Advances in

VIRUS RESEARCH

VOLUME 68

ADVISORY BOARD

Advances in
VIRUS RESEARCH

Edited by

KARL MARAMOROSCH
Department of Entomology
Rutgers University
New Brunswick, New Jersey

AARON J. SHATKIN
Center for Advanced Biotechnology
and Medicine
Piscataway, New Jersey

VOLUME 68
Insect Viruses: Biotechnological Applications

Edited by

BRYONY C. BONNING
Department of Entomology
Iowa State University
Ames, Iowa

ELSEVIER

AMSTERDAM • BOSTON • HEIDELBERG • LONDON
NEW YORK • OXFORD • PARIS • SAN DIEGO
SAN FRANCISCO • SINGAPORE • SYDNEY • TOKYO
Academic Press is an imprint of Elsevier

Academic Press is an imprint of Elsevier
525 B Street, Suite 1900, San Diego, California 92101-4495, USA
84 Theobald's Road, London WC1X 8RR, UK

This book is printed on acid-free paper. ∞

For information on all Elsevier Academic Press publications
visit our Web site at www.books.elsevier.com

ISBN-13: 978-0-12-039868-3
ISBN-10: 0-12-039868-0

PRINTED IN THE UNITED STATES OF AMERICA
06 07 08 09 9 8 7 6 5 4 3 2 1

CONTENTS

Baculovirus Display: A Multifunctional Technology for Gene Delivery and Eukaryotic Library Development

Anna R. Mäkelä and Christian Oker-Blom

Stably Transformed Insect Cell Lines: Tools for Expression of Secreted and Membrane-Anchored Proteins and High-Throughput Screening Platforms for Drug and Insecticide Discovery

Vassilis Douris, Luc Swevers, Vassiliki Labropoulou, Evi Andronopoulou, Zafiroula Georgoussi, and Kostas Iatrou

Protein *N*-Glycosylation in the Baculovirus–Insect Cell Expression System and Engineering of Insect Cells to Produce "Mammalianized" Recombinant Glycoproteins

Robert L. Harrison and Donald L. Jarvis

Vaccines for Viral and Parasitic Diseases Produced with Baculovirus Vectors

Monique M. van Oers

Baculoviruses and Mammalian Cell-Based Assays for Drug Screening

J. Patrick Condreay, Robert S. Ames, Namir J. Hassan, Thomas A. Kost, Raymond V. Merrihew, Danuta E. Mossakowska, David J. Pountney, and Michael A. Romanos

Baculovirus Vectors for Gene Therapy

Yu-Chen Hu

Virus-Derived Genes for Insect-Resistant Transgenic Plants

SIJUN LIU, HUARONG LI, S. SIVAKUMAR, AND BRYONY C. BONNING

Small RNA Viruses of Insects: Expression in Plants and RNA Silencing

KARL H. J. GORDON AND PETER M. WATERHOUSE

PREFACE

Baculoviruses are perhaps unique among viruses in the breadth of their biotechnological applications. These insect-specific viruses are used not only for insect pest management purposes but also as laboratory research tools for production of recombinant proteins and for protein display and as potential vectors for human gene therapy. Although the biotechnological use of insect viruses has historically been dominated by the baculoviruses, the development of technologies from other insect virus families, such as the *Polydnaviridae* and *Tetraviridae*, is gaining momentum. In addition to highlighting recent advances, this volume provides a comprehensive review of the biotechnological applications of these and other insect viruses.

Although there is some overlap, the chapters in this volume are divided into the following sections: (1) Insect Viruses as Laboratory Research Tools, (2) Applications to Human and Animal Health, and (3) Insect Pest Management. While chapters on stably transformed insect cell lines (Douris *et al.*, pp. 113–156) and mammalian glycosylation in insect cells (Harrison and Jarvis, pp. 159–191) describe "spin-off" technologies from baculovirus research, all other chapters deal with application of the insect virus itself or virus-derived genes for technological purposes.

Given that baculoviruses comprise the most widely used of the insect viruses within the biotechnological arena, it is appropriate that this volume begins with a personal chronicle of the milestones leading to the use of baculoviruses for recombinant protein expression and as genetically optimized insecticides (Max D. Summers, pp. 3–73). This chronicle was written by Max Summers, a leader in the development and implementation of the baculovirus expression vector system (BEVS), and provides some historical background for many of the subsequent chapters. Angelika Fath-Goodin *et al.* describe a cutting-edge technology for enhancement of the BEVS by using a polydnavirus gene that prolongs the longevity of baculovirus-infected cells (Fath-Goodin *et al.*, pp. 75–90), and Anna Mäkelä and Christian Oker-Blom provide an overview of baculovirus display technologies for display of proteins either on the budded virus or on the surface of baculovirus-infected cells (Mäkelä and Oker-Blom, pp. 91–112). Applications of these display techniques range from production of monoclonal antibodies, to study of protein–protein interactions, and targeting of gene therapy vectors to specific cell types. Vassilis Douris *et al.* describe the use of stably transformed cell lines for recombinant protein production and for high-throughput screening for drug and insecticide discovery, including a detailed description of expression cassettes available for specific applications that employ various baculovirus-derived promoter and enhancer elements (Douris *et al.*, pp. 113–156).

The differences between mammalian and insect glycosylation compromise the biomedical value of glycoproteins produced by using the BEVS. In their chapter, Robert Harrison and Donald Jarvis (pp. 159–191) describe insect glycosylation pathways and efforts to engineer both baculovirus vectors and insect cell lines to remedy the problem of inappropriate glycosylation of mammalian proteins. Monique van Oers (pp. 193–253) describes different approaches and tailoring of the BEVS for production of vaccines against viral and parasitic disease. Descriptions of the first vaccine for animal use produced in insect cells, and the development of tests that distinguish between immunized and infected individuals, highlight the potential for application of the BEVS toward development of effective vaccines.

The broad range of mammalian cells that are permissive to baculovirus transduction provides the foundation for research described in chapters by Condreay *et al.* (pp. 255–286) and Andy Hu (pp. 287–320). Pat Condreay *et al.* (pp. 255–286) address the use of baculoviruses for automated screening of chemical libraries in mammalian cell-based assays. This chapter describes the use of the BacMam system both for basic research and therapeutic intervention with a focus on nuclear receptors, transporters, G-protein–coupled receptors, ion channels, and viral targets. Andy Hu (pp. 287–320) provides an overview of the use of baculoviruses as potential vectors for both *in vitro* and *in vivo* gene therapy including the benefits and limitations of this system. Baculovirus gene therapy vectors have potential for use in hybrid vectors for stable integration of expression cassettes into the host genome and for delivery of dsRNA for RNA interference-based gene silencing.

In the final section, which deals with biotechnological application of insect viruses for the management of insect pests, Bora Inceoglu *et al.* (pp. 323–360) provide a historical overview of the development of recombinant baculovirus insecticides and an assessment of the prospects for their use. Genetically enhanced baculovirus insecticides are now competitive with pyrethroid insecticides under field conditions, and, as public acceptance of genetically modified organisms increases, could be adopted as an integral part of modern pest management programs. In their chapter, John Carlson *et al.* (pp. 361–392) describe the potential use of densoviruses as transducing vectors to directly combat mosquito-borne disease agents and also highlight the potential use of densoviruses for management of mosquitoes. With the emergence of mosquito-borne diseases, such as that caused by *West Nile virus* and dengue hemorrhagic fever, management of both the disease agent and the insect vectors are increasingly important goals.

In the final three chapters, Gill, *et al.* (pp. 393–426), Liu *et al.* (pp. 427–457), and Gordon and Waterhouse (pp. 459–502) address transgenic plant-mediated delivery of insect virus-derived genes and insect viruses themselves for pest management purposes. Torrence Gill *et al.* describe the intricate biology of the parasitoid wasp–polydnavirus–host insect interactions, and the potential use of both polydnavirus and wasp genes that disrupt the physiology of the host insect, for pest control purposes. Of particular note is the use of the Cys-motif genes derived from the *Campoletis sonorensis* ichnovirus and wasp teratocytes as transgenes for plant protection against lepidopteran pests. Sijun Liu *et al.* (pp. 427–457) describe the potential for use of other virus-derived genes for production of insect-resistant transgenic plants, particularly virus genes that have evolved to counter the physiological defenses of the host insect. These authors also describe the use of plant lectins for appropriate delivery of toxins that act within the insect hemocoel, from a transgenic plant. This delivery system allows for exploitation of intrahemocoelic toxins that have effectively been screened for efficacy during the optimization of recombinant baculovirus insecticides (described in chapter by Bora Inceoglu *et al.*, pp. 323–360). In the final chapter, Karl Gordon and Peter Waterhouse (pp. 459–502) describe the potential for use of the relatively little studied, small RNA viruses of insects for pest management. In addition to their classical use as insecticidal agents, the authors describe efforts to develop transgenic plants that express the *Helicoverpa armigera* stunt virus and the resulting encounters with the RNA-based antiviral immune response of plants. However, the virus-like particles of small RNA viruses assembled in plants have potential for delivery of dsRNAs for silencing of genes in target insect pests. This novel biotechnological application of small RNA viruses that does not require plant expression of a viable virus is likely to stimulate a good deal of interest for plant protection purposes.

The assembly of the 13 chapters within this volume highlights the remarkable versatility of insect viruses for a diverse range of applications. Not covered in this volume are a number of other important insect virus-derived tools, which include (1) the baculovirus gene *P35*, an inhibitor of apoptosis identified by Lois Miller and colleagues that is the most broadly acting caspase inhibitor protein known (Clem, 2001; Clem *et al.*, 1991), and (2) the transposon *piggyBac*, which was isolated from a baculovirus and which, among other applications, has been widely adopted for production of transgenic insects (Handler, 2002). Aside from the myriad applications of baculoviruses and baculovirus-derived genes, the internal ribosome entry site (IRES) sequences of the *Dicistroviridae* have been extensively studied both for their novelty (Jan, 2005) and for construction of dicistronic expression vectors (Royall *et al.*, 2004).

This volume on the biotechnological applications of insect viruses was inspired in part by a Virus Division symposium on “Insect Expression Systems, Gene Therapy, and Vaccine Development” held at the annual meeting of the Society for Invertebrate Pathology in Alaska, Anchorage in 2005. I would like to thank the participants of that symposium, in addition to all other authors who have contributed to this volume. With the exponential increase in insect genomic efforts, growth in the study of insect virus–host interactions, and discovery of new insect viruses, the continued development of biotechnological products from these viruses is assured.

Bryony Bonning

References

Clem, R. J. (2001). Baculoviruses and apoptosis: The good, the bad, and the ugly. *Cell Death Differ.* **8**(2)**:**137–143.

Clem, R. J., Fechheimer, M., and Miller, L. K. (1991). Prevention of apoptosis by a baculovirus gene during infection of insect cells. *Science* **254:**1388–1389.

Handler, A. M. (2002). Use of the piggyBac transposon for germ-line transformation of insects. *Insect Biochem. Mol. Biol.* **32**(10)**:**1211–1220.

Jan, E. (2005). Divergent IRES elements in invertebrates. *Virus Res.* **119**(1):16–28.

Royall, E., Woolaway, K. E., Schacherl, J., Kubick, S., Belsham, G. J., and Roberts, L. (2004). The *Rhopalosiphum padi* virus 5′ internal ribosome entry site is functional in *Spodoptera frugiperda* 21 cells and in their cell-free lysates: Implications for the baculovirus expression system. *J. Gen. Virol.* **85**(6)**:**1565–1569.

SECTION I

INSECT VIRUSES AS LABORATORY RESEARCH TOOLS

MILESTONES LEADING TO THE GENETIC ENGINEERING OF BACULOVIRUSES AS EXPRESSION VECTOR SYSTEMS AND VIRAL PESTICIDES

Max D. Summers

Department of Entomology, Texas A&M University, College Station, Texas 77843
Department of Biochemistry and Biophysics, Texas A&M University
College Station, Texas 77843

Abstract

The baculovirus expression vector system (BEVS) is widely established as a highly useful and effective eukaryotic expression system. Thousands of soluble and membrane proteins that, in general, are correctly folded, modified, sorted and assembled to produce highly authentic recombinant proteins have been cloned and expressed.

0065-3527/06 $35.00
DOI: 10.1016/S0065-3527(06)68001-9

This historical chronology and perspective will focus on the original, peer-reviewed discoveries that were pioneering and seminal to the development of the BEVS and that provided the basis for subsequent and more recent developments and applications.

I. Introduction

The modern era for baculovirus research spans the course of several decades of which developments since the early 80s have been rapid and notable. Why? Initially, baculoviruses were studied primarily for agricultural applications involving pest control. At first, attention was drawn to the unique structure of baculoviruses and their natural process of infection, which by its nature stimulated significant curiosity about the biology and molecular basis of the viral infection pathway. This led to fundamental discoveries of the structure of baculoviruses, the structure and function of virus-encoded proteins, and the nature of virus-host specificity; in particular, the cellular and molecular basis of the virus infection pathway, *in vivo* and *in vitro*. As the era encompassing the fundamental development of recombinant DNA technologies and genetic manipulation of DNA had already arrived, baculovirologists were ready to apply those discoveries and innovations. As such, biomedical research and commercial human health needs were timely for the rapid development and cost-effective production of structurally complex and functionally authentic recombinant proteins for research and drug discovery, vaccines, therapeutics, and diagnostics. The genetic engineering of the baculovirus polyhedrin gene promoter and the enabling technology for the baculovirus expression vector system (BEVS) became an important tool in biomedical research.

The BEVS has taken its place among the prokaryotic and eukaryotic expression systems. It has been, and currently is, widely used for the routine cloning, expression, and production of thousands of soluble and membrane proteins that, in general, are correctly folded, modified, sorted to the correct cellular location and assembled to produce highly authentic recombinant proteins. Recombinant proteins from a very diverse range of organisms have been expressed, usually in quite abundant quantities (several micrograms to milligrams per liter of infected cell culture or individual infected insect) for experimental or practical applications. The ability to cost effectively generate abundant quantities of functionally authentic proteins, coupled with the development of large-scale production systems, has served as a powerful force to promote baculovirus research at all levels, especially in medical research and human health applications. Interest in the development of viral pesticides has proceeded with varying levels of popularity and acceptance during the past decades, and the success of the BEVS coupled with environmental concerns for the use of chemical pesticides has kept this potential application active in research development.

The objective of this chapter is to list from the original refereed baculovirus literature the sequence and discovery of seminal and first discoveries with focus on the development of the BEVS. This is done in order to place in perspective developments relative to the seminal discoveries providing the groundwork for the evolution and development of the BEVS field and its related technologies and applications; and, from that, the progress and developments in pursuit of genetically engineered viral pesticides. It must be noted that to date only the BEVS has been a truly remarkable success story; the commercial use of natural or genetically engineered baculoviruses as cost effective and routinely used pest control agents is still to become a practical reality.

This chapter is not to be a comprehensive compendium of all the related literature during the development of the BEVS. The purpose for focusing on the original, peer-reviewed publications and presentations in the correct chronology is that it is often too convenient for an author(s), especially those not working directly in the field, to cite reviews representing a collective body of literature; as such, for scientists and lay people not directly involved in baculovirus research it is difficult to discern those discoveries that are pioneering and seminal relative to the rather vast literature of the many related disciplines and cross-cutting developments that have occurred during the development of the BEVS. The papers cited in Table I represent primarily the original peer-reviewed reports with only a few pertinent non-reviewed publications. To record the correct sequence of the chronology, the dates are recorded giving the month and year of both submission and publication. Given that it is too soon to assess the long-term impact of the more recent developments, the primary original literature addressed is up to the year 2000.

II. Baculovirus Advances Before the Application of Recombinant DNA

A. *Virus Structure, Composition, and Infectivity*

Because of their size (0.5–5 μm diameter), highly refractile (by light and dark field microscopy) polyhedral bodies were easily observed in infected insect tissues by optical microscopy as early as 1856.

TABLE I
MILESTONES

Authors	Citation	Submitted	Published
	Polyhedron-shaped bodies are associated with disease of silkworms		
Cornalia, E.	*Memorie dell' I.R. Instituo Lombardo di Scienze, Lettere ed Arti* **6:**3–387 [Parte quarta: Patologia del baco., pp. 332–336]: Monografia del bombice del gelso (*Bombyx mori* Linneo).		1856
Maestri, A.	Frammenti anatomici, fisiologici e patologici sul baco da seta (*Bombyx mori* Linn). Fratelli Fusi, Pavia, p. 172		1856
Bolle, J.	**Correctly associated polyhedral bodies observed in the disease, silkworm jaundice, with the disease**		1894
	Discovered that polyhedra are alkali-sensitive		
	Atti e Mem. dell' i.R. Soc. Agr. Gorizia **34:**133–136: Il giallume od il mal del grasso del baco da seta. Communicaizone preliminare.		
Fischer, E.	**Proposed the name polyhedrosis**		1906
	Biol. Zentr. **26:**448–463; 534–544: Über die Ursachen der Disposition und über Frühsymptome der Raupenkrankheiten.		
Goldschmidt, R.	**Culture of insect tissues *in vitro*: explant of *Cecropia* moth spermatozoa**		1915
	Pro. Nat. Acad. Sci. A **1:**220–222: Some experiments on spermatogenesis *in vitro*.		
Glaser, R. W.	**Demonstrated *in vitro* formation of NPV in harvested blood cells from infected insects, observed in hanging drops**		1917
	Psyche. **24:**1–7: The growth of insect blood *in vitro*.		
Komárek, J., Breindl, V.	**Proposed that virions are occluded within polyhedra**		1924
	Z. Angew. Entomol. **10:**99–162: Die Wipfel-Krankheit der Nonne und der Erreger derselben.		
	European field introduction of baculovirus for insect control		
Klöck	*Fortwiss. Cent.* **47:**241–245: Zur Lösung der Nonnenbekampfungsfrage Auf biologischem Wege.		1925

Ruzicka, J.	*Forstwiss. Zentr.* **47:**537–538: Einige Bemerkungen über die Nonnenbekämpfung auf biologischen Wege.	1925
Paillot, A.	**Discovery of granulosis virus** *Compt. Rend. Acad. Sci.* **182:**180–182: Sur une nouvelle maladie du noyau ou grasserie* des chenilles de *P. brassicae* et un nouveau groupe de microorganismes parasites.	1926
	First (modern) treatise on diseases of insects	
Paillot, A.	*Traité des maladies du ver à soie.* G. Doin et Cie, Paris, p. 279	1930
Paillot, A.	*L'Infection chez les insectes.* G. Patissier, Trévoux, p. 535	1933
Trager, W.	**Demonstrated baculovirus infection in cultured insect tissue and the subculture of virus infectivity to healthy cultures** *J. Exp. Med.* **61:**501–513: Cultivation of the virus of grasserie in silkworm tissue cultures.	1935
Paillot, A., Gratia, A.	**Demonstrated virions by alkali dissolution of inclusion bodies** *Arch Gesell. Virusforschung* **1:**120–129: Essai d'isolement du virus de la grasserie des vers à soie par l'ultracentrifugation.	1939
Balch, R. E., Bird, F. T.	**North American field introduction of baculovirus for insect control** *Sci. Agr.* **25:**65–80: A disease of the European spruce sawfly, *Gilpiniia hercyniae* (Htg.), and its place in natural control.	1944
Bergold, G. H.	**Comprehensive early studies on purification of baculovirus virions, confirmed that virus-like particles were occluded in polyhedra** *Z. f. Naturforsch.* **2b:**122–143: Die Isolierung des Polyeder-Virus und die Natur der Polyeder.	1947
Steinhaus, E. A.	**Landmark book and review of pathogens of insects including occluded viruses and their potential for insect control** *Principles of Insect Pathology.* McGraw-Hill, New York. p. 757	1949

(continues)

TABLE I (*continued*)

Authors	Citation	Submitted	Published
Bergold, G. H.	**First comprehensive review of baculovirus chemistry and biochemistry** *Insect Viruses*. *In* Advances Virus Research, Vol. I, pp. 91–139.		1953
Gershenzon, S.	**Differences in polyhedron shapes are determined by the virus causing the disease: correlation of mutant NPV strains with shape of the polyhedron** *Mikrobiologiya* **24:**90–98: On the species specificity of viruses of the polyhedral disease of insects.		1955
Morgan, C., Bergold, G. H., Moore, D. H., Rose, H. M.	**The macromolecular paracrystalline lattice of viral polyhedral bodies as examined in the electron microscope** *J. Biophys. Biochem. Cyto.* **1:**187–190: The macromolecular paracrystalline lattice of insect viral polyhedral bodies demonstrated in ultrathin sections examined in the electron microscope.		1955
Bergold, G. H.	*Viruses of Insects*. pp. 60–142. *In* Handbuch der Virusforschung. Doerr and Hallauer, eds., Springer-Verlag, Wien		1958
Yamafuji, K., Yoshinara, Y., Hirayama, K.	**The NPV of the silkworm contains DNA and there is protease activity associated with polyhedra** *Enzymol. Biol. Clin.* **19:**53–58: Protease and desoxyribonuclease in viral polyhedral crystal.		1958
Gaw, Z. Y., Liu, N. T., Zia, T. U.	**First cultivation of insect cells in continuous monolayer culture: *Bombyx mori* gonad epithelial cells** *Acta Virol.* **3:**55–60: Tissue culture methods for cultivation of virus grasserie.		1959
Martignoni, M. E., Scallion, R. J.	**Baculovirus infection *in vitro* of insect primary blood cell cultures** *Nature (London)* **190:**1133–1134: Establishment of strains of cells from insect tissue cultured *in vitro*.		1961

Grace, T. D. C.	**Established cell strains from tissue cultured *in vitro*** *Nature* **195:**788–799: Establishment of four strains of cells from insect tissues grown *in vitro.*		1962
Caspar, D. L. D., Dulbecco, R., Klug, A., Lwoff, A., Stoker, M. G. P., Tournier, P., Wildy, P.	Cold Spring Harbor Symposium Quantitative Biology **27:**49–50: *Proposals*		1962
Vaughn, J. L., Faulkner, P.	**Hemolymph from an infected insect will infect cells cultured *in vitro*, but virus purified from polyhedra is not infectious** *Virol.* **20:**484–489: Susceptibility of an insect tissue culture to infection by virus preparations of the nuclear polyhedrosis of the silkworm (*Bombyx mori*).		1963
Steinhaus, E. A., ed.	*Academic Press*, New York and London. Vol. I, p. 661, Vol. II, p. 689: Insect pathology, An Advanced Treatise		1963
Steinhaus, E. A.	*Insect Microbiology:* Hafner Publishing Co., New York and London, p. 763		1967
Harrap, K. A., Robertson, J. S.	**NPV replication occurs in the gut cell without being occluded** **Viral replication in gut cells is responsible for secondary infection "Short projections" on the surface of the envelope of the viruses in the basal cytoplasm of the columnar cell are not present on virus envelopes in fat body cells** *J. Gen. Virol.* **3:**221–225: A possible infection pathway in the development of a nuclear polyhedrosis virus.	Mar 1968	Sept 1968
Shvedchikova, N. G., Ulanov, V. P., Tanasevich, L. M.	**By electronmicroscopy the DNA of a granulosis virus is high molecular weight (80 $\times$ 10^{6}), observed as linear and circular forms** *Molekulyarwaya Biologiya* **3:**361–365: Structure of the granulosis virus of Siberian silkworm *Dendrolinus sibiricus* Tschetw.	Oct 1968	May 1969

(continues)

TABLE I (*continued*)

Authors	Citation	Submitted	Published
Summers, M. D.	**Baculovirus (granulosis virus) enters the gut cell by fusion of viral envelope with microvillar membrane** **Nucleocapsid uncoating occurs by nuclear pore interaction** **Viral replication confined to nucleus of gut cell but occurs throughout nuclear and cytoplasmic regions of fat body cells** *J. Virol.* **2:**188–190: Apparent *in vivo* pathway of granulosis virus invasion and infection.	Apr 1969	Aug 1969
Hink, W. F.	**Continuous culture of the cabbage looper cell line** *Nature* **226:**466–467: Established insect cell line from the cabbage looper, *Trichoplusia ni.*	Sept 1969	May 1970
Kawanishi, C. Y., Paschke, J. D.	**Application of standard virological terminology to baculovirus structure using the convention proposed by Caspar, *et al.* 1962** **Rate zonal banding of ODV separates virions with 1, 2, 3, etc., nucleocapsids for virus purification** *J. Inverteb. Pathol.* **16:**89–92: Density gradient centrifugation of the virions liberated from *Rachoplusia ou* nuclear polyhedra.	Oct 1969	July 1970
Summers, M. D., Paschke, J. D.	*J. Inverteb. Pathol.* **16:**227–240: Alkali-liberated granulosis virus of *Trichoplusia ni* I. density gradient purification of virus components and some of their *in vitro* chemical and physical properties.	Jan 1970	Sept 1970
Goodwin, R. H., Vaughn, J. L., Adams, J. R., Louloudes, S. J.	**NPV replication in continuous cell culture** *J. Inverteb. Pathol.* **16:**284–288: Replication of a nuclear polyhedrosis virus in an established cell line.	Apr 1970	Sept 1970
Vail, P. V., Sutter, G., Jay, D. L., Gough, D.	**Discovery and identification of the baculovirus, *Autographa californica* nuclear polyhedrosis virus**	Nov 1970	May 1971

	J. Inverteb. Pathol. **17:**383–388: Reciprocal infectivity of nuclear polyhedrosis viruses of the cabbage looper and alfalfa looper.		
Zlotkin, E., Rochart, H., Kopeyan, C., Miranda, F., Lissitzky, S.	**The demonstration of insect specific toxins in the venom of scorpions, amino acid analysis, N-terminal sequence**		1971a
	Biochimie. **53:**1073–1078: Purification and properties of the insect toxin from the venom of the scorpion, *Androctonus australis* Hector.		
	Toxicon. **9:**1–8: The effect of scorpion venom on blowfly larvae—a new method for the evaluation of scorpion venoms potency.		1971b
	Toxicon. **9:**9–13: A new toxic protein in the venom of the scorpion *Androctonus australis* Hector.		1971c
Summers, M. D.	**Detailed EM analysis of nuclear pore interaction**		1971
	J. Ultrastruct. Res. **35:**606–625: Electron microscopic observations on granulosis virus entry, uncoating and replication processes during infection of the midgut cells of *Trichoplusia ni.*		
Summers, M. D., Anderson, D. L.	**The purification, isolation and sedimentation of baculovirus DNA as three forms: ds linear, ds relaxed circular, ds covalently closed**	Oct 1971	Apr 1972
	J. Virol. **9:**710–713: Granulosis virus deoxyribonucleic acid: a closed, double-stranded molecule.		
Egawa, K., Summers, M. D.	**The kinetics of polyhedrin solubilization, established methods for solubilization at neutrality, estimate protein monomers of 20,000 to 40,000 daltons**	Jan 1971	May 1972
	J. Inverteb. Pathol. **19:**395–404: Solubilization of *Trichoplusia ni* granulosis virus proteinic crystal.		
Kawanishi, C. Y., Summers, M. D., Stoltz, D. B., Arnott, H. J.	**NPV entry to gut cells by fusion of viral envelope**	Jan 1972	July 1972
	J. Inverteb. Pathol. **20:**104–108: Entry of an insect virus *in vivo* by fusion of viral envelope and microvillus membrane.		
Harrap, K. A.	**Comprehensive ultra-structural analysis of the viral occlusion, the virion and virus assembly**		
	Virol. **50**:114–123: The structure of Nuclear Polyhedrosis Viruses I. The inclusion body.	July 1972	Oct 1972a

(*continues*)

TABLE I (*continued*)

Authors	Citation	Submitted	Published
	Virol. **50:**124–132: The Structure of Nuclear Polyhedrosis Viruses II. The virus particle.	July 1972	Oct 1972b
	Virol. **50:**133–139: The Structure of Nuclear Polyhedrosis Viruses III. Virus assembly.	July 1972	Oct 1972c
Faulkner, P., Henderson, J. F.	**Serial passage of infectious virus propagated in continuous cell culture**	Sept 1972	Dec 1972
	Virol. **50:**920–924: Serial passage of a nuclear polyhedrosis disease virus of the cabbage looper (*Trichoplusia ni*) in a continuous tissue culture cell line.		
Jackson, D. A., Symons, R. H., Berg, P.	**Gene cloning, manipulation of DNA**		1972
	Pro. Nat. Acad. Sci. A **69:**2904–2909; Biochemical method for inserting new genetic information into DNA of simian virus 40: circular SV40 DNA molecules containing Lambda Phage genes and the galactose operon of *Escherichia coli.*		
Cohen, S. N., Chong, A. C. Y., Boyer, H. W., Helling, R. B.	*Pro. Nat. Acad. Sci. A* **70:**3240–3244; Construction of biologically functional bacterial plasmids *in vitro.*		1973
Stoltz, D. B., Pavan, C., da Cunha, A. B.	**Describes the process of *de novo* membrane morphogenesis as the source of intranuclear membranes for occlusion-derived virus (ODV) envelopes.**	Sept 1972	Apr 1973
	J. Gen. Virol. **19:**145–150: Nuclear polyhedrosis virus: a possible example of *de novo* intranuclear membrane morphogenesis.		
Vail, P. V., Jay, D. L., Hink, W. F.	**Replication of *Ac*MNPV in cell culture**	Dec 1972	Sept 1973
	J. Inverteb. Pathol. **22:**231–237: Replication and infectivity of the nuclear polyhedrosis virus of the alfalfa looper, *Autographa californica*, produced in cells grown *in vivo.*		

Kozlov, E. A., Levitina, T. L., Radavskii, Y. L., Sogulyaeva, V. M. Sidorova, N. M., Serebryanyi, S. B.	**The molecular weight of *Bombyx mori* polyhedrin is 28,000 daltons as determined by PAGE.** *Biokhimiya* **38:**1015–1019: A determination of the molecular weight of the inclusion body protein of the nuclear polyhedrosis virus of the mulberry silkworm *Bombyx mori*.	Mar 1972	Sept 1973
Hink, W. F., Vail, P. V.	**The first baculovirus plaque assay (methycellulose)** **Observed and documented FP(Few Polyhedra) and MP (Many polyhedra) plaques in the overlay** *J. Inverteb. Pathol.* **22:**168–174: A plaque assay for titration of alfalfa looper nuclear polyhedrosis virus in a cabbage looper (TN-368) cell line.	Feb 1973	Sept 1973
Summers, M. D., Egawa, K.	**The terms "polyhedrin" and "granulin" are designated** *J. Virol.* **12:**1092–1103: Physical and chemical properties of *Trichoplusia ni* granulosis virus granulin.	July 1973	Nov 1973
Tinsley, T. W., Melnick, J. L.	**The safety and potential of pesticidal viruses** *Intervirology* **2:**206–208: Potential ecological hazards of pesticidal viruses.		1973/74
Vago, C., Aizawa, K., Ignoffo, C., Martignoni, M. E., Tarasevitch, L., Tinsley, J. W.	**The genus baculovirus is adopted for the NPVs and GVs. Proposed by M. Martignoni** *J. Inverteb. Pathol.* **23:**133–134: Editorial: Present status of the nomenclature and classification of invertebrate viruses.		1974
Henderson, J. F., Faulkner, J., MacKinnon, E.	**There is an infectious viral form in cell culture extracellular media which suggests that non-occluded virus particles are responsible for systemic infection of the insect** **Derived the term "non-occluded" (NOV) viral form: later to be called the "budded virus" form (BV)** **Suggested that the NOV form consists of fragile enveloped particles.** *J. Gen. Virol.* **22:**143–146: Some biophysical properties of virus present in tissue cultures infected with the nuclear polyhedrosis virus of *Trichoplusia ni*.	Aug 1973	Jan 1974

(continues)

TABLE I (*continued*)

Authors	Citation	Submitted	Published
Ramoska, W. A., Hink, W. F.	**First *in vitro* (cell culture)demonstration of genetic differences (plaque variants) in a baculovirus isolate polyhedra morphology and nucleocapsid envelopment** *J. Inverteb. Pathol.* **23:**197–201: Electron microscope examination of two plaque variants from a nuclear polyhedrosis virus of the alfalfa looper, *Autographa californica*.	Aug 1973	Mar 1974
Knudson, D. L., Tinsley, T. W.	**Insect cell culture systems represent a feasible means for investigating the replication of baculoviruses** **Physical particle: Infectious particle ratios** Viral growth cycle—virus released by 12 h p.i., maximal titers at 4 days *J. Virol.* **14:**934–944: Replication of a nuclear polyhedrosis virus in continuous cell culture of *Spodoptera frugiperda*: purification, assay of infectivity, and growth characteristics of the virus.	Apr 1974	Oct 1974
Steinhaus, E. A.	**Reflections on the history of insect pathology, yet a visionary excursion into the world of diseases of insects and other arthropods** *Disease in a Minor Chord*. Ohio State University Press. 488 p: Edited collection of leading expert opinions on principal developments in insect pathology.		1975
Brown, M., Faulkner, P.	**Clonal isolates of *Trichoplusia ni* cells** *J. Invert. Pathol.* **26:**251–257: Factors affecting the yield of virus in a cloned cell line of *Trichoplusia ni* infected with a nuclear polyhedrosis virus.	Dec 1974	Sept 1975
Volkman, L. E., Summers, M. D.	**Clonal cell isolates show differences in virus growth curves and polyhedra production** **Cells should be in log growth for optimal polyhedra production**	July 1975	Dec 1975

	J. Virol. **16:**1630–1637: Nuclear polyhedrosis virus detection: relative capabilities of clones developed from *Trichoplusia ni* ovarian cell line TN-368 to serve as indicator cells in a plaque assay.		
Summers, M. D., Engler, R., Falcon, L. A., Vail, P.	**Baculoviruses for Insect Pest Control: Safety Considerations** *Am. Soc. Microbiol.*, 1913 I Street, N.W. Washington, D.C. 20006. 186 p		1975
Summers, M. D., Smith, G. E.	**Because each polyhedrin is similar, yet different to some extent in primary amino acid sequence, postulate that polyhedrin/granulin is encoded by the viral genome** *Intervirology* **6:**168–180: Comparative studies of baculovirus granulins and polyhedrins.	Nov 1975	Jan 1976
Summers, M. D. Volkman, L. D.	**Non-occluded viral forms (budded viruses) of NPV from tissue culture and the infected insect hemolymph are enveloped but, physically different from occlusion derived virus, ODV** **The budded virus form has distinct peplomers localized to one end of the virion** *J. Virol.* **17:**962–972: Comparison of biophysical and morphological properties of occluded and extracellular nonoccluded baculovirus from *in vivo* and *in vitro* host systems.	Sept 1975	Mar 1976
Potter, J. N., Faulkner, P., MacKinnon, E. A.	**Virus plaque purification** **Demonstration that new strains of virus with FP phenotype occur during serial passage** **Replication of MP after plaque purification** *J. Virol.* **18:**1040–1050: Strain selection during serial passage of *Trichoplusia ni* nuclear polyhedrosis virus.	Dec 1975	Jun 1976
Volkman, L. E., Summers, M. D., Hsieh, C.-H.	**BV and ODV have different neutralization antigens** **Temporal relationship of BV production and polyhedrin synthesis: BV shuts down with onset of polyhedrin synthesis** **BV is 1700-fold more infectious than ODV in cell culture**	May 1975	Sept 1976

(continues)

TABLE I (*continued*)

Authors	Citation	Submitted	Published
	J. Virol. **19:**820–832: Occluded and nonoccluded nuclear polyhedrosis virus grown in *Trichoplusia ni*: Comparative neutralizations, comparative infectivity and *in vitro* growth studies.		
Wood, H. A.	**Plaque assay improvement, use of Sea plaque agarose** *J. Invert. Pathol.* **29:**304–307: An agar overlay plaque assay method for *Autographa californica* nuclear-polyhedrosis virus.	July 1976	May 1977
Volkman, L. E., Summers, M. D.	**Quantitative infectivity: BV and ODV are different in their ability to infect the insect and tissue culture cells** *J. Invert. Pathol.* **30:**102–103: *Autographa californica* nuclear polyhedrosis virus: comparative infectivity of the occluded, alkali-liberated and non-occluded forms.	Aug 1976	July 1977
Tinsley, T. W.	Viruses and the Biological Control of Insect Pests *BioScience* **27:**659–661: Viruses and the biological control of insect pests	1977	Oct 1977
Serebryani, S. B., Levitina, T. L., Kautsman, M. L., Radavski, Y. L., Gusak, N. M., Ovander, M. N., Sucharenko, N. V., Kozlov, E. A.	**The primary amino acid sequence of polyhedrin** *J. Invert. Pathol.* **30:**442–443: The primary structure of the polyhedrin protein of nuclear polyhedrosis virus (NPV) of *Bombyx mori*.	Apr 1977	Nov 1977
Rohrmann, G. F., Beaudreau, G. S.	**Restriction endonuclease (REN) analysis of baculovirus (*Orgyia pseudotsugata*) DNA: *Eco*RI digests** *Virol.* **83:**474–478: Characterization of DNA from polyhedral inclusion bodies of the nucleopolyhedrosis single-rod virus pathogenic for *Orgyia pseudotsugata*.	Aug 1977	Dec 1977

Rohrmann, G. F., McParland, R. H., Martignoni, M. E., Beadreau, G. S.	**Use of REN to compare and identify baculoviruses** **Comparison of homology by DNA-DNA hybridization** *Virol.* **84:**213: Genetic relatedness of two nucleopolyhedrosis viruses pathogenic for *Orgyia pseudotsugata.*	Sept 1977	Jan 1978
Miller, L. K., Dawes, K. D.	**Virus passage through alternate hosts does not alter REN patterns** **Submolar fragments in REN digests suggest heterogeneity in the virus strain due to possible a) contaminating virus, b) a genetic variant, c) defective virus particles** *Appl. Environ. Microbiol.* **35:**411: Restriction endonuclease analysis for the identification of baculovirus pesticides.	July 1977	Feb 1978
Summers, M. D., Smith, G. E.	**ODV *Ac*MNPV purified single enveloped nucleocapsids have a different structural polypeptide composition as compared to purified multiple enveloped nucleocapsids.** **Removal of viral envelopes, analysis of capsid proteins** *Virol.* **84:**390–402: Baculovirus structural polypeptides.	Sept 1977	Feb 1978
Summers, M. D., Kawanishi, C. Y.	**Viral Pesticides: Present knowledge and potential effects on public and environmental health** *EPA-600/9-78-026*. September, 1978 pp. 311		Sept 1978
Summers, M. D., Volkman, L. E., Hsieh, C.-H.	**Polyhedrin expression is not directly proportional to virus titers** **Polyhedrin expression differs according to cell type** **Assays more direct and quantitative for virus replication are needed of which immunoperoxidase detection is an example** *J. Gen. Virol.* **40:**545–557: Immunoperoxidase detection of baculovirus antigens in insect cells.	Mar 1978	Sept 1978
Lee, H. H., Miller, L. K.	**Infectious genotypic variants occur in plaque-purified *Ac*MNPV budded virus isolates from infected cell cultures** *J. Virol.* **27:**754–767: Isolation of genotypic variants of *Autographa californica* nuclear polyhedrosis virus.	Mar 1978	Sept 1978
Smith, G. E., Summers, M. D.	**Genotypic variants occur in both BV and ODV isolates of *Ac*MNPV from infected cell cultures**	Jun 1978	Sept 1978

(*continues*)

TABLE I (*continued*)

Authors	Citation	Submitted	Published
	Demonstration of genotypic variation in plaque purified isolates from purified BV, SNPV, MNPV **Potential significance of natural or genotypic variation relative to virus-host specificity** **BV and ODV have the same genome, but different structural polypeptides** *Virol.* **89:**517–527: Analysis of baculovirus genomes with restriction endonucleases.		
Granados, R. R.	**NPV nucleocapsid uncoating is different from that of GV, uncoating occurs after entry through gut cell nuclear pore (NPV) and not at the nuclear pore (GV)** *Virol.* **90:**170–174: Early events in the infection of *Heliothis zea* midgut cells by a baculovirus.	Jun 1978	Oct 1978
Miller, L. K., Dawes, K. P.	**Preliminary physical map of the *Ac*MNPV genome** **Order of *Bam*HI and *Xma*I REN fragments** **Partial order of *Eco*RI and *Hin*dIII** Preliminary comparison of genotypic variants *J. Virol.* **29:**1044–1055. Physical map of the DNA genome of *Autographa californica* nuclear polyhedrosis virus.	Oct 1978	Mar 1979
	Protease activity is not associated with viral occlusions isolated from cell culture when compared to occlusions purified from insects		
Maruniak, J. E., Summers, M. D., Falcon, L. A., Smith, G. E.	*Intervirology* **11:**82–88; *Autographa californica* nuclear polyhedrosis virus structural proteins compared from *in vivo* and *in vitro* sources.	Jan 1978	May 1979
Zummer, M., Faulkner, P.	*J. Inverteb. Pathol.* **33**, 382–384; Absence of protease in baculovirus polyhedral bodies propagated *in vitro.*	Jun 1978	May 1979
Smith, G. E., Summers, M. D.	**Physical mapping and order of sixty-one REN for *Ac*MNPV** **Physical maps of clonal variants**	Jan 1979	Jun 1979

	J. Virol. **30:**828–838: Restriction maps of five *Autographa californica* MNPV variants, *Trichoplusia ni* MNPV, and *Galleria mellonella* MNPV DNAs with endonucleases *Sma*I, *Kpn*I, *Bam*HI, *Sac*I, *Xho*I, and *Eco*RI.		
Brown, M., Crawford, M., Faulkner, P.	**ts mutants** **Report patterns of virus-specific DNA synthesis** *J. Virol.* **31:**190–198: Isolation of temperature sensitive mutants and assortment into complementation groups.	Jan 1979	July 1979
Lee, H. H., Miller, L. K.	**ts mutants** **Detectable viral DNA at 6–9 h p.i.** *J. Virol.* **31:**240–252: Isolation, complementation, and initial characterization of temperature-sensitive mutants of the baculovirus *Autographa californica* nuclear polyhedrosis virus.	Mar 1979	July 1979
Rohrmann, G. F., Bailey, T. J., Brimhall, B., Becker, R. R., Beaudreau, G. S.	**The N-terminal amino acids of polyhedrins are highly conserved** *Pro. Nat. Acad. Sci. A* **76:**4976–4980: Tryptic peptide analysis and NH_2-terminal amino acid sequences of polyhedrins of two baculoviruses from *Orgyia pseudotsugata.*	July 1979	Oct 1979
Carstens, E. B., Tjia, S. T., Doerfler, W.	**Study of infected cell specific proteins (ICSP) during AcMNPV infection** *Virol.* **99:**386–398: Infection of *Spodoptera frugiperda* cells with *Autographa californica* nuclear polyhedrosis virus. I. Synthesis of intracellular proteins after infection.	July 1979	Dec 1979
Tjia, S. T., Carstens, E. B., Doerfler, W.	**Viral DNA replication initiates at 5 hours post infection** *Virol.* **99:**399–409: Infection of *Spodoptera frugiperda* cells with *Autographa californica* nuclear polyhedrosis virus. II. The viral DNA and the kinetics of its replication.	July 1979	Dec 1979
Tinsley, T. W.	**The safety and potential of insecticidal viruses** *Ann. Rep. Entomol. Soc. Ont.* **24:**63–87: The potential of insect pathogenic viruses as pesticidal agents.		1979

(*continues*)

TABLE I *(continued)*

Authors	Citation	Submitted	Published
van der Beek, C. P., Saaijer-Riep, J. D., Vlak, J. M.	**Hybridization selection, *in vitro* translation and immunoprecipitation: polyhedrin is encoded by the virus** *Virol.* **100:**326–333: On the origin of the polyhedral protein of *Autographa californica* nuclear polyhedrosis virus.	Sept 1979	Jan 1980
Burand, J. P., Summers, M. D., Smith, G. E.	**Transfection with baculovirus DNA** *Virol.* **101:**286–290: Transfection with baculovirus DNA.	Oct 1979	Feb 1980
Carstens, E. B., Tjia, S. T., Doerfler, W.	*Virol.* **101:**311–314: Infectious DNA from *Autographa californica* nuclear polyhedrosis virus.	Nov 1979	Feb 1980
Brown, M., Faulkner, P.	**Preliminary genetic map for *Ac*MNPV** **Demonstration of recombination between ts mutants: project correlations with a physical mapping, marker rescue and heteroduplex mapping.** *J. Gen. Virol.* **48:**247–251: A partial genetic map of the baculovirus, *Autographa californica* nuclear polyhedrosis virus, based on recombination studies with ts mutants.	Dec 1979	May 1980
Summers, M. D., Smith, G. E., Knell, J. D., Burand, J. P.	**A method for the physical mapping of viral genetic markers, virus-induced polypeptides, physical mapping of gene loci by intertypic marker rescue** **Polyhedrin maps within 70–89 map units of the *Ac*MNPV genome** **Recombination mapping between two closely related viruses (*Ac*MNPV, *Ro*MNPV)** *J. Virol.* **34:**693–703: Physical maps of *Autographa californica* and *Rachoplusia ou* nuclear polyhedrosis virus recombinants.		Jun 1980
Dobos, P., Cochran, M. A.	**Viral protein synthesis is sequentially ordered in cascades** *Virol.* **103:**446–464: Protein synthesis in cells infected with *Autographa californica* nuclear polyhedrosis virus (Ac-NPV): the effect of cytosine arabinoside.	Feb 1980	Jun 1980

Wilkie, G. E., Stockdale, H., Pirt, S. V.	**A serum-free medium for insect cells** *Dev. Biol. Stand.* **46:**29–37: Chemically-defined media for production of insect cells and viruses *in vitro.*		1980
Granados, R. R., Lawler, K. A.	**Infectious parental virus may penetrate directly through the gut cell to infect the insect** *Virol.* **108:**297–308: *In vivo* pathway of *Autographa californica* baculovirus invasion and infection.	Aug 1980	Jan 1981
Smith, G. E., Summers, M. D.	**SDS-Page gels comparing baculoviruses from all subgroups Immunological cross-sections among baculoviruses** *J. Virol.* **39:**125–137: Application of a novel radioimmunoassay to identify baculovirus structural proteins that share interspecies antigenic determinants.	Dec 1980	July 1981
Rohrmann, G. F., Pearson, M. N., Bailey, T. J., Becker, R. R., Beaudreau, G. S.	**The amino acid sequences are conserved among NPV polyhedrins** *J. Mol. Evol.* **17:**329–333: N-terminal polyhedron sequences and occluded baculovirus evolution.	Dec 1980	Sept 1981
Miller, L. K.	**Marker rescue and physical map positions of seven ts mutants on the *Ac*NPV genome.** *J. Virol.* **39:**973–976: Construction of a genetic map of the baculovirus *Autographa californica* nuclear polyhedrosis virus by marker rescue of temperature-sensitive mutants.	Mar 1981	Sept 1981
Vlak, J. M., Smith, G. E., Summers, M. D.	**The Polyhedrin gene is located in *Eco*RI-I between map units 0 and 0.045** *J. Virol.* **40:**762–771: Hybridization selection and *in vitro* translation of *Autographa californica* nuclear polyhedrosis virus mRNA.	Mar 1981	Dec 1981
Lubbert, H., Kruczek, I., Tjia, S., Doerfler, W.	**Construction of a genomic library of 21 of the 24 *Ac*MNPV *Eco*RI fragments** *Gene* **16:**343–345: The cloned *Eco*RI fragments of *Autographa californica* nuclear polyhedrosis virus DNA.	Aug 1981	Dec 1981

(continues)

TABLE I (*continued*)

Authors	Citation	Submitted	Published
Kozlov, E. A., Levitina, T. L., Gusak, N. M., Ovander, M. N., Serebryany, S. B.	**Comparison of polyhedrin protein amino acid sequences** *Bioorgan Chimija* **7:**1008–1015: Comparison of amino acid sequences of inclusion body proteins of nuclear polyhedrosis viruses of *Bombyx mori, Porthetria dispar* and *Galleria mellonella.*		1981
Vlak, J. M., Smith, G. E.	**A consensus orientation physical map for *Ac*MNPV** *J. Virol.* **41:**1118–1121: Orientation of the genome of *Autographa californica* nuclear polyhedrosis virus: a proposal.	Sept 1981	Mar 1982
Volkman, L. E., Goldsmith, P. A.	**Establishment of a rapid virus titer assay (40 hours) not dependant on polyhedra production** *Appl. Environ. Microbiol.* **44:**227–233: Generalized immunoassay for *Ac*NPV infectivity *in vitro.*	Nov 1981	July 1982
Rohrmann, G. F., Leisy, D. J., Chou, K.-C., Pearson, G. D., Beaudreau, G. S.	**cDNA mapping of a baculovirus (polyhedrin) mRNA** **Comparison of determined amino acid sequence of *Op*MNPV polyhedrin with the 5′ nucleotide coding sequences** **Orientation of gene and region of the insert encoding the N-terminus of the polyhedrin protein were determined by DNA sequencing; R-loop mapping indicated mRNA is 980 ± 75 bases; no observable intron** **Polyhedrin mRNA is not spliced** *Virol.* **121:**51–60: Identification, cloning, and R-loop mapping of the polyhedrin gene from the multicapsid nuclear polyhedrosis virus of *Orgyia pseudotsugata.*	May 1982	Aug 1982
Smith, G. E., Vlak, J. M., Summers, M. D.	**Preliminary translational map of *Ac*MNPV genome by hybrid selection, *in vitro* translation: mapping of 19 translation products for virus specific proteins in early and late infection** ***Hin*dIII-V to a large extent contains the coding sequence for polyhedrin, actual gene extends into *Hin*dIII-F and possibly *Hin*d-T**	Mar 1982	Oct 1982

	Temporal and abundant levels of p10 and polyhedrin proteins are different and possibly under the control of separate promoters **p10 maps to *Hin*dIII-P** *J. Virol.* **44:**199–208: *In vitro* translation of *Autographa californica* nuclear polyhedrosis virus early and late mRNAs.		
Miller, D. W., Miller, L. K.	**A transposable element is integrated into a baculovirus viral genome** *Nature* **299:**562–564: A virus mutant with an insertion of a *Copia*-like transposable element.	July 1982	Oct 1982
Darbon, H., Zlotkin, E., Kopeyan, C., van Rietschoten, J., Rochart, H.	**The amino acid sequence of the insect-specific scorpion toxin AaIT** *Int. J. Pept. Protein. Res.* **20:**320–330: Covalent structure of the insect toxin of the North African scorpion *Androctonus australis* Hector.		Oct 1982
Adang, M. J., Miller, L. K.	**Transcription map of cDNA's for late gene products** *J. Virol.* **44:**782–793: Molecular cloning of DNA complementary to mRNA of the baculovirus *Autographa californica* nuclear polyhedrosis virus: location and gene products of RNA transcripts found late in infection.	Apr 1982	Dec 1982
Esche, H., Lubbert, H., Siegmann, B., Doerfler, W.	**A preliminary map of early and late *Ac*MNPV gene products mapped by cell-free translation of virus specific mRNA** *EMBO J.* **1:**1629–1633: The translational map of the *Autographa californica* nuclear polyhedrosis virus (AcNPV) genome.	Nov 1982	Dec 1982
Fraser, M. J., Smith, G. E., Summers, M. D.	**Acquisition of host cell DNA sequences by baculoviruses** *J. Virol.* **47:**287–300: Relationship between host DNA insertions and FP mutants of *Autographa californica* and *Galleria mellonella* nuclear polyhedrosis viruses.	Feb 1983	Aug 1983
Rohel, D. Z., Cochran, M. A., Faulkner, P.	**The p10 gene maps to *Ac*MNPV map units 87.35–89.55**	July 1982	Jan 1983

(*continues*)

TABLE I *(continued)*

Authors	Citation	Submitted	Published
	Virol. **124:**357–365: Characterization of two abundant mRNA's of *Autographa californica* nuclear polyhedrons virus present late in infection.		
Smith, G. E., Vlak, J. M., Summers, M. D.	**S1 mapping of 5′ and 3′ ends of the *Ac*NPV polyhedrin gene** **p10 gene maps to *Hin*dIII-P** **Physical map of polyhedrin and p10 mRNA locations** *J. Virol.* **45:**215–225: Physical analysis of *Autographa californica* nuclear polyhedrosis virus transcripts for polyhedron and 10,000-molecular-weight protein.	Aug 1982	Jan 1983a
Miller, L. K., Lingy, A. J., Bulla, L. A.	**Bacterial, Viral and Fungal Insecticides** Science **219**:715–721: Bacterial, Viral and Fungal Insecticides.		Feb 1983
Cochran, M. A., Faulkner, P.	**Discovery of the *Ac*MNPV hr1–5 regions; postulated a potential role for hrs as origins of replication** *J. Virol.* **45:**961–970: Location of homologous DNA sequences interspersed at five regions in the baculovirus AcNPV genome.	Aug 1982	Mar 1983
Smith, G. E., Fraser, M. J., Summers, M. D.	**The polyhedrin gene is not essential for infection and therefore is not essential in cell culture** **Introduction of site-specific mutations into the polyhedrin gene** **Exchange of mutated gene for wild-type polyhedrin gene by co-transfection and homologous recombination** **Screening and selection for recombinant virus by occlusion negative plaques** *J. Virol.* **46:**584–593: Molecular engineering of the *Autographa californica* nuclear polyhedrosis virus genome: deletion mutations within the polyhedrin gene.	Nov 1982	May 1983b
Smith, G. E., Summers, M. D., Fraser, M. J.	**Genetic engineering of *Ac*MNPV for foreign gene expression** **DNA sequence of the transcriptional promoter and 5′end of polyhedrin open reading frame**	May 1983	Dec 1983c

	Abundant production of nonfused and fusion recombinant proteins in insect cells *Mol. Cell. Biol.* **3:**2156–2165: Production of human beta interferon in insect cells infested with a baculovirus expression vector.		
Hooft van Iddekinge, B. J. L., Smith, G. E., Summers, M. D.	**The nucleotide sequence of the *Ac*MNPV polyhedrin gene** *Virol.* **131:**561–565: Nucleotide sequence of the polyhedrin gene of *Autographa californica* nuclear polyhedrosis virus.	July 1983	Dec 1983
Pennock, G. D., Shoemaker, C., Miller, L. K.	*Mol. Cell. Biol.* **4:**399–406: Strong and regulated expression of *Escherichia coli* β-galactosidase in insect cells with a baculovirus vector.	Sept 1983	Mar 1984
Volkman, L. E., Goldsmith, P. A., Hess, R. T., Faulkner, P.	**Preliminary identification and demonstration that gp64 is a "neutralizing antigen" of the budded virus** *Virol.* **133:**354–362: Neutralization of budded *Autographa californica* NPV by a monoclonal antibody: identification of the target antigen.	Oct 1983	Mar 1984
Lubbert, H., Doerfler, W.	**Viral mRNAs exist as overlapping sets with common 3′ or 5′ termini** *J. Virol.* **52:**255–265: Transcription of overlapping sets of RNAs from the genome of *Autographa californica* nuclear polyhedrosis viruses: a novel method for mapping RNAs.	Apr 1984	Oct 1984
Kuzio, J., Rohel, D. Z., Curry, C. J., Krebs, A., Carstens, E. B., Faulkner, P.	**First step for construction of *Ac*MNPV-p10 vectors for foreign gene expression or use in viral pesticides** *Virology* **139:**414–418: Nucleotide sequence of the p10 polypeptide gene of *Autographa californica* nuclear polyhedrosis virus.	July 1984	Dec 1984
Maeda, S., Kawai, T., Obinata, M., Chika, T., Horiuchi, T., Maekawa, K., Nakasuji, K., Saeki, Y., Sato, Y., Yamada, K., Furusawa, M.	**Foreign proteins can be abundantly produced by recombinant baculoviruses in infected insect larvae** *Proc. Jpn. Acad.* **60(Ser. B):**423–426: Characteristics of human interferon-α produced by a gene transferred by a baculovirus vector in the silkworm, *Bombyx mori*.	Dec 1984	Dec 1984

(*continues*)

TABLE I (*continued*)

Authors	Citation	Submitted	Published
Volkman, L. E., Goldmith, P. A.	***Ac*MNPV gp64 is responsible for fusogenic activity (hemolysis) and increased infection over ODV in cell culture** **BV infects via pH sensitive pathway (endosome) *in vitro* while ODV does not** *Virol.* **143:**185–195: Mechanism of neutralization of budded *Autographa californica* nuclear polyhedrosis virus by a monoclonal antibody: inhibition of entry by adsorptive endocytosis.	Dec 1984	May 1985
Knebel, D., Lubbert, H., Doerfler, W.	**Engineering and transient expression of the *Ac*MNPV p10 gene promoter for foreign gene expression** *EMBO J.* **4:**1301–1306: The promoter of the late p10 gene in insect nuclear polyhedrosis virus *Autographa californica*: activation by viral gene products and sensitivity to DNA methylation.	Mar 1985	May 1985
Maeda, S., Kawai, T., Obinata, M., Fujiwara, H., Horiuchi, T., Saeki, Y., Sato, Y., Furusawa, M.	*Nature* **315:**592–594: Production of human α-interferon in silkworm using a baculovirus vector.	Oct 1984	Jun 1985
Fraser, M. J., Brusca, J. S., Smith, G. D., Summers, M. D.	*Virol.* **145:**356–361: Transposon-mediated mutagenesis of a baculovirus.	May 1984	Sept 1985
Carbonell, L. F., Klowden, M. J., Miller, L. K.	**Expression of two foreign genes with the baculovirus vector using the polyhedrin promoter and the heterologous viral promoter of RSV-LTR** **Baculoviruses can enter and express genes in mammalian cells** **Considered the use of baculovirus early promoters to express in replication-refractive cells**	Mar 1985	Oct 1985

	Importance to widening host range with insect specific neurotoxins *J. Virol.* **56:**153–160: Baculovirus-mediated expression of bacterial genes in dipteran and mammalian cells.		
Tramper, J., Williams, J. B., Joustra, D., Vlak, J. M.	**Hallmark for development of bioreactor production using insect cells: shear sensitivity is a major limitation** *Enzyme Microb. Technol.* **8:**33–36: Shear sensitivity of insect cells in suspension.	Mar 1985	Jan 1986
Guarino, L. A., Summers, M. D.	**Development of the transient expression assay for Sf9 cells using *Ac*MNPV gene promoters not regulated by baculovirus infection** **A baculovirus immediate early gene (IE1) which transactivates delayed early (39K) gene expression** *J. Virol.* **57:**563–571: Functional mapping of a *trans*-activating gene required for expression of a baculovirus delayed early gene.	July 1985	Feb 1986a
Rohrmann, G. F.	**Prediction of the functional significance of the 12mer AATAAGTATTTT in the initiation of late mRNA synthesis and polyhedrin expression** *J. Gen. Virol.* **67:**1499–1513: Review Article: Polyhedrin Structure.		Aug 1986
Matsuura, Y., Possee, R. D., Bishop, D. H. L.	**The site of foreign gene insertion in the *polyhedrin* 5′ untranslated leader is important for high level expression** *J. Gen. Virol.* **67:**1515–1529: Expression of the S-coded genes of *Lymphocytic choriomengitis* arena virus using a baculovirus vector.	Apr 1986	Aug 1986
Guarino, L. A., Summers, M. D.	**AcMNPV hrs function as transcriptional enhancers in transfected cells** *Virol.* **60:**215–223: Homologous DNA of *Autographa californica* nuclear polyhedrosis virus enhances delayed-early gene expression.	Apr 1986	Oct 1986b
Bishop, D. H.	**Field testing of a genetically marked baculovirus-strategy** **Assessment of risk and environmental consequences of genetically engineered baculoviruses** *Nature* **323:**496: UK release of genetically marked virus.		Oct 1986

(*continues*)

TABLE I (*continued*)

Authors	Citation	Submitted	Published
Wilson, M. E., Mainprize, T. H., Friesen, P. D., Miller, L. K.	**Identification and DNA sequence of basic AcMNPV DNA binding protein (p6.9)** *J. Virol.* **61:**661–666: Location, transcription, and sequence of a baculovirus gene encoding a small arginine-rich polypeptide.	Aug 1986	Mar 1987
Summers, M. D., Smith, G. E.	**A BEVS (Baculovirus Expression Vector System) manual** A manual of methods for baculovirus vectors and insect cell culture procedures. Texas Agric. Exp. Station Bulletin No. 1555. p. 57		Apr 1987
Jeang, K.-T., Holmgren-Konig, M., Khoury, G.	**Correct mRNA splicing can occur in baculovirus-infected cells** *J. Virol.* **61:**1761–1764: A baculovirus vector can express intron-containing genes.	Oct 1986	May 1987
Matsuura, Y., Possee, R. D., Overton, H. A., Bishop, D. H.	**Optimal foreign gene expression relative to the site of insertion (+1 to −60 nt) in the polyhedrin gene leader sequence** *J. Gen. Virol.* **68:**1233–1250: Baculovirus expression vectors: the requirements for high level expression of proteins, including glycoproteins.	Feb 1987	May 1987
Estes, M. K., Crawford, S. E., Penaranda, M. E., Petrie, B. L., Burns, J. W., Chan, W.-K., Ericson, B., Smith, G. E., Summers, M. D.	**Recombinant rotavirus VP6 assembles into tubules** *J. Virol.* **61:**1488–1494: Synthesis and immunogenicity of the rotavirus major capsid antigen using a baculovirus expression system.	Oct 1986	May 1987
Guarino, L. A., Summers, M. D.	**Nucleotide sequence of an *Ac*MNPV immediate early gene Demonstration that IE1 is expressed immediate early and through late in infection** *J. Virol.* **61:**2091–2099: Nucleotide sequence and temporal expression of a baculovirus regulatory gene.	Feb 1987	July 1987

Emery, V. C., Bishop, D. H. L.	**Multigene expression vectors: Engineering of the polyhedrin gene promoter in opposite orientations for an occ$^+$ vector (pAcVC2)** *Protein Eng.* **1:**359–366: The development of multiple expression vectors for high level synthesis of *Ac*NPV polyhedrin protein by a recombinant baculovirus.	Jun 1987	Aug–Sep 1987
Vlak, J. M., Klinkenberg, F. A., Zaal, K. J., Usmany, M., Klinge-Roode, E. C., Geervliet, J. B., Roosien, J., van Lent, J. W. M.	**Foreign gene expression with the *Ac*MNPV p10 gene promoter** *J. Gen. Virol.* **69:**765–776: Functional studies on the p10 gene of *Autographa californica* nuclear polyhedrosis virus using a recombinant expressing a p10-β-galactosidase fusion gene.	Dec 1987	Apr 1988
Chisholm, G. E., Henner, D. J.	**Splicing of baculovirus genes** *J. Virol.* **62:**3193–3200: Multiple early transcripts and splicing of the *Autographa californica* nuclear polyhedrosis virus IE-1 gene.	Mar 1988	Sept 1988
Carbonell, C. F., Hodge, M. R., Tomalski, M. D., Miller, L. K.	**Attempt to express an insecticidal insect toxin-1 gene of the scorpion *Buthus eupeus:* no effect** *Gene* **73:**409–418: Synthesis of a gene coding for an insect-specific scorpion neurotoxin and attempt to express it using baculovirus vectors.	Feb 1988	Dec 1988
Pearson, M. N., Quant-Russell, R. L., Rohrmann, G. F., Beaudreau, G. S.	**Identification of the major baculovirus capsid protein p39** *Virol.* **167:**407–413: P39, a major baculovirus structural protein: Immunocytochemical characterization and genetic location.	Jun 1988	Dec 1988
Jarvis, D. L., Summers, M. D.	**Secretion of recombinant proteins is compromised during late stages of infection** *Mol. Cell. Biol.* **9:**214–223: Glycosylation and secretion of human tissue plasminogen activator in recombinant baculovirus-infected insect cells.	July 1988	Jan 1989

(*continues*)

TABLE I (*continued*)

Authors	Citation	Submitted	Published
Blissard, G., Quant-Russell, R. L., Rohrmann, G. F., Beaudreau, G. S.	*Virol.* **168:**354–362: Nucleotide sequence, transcriptional mapping, and temporal expression of the gene encoding P39, a major structural protein of the multicapsid nuclear polyhedrosis virus of *Orgyia pseudotsugata.*	Aug 1988	Feb 1989
Whitford, M., Stewart, S., Kuzio, J., Faulkner, P.	**Cloning and sequencing of *gp64*, the major envelope protein of BV** *J. Virol.* **63:**1393–1399: Identification and sequence analysis of a gene encoding *gp67,* an abundant envelope glycoprotein of the baculovirus *Autographa californica* nuclear polyhedrosis virus.	Aug 1988	Mar 1989
Devlin, J. J., Devlin, P. E., Clark, R., O'Rourke, E. C., Levenson, C., Mark, D. F.	**Substitution of the signal peptide of a secreted recombinant protein to enhance and direct the more efficient secretion** *Bio / Technology* **7:**286–292: Novel expressions of chimeric plasminogen activators in insect cells.	Oct 1988	Mar 1989
Keddie, B. A., Aponte, G. W., Volkman, L. E.	**Studies describing the *in vivo* pathway of baculovirus infection providing new insights for strategies involving the use of genetically engineered baculovirus pesticides** *Science* **243:**1728–1730: The pathway of infection of *Autographa californica* nuclear polyhedrosis virus in an insect host.	Nov 1988	Mar 1989
Thiem, S., Miller, L. K.	*J. Virol.* **63:**2008–2018: Identification, sequence, and transcription mapping of the major capsid protein gene of the baculovirus *Autgrapha californica* nuclear polyhedrosis virus.	Nov 1988	May 1989
Dolin, R., Graham, B. S., Greenberg, S. B., Tacket, C. O., Belseh, R. B., Midthun, K., Clements, M. L., Gorse, G. J., Horgan, B. W., Atmar, R. L.	**First report of a baculovirus recombinant protein tested in humans: HIV-gp160 envelope protein was safe and immunogenic** *Ann. Intern. Med.* **114:**119–127: The safety and immunogenicity of a human immuno deficiency virus type 1 (HIV-1) recombinant gp160 candidate vaccine in humans.	Oct 1988	Mar 1989

Urakawa, T., Ferguson, M., Minor, P. D., Cooper, J., Sullivan, M., Almond, J. W., Bishop, D. H. L.	**The development and use of baculovirus multigene expression vectors for studies of the structure and assembly of heteroligomer protein particles, virus-like particles and multiprotein complexes** *J. Gen. Virol.* **70:**1453–1463: Synthesis of immunogenic, but non-infectious, poliovirus particles in insect cells by a baculovirus expression vector	Nov 1988	Jun 1989
Cary, L. C., Goegel, M., Corsaro, B. G., Wang, H.-G., Rosen, E., Fraser, M. J.	*Virology* **172:**156–169; Transposon mutagenesis of baculoviruses: analysis of *Trichoplusia ni* transposon IFP2 insertions within the FP locus of nuclear of polyhedrosis viruses.	Oct 1988	Sept 1989
Tomalski, M. D., Kutney, R., Bruce, W. A., Brown, M. R., Blum, M. S., Travis, J.	**Identification a of potential insecticidal toxin purified from the mite** *Toxicon* **27:**1151–1167: Purification and characterization of insect toxins derived from the mite, *Pyemotes tritici.*	Apr 1989	Oct 1989
Zuidema, D., Klinge Roode, E. C., van Lent, J. W. M., Vlak, J. M.	**A recombinant virus with improved virulence (LD_{50}) by deleting the gene for the polyhedral envelope** *Virol.* **173:**98–108: Construction and analysis of an *Autographa californica* nuclear polyhedrosis virus mutant lacking the polyhedral envelope.	May 1989	Nov 1989
Maeda, S.	**A recombinant baculovirus exhibiting an increase in "insecticidal" activity by expression of a diuretic hormone gene using a heterologous signal sequence of the *Drosophila* CP2 cuticle protein** *Biochem. Biophys. Res. Commun.* **165:**1177–1183: Increased insecticidal effect by a recombinant baculovirus carrying a synthetic diuretic hormone gene.	Nov 1989	Dec 1989
Merryweather, R. A., Weyer, U., Harris, M. P., Hirst, M., Booth, T., Possee, R. D.	**Use of dual promoters for *Bacillus thuringiensis* delta endotoxin expression with an occluding positive baculovirus pesticide: no effect** *J. Gen. Virol.* **71:**1535–1544: Construction of genetically engineered baculovirus insecticides containing the *Bacillus thuringiensis* subsp. *kurstaki* HD-73 delta endotoxin.	Oct 1989	Feb 1990

(*continues*)

TABLE I (*continued*)

Authors	Citation	Submitted	Published
Hammock, B. D., Bonning, B. C., Possee, R. D., Hanzlik, T. N., Maeda, S.	Expression of juvenile hormone esterase *Nature* **344:**458–461: Expression and effects of the juvenile hormone esterase in a baculovirus vector.	Oct 1989	Mar 1990
French, T. J., Roy, P.	*J. Virol.* **64:**1530–1536: Synthesis of bluetongue virus (BTV) core-like particles by a recombinant baculovirus expressing the two major structural core proteins of BTV.		Apr 1990a
Dee, A., Belagaje, R. M., Ward, K., Chio, E., Lai, M. H.	**The synthetic gene for AaIT has "insecticidal" activity** *Bio/Technology* **8:**339–342: Expression and secretion of a functional scorpion insecticidal toxin in cultured mouse cells.	Aug 1989	Apr 1990
Orentas, R. J., Heldreth, J. E. K., Obah, B., Polydefkis, M., Smith, G. E., Clements, M. L., Siliciano, R. F.	**A baculovirus expressed subunit antigen can induce T-cell response in humans** *Science* **248:**1234–1236: Induction of $CD4^+$ human cytolytic T cells specific for HIV-infected cells by a gp160 subunit vaccine.	Feb 1990	Jun 1990
Jarvis, D. L., Fleming, J. G. W., Kovacs, G. R., Summers, M. D., Guarino, L. A.	**The Sf9 cell line is stably transformed with the AcMNPV IE1 promoter for continuous expression** *Bio/Technology* **8:**950–955: Use of early baculovirus promoters for continuous expression and efficient processing of foreign gene products in stably transformed lepidopteran cells.	Mar 1990	Oct 1990
Kitts, P. A., Ayres, M. A., Possee, R. D.	**Major improvement for a higher efficiency selection (30%) of recombinant baculoviruses** *Nuc. Acid Res.* **18:**5667–5672: Linearization of baculovirus DNA enhances the recovery of recombinant virus expression vectors.	July 1990	Oct 1990
French, T. J., Marshall, J. J. A., Roy, P.	*J. Virol.* **64:**5696–5700: Assembly of double-shelled virus-like particles of bluetongue virus by the simultaneous expression of four structural proteins.	Oct 1989	Dec 1990b

Tessier, D. C., Thomas, D. Y., Khouri, H. E., Laliberts, F., Vernet, T.	*Gene.* **98:**177–183: Enhanced secretion from insect cells of a foreign protein fused to the honeybee mellittin signal peptide.	July 1990	Feb 1991
Wang, X., Ooi, B. G., Miller, L. K.	**The development of a variety of hybrid promoter constructs (*polyhedrin*, p10, capsid, basic core protein) in an occlusion positive vector for foreign gene expression** *Gene* **100:**131–137: Baculovirus vectors for multiple gene expression and for occluded virus production.	Oct 1990	Apr 1991
Loudon, P. T., Hirasawa, T., Oldfield, S., Murphy, M., Roy, P.	*Virol.* **182:**793–801: Expression of outer capsid protein VP5 of two bluetongue viruses and synthesis of chimeric double-shelled virus-like particles using combination of recombinant baculovirus.	Jan 1991	Jun 1991
Tomalski, M. D., Miller, L. K.	*Nature* **352:**82–85: Insect paralysis by baculovirus-mediated expression of a mite neurotoxin gene.	Feb 1991	July 1991
Stewart, L. M. D., Hirst, M., Ferber, M. L., Merryweather, A. T., Cayley, J., Possee, R. D.	**Use of a heterologous secretory signal sequence to improve insecticidal effects** *Nature* **352:**85: Construction of an improved baculovirus insecticide containing an insect-specific toxin gene.	Apr 1991	July 1991
McCutchen, B. F., Chandary, V., Crenshaw, R., Maddox, D., Kamita, S. G., Palekar, N., Volrath, S., Fowler, E., Hammock, B. D., Maeda, S.	*Bio/Technology* **9:**848: Development of a recombinant baculovirus expressing an insect-selective neurotoxin: Potential for pest control.	Apr 1991	Sept 1991
Kool, M., Voncken, F. J. L., VanLier, T., Vlak, J. M.	**Discovered and documented defective interference in baculovirus replication in cell culture** *Virol.* **183:**739–746: Detection and analysis of *Autographa californica* nuclear polyhedrosis virus mutants with defective interfering properties.	Mar 1991	Aug 1991
Maeda, S., Volrath, S. L., Hanzlik, T. N., Harper, S. A., Majima, K., Maddox, D. W., Hammock, B. D., Fowler, E.	*Virol.* **184:**777: Insecticidal effects of an insect specific neurotoxin expressed by a recombinant baculovirus.	Apr 1991	Oct 1991

(*continues*)

TABLE I (*continued*)

Authors	Citation	Submitted	Published
Clem, R. J., Fechheimer, M., Miller, L. K.	**Discovery of an anti-apoptosis gene in baculovirus** *Science* **254:**1388–1390: Prevention of apoptosis by a baculovirus gene during infection of insect cells.	Jun 1991	Nov 1991
Tomalski, M. D., Miller, L. K.	**Use of hybrid promoters to express toxin** ***Bio/Technology* 10:545: Expression of a paralytic neurotoxin gene to improve insect baculoviruses as biopesticides.**	Dec 1991	May 1992
Pearson, M., Bjornson, R., Pearson, G., Rohrmann, G. F.	**hr sequences function to enhance baculovirus replication** *Science* **257:**1382–1384: The *Autographa californica* baculovirus genome: evidence for multiple replication origins.	Apr 1992	Sept 1992
Basak, A. K., Stuart, D. I., Roy, P.	**Crystallographic structure of a protein produced with the BEVS** *J. Mol. Biol.* **228:**687–689: Preliminary crystallographic study of bluetongue virus capsid protein, VP7.	Jan 1992	Nov 1992
Kitts, P. A., Possee, R. D.	**BEVS vectors improved by high efficiency selection (approaching 100%) of recombinant baculoviruses** *BioTechniques* **14:**810–817: A method for producing recombinant baculovirus expression vectors at high frequency.		May 1993
Luckow, V. A., Lee, S. C., Barry, G. F., Olins, P. O.	**Production of a recombinant baculovirus in *Escherichia coli* by site-specific transposition *in vivo* of a foreign gene: The BAC-to-BAC Expression System** *J. Virol.* **67:**4566–4579: Efficient generations of infectious recombinant baculoviruses by site-specific transposition mediated insertion of foreign genes into a baculovirus genome propagated in *Escherichia coli.*	Oct 1992	Aug 1993
Wood, H. A., Hughes, P. R., Shelton, A.	**Field release of recombinant virus in the U.S.** *Environ. Entomol.* **23:**211: Field studies of the co-occlusion strategy with a genetically altered isolate of the *Autographa californica* nuclear polyhedrosis virus.	Apr 1993	Apr 1994

Engelhard, K. K., Kam-Morgan, L. N. W., Washburn, J. O., Volkman, L. E.	*Proc. Nat. Acad. Sci. A.* **91:**3224–3227: The insect tracheal system: A conduit for the systemic spread of *Autographa californica* M nuclear polyhedrosis virus.	Dec 1993	Apr 1994
Braunagel, S. C., Summers, M. D.	**Comprehensive comparisons of ODV and BV envelope and nucleocapsid proteins, antigenicity, and lipid and fatty acid compositions** *Virol.* **202:**315–328: *Autographa californica* nuclear polyhedrosis virus PDV, and ECV viral envelopes and nucleocapsids: structural proteins, antigens, lipid and fatty acid profiles.	Dec 1993	July 1994
Cory, J. S., Hirst, M. L., Williams, T., Hails, R. S., Goulson, D., Green, B. M., Carty, T. M., Possee, R. D., Cayley, P. J., Bishop, D. H. L.	**Field trial of genetically engineered improved viral pesticide** *Nature* **370:**138–140: field trial of a genetically improved baculovirus insecticide.	Jan 1994	July 1994
Ayres, M. D., Howard, S. C., Kuzio, J., Lopez-Ferber, M., Possee, R. D.	**The sequence of the *Ac*MNPV genome** *Virol.* **202:**586–605: The complete DNA sequence of *Autographa californica* nuclear polyhedrosis virus.	Jan 1994	Aug 1994
Hsu, T.-A., Eiden, J. J., Bourgarel, P., Meo, T., Betenbaugh, M. J.	**Co-expressed chaperone can increase intra-cellular soluble and functional antibody yields** *Pro. Exp. Purif.* **5:**595–603: Effect of co-expressing chaperone BiP on functional antibody production in the baculovirus systems	May 1994	Aug 1994
Martens, J.	*Thesis Wageningen*, p. 135 ISBN 90-5485-241-7; Development of a baculovirus insecticide exploring the *Bacillus thuringiensis* insecticidal crystal protein.		1994
Grimes, J., Basak, A. K., Roy, P., Stewart, I.	*Nature* **373:**167–170: The crystal structure of bluetongue virus VP7.	Sept 1994	Jun 1995

(*continues*)

TABLE I (*continued*)

Authors	Citation	Submitted	Published
Powers, D. C., Smith, G. E., Anderson, E. L., Kenney, D. J., Hanchett, C. S., Wilkinson, B. E., Volvovitz, F., Belshe, R. B., Treanor, J. J.	**Influenza vaccine containing baculovirus expressed, purified recombinant uncleaved hemagglutinin from influenza A virus was equal or better to natural flu vaccine in safety and protection** *J. Infect. Dis.* **171:**1595–1599: Influenza A virus vaccines containing purified recombinant H3 hemagglutinin are well tolerated and induce protective immune responses in healthy adults.	Oct 1994	Jun 1995
Ignoffo, C. M., Garcia, C., Zuidema, D., Vlak, J. M.	*J. Inverteb. Pathol.* 66:212–213: Relative *in vivo* activity and simulated sunlight-UV stability of inclusion bodies of a wild-type and an engineered polyhedral envelope-negative isolate of the nucleopolyhedrosis virus of *Autographa californica.*	Mar 1995	Sept 1995
Mori, H., Yamao, M., Nakazawa, H., Sugahara, Y., Shirai, N., Matsubara, F., Sumida, M., Imamura, T.	**Recombinant baculoviruses can be used as a vector for transovarian transmission of foreign genes in the silkworm** *Nature Biotechnology* **13:**1005–1007: Transovarian transmission of a foreign gene in the silkworm, *Bombyx mori*, by *Autographa californica* nuclear polyhedrosis virus.	May 1995	Sept 1995
Hoffman, C., Sandig, V., Jennings, G., Rudolph, M., Schleg, P., Strauss, M.	**Baculovirus–Mediated gene delivery in mammalian cells** *Pro. Nat. Acad. Sci. A* **92:**10099–10103: Efficient gene transfer into human hepatocytes by baculovirus vectors.	Jun 1995	Oct 1995
Monsma, S. A., Oomens, A. G. P., Blissard, G. W.	**Use of a stably transformed cell line to complement the deletion of an essential baculovirus gene** *J. Virol.* **70:**4607–4616: The GP64 envelope fusion protein is an essential baculovirus protein required for cell to cell transmission of infection.	Dec 1995	July 1996
Roy, P., Mikhailow, M., Bishop, D. H. L.	*Gene* **190:**119–129: Baculovirus multigene expression vectors and their use for understanding the assembly process of architecturally complex virus particles.	Mar 1996	Apr 1997

Wang, P., Granados, R. R.	**Demonstration of how *Ac*NPV overcomes the intestinal barrier in the host organism** *Pro. Nat. Acad. Sci. A* **94:**6977–6982: An intestinal mucin is the target substrate for a baculovirus enhancin.	Oct 1996	Jun 1997
Murges, D., Kremer, A., Knebel-Moensdorf, D.	***Ac*MNPV IE1 is functional in mammalian cells** *J. Gen. Virol.* **78:**1507–1510: Baculovirus transactivator IE1 is functional in mammalian cells.	Oct 1996	Jun 1997
Hawtin, R. E., Zarkowska, T., Arnold, K., Thomas, C. J., Gooday, G. W., King, L. A., Kuzio, J. A., Possee, R. D.	*Virol.* **238:**243–253: Liquefaction of *Autographa californica* nucleopolyhedorvirus-infected insects is dependent on the integrity of virus-encoded chitinase and cathepsin genes.	May 1997	Nov 1997
Handler, A. M., McCombs, S. D., Fraser, M. J., Saul, S. H.	**Stable germ-line transformation of non-host insects by a terminal repeat transposable element (*piggy-bac*) discovered in *Ac*MNPV FP mutants** *Pro. Nat. Acad. Sci. A* **95:**7520–7525: The lepidopteran transposon vector *piggybac,* mediates germ-line transformation in the Mediterranean fruit fly.	Oct 1997	Jun 1998
Yamao, M., Katayama, N., Nakazawa, H., Yamakawa, M., Hayashi, Y., Hara, S., Kamei, K., Hajime, M.	**Gene targeting and transgenesis in the silkworm by a baculovirus** *Genes and Development* **13:**511–516: Gene targeting in the silk worm by use of a baculovirus.	Nov 1998	Mar 1999
Toshiki, T., Chantal, T., Corinne, R., Toshio, K., Eappen, A., Kamba, M., Natus, K., Jean-Luc, T., Manchamp, B., Gerard, C., Shirk, P., Fraser, M. N., Prudhomme, J.-C., Couble, P.	*Nature Biotechnology* **18:**81–84: Germline transformation of the silkworm *Bombyx mori* L. using a *piggybac* transposon-derived vector.	July 1999	Jan 2000

(*continues*)

TABLE I *(continued)*

Authors	Citation	Submitted	Published
Hom, L. G., Volkman, L. E.	*Virol.* **277:**178–183: *Autographa californica* M nucleopolyhedrovirus chiA is required for processing V-CATH.	Jun 2000	Nov 2000
Zhao, Y., Chapman, D. A., Jones, I. M.	**Direct or "ET cloning" *in vitro* by placing a foreign gene under the regulation of the polyhedrin gene promoter to produce 100% recombinant viruses.** *Nuc. Acid. Res.* **31:** e6: Improving baculovirus recombination.	Jun 2002	Jan 2003
Braunagel, S. C., Russell, W. K., Rosas-Acosta, G., Russell, D. H., Summers, M. D.	**Proteomics analysis of ODV proteins** *Pro. Nat. Acad. Sci. A* **100:**9797–9802: Determination of the protein composition of the occlusion-derived virus of *Autographa californica* nucleopolyhedrovirus.	Jun 2003	Aug 2003
Kaba, S. A., Adriana, M. S., Wafula, P. O., Vlak, J. M., Van Oers, M. M.	**Deletion of viral genes that facilitate proteolysis of recombinant proteins provide improved vectors for foreign gene expression** *J. Virol. Meth.* **122:**113–118: Development of a chitinase and v-cathepsin negative bacmid for improved integrity of secreted recombinant proteins.	Mar 2004	Sept 2004
Kost, T. A., Condreay, J. P., Jarvis, D. L.	*Nature Biotechnology* **23:**567–575: Baculovirus as versatile vectors for protein expression in insect and mammalian cells.		May 2005
Hu, Yu-Chen	*Acta Pharmacol. Sin.* **26:**405–416: Baculovirus as a highly efficient expression vector in insect and mammalian cells.		May 2005

Both Cornalia (1856) and Maestri (1856) were able to associate the refractile bodies with a disease of silkworms. Fischer (1906) derived the name "polyhedrosis" to describe this disease now known to be caused by nucleopolyhedroviruses (NPVs). Compared to plant and vertebrate viruses, several insect pathogenic viruses, like baculoviruses, are unique in that the virion becomes embedded in a highly ordered protein crystal called the polyhedron (plural, polyhedra). The granuloviruses or granulosis viruses (GVs) are occluded baculoviruses also embedded in a protein crystal but with only one virion per occlusion of a much smaller size; GVs were first discovered by Paillot (1926). The ultrastructure of the NPV polyhedron was first described by Morgan *et al.* (1955), who also revealed that the virion was an enveloped nucleocapsid incorporated randomly in the protein crystal without apparent perturbation of crystal lattice structure and that many enveloped nucleocapsids were embedded in a single polyhedron. For decades these observations stimulated curiosity and prompted the search for, and understanding of, the origin of the protein crystals called "polyhedra" (now referred to as viral occlusions). Early studies of polyhedra included a search for the identity and source of the protein forming the crystalline lattice, and an understanding of the apparent ability of polyhedra to incorporate enveloped virus particles without disturbing the crystal lattice structure.

In order to characterize baculoviruses as infectious agents, it is important to place in perspective what was known about the structure of viruses. In the 1940s and early 1950s, viruses were understood only as filterable agents composed of protein and perhaps nucleic acid. Thus, the demonstration that polyhedra were dissolved by high pH (Bolle, 1894) still leaving infectious material must have been particularly intriguing. Komárek and Breindl (1924) suggested that infectious virus was occluded within the polyhedra and this proposal was later supported by Paillot and Gratia (1939). These observations were confirmed by the elegant biochemistry and virus purification techniques using the analytical ultracentrifuge by Bergold (1947). Although the first modern treatises on diseases of insects were published by Paillot (1930, 1933), it was Bergold's pioneering studies (1947; reviewed in 1953, 1958) and his comprehensive treatises on the biology, chemistry, and biochemistry of baculovirus that set the standards and established experimental protocols leading to the basic knowledge of baculovirus structure and composition. The result of these studies placed baculoviruses into the modern taxonomic structure for classification as NPVs and GVs. As a result of his contributions, Bergold is considered the father of modern day baculovirus molecular biology.

B. In Vivo *and* In Vitro *Developments for Virus Infection and Replication Studies, Preliminary Studies of Protein and Nucleic Acid Composition*

The next requirement to advance baculovirology into a modern context was the development and study of virus infection and assembly processes in insect tissues and cells *in vitro*. Goldschmidt (1915) cultured explants of *Cecropia* moth spermatozoa and Glaser (1917) demonstrated the *in vitro* formation of NPV by observing infected insect blood cells in hanging drops. Trager (1935) demonstrated baculovirus infection of cultured silkworm tissues and the subculture of virus infectivity to healthy tissues. Although not widely known, Gaw *et al.* (1959) were the first to report a monolayer, continuous culture of *Bombyx mori* cells (22 passages). Following rapidly on these seminal developments was the study of baculovirus infection and replication in primary blood cell cultures of *Peridroma saucia* by Martignoni and Scallion (1961), and the development of tissue culture media for several cell lines maintained in continuous culture (Grace, 1962).

The application and standardization of virus purification techniques coupled with the potential to propagate virus in continuous insect cell cultures under controlled and standardized conditions greatly advanced the studies of virus structure, infection, and host–cell interactions. The ability to purify virus, coupled with preliminary knowledge of the structure of infectious virions, stimulated curiosity of how the baculovirus penetrated the gut barrier during normal host infection to produce viral occlusions in a variety of insect tissues. This led Harrap and Robertson (1968) to show that during invasion of the host insect baculovirus replication occurs initially in the midgut columnar cell without the production of viral occlusions; yet viral occlusions in large numbers were easily observed in the nuclei of many other infected tissues of the insect. Prior to this discovery, viral replication in the host insect gut cell without the production of polyhedra was not detected by routine light microscopy and therefore it was assumed for some baculoviruses that the gut cells were not infected. Harrap and Robertson (1968) also noted that progeny virus replication in gut cells precedes infection of other cells and tissues in the host insect and was likely responsible for secondary infection. Electron microscopy (EM) observations further revealed "short projections" on the surface of progeny viral envelopes in the basal cytoplasm of the columnar cell. These projections had not been observed on the envelopes of virions assembled in the nuclei of fat body cells or other tissues. The nature of the structural differences for these viral forms was not understood at the time, but this was insight and partial confirmation of an earlier study by Vaughn and Faulkner (1963) who demonstrated the presence of two infectious baculovirus forms in an infected insect. They determined that while hemolymph from an infected insect would infect tissues cultured *in vitro*, virus purified from polyhedra would not. Summers (1969) confirmed the nature of host midgut cell infection with a GV and extended understanding of the mechanisms of virus entry and uncoating in the gut cell. His EM observations revealed that the GV entered the gut cell by fusion of the viral envelope with the columnar cell microvillar membrane and that GV nucleocapsid uncoating and release of the viral genome into the nucleus occurred by interaction with the nuclear pore. In 1978, Granados discovered an important difference in the uncoating of NPV nucleocapsids as compared to GV nucleocapsids; the NPV nucleocapsid passes through the nuclear pore before releasing its DNA genome into the nucleoplasm. Summers also noted that GV replication in midgut cells as compared to the fat body had a unique cell biology: viral replication occurred only in the nucleus of the midgut cell during invasion of the host insect, yet replication and viral assembly occurred throughout the nucleus and cytoplasm of infected fat body cells in which the nuclear envelope had apparently disassembled. The details of virus entry, nuclear pore interactions, and virus uncoating and penetration into the hemocoel were more comprehensively detailed in a subsequent study (Summers, 1971). Kawanishi *et al.* (1972) documented and confirmed that NPV entry to gut cells occurred by fusion of the viral envelope with the columnar cell microvillar membrane.

Concomitant with studies of the pathways of viral invasion and infection were the initial studies of the macromolecular structure of baculoviruses. Because of their size and abundance in the nuclei, initial attention was focused on the molecular structure of polyhedra (viral occlusions). Bolle (1894) discovered that the polyhedral bodies were composed of protein. By analytical ultracentrifugation Bergold (1947) demonstrated that the polyhedral protein had a molecular weight (MW) of 267.0–378.0 kDa, whereas on addition of alkali the most elementary subunit had a size of 20.3 kDa. Although Komárek and Breindl (1924) using histological methods, demonstrated numerous small particles in the polyhedra which they believed to be the infectious viral agents, Bergold (1947) isolated the virus particles and demonstrated by EM that they were rod shaped and occurred in bundles contained within a membrane (Bergold, 1947, 1953).

The next advances used modern purification techniques. Kawanishi and Paschke (1970) and Summers and Paschke (1970) purified occlusions of NPV and GV and alkali-released virions, respectively, by sedimentation through linear sucrose gradients. Unexpected multiple banding patterns were observed for the purified virions and it was Kawanishi who demonstrated that the multiple banding of occlusion-derived virus (ODV) was due to the number of virions (or nucleocapsids 1, 2, 3, and so on) per viral envelope. Kawanishi implemented the use of standard virus terminology for baculovirus structure after the conventions proposed by Caspar *et al.* (1962). Shvedchikova *et al.* (1969) visualized GV DNA as double stranded, linear, and circular molecules of high MW (80×10^6 bp). Using rate-zonal ultracentrifugation and sucrose and CsCl gradients, Summers and Anderson (1972) purified baculovirus DNA and determined that the sedimentation profile of the high-molecular-weight baculovirus DNA observed was due to the presence of double-stranded linear, relaxed circular, and covalently closed (superhelical) DNA molecules purified from the virus.

In 1953, Bergold proposed the first formal classification of insect viruses in a comprehensive treatise using the conventions established by the 5th International Congress for Microbiology. This was revised in the First Report of the International Committee on Nomenclature of Viruses at which time the taxonomic genus *Baculovirus* was formally established as initially proposed by Martignoni in 1969 (Vago *et al.*, 1974).

With established knowledge of polyhedra, virus and DNA purification techniques, and a preliminary understanding of the cellular basis for virus host invasion and replication, attention was drawn to the identity of virus structural proteins and their functions. Because of its tremendous abundance and ease of purification, the major structural protein of the viral occlusion, polyhedrin (*polyhedrin* from the NPVs and *granulin* from the GVs; terms derived by Summers and Egawa, 1973), was the initial focus. Gram quantities of polyhedrin or granulin could be purified from viral occlusions produced and purified from insects (one to two milligrams could be purified from an individual cabbage looper, *Trichoplusia ni*). Egawa and Summers (1972) established the neutral conditions to solubilize GV occlusions and estimated the dissociated granulin polymer subunit to be approximately 20–40 kDa in size. Kozlov *et al.* (1973) used sodium dodecyl sulfate polyacrylamide gel electrophoresis (SDS-PAGE) to demonstrate that the relative molecular weight of *B. mori* polyhedrin was 28 kDa. In these early studies, there was heterogeneity in the SDS-PAGE protein-banding profiles suggesting that occlusions were composed of one major and several minor structural protein subunits. This was partially resolved by inactivation of the alkali protease associated with viral occlusions purified from infected insects that was originally reported by Yamafuji *et al.* (1958). Inactivation of the protease in purified larval-derived occlusions resulted in SDS-PAGE resolution of one major band, the polyhedrin protein. Later, Maruniak *et al.* (1979) and Zummer and Faulkner (1979) reported independently that the alkali protease activity in viral occlusions was a property unique to occlusions derived from infected insects and was not present in occlusions purified from infected cells cultured *in vitro*. Subsequent comparison (utilizing peptide mapping) of the primary structures of polyhedrins and granulins from different host insects led Summers and Smith (1976) to speculate that the protein was virus encoded. Later, this was confirmed when van der Beek *et al.* (1980) isolated RNA from infected cells and translated polyhedrin *in vitro* by hybridization selection with total viral DNA. In retrospect, these results are not surprising since Gershenzon (1955) first correlated differences in the shapes of polyhedra with mutant strains of NPVs. Using classical techniques for amino acid sequencing, Serebryani *et al.* (1977) reported the first primary amino acid sequence for polyhedrin.

C. Insect Cell Culture Advances; Budded Virus (BV) and Occlusion-Derived Virus (ODV) Structure and Role in Infection

The development of new tools fundamental to standardized scientific inquiry significantly advanced baculovirus cell and molecular biology. Most significant were: (1) standardized purification procedures for viruses (Kawanishi and Paschke, 1970), viral proteins (Kozlov *et al.*, 1973), and viral DNA (Summers and Anderson, 1972); (2) the development of tissue culture media (Grace, 1962), establishment of continuous lepidopteran cell cultures (Gaw *et al.*, 1959; Hink, 1970), demonstration that baculovirus can replicate in these cells (Goodwin *et al.*, 1970), and the ability to propagate a baculovirus in continuous cell culture (Faulkner and Henderson, 1972); (3) discovery of *Autographa californica* multiple nucleopolyhedrovirus (AcMNPV), a baculovirus with a wide host range and ability to infect cells *in vitro* (Vail *et al.*, 1971); (4) comprehensive studies of polyhedra and virus ultrastructure, replication, and assembly (Harrap, 1972a,b,c); and (5) the development of baculovirus plaque assays (methylcellulose overlay, Hink and Vail, 1973 and Vail *et al.*, 1973; solid Seaplaque overlay, Wood, 1977). These seminal developments facilitated the next level of advances in baculovirus cell and molecular biology.

With the ability to isolate infectious foci from plaques in cell monolayers, Ramoska and Hink (1974) observed phenotypic differences and described unique "plaque variants" in a field of plaques resulting from infection with a wild-type (wt) virus. They noted distinct differences in polyhedra morphology and nucleocapsid envelopment in the nucleus and established the MP (many polyhedra) and FP (few polyhedra) terminology to describe these genetic variants. The FP plaque variants contained occlusions with a few enveloped, single nucleocapsids with the total number of enveloped virions significantly reduced to the point that many occlusions appeared to be devoid of them. After plaque purification, Potter *et al.* (1976) demonstrated that new strains of virus with FP phenotype develop rapidly on continuous passage in cell culture. They demonstrated a selective advantage for the rapid development of FP variants *in vitro* and introduced the use of plaque neutralization assays. The molecular significance of the FP genetic variants was not known at this time, but the nature of transposon insertion at a unique site for the AcMNPV *FP25K* gene relative to the FP phenotype was a seminal discovery ultimately leading to the identification and isolation of transposable element sequences for the development of popular and highly efficient *piggyBac* vectors for stable germ line modifications of insects (Fraser *et al.*, 1983, 1985; Handler *et al.*, 1998).

The significant differences observed by Vaughn and Faulkner (1963) in the infectivity of virus purified from viral occlusions and the "infectious" hemolymph from infected insects stimulated interest in the identity of the "unusual" infectious viral form present in insect blood. Henderson *et al.* (1974) established the term for this form of the virus, "*n*on*o*ccluded" *v*irus (NOV), and demonstrated that NOV [now called the budded virus (BV)] in the blood banded on sucrose gradients, and its physical structure and infectivity was abolished by detergent. They proposed that BV in tissue culture supernatant consisted of fragile enveloped virions and this proposal was confirmed by Knudson and Tinsley (1974). Summers and Volkman (1976) extended the characterization of the two viral forms and determined that BV from both cell culture supernatant and infected insect hemolymph consisted of enveloped single nucleocapsids and that BV and ODV [then termed *p*olyhedra-*d*erived *v*irus (PDV)] were physically different based on their sedimentation and density separation profiles. They also showed that BV and ODV had the same nucleocapsid morphology but that the envelopes for each viral form had a different physical structure. The BV envelope was loosely associated with the nucleocapsid and the envelope contained distinct peplomers localized to one end: the unique peplomer structure was initially observed by Harrap and Robertson (1968). In contrast, the envelope of the ODV uniformly associated with the nucleocapsid and without distinct surface structure.

With basic knowledge of structures of these two infectious viral forms, the major differences for the two forms in virus infection and maturation pathways *in vivo* and *in vitro* now became an obvious target for study. Stoltz *et al.* (1973) studied the ultrastructure of the intranuclear envelope maturation processes for the ODV showing the abundant presence of intranuclear "unit membrane" structures through which the nucleocapsids might bud to obtain an envelope prior to incorporation within the polyhedrin protein crystal as it assembled. He noted that the abundant presence of these unique intranuclear membranes occurred in a nucleus with an apparently intact nuclear envelope. Thus, he deduced that the viral-induced membranes were assembled "*de novo*" and established the term *de novo* intranuclear membrane morphogenesis. Intuitive to these observations was that the source of the ODV envelope and BV envelope was different and therefore the two viral envelopes were likely different in biochemical composition. If BV and ODV were different in site of maturation and therefore composition, then they were likely differences in the role and function of the two forms of virus progeny.

Knowing that ODV and BV envelopes appeared structurally different and likely functioned in different roles in the infection processes, Volkman *et al.* (1976) and Volkman and Summers (1977) demonstrated differences in infectivity, neutralization antigens, and the temporal production of the two viral forms. BV was shown to be 1700 times more infectious than ODV in cell culture, while ODV was 2500 times more infectious by *per os* (feeding) in the host insect. It was shown that during infection *in vitro*, BV is formed early and obtains its envelope by budding from the cell surface, whereas later, ODV assembles and acquires its envelope in the nucleus (Stoltz *et al.*, 1973) and is then incorporated in the highly ordered crystal of polyhedrin protein. Effort was also directed to understand the biochemical basis for the differences demonstrated by BV and ODV. Summers and Smith (1978) performed the initial comparison of the structural proteins and genomes of BV and ODV using SDS-PAGE and restriction endonuclease (REN) enzyme analyses. They showed that BV and ODV have some similar yet different proteins (Smith and Summers, 1978) but the same REN fragment profiles. SDS-PAGE and Western blot analyses also showed different structural proteins and antigens in both the nucleocapsids and viral envelopes and identified immunological cross-reaction existing among several specific baculovirus proteins (Smith and Summers, 1981). Studies to understand the composition of BV and ODV have continued to more recent times. In 1994, Braunagel and Summers showed that protein and lipid compositions of BV and ODV envelopes are significantly different; and as recently as 2003, Braunagel *et al.* identified the proteins comprising ODV using mass spectrometry proteomics. Knowing the structure and functions of the virion envelope proteins and their processing during entry to susceptible cells is basic to defining mechanisms of infection and host-range specificity.

The selection of clonal cell lines and the isolation of baculovirus genetic variants further advanced baculovirus genetics. Brown and Faulkner (1975) selected three cell lines derived from *T. ni* and reported for each that there was little difference in yield of polyhedra on infection. They concluded that the variability in number of occlusions produced per cell was not due to genetic variability in cells but more likely due to virus strain variant or stage of the cell cycle during infection. Volkman and Summers (1975) and Volkman *et al.* (1976), however, showed that there were clonal cell line-specific responses to baculovirus infection: (1) clonal cell isolates differed from the parent cell line as plaque assay indicators and in their susceptibility to infection, (2) clonal cell isolates demonstrated different capacities for occlusion production and that optimal occlusion production occurred when cells were in log growth phase, (3) clonal isolates displayed differing temporal patterns of polyhedrin synthesis and there was a correlation between cessation of BV production with the onset of polyhedrin synthesis, and (4) the optimal time to infect cells in culture is in log phase growth. These studies led to the observation that polyhedrin expression is not only different for cell isolates but also that polyhedrin expression and the steady state levels produced are not directly proportional to virus titer (Summers *et al.*, 1978). The need for a quantitative assay to directly titer virus was realized and resulted in the development of rapid titer kits using baculovirus-specific antisera (Volkman and Goldsmith, 1982).

III. Recombinant DNA Technologies

A. Virus Identification, Genotypic Variation, Physical Mapping of Genomes

With the fundamentals of recombinant DNA technologies for gene cloning and DNA manipulation established for animal viruses (Jackson *et al.*, 1972) and *Escherichia coli* (Cohen *et al.*, 1973), Rohrmann pioneered the use of REN analysis for the identification and comparison of baculovirus DNAs (Rohrmann and Beaudreau, 1977; Rohrmann *et al.*, 1978). Miller and Dawes (1978) demonstrated that passage of virus through alternate hosts did not alter REN fragment profiles but did notice submolar fragment heterogeneity suggesting contaminating virus, genetic variants, or the production of defective interfering virus particles. With the application of REN for DNA analysis of plaque-purified viral isolates, Lee and Miller (1978) and Smith and Summers (1978) independently reported that genotypic variants could be identified after plaque-purifying clonal isolates of BV from cells infected with wt AcMNPV. Lee and Miller reported that the plaque-purified genotype could be maintained on serial passage, while Smith and Summers extended the comparison of genotypic variation and identified different genotypic variants in plaque-purified ODV for virions with a single nucleocapsid per envelope (SNPV) and multiple nucleocapsids per envelope (MNPV). They both speculated on the significance of natural genotypic variants in the wt virus population with regard to potential virus–host cell interactions relative to individual cell type, virulence, and/or natural host range. One impact of the combined effects of these studies was to redirect the attention of several labs in the search of more virulent viral strains for insect pest control.

The use of REN led to the first physical maps for the prototype baculovirus AcMNPV and several of its related strains. Miller and Dawes (1979) published a physical map of the AcMNPV-L1 strain showing the order of 11 *Bam*HI and *Xma*I fragments, a partial order for the *Eco*RI and *Hin*dIII fragments, and a preliminary comparison of AcMNPV genotypic variants and *T. ni* NPV. Smith and Summers (1979) extended the physical mapping to six AcMNPV-E2 genotypic variants by mapping the fragments generated by *Eco*RI, *Xho*I, *Sae*I, *Kpn*I, and *Sma*I digestion; the physical maps of plaque-purified viruses of AcMNPV-E2 variants representing BV and ODV single (SNPV) and ODV multiple (MNPV) were compared with the closely related viruses of *Rachiplusia ou* (R9 strain), *Galleria mellonella* and *T. ni*. Vlak and Smith (1982) led the organized effort within the baculovirus community to develop a consensus map for the AcMNPV genome.

B. Functional Mapping, Gene Identification, Virus Protein Structure and Function, Regulation of Viral Gene Expression

The development of physical maps for baculovirus genomes concomitant with the development of viral genomic libraries led to an explosion of studies to map the functional organization of the viral genome and the systematic identification of viral genes and their encoded proteins and functions. Requisite for such experiments was the ability to transfect insect cells with baculovirus DNA, and this was independently reported by Burand *et al.* (1980) and Carstens *et al.* (1980).

The selection of temperature-sensitive (ts) mutants for the studies of viral genetics was reported independently by both Brown *et al.* (1979) and Lee and Miller (1979). A preliminary genetic map for AcMNPV and the demonstration that recombination occurred between ts mutants was reported by Brown and Faulkner (1980) who projected the use of ts mutants and correlation of the genetic map with physical mapping, marker rescue, and heteroduplex mapping to explore the functional organization of the baculovirus genome.

Using EM and SDS-PAGE, Carstens *et al.* (1979) performed a comprehensive study of the temporal pattern of virus-specific protein expression in infected cells. He noted that some viral proteins were synthesized before viral DNA replication and that virus-induced polypeptides did not appear until after 6 hours postinfection (h p.i.). He also observed that polyhedrin appeared late in infection with increased steady state levels of synthesis until at least 65 h p.i. He concluded that viral protein expression was temporally regulated (both early and late) and as such polyhedrin was expressed as a very late protein. Concurrently, Tjia *et al.* (1979) showed that AcMNPV DNA replication was detected as early as 5 h p.i. with maximum levels occurring at 18 h p.i. Although Tjia and colleagues are usually given credit for the first detailed analysis of AcMNPV DNA replication, Brown *et al.* (1979) also reported patterns of viral DNA synthesis starting at 6 h p.i. (as did Lee and Miller, 1979), with maximum synthesis occurring around 15 h p.i. By incorporating the use of specific inhibitors of DNA synthesis, Dobos and Cochran (1980) refined understanding of the sequentially ordered cascade of early, middle, and late genes and extended these studies to examine viral proteins for evidence of post-translational modifications involving glycosylation and phosphorylation. Collectively, these studies defined the temporal postinfection times for early and late gene expression and provided the basis for studies on the mechanisms fundamental to the regulated expression of viral proteins.

Rohrmann *et al.* (1979) observed that the amino terminal amino acids of several NPV polyhedrins are highly conserved. Because of its considerable abundance in infected cells, its unique role in the baculovirus morphology, and apparent role in providing the baculovirus with environmental stability, priority was now directed to the identification and location of the polyhedrin gene, and the function of its encoded protein. These studies were not performed in isolation; many were done in conjunction with attempts to characterize, identify, and understand the regulated expression of the many viral-encoded proteins.

C. *Search for the Polyhedrin Gene and Development of the BEVS*

Extensive efforts by several laboratories were fundamental to discovering the identity, role, and function of the *polyhedrin* gene and its encoded protein. Again, these included comparisons of the primary amino acid sequences of polyhedrins (Kozlov *et al.*, 1981; Serebryani *et al.*, 1977), and Rohrmann *et al.* (1979) who observed that the N-terminal sequences of polyhedrin proteins were highly conserved. Now, however, the stage was set to map the location of the polyhedrin gene. Researchers had the ability to purify (Miller, 1981; Vlak *et al.*, 1981) and clone specific REN fragments (Lubbert *et al.*, 1981), and fundamental techniques like Southern hybridization, nucleotide radiolabeling, and so on were established. Summers *et al.* (1980) roughly located the AcMNPV polyhedrin gene to map units 70–89 (*Eco*RI-I) using the technique of physical mapping of virus-specific polypeptides to genomic locations by intertypic marker rescue between two closely related viruses, AcMNPV and *Rachiplusia ou* MNPV (RoMNPV). In 1981, Vlak *et al.* defined the location of the polyhedrin gene within the *Eco*RI-I fragment. The search for spliced mRNAs led Rohrmann *et al.* (1982) to conduct the first cDNA mapping of polyadenylated mRNAs from infected larvae and from this he determined the orientation of the polyhedrin gene and region encoding the amino terminus. R-loop mapping indicated the mRNA to be 980 ± 75 bases with no detectable introns. During this time, Smith *et al.* (1982) and Esche *et al.* (1982) reported preliminary translational maps of the AcMNPV genome. Using 18 REN fragments from a genomic library to hybrid-select infected cell mRNA from early and late times postinfection, Smith *et al.* (1982) identified and mapped regions on the genome for 6 early and 13 late polypeptides. The polyhedrin gene was more precisely located to the *Hin*dIII-F and possibly *Hin*dIII-V fragments. These results were consistent with the placement on the genome previously predicted by Vlak *et al.* (1981). Transcriptional mapping of late viral mRNA by Adang and Miller (1982) subsequently indicated the location of the AcMNPV polyhedrin gene to be within the *Hin*dIII-V fragment. Smith *et al.* (1983a) then more precisely determined the location of the polyhedrin gene by S1 mapping of the 5′ and 3′ ends to a region of the *Eco*RI-I fragment and the region encompassing the *Hin*dIII-F and *Hin*dIII-V REN cleavage site. Both Rohel *et al.* (1983) and Smith *et al.* (1983a) located a second, highly expressed late gene, *p10*, within 87.35

and 89.55 map units and *Hin*dIII-P, respectively. With these discoveries, the foundation had been laid for the development of the BEVS.

Several pieces of data led Smith *et al.* (1983c) to the enabling research that established and developed the BEVS as a routine tool for the cloning and expression of foreign genes: (1) the results of Serebryani *et al.* (1977), Rohrmann *et al.* (1979), and Kozlov *et al.* (1981); (2) the location of the 5′ ends of the *polyhedrin* genes for *Orgyia pseudotsugata* MNPV (OpMNPV) polyhedrin as determined by Rohrmann *et al.* (1982); and (3) S1 mapping of the 5′ and 3′ ends of the AcMNPV polyhedrin gene (Smith *et al.* 1983a). These discoveries allowed Smith *et al.* (1983b) to develop strategies for directed mutations and engineering of the polyhedrin gene promoter for the insertion of foreign genes. This was done initially by constructing a series of deletions within the polyhedrin open reading frame encompassed within the AcMNPV *EcoR*I-I cloned fragment, and then transferring the mutated fragments into the viral genome by cotransfecting them into cells along with AcMNPV DNA for homologous recombination to occur. Recombinant virus was identified by an occlusion negative (occ^-) phenotype and plaque purified. The insertion of directed mutations into the polyhedrin gene locus demonstrated the molecular engineering of a baculovirus gene. It also demonstrated the final criterion needed for the development of the BEVS, that the polyhedrin gene was not essential and the ability to produce BV from an occ^- genotype that was capable of efficiently infecting cells *in vitro* or host insects on injection into the hemocoel.

By demonstrating that the polyhedrin gene was not essential for viral infection and the ability to specifically modify DNA sequences within the polyhedrin gene and under the transcriptional regulation of the polyhedrin gene promoter, Smith *et al.* (1983c) then cloned and expressed recombinant human β-interferon as both fusion and nonfused constructs, the latter of which was secreted, and reported the enabling details for AcMNPV polyhedrin promoter-regulated foreign gene expression. At the same time, the nucleotide sequence of the polyhedrin gene was reported (Hooft van Iddekinge *et al.*, 1983). Pennock *et al.* (1984) subsequently confirmed the approach to generate recombinant virus and produce foreign proteins by selecting a recombinant virus that expressed a polyhedrin–β-galactosidase fusion protein.

The next logical extension of the BEVS technology was to express such proteins directly in insects. Using similar techniques, Maeda *et al.* (1984, 1985) reported the expression of polyhedrin promoter-directed human α-interferon in silkworm larvae using a recombinant *B. mori* NPV.

D. BEVS: Strategies for Optimizing Recombinant Protein Expression, Expression of Multiple Recombinant Proteins, Enhancement of Foreign Gene Expression

The developments for advancing and enhancing a variety of applications for baculovirus-directed recombinant protein expression involved the development and use of other baculovirus or viral gene promoters for the regulated expression of foreign genes and the discovery of other nonessential regions of the AcMNPV genome. It was discovered that baculovirus mRNAs can exist as overlapping sets with common 3′ or 5′ termini (Lubbert and Doerfler, 1984) and that splicing of AcMNPV IE1 (Chisholm and Henner, 1988) and correct splicing of foreign gene mRNA can occur in baculovirus-infected cells (Jeang *et al.*, 1987). As important was the rapidly expanding knowledge of baculovirus gene promoters that could be used alone or in combination with the polyhedrin promoter to potentially manipulate the temporal pattern(s) of recombinant protein(s) expression and/or express multiple genes using the same recombinant virus. These were important for the next round of the development of baculoviruses as expression vectors.

With knowledge of the approximate genomic location of another AcMNPV gene that was also expressed at high levels, *p10* (Rohel *et al.*, 1983; Smith *et al.*, 1982, 1983a) and the *p10* gene sequence (Kuzio *et al.*, 1984), the stage was set for similar construction of AcMNPV-*p10* promoter-regulated expression of foreign genes. Knebel *et al.* (1985) engineered and tested the viability of using the *p10* promoter for foreign gene expression by expression of chloramphenicol acetyl transferase (CAT) fusion protein. However, it was Vlak *et al.* (1988) who demonstrated that *p10* was not essential for virus infection and expressed β-galactosidase fusion proteins placed in the p10 locus (AcMNPV) and regulated by the *p10* promoter. Carbonell *et al.* (1985) demonstrated the ability of a recombinant baculovirus to enter and express viral DNA in dipteran and mammalian cell lines that were considered refractory to baculovirus replication. This was done by use of a recombinant baculovirus expressing the CAT gene under the Rous sarcoma LTR (long terminal repeat) promoter, and β-galactosidase under the polyhedrin promoter. With this discovery they postulated on the use of baculovirus early promoters and heterologous promoters to express insect-specific neurotoxins in replication-refractive cells, and the potential of such applications to expand host range for viral pesticides. They further considered the potential role of baculovirus in facilitating interorganismal movement of transposable elements (Fraser *et al.*, 1983; Miller and Miller, 1982).

During the discovery of temporally regulated AcMNPV protein expression, Guarino and Summers (1986a) identified and functionally mapped a gene regulating early gene expression. Using an assay based on transient expression with a reporter plasmid, *39K* was identified as a delayed early gene and the gene responsible for its expression was located between 95.0 and 97.5 map units. The immediate early gene responsible for activation of delayed early genes was identified and named immediate early gene-1 (*IE1*). Guarino and Summers (1987) studied the temporal expression of IE1 and showed that it was expressed very early after viral entry into *Spodoptera* cells and that its expression continued late through the infection. These studies identified the first baculovirus gene with a dual phase promoter; moreover the *IE1* promoter was recognized and transcribed by host cell polymerase. *IE1* transcription was further enhanced by the use of the AcMNPV homologous repeat sequences (*hrs*) discovered and described by Cochran and Faulkner (1983) who predicted a role of the *hrs* as origins of DNA replication.

Knowledge of IE1, its promoter, and its role in transactivating late genes now presented the option of expressing foreign gene(s) in recombinant viruses within a different temporal context than that previously provided by the *p10* or *polyhedrin* promoters. Thus, an immediate early or late gene promoter could be used to supplement, replace, or add to the combinational effects of recombinant protein expression during infection. Moreover, the discovery that the IE1 promoter that was recognized by cell polymerase now resulted in the use of the IE1 promoter for transient expression of gene products in uninfected cells and in the generation of vectors for stable transformation of cells lines. Demonstration that *IE1* was functional in mammalian cells further extended the potential use of this promoter for transient expression or stable germ line transformation (Murges *et al.*, 1997).

To develop baculovirus vectors with multiple promoters, one approach was to define the functional limits of the polyhedrin promoter, and the identity of the genes juxtaposed upstream and downstream of the polyhedrin gene. Rohrmann (1986) identified a 12-mer nucleotide sequence (AATAAGTATTTT) in the *polyhedrin* 5′ untranslated leader and predicted the functional significance of this highly conserved motif in the initiation of late mRNA synthesis and polyhedrin expression. In recognition of this discovery, the term "Rohrmann Box" was routinely used to describe this motif. Matsuura *et al.* (1986, 1987) investigated essential functional features of the polyhedrin promoter by directed deletions in the leader from positions −60 to +1. They determined that interruption of the sequence in the leader resulted in lower levels of gene expression and demonstrated the highest level of expression with a vector constructed for insertion at +1 relative to the polyhedrin translation initiation codon (+1 ATG).

Emery and Bishop (1987) engineered the first multiple polyhedrin gene promoter recombinant baculoviruses by duplicating the polyhedrin gene promoter in opposite transcription orientation inserted into the unique *Eco*RV site situated upstream of the polyhedrin gene. This discovery expanded the horizon for several possibilities to develop versatility with the BEVS: (1) the identification of several nonessential loci in the AcMNPV genome (*Eco*RV site, *p10, polyhedrin*) in which to place foreign constructs; (2) the availability or potential use of promoters that, in addition to the *polyhedrin* and *p10* (Kuzio *et al.*, 1984) promoters, could express immediately early to very late in the infection process; these included *IE1* (Guarino and Summers, 1987) and the gene promoter for the major envelope protein and neutralization antigen for BV (Volkman *et al.,* 1984), gp64 (Whitford *et al.*, 1989). The availability of late gene promoters was now expanded to include the major capsid protein 39K (Blissard *et al.*, 1989; Pearson *et al.*, 1988; Thiem and Miller, 1989) and the highly expressed basic DNA-binding protein p6.9 (Wilson *et al.*, 1987); (3) the potential use of heterologous gene promoters (Carbonell *et al.*, 1985). From all the above, vector construction was expanded by the ability to clone or make synthetic promoters to allow the development of hybrid promoter constructs using minimal promoter sequences for polyhedrin, p10, capsid, and p6.9 in tandem arrays and in occlusion-positive (occ^+) vectors (Wang *et al.*, 1991). Collectively all of these potential developments not only became reality in whole or part leading to a variety of potential cloning and expression strategies that established the basis for rapidly expanding the potential uses for BEVS in medicine and human health but were also the basis for agricultural applications involving genetically engineered baculoviruses.

The BEVS has been used successfully for the production of virus-like particles (VLPs) to study viral assembly processes, to produce VLP antigens for immunization, and for diagnostic assays (Kost *et al.*, 2005; Roy *et al.*, 1997). Estes *et al.* (1987) first reported the spontaneous tubule assembly of baculovirus expressed simian rotavirus expressed VP6. Urakawa *et al.* (1989) and Roy and colleagues (1990a,b, 1991) creatively expanded the use of baculovirus multigene vectors and very elegantly utilized these vectors to study the structure and assembly of viruses and heterooligomer protein particles and protein complexes (Roy *et al.*, 1997). These vectors were also used to generate the first crystal structures of virus proteins produced in insect cells

(Basak *et al.*, 1992; Grimes *et al.*, 1995). The potential use of baculovirus-expressed antigens as subunit vaccines was successfully tested early. Dolin *et al.* (1989) and Orentas *et al.* (1990) demonstrated that baculovirus-expressed HIV-gp160 envelope protein was safe and immunogenic and was a candidate subunit human vaccine. Powers *et al.* (1995) reported that a subunit vaccine antigen of recombinant hemagglutinin from influenza A virus was equal to or better in both safety and protection than the natural vaccine. As a number of products are currently in the vaccine cue, and awaiting US Food and Drug Administration approvals, it will be exciting to see what baculovirus-expressed recombinant products become successful in the commercial arena.

By 1994, approximately 60–70% of the baculovirus genome had been sequenced. These sequences were derived from many different laboratories and entered into the databases as separate entries. An important piece of the AcMNPV genome puzzle was finally put in place with the sequence and annotation of the AcMNPV genome (Ayres *et al.* 1994).

Fundamental to rapidly advancing the worldwide acceptance, use, and the further conceptual development of the BEVS by scientists outside of the baculovirus community was the policy of M. D. Summers to voluntarily distribute several thousand BEVS kits free of charge containing the component reagents, along with the *Baculovirus Expression Vector Manual* (Summers and Smith, 1987). This was done without regard to the potential effect or compromise to the development of intellectual property, and this policy was very kindly supported and funded by the Texas Agricultural Experiment Station and Texas A&M University.

IV. Quality Improvements: Expression Vectors, Enhanced Expression, Secretion, and Protein Integrity

A major improvement of the BEVS that allowed for the more efficient selection of recombinant viruses was developed by Kitts *et al.* (1990). They discovered that linearized baculovirus DNA resulted in a higher frequency of vector recombinants (~25–30%), a significant improvement compared to the 0.01% produced by the original transfection technique (Burand *et al.*, 1980; Carstens *et al.*, 1980). Kitts and Possee (1993) further improved the selection of recombinant baculoviruses to greater than 90% by engineering unique restriction sites within an essential gene thus allowing selection of recombinant viruses that only contained the inserted foreign gene. The ability to select for recombinant viruses bypassing the plaque assay would be ideal; and this was accomplished by Luckow *et al.* (1993) when they developed a method to insert a foreign gene by site-specific transposition into a baculovirus genome propagated in *E. coli*. The recombinant baculovirus DNA is then purified from *E. coli* and transfected into host cells to produce a recombinant baculovirus (Bac-to-Bac System). Zhao *et al.* (2003) developed a targeted gene knockout technology to inactivate an essential gene adjacent to the locus for recombination. The viral DNA from the knockout can only be rescued by recombination with a baculovirus vector and this results in 100% recombinant virus selection. These were major achievements facilitating the use of baculovirus vectors especially for that community of scientists not familiar with basic baculovirus techniques.

During the 1980s and 1990s, there were several other discoveries which merit comment that enhanced the utility of BEVS, the efficacy of expression, and the quality and integrity of recombinant protein products.

Wilkie *et al.* (1980) developed a serum-free medium for insect cells. The discovery that shear sensitivity was a major limitation and required the use of chemical agents to minimize it was a hallmark for the development of bioreactor production of recombinant proteins (Tramper *et al.*, 1986), as was the discovery of Kool *et al.* (1991) of the conditions to minimize defective interfering particles during bioreactor scale-up.

Cochran and Faulkner (1983) discovered the multiple *hrs* in the AcMNPV genome and postulated their potential to function as origins of replication. Pearson *et al.* (1992) demonstrated that *hrs* function to enhance viral DNA replication. The *hrs* were also shown to function as transcriptional enhancers (Guarino and Summers, 1986b). Selected AcMNPV *hrs* are used routinely in a variety of vectors to enhance expression.

Jarvis and Summers (1989) discovered that the secretory pathway was compromised late in infection in Sf9 cells. In an attempt to resolve this, attention was first given to the type of signal peptide and the efficiency of its processing during translocation in the endoplasmic reticulum. Devlin *et al.* (1989) used a synthetic signal peptide designed to be optimal for codon usage but without effect. However, Tessier *et al.* (1991) reported the enhanced secretion of a plant protein using the honeybee mellittin signal peptide. For both recombinant protein production and viral pesticides, the use of heterologous signal sequences for improved secretion has been used extensively in foreign gene expression and viral pesticides in attempts to optimize insecticidal protein delivery. The success of this approach has given variable results.

Baculoviruses expressing chaperones to enhance and facilitate correct protein folding during secretion and thus potentially improve the quality of recombinant protein folding along with levels of secretion was first examined by Hsu *et al.* (1994). Hsu demonstrated that coexpressed BiP increased intracellular soluble and functional recombinant immunoglobulin IgG levels but did not improve secretion. At this point, I will depart from the established policy of not citing reviews and, in addition to chapters in this volume, recommend Kost *et al.* (2005) and Hu (2005) for a more comprehensive and current status of this literature and the use of baculoviruses for stable integration and expression in mammalian cells. Finally, the potential roles of viral-encoded chitinase and cathepsin (Hawtin *et al.*, 1997) as competitive factors during secretion or in recombinant protein degradation (Hom and Volkman, 2000) led to the development of a baculovirus vector in which these genes were deleted for the improved integrity of recombinant protein expression (Kaba *et al.*, 2004).

V. Pathways of Baculovirus Invasion and Infection

Knowledge of the molecular and cellular basis and factors involved in host range and the infection process in the insect and, of necessity, in cultured cells were essential to BEVS and viral pesticide development. Harrap and Robertson (1968) revealed that ODV invaded the host insect by entering through the microvilli and replicating in the midgut cells. Summers (1969, 1971) showed that nucleocapsid entry occurred by fusion of the ODV envelope and microvillus membrane (clearly implied by Harrap and Robertson's observations), with subsequent GV nucleocapsid uncoating by specific association with the host cell nuclear pore complex. Granados (1978) discovered that there were differences in the uncoating of baculovirus nucleocapsids by showing that the nucleocapsid of *Heliothis zea* baculovirus passed intact through the nuclear pore complex before uncoating. These discoveries had set the dogma for pathways of infection until Granados and Lawler (1981) showed that progeny baculovirus could penetrate through the midgut cell barrier to directly infect other cells and tissues via the hemocoel. With discovery of the role of gp64, the observation that BV enters cells and uncoats differently from ODV (Volkman and Goldsmith, 1985) and that baculovirus invasion and infection through the insect midgut cell very likely involves cell-to-cell transmission through insect tracheoblast cells to infect other tissues (Engelhard *et al.*, 1994; Keddie *et al.*, 1989), comes the realization that baculoviruses have evolved with a remarkable menu of invasion and infection strategies. It was no surprise then to learn that baculovirus encodes an enhancing enzyme to facilitate penetration through the peritrophic membrane (Wang and Granados, 1997). Another most notable discovery of how the baculovirus manipulates its host cell to optimize infection was that AcMNPV encodes antiapoptosis genes (Clem *et al.*, 1991). As time and understanding of the functions of virus-encoded genes progresses, I am confident that much more will be revealed of the ability of baculoviruses to acquire and adapt host genes to leverage the virus's selective advantage for infection and replication.

VI. Genetically Engineered Viral Pesticides

The experimental use of baculoviruses and other insect viruses to explore their potential as environmentally safe natural agents for crop protection and pest control has a distinguished history. Klöck (1925) and Ruzicka (1925) reported the first European studies using baculoviruses for pest control followed by those of Balch and Bird (1944) in North America. Unfortunately, much of the current refereed literature ignores the excellent scholarship, science, and pioneering contributions of many early outstanding visionaries, who established the historic foundations for much of the current applications in microbial pest control and genetic engineering. These include Bergold's pioneering biochemistry efforts and the edited contributions and texts by the father of insect pathology, E. A. Steinhaus (1949, 1963, 1967, and 1975) who accurately, comprehensively, and objectively recorded the history and progress of insect pathology and microbiology from the earliest records through the 1960s. As the development of baculoviruses as naturally occurring and environmentally safe alternatives for augmenting or replacing chemical pest control strategies increased in popularity, the mass production and deliberate release of biologically viable and replicating virus into the world ecosystems became an important concern (Tinsley and Melnick, 1973/74). It was primarily their concern for the lack of sensitive and specific diagnostic technology to monitor the deliberate release and use of pathogens and their fate in the environment and nonhost systems (summarized by Tinsley, 1977, 1979) that led to a careful scrutiny of the use of baculoviruses for pest control by an international community of virologists who comprehensively examined and evaluated those environmental and human health concerns (Summers and Kawanishi 1978; Summers *et al.*, 1975). These activities generated an understanding and awareness within the virology community at large leading to consensus and acceptance of the use of insect pathogens for insect control contingent on the implementation of necessary basic research fundamental to the ability to assess safe environmental and human use of mass-produced insect pest pathogens. This stimulated a revolution in basic research initially led by Tom Tinsley (Unit of Invertebrate Virology, Commonwealth Forestry Institute, Oxford), which facilitated fundamental research worldwide advancing studies and discoveries of the molecular biology and genetics of baculovirus, in general, and their development as expression vectors and exploration of their potential as viral pesticides.

There were several factors that facilitated renewed attention to the development of baculovirus pesticides: (1) The extant success of the BEVS and its potential and broad acceptance for drug discovery and the cloning, study, and production of medically important recombinant gene products; (2) the broad awareness and acceptance of the potential applications resulting from the genetic engineering of baculoviruses concomitant with rapid advances in baculovirus molecular biology and genetics from several laboratories expert in the field during the late 1970s and early 1980s; (3) the development of highly specific and sensitive tools basic to pathogen identification, detection, and safety assessment for baculoviruses; (4) the development of insect pest resistance to chemical pesticides and the resulting pollution of the biosphere; and (5) the potential for bioreactor production of recombinant viruses and proteins. Collectively these provided the basic background that generated enthusiastic "cross talk" within the insect virology and pathology scientific communities during the late 1970s and early 1980s of the potential for genetic engineering of pathogens of insects (Miller *et al.*, 1983). The basics were essentially in place and ready for application of BEVS as a tool for pest control. Needed were strategies for the environmental delivery and effective use of baculovirus-expressed insecticidal products.

Major questions to consider were what "insecticidal" product should be used, the target in the insect (behavioral, developmental, metabolic, host range, or other), and/or unique property of the virus or host to exploit or manipulate in order to achieve a significant insecticidal effect. An insecticidal effect in the order of magnitude of 100–1000-fold relative to natural pathogenicity would be ideal for a commercially viable product. One clue for a testable product was already published.

Zlotkin and colleagues (Darbon *et al.* 1982; Zlotkin *et al.* 1971a,b,c) had identified and sequenced an insect-specific protein toxin (AaIT, 70 amino acids) from the venom of the scorpion, *Androctonus australis*. The concept of an insect-specific, highly potent neurotoxin was a reality begging to be tested. Needed was an effective and efficient occ^+ expression vector and the insecticidal product for delivery. This engaged several laboratories in the baculovirus community.

A. *Development of Genetically Engineered Baculovirus Pesticides*

The basic molecular requirements for optimal polyhedrin gene promoter–directed expression were reasonably well known. A representative selection of baculovirus gene promoters and nonessential sites for foreign gene insertion were also available and could be engineered as needed. The know-how was in place to initiate these studies.

Carbonell *et al.* (1988) were not successful in the first test of a recombinant virus expressing a synthetic gene encoding a 4 kDa insect-specific neurotoxin (insectotoxin-1) from the scorpion, *Buthus eupeus*. It was Maeda (1989) who reported the first study showing a 20% increase in "insecticidal" effects of a recombinant *B. mori* NPV infection of the silkworm by expressing the diuretic hormone gene of *Manduca sexta* fused to the signal sequence of the *Drosophila* CP2 cuticle protein. Just prior to Maeda's report, Zuidema *et al.* (1989) published the details of an engineered AcMNPV mutant with the gene for the polyhedral envelope deleted. It was not until 1995, however, that bioassay results showed that this mutant virus was six times more infectious than wt virus (Ignoffo *et al.*, 1995). So, in reality Zuidema *et al.* (1989) reported the first recombinant baculovirus with improved insecticidal effects.

The marginal successes of Maeda (1989) and Zuidema *et al.* (1989) were quickly followed by Merryweather *et al.* (1990) and Hammock *et al.* (1990) with recombinant viruses expressing the *Bacillus thuringiensis* delta endotoxin and juvenile hormone esterase, respectively, but without significant insecticidal effects. At this time, Dee *et al.* (1990) expressed a functional synthetic AaIT gene [the scorpion-derived insect-specific neurotoxin discovered by Zlotkin *et al.* (1971a, b,c)] coupled to the coding sequences for the interleukin-signal sequence. The recombinant AaIT protein was tested on the larvae of *Aedes aegypti* mosquitoes and they predicted the potential use for this insecticidal toxin.

In these early studies, Tomalski and Miller (1991) reported on the insecticidal effects of a baculovirus-expressed mite toxin, TxP-1, that

Tomalski *et al.* discovered in 1989. At the same time, Stewart *et al.* (1991) reported insecticidal effects with the expression of scorpion AaIT neurotoxin fused to the AcMNPV gp64 secretory signal, as did McCutchen *et al.* (1991) and Maeda *et al.* (1991) but with the AaIT fused to the bombyxin secretory signal. Tomalski and Miller (1992) compared a series of different hybrid promoters representative of early to late-regulated promoters to express the mite toxin TxP-1 that also demonstrated insecticidal effects.

At this point it must be emphasized that it is not the intent to comprehensively detail the literature of this era in this paper but to highlight the pioneering studies relative to the current state of genetically engineered baculovirus pesticides. Further in these studies it is difficult to directly compare the similarities or differences in insecticidal activity of these recombinant viruses when evaluating the data from individual reports using different BEVS, especially without a standard reference with which to compare. It is beyond the scope of this chapter to address this issue, but for the interested reader a reasonably comprehensive comparison of the studies regarding the testing of insecticidal products referenced in this chapter and other studies of this time period please consult Martens (1994).

Public acceptance of the environmental release of genetically engineered organisms was quite problematical and remains controversial. In anticipation of this, Bishop (1986) described their experimental implementation for the environmental testing and release of a genetically marked baculovirus emphasizing a series of tests with occ$^-$ recombinant baculoviruses leading to occ$^+$ genetically engineered vectors. The first field test in the United Kingdom to test the efficacy of a recombinant baculovirus-expressing AaIT showed a marked improvement of the recombinant virus compared to wt: a 12% reduction in the field compared to 25% in the laboratory, and a 29% reduction in feeding damage (Cory *et al.*, 1994). Wood *et al.* (1994) also conducted an environmental release in the United States with a genetically altered baculovirus isolate.

Research and development of recombinant baculovirus insecticides are still underway but have not reached the point of successful commercial use. Even the most efficacious recombinant product of the pioneering research and that of subsequent developments has not been produced by the commercial or public sectors for routine pest control. Because of this, other than the basic discoveries leading to the BEVS it is not yet possible to identify the seminal "insecticidal" development leading to the successful practical and routine application of a genetically engineered viral pesticide.

VII. Other Notable Developments Emerging from Baculovirus Molecular Biology and the BEVS

A. *Stable Transformation of Insect Cell Lines and Insects*

Jarvis *et al.* (1990) pioneered the development of transgenic lepidopteran cell lines for the continuous and stable expression of the genes for neomycin-resistance, human tissue plasminogen activator, and β-galactosidase; these were expressed under the transcriptional control of the IE1 gene promoter. Transformed lepidopteran cells have shown promise for functional studies involving the expression of factors to complement the deletion of essential viral genes (Monsma *et al.*, 1996).

Maeda *et al.* (1984, 1985) first demonstrated the cost-effective potential for production of recombinant proteins in the silkworm, *B. mori*, using the BEVS. Exploiting the unique character of AcMNPV replication in silkworm larvae, which can continue to grow without symptoms, Mori *et al.* (1995) demonstrated the ability to use recombinant virus as a vector for transovarian transmission and expression of luciferase under the control of a heat shock promoter in subsequent silkworm generations suggesting that the luciferase gene had been vertically transmitted. Yamao *et al.* (1999) were able to demonstrate integration into the *B. mori* genome by homologous recombination of a fibroin *light chain*-green fluorescent protein (GFP) chimera using a recombinant *Autographa californica* M nuclear polyhedrosis virus (AcMNPV) containing the foreign gene construct inserted into the *polyhedrin* locus. The GFP reporter was expressed in the targeted posterior silk gland and incorporated into the cocoon layer.

The lepidopteran transposon vector, *piggyBac*, was an unexpected spin-off of research to characterize the FP mutants of AcMNPV and *G. mellonella* NPV. It was discovered that the FP mutation and resulting plaque phenotype is a result of spontaneous host cell DNA insertions within the viral FP25K gene locus (Fraser *et al.*, 1983, 1985). The *piggyBac* element is a short inverted terminal repeat (ITR) transposable element and part of a subclass of ITR elements thus far found only in lepidopteran insects. These transposable elements insert exclusively into TTAA target sites (Cary *et al.*, 1989), which on insertion duplicate the target site. *PiggyBac* was tested for gene transfer in the Mediterranean fruit fly by Handler *et al.* (1998) who demonstrated efficient and stable germ line transformation. Toshiki *et al.* (2000) also stably transformed the silkworm for expression of GFP. The use of the baculovirus and insect-derived *piggyBac* has considerable implications for not only the mass production of recombinant proteins but also the potential for genetic engineering of insect vectors of plant, animal, and human diseases.

B. Baculovirus-Mediated Gene Delivery in Mammalian Cells

In 1995, Hoffman *et al.* demonstrated that a recombinant baculovirus can efficiently infect human hepatocytes and can deliver functional genes to the nucleus using an immediate early cytomegalovirus gene promoter. Thus, a new type of vector for liver-directed gene therapy was pioneered. Although baculoviruses had been extensively studied for their ability to infect mammalian cells (reviewed by Hoffman *et al.* 1995), Hoffman *et al.* brought attention to the fact that their results, and those of Carbonell and Miller (1985), suggest a more careful look at the possible hazards for humans in the unrestricted use of BEVS but did emphasize that properly designed vectors should not be harmful. The development of the BacMam™ System for gene delivery into mammalian cells has accomplished such a design. There has been significant use of this system (reviewed by Kost *et al.*, 2005; Hu, 2005) in which the vector has been shown to be efficient in delivery of genes into many cell types.

VIII. Conclusions

The purpose of this chronicle of the seminal refereed literature was to articulate the basic discoveries resulting in the development of the BEVS and emphasize some of the major developments, expected and unexpected, from the prolific basic research that was stimulated by the potential use of the BEVS for the expression of recombinant proteins. I am certain that I have not cited several discoveries that some feel important for this historical record. For that I apologize, but the literature and developments from the discovery of the BEVS is vast and there are dozens of areas and topics for which comprehensive reviews can be written, each with an equal or greater literature base than this chapter. I will not predict the future for the BEVS and viral pesticides except to quote Kost *et al.* (2005): *"Yet in addition to its value in producing recombinant proteins in insect cells and larvae, this viral system continues to evolve in new and unexpected ways."*

Finally, I wish to give special recognition and thanks to Gale Smith without whose special intuition and creativity at the bench the reality of the BEVS would not have been pioneered in my laboratories with basic research starting at the University of Texas at Austin and coming to fruition at Texas A&M University and The Texas Agricultural Experiment Station. I also wish to thank all of those talented individuals who worked in my laboratories without whom the competitive edge for BEVS discovery would also not have been possible.

For me it has been an exceptional privilege to have been at the core of the leadership for the pioneering developments and implementation of the BEVS. It was very rewarding to actively promote awareness and use of the BEVS by working with all those who asked and by providing the basic reagents and technical basis for their successful use. The very considerable success of the BEVS and its worldwide acceptance has been most gratifying. Through the development and application of the BEVS, I have met and made many friends, professional and personal, and I am privileged and honored to have played a role in advancing those BEVS applications that extended beyond the baculovirus community. This has been personally and professionally gratifying, and I can best describe the emotional effects during this period as having been "one-hell-of-a rush!"

Acknowledgments

I had written a draft of Table I for this chronology about 5 or 6 years ago and had distributed it to the following experts in the baculovirus field for their review. I am especially grateful for their constructive criticisms and advice, which I have endeavored to include: D. H. L. Bishop, G. Blissard, E. Carstens, W. Doerfler, M. J. Fraser, Jr., R. R. Granados, D. Jarvis, C. Y. Kawanishi, D. Knebel-Moersdorf, C. Y. Kang, S. Maeda, M. E. Martignoni, R. D. Possee, G. Rohrmann, P. Roy, J. M. Vlak, L. E. Volkman. I also wish to thank B. Bonning for the chance and motivation to finish this chronology. And I am forever grateful to S. C. Braunagel for the continuing advice and due diligence in helping keep my focus and intensity, and for her constructive review and editing of this chapter. This study was funded in part by the Texas Agricultural Experiment Station (Hatch Project No. 8087) and Texas A&M University.

References

Adang, M. J., and Miller, L. K. (1982). Molecular cloning of DNA complementary to mRNA of the baculovirus *Autographa californica* nuclear polyhedrosis virus: Location and gene products of RNA transcripts found late in infection. *J. Virol.* **44:**782–793.

Ayres, M. D., Howard, S. C., Kuzio, J., Lopez-Ferber, M., and Possee, R. D. (1994). The complete DNA sequence of *Autographa californica* nuclear polyhedrosis virus. *Virology* **202:**586–605.

Balch, R. E., and Bird, F. T. (1944). A disease of the European spruce sawfly, *Gilpiniia hercyniae* (Htg.), and its place in natural control. *Sci. Agr.* **25:**65–80.

Basak, A. K., Stuart, D. I., and Roy, P. (1992). Preliminary crystallographic study of bluetongue virus capsid protein, VP7. *J. Mol. Biol.* **228:**687–689.

Bergold, G. H. (1947). Die Isolierung des Polyeder-Virus und die Natur der Polyeder. *Z. f. Naturforsch.* **2b:**122–143.

Bergold, G.H (1953). Insect viruses. *In* "Advances Virus Research" Vol. 1, pp. 91–139.

Bergold, G. H. (1958). Viruses of insects. *In* "Handbuch der Virusforschung" (R. Doerr and C. Hallauer, eds.), pp. 60–142. Springer-Verlag, Wien.

Bishop, D. H. (1986). UK release of genetically marked virus. *Nature* **323:**496.

Blissard, G., Quant-Russell, R. L., Rohrmann, G. F., and Beaudreau, G. S. (1989). Nucleotide sequence, transcriptional mapping, and temporal expression of the gene encoding P39, a major structural protein of the multicapsid nuclear polyhedrosis virus of *Orgyia pseudotsugata*. *Virology* **168:**354–362.

Bolle, J. (1894). Il giallume od il mal del grasso del baco da seta. Communicaizone preliminare. *Atti e Mem. dell' i.R. Soc. Agr. Gorizia* **34:**133–136.

Brown, M., and Faulkner, P. (1975). Factors affecting the yield of virus in a cloned cell line of *Trichoplusia ni* infected with a nuclear polyhedrosis virus. *J. Invertebr. Pathol.* **26:**251–257.

Brown, M., and Faulkner, P. (1980). A partial genetic map of the baculovirus, *Autographa californica* nuclear polyhedrosis virus, based on recombination studies with ts mutants. *J. Gen. Virol.* **48:**247–251.

Brown, M., Crawford, M., and Faulkner, P. (1979). Isolation of temperature sensitive mutants and assortment into complementation groups. *J. Virol.* **31:**190–198.

Burand, J. P., Summers, M. D., and Smith, G. E. (1980). Transfection with baculovirus DNA. *Virology* **101:**286–290.

Carbonell, C. F., Hodge, M. R., Tomalski, M. D., and Miller, L. K. (1988). Synthesis of a gene coding for an insect-specific scorpion neurotoxin and attempt to express it using baculovirus vectors. *Gene* **73:**409–418.

Carbonell, L. F., Klowden, M. J., and Miller, L. K. (1985). Baculovirus-mediated expression of bacterial genes in dipteran and mammalian cells. *J. Virol.* **56:**153–160.

Carstens, E. B., Tjia, S. T., and Doerfler, W. (1979). Infection of *Spodoptera frugiperda* cells with *Autographa californica* nuclear polyhedrosis virus. I. Synthesis of intracellular proteins after infection. *Virology* **99:**386–398.

Carstens, E. B., Tjia, S. T., and Doerfler, W. (1980). Infectious DNA from *Autographa californica* nuclear polyhedrosis virus. *Virology* **101:**311–314.

Cary, L. C., Goegel, M., Corsaro, B. G., Wang, H-G, Rosen, E., and Fraser, M. J. (1989). Transposon mutagenesis of baculoviruses: Analysis of *Trichoplusia ni* transposon IFP2 insertions within the FP locus of nuclear of polyhedrosis viruses. *Virology* **172:** 156–169.

Caspar, D. L. D., Dulbecco, R., Klug, A., Lwoff, A., Stoker, M. G. P., Tournier, P., and Wildy, P. (1962). Proposals. *Cold Spring Harb. Symp. Quant. Biol.* **27:**49–50.

Chisholm, G. E., and Henner, D. J. (1988). Multiple early transcripts and splicing of the *Autographa californica* nuclear polyhedrosis virus IE-1 gene. *J. Virol.* **62:**3193–3200.

Clem, R. J., Fechheimer, M., and Miller, L. K. (1991). Prevention of apoptosis by a baculovirus gene during infection of insect cells. *Science* **254:**1388–1390.

Cochran, M. A., and Faulkner, P. (1983). Location of homologous DNA sequences interspersed at five regions in the baculovirus AcNPV genome. *J. Virol.* **45:**961–970.

Cohen, S. N., Chong, A. C. Y., Boyer, H. W., and Helling, R. B. (1973). Construction of biologically functional bacterial plasmids *in vitro*. *Proc. Natl. Acad. Sci. USA* **70:** 3240–3244.

Cornalia, E. (1856). Monografia del bombice del gelso (*Bombyx mori* Linneo). *Memorie dell' I.R. Instituo Lombardo di Scienze, Lettere ed Arti* **6:**3–387 (Parte quarta: Patologia del baco. pp. 332–336).

Cory, J. S., Hirst, M. L., Williams, T., Hails, R. S., Goulson, D., Green, B. M., Carty, T. M., Possee, R. D., Cayley, P. J., and Bishop, D. H. L. (1994). Field trial of a genetically improved baculovirus insecticide. *Nature* **370:**138–140.

Darbon, H., Zlotkin, E., Kopeyan, C., van Rietschoten, J., and Rochart, H. (1982). Covalent structure of the insect toxin of the North African scorpion *Androctonus australis* Hector. *Int. J. Pept. Protein Res.* **20:**320–330.

Dee, A., Belagaje, R. M., Ward, K., Chio, E., and Lai, M. H. (1990). Expression and secretion of a functional scorpion insecticidal toxin in cultured mouse cells. *Biotechnology* **8:**339–342.

Devlin, J. J., Devlin, P. E., Clark, R., O'Rourke, E. C., Levenson, C., and Mark, D. F. (1989). Novel expressions of chimeric plasminogen activators in insect cells. *Biotechnology* **7:**286–292.

Dobos, P., and Cochran, M. A. (1980). Protein synthesis in cells infected with *Autographa californica* nuclear polyhedrosis virus (Ac-NPV): The effect of cytosine arabinoside. *Virology* **103:**446–464.

Dolin, R., Graham, B. S., Greenberg, S. B., Tacket, C. O., Belseh, R. B., Midthun, K., Clements, M. L., Gorse, G. J., Horgan, B. W., and Atmar, R. L. (1989). The safety and immunogenicity of a human immuno deficiency virus type 1 (HIV-1) recombinant gp160 candidate vaccine in humans. *Ann. Intern. Med.* **114:**119–127.

Egawa, K., and Summers, M.D (1972). Solubilization of *Trichoplusia ni* granulosis virus proteinic crystal. *J. Invertebr. Pathol.* **19:**395–404.

Emery, V. C., and Bishop, D. H. L. (1987). The development of multiple expression vectors for high level synthesis of *Ac*NPV polyhedrin protein by a recombinant baculovirus. *Protein Eng.* **1:**359–366.

Esche, H., Lubbert, H., Siegmann, B., and Doerfler, W. (1982). The translational map of the *Autographa californica* nuclear polyhedrosis virus (AcNPV) genome. *EMBO J.* **1:**1629–1633.

Estes, M. K., Crawford, S. E., Penaranda, M. E., Petrie, B. L., Burns, J. W., Chan, W-K, Ericson, B., Smith, G. E., and Summers, M. D. (1987). Synthesis and immunogenicity of the rotavirus major capsid antigen using a baculovirus expression system. *J. Virol.* **61:**1488–1494.

Faulkner, P., and Henderson, J. F. (1972). Serial passage of a nuclear polyhedrosis disease virus of the cabbage looper (*Trichoplusia ni*) in a continuous tissue culture cell line. *Virology* **50:**920–924.

Fischer, E. (1906). Über die Ursachen der Disposition und über Frühsymptome der Raupenkrankheiten. *Biol. Zentr.* **26:**448–463; 534–544.

Fraser, M. J., Smith, G. E., and Summers, M. D. (1983). Relationship between host DNA insertions and FP mutants of *Autographa californica* and *Galleria mellonella* nuclear polyhedrosis viruses. *J. Virol.* **47:**287–300.

Fraser, M. J., Brusca, J. S., Smith, G. D., and Summers, M. D. (1985). Transposon-mediated mutagenesis of a baculovirus. *Virology* **145:**356–361.

Gaw, Z. Y., Liu, N. T., and Zia, T. U. (1959). Tissue culture methods for cultivation of virus grasserie. *Acta Virol.* **3:**55–60.

Gershenzon, S. (1955). On the species specificity of viruses of the polyhedral disease of insects. *Mikrobiologiya* **24:**90–98.

Glaser, R. W. (1917). The growth of insect blood *in vitro*. *Psyche* **24:**1–7.

Goldschmidt, R. (1915). Some experiments on spermatogenesis *in vitro*. *Proc. Natl. Acad. Sci. USA* **1:**220–222.
Goodwin, R. H., Vaughn, J. L., Adams, J. R., and Louloudes, S. J. (1970). Replication of a nuclear polyhedrosis virus in an established cell line. *J. Invertebr. Pathol.* **16:**284–288.
Grace, T. D.C (1962). Establishment of four strains of cells from insect tissues grown. *in vitro*. *Nature* **195:**788–799.
Granados, R. R. (1978). Early events in the infection of *Heliothis zea* midgut cells by a baculovirus. *Virology* **90:**170–174.
Granados, R. R., and Lawler, K. A. (1981). *In vivo* pathway of *Autographa californica* baculovirus invasion and infection. *Virology* **108:**297–308.
Grimes, J., Basak, A. K., Roy, P., and Stewart, I. (1995). The crystal structure of bluetongue virus VP7. *Nature* **373:**167–170.
Guarino, L. A., and Summers, M. D. (1986a). Functional mapping of a *trans*-activating gene required for expression of a baculovirus delayed early gene. *J. Virol.* **57:**563–571.
Guarino, L. A., and Summers, M. D. (1986b). Homologous DNA of *Autographa californica* nuclear polyhedrosis virus enhances delayed-early gene expression. *Virology* **60:** 215–223.
Guarino, L. A., and Summers, M. D. (1987). Nucleotide sequence and temporal expression of a baculovirus regulatory gene. *J. Virol.* **61:**2091–2099.
Hammock, B. D., Bonning, B. C., Possee, R. D., Hanzlik, T. N., and Maeda, S. (1990). Expression and effects of the juvenile hormone esterase in a baculovirus vector. *Nature* **344:**458–461.
Handler, A. M., McCombs, S. D., Fraser, M. J., and Saul, S. H. (1998). The lepidopteran transposon vector *piggyBac*, mediates germ-line transformation in the Mediterranean fruit fly. *Proc. Natl. Acad. Sci. USA* **95:**7520–7525.
Harrap, K. A. (1972a). The structure of nuclear polyhedrosis viruses. I. The inclusion body. *Virology* **50:**114–123.
Harrap, K. A. (1972b). The structure of nuclear polyhedrosis viruses. II. The virus particle. *Virology* **50:**124–132.
Harrap, K. A. (1972c). The Structure of nuclear polyhedrosis viruses. III. Virus assembly. *Virology* **50:**133–139.
Harrap, K. A., and Robertson, J. S. (1968). A possible infection pathway in the development of a nuclear polyhedrosis virus. *J. Gen. Virol.* **3:**221–225.
Hawtin, R. E., Zarkowska, T., Arnold, K., Thomas, C. J., Goodays, G. W., King, L. A., Kuzio, J. A., and Possee, R. D. (1997). Liquefaction of *Autographa californica* nucleopolyhedorvirus-infected insects is dependent on the integrity of virus-encoded chitinase and cathepsin genes. *Virology* **238:**243–253.
Henderson, J. F., Faulkner, J., and MacKinnon, E. (1974). Some biophysical properties of virus present in tissue cultures infected with the nuclear polyhedrosis virus of *Trichoplusia ni*. *J. Gen. Virol.* **22:**143–146.
Hink, W. F. (1970). Established insect cell line from the cabbage looper. *Trichoplusia ni*. *Nature* **226:**466–467.
Hink, W. F., and Vail, P. V. (1973). A plaque assay for titration of alfalfa looper nuclear polyhedrosis virus in a cabbage looper (TN-368) cell line. *J. Invertebr. Pathol.* **22:**168–174.
Hoffman, C., Sandig, V., Jennings, G., Rudolph, M., Schleg, P., and Strauss, M. (1995). Efficient gene transfer into human hepatocytes by baculovirus vectors. *Proc. Natl. Acad. Sci. USA* **92:**10099–10103.
Hom, L. G., and Volkman, L. E. (2000). *Autographa californica* M nucleopolyhedrovirus chiA is required for processing V-CATH. *Virology* **277:**178–183.

Hooft van Iddekinge, B. J. L., Smith, G. E., and Summers, M. D. (van Iddekinge 1983). Nucleotide sequence of the polyhedrin gene of *Autographa californica* nuclear polyhedrosis virus. *Virology* **131:**561–565.

Hu, Y.-C. (2005). Baculovirus as a highly efficient expression vector in insect and mammalian cells. *Acta Pharmacol. Sin.* **26:**405–416.

Ignoffo, C. M., Garcia, C., Zuidema, D., and Vlak, J. M. (1995). Relative *in vivo* activity and simulated sunlight-UV stability of inclusion bodies of a wild-type and an engineered polyhedral envelope-negative isolate of the nucleopolyhedrosis virus of *Autographa californica*. *J. Invertebr. Pathol.* **66:**212–213.

Jackson, D. A., Symons, R. H., and Berg, P. (1972). Biochemical method for inserting new genetic information into DNA of simian virus 40: Circular SV40 DNA molecules containing Lambda Phage genes and the galactose operon of *Escherichia coli*. *Proc. Natl. Acad. Sci. USA* **69:**2904–2909.

Jarvis, D. L., and Summers, M. D. (1989). Glycosylation and secretion of human tissue plasminogen activator in recombinant baculovirus-infected insect cells. *Mol. Cell Biol.* **9:**214–223.

Jarvis, D. L., Fleming, J. G. W., Kovacs, G. R., Summers, M. D., and Guarino, L. A. (1990). Use of early baculovirus promoters for continuous expression and efficient processing of foreign gene products in stably transformed lepidopteran cells. *Biotechnology* **8:**950–955.

Kaba, S. A., Adriana, M. S., Wafula, P. O., Vlak, J. M., and Van Oers, M. M. (2004). Development of a chitinase and v-cathepsin negative bacmid for improved integrity of secreted recombinant proteins. *J. Virol. Methods* **122:**113–118.

Kawanishi, C. Y., and Paschke, J. D. (1970). Density gradient centrifugation of the virions liberated from *Rachoplusia ou* nuclear polyhedra. *J. Invertebr. Pathol.* **16:** 89–92.

Kawanishi, C. Y., Summers, M. D., Stoltz, D. B., and Arnott, H. J. (1972). Entry of an insect virus *in vivo* by fusion of viral envelope and microvillus membrane. *J. Invertebr. Pathol.* **20:**104–108.

Keddie, B. A., Aponte, G. W., and Volkman, L. E. (1989). The pathway of infection of *Autographa californica* nuclear polyhedrosis virus in an insect host. *Science* **243:** 1728–1730.

Kitts, P. A., and Possee, R. D. (1993). A method for producing recombinant baculovirus expression vectors at high frequency. *Biotechniques* **14:**810–817.

Kitts, P. A., Ayres, M. A., and Possee, R. D. (1990). Linearization of baculovirus DNA enhances the recovery of recombinant virus expression vectors. *Nucleic Acid Res.* **18:**5667–5672.

Klöck (1925). Zur Lösung der Nonnenbekampfungsfrage Auf biologischem Wege. *Fortwiss Cent.* **47:**241–245.

Knebel, D., Lubbert, H., and Doerfler, W. (1985). The promoter of the late p10 gene in insect nuclear polyhedrosis virus *Autographa californica*: Activation by viral gene products and sensitivity to DNA methylation. *EMBO J.* **4:**1301–1306.

Knudson, D. L., and Tinsley, T.W (1974). Replication of a nuclear polyhedrosis virus in continuous cell culture of *Spodoptera frugiperda*: Purification, assay of infectivity, and growth characteristics of the virus. *J. Virol.* **14:**934–944.

Komárek, J., and Breindl, V. (1924). Die Wipfel-Krankheit der Nonne und der Erreger derselben. *Z. Angew. Entomol.* **10:**99–162.

Kool, M., Voncken, F. J. L., VanLier, T., and Vlak, J. M. (1991). Detection and analysis of *Autographa californica* nuclear polyhedrosis virus mutants with defective interfering properties. *Virology* **183:**739–746.

Kost, T. A., Condreay, J. P., and Jarvis, D. L. (2005). Baculovirus as versatile vectors for protein expression in insect and mammalian cells. *Nat. Biotechnol.* **23:**567–575.

Kozlov, E. A., Levitina, T. L., Radavskii, Y. L., Sogulyaeva, V. M., Sidorova, N. M., and Serebryanyi, S. B. (1973). A determination of the molecular weight of the inclusion body protein of the nuclear polyhedrosis virus of the mulberry silkworm *Bombyx mori*. *Biokhimiya* **38:**1015–1019.

Kozlov, E. A., Levitina, T. L., Gusak, N. M., Ovander, M. N., and Serebryany, S. B. (1981). Comparison of amino acid sequences of inclusion body proteins of nuclear polyhedrosis viruses of *Bombyx mori*, *Porthetria dispar* and *Galleria mellonella*. *Bioorgan. Chimija* **7:**1008–1015.

Kuzio, J., Rohel, D. Z., Curry, C. J., Krebs, A., Carstens, E. B., and Faulkner, P. (1984). Nucleotide sequence of the p10 polypeptide gene of *Autographa californica* nuclear polyhedrosis virus. *Virology* **139:**414–418.

Lee, H. H., and Miller, L. K. (1978). Isolation of genotypic variants of *Autographa californica* nuclear polyhedrosis virus. *J. Virol.* **27:**754–767.

Lee, H. H., and Miller, L. K. (1979). Isolation, complementation, and initial characterization of temperature-sensitive mutants of the baculovirus *Autographa californica* nuclear polyhedrosis virus. *J. Virol.* **31:**240–252.

Lubbert, H., and Doerfler, W. (1984). Transcription of overlapping sets of RNAs from the genome of *Autographa californica* nuclear polyhedrosis viruses: A novel method for mapping RNAs. *J. Virol.* **52:**255–265.

Lubbert, H., Kruczek, I., Tjia, S., and Doerfler, W. (1981). The cloned *Eco*RI fragments of *Autographa californica* nuclear polyhedrosis virus DNA. *Gene* **16:**343–345.

Luckow, V. A., Lee, S. C., Barry, G. F., and Olins, P. O. (1993). Efficient generations of infectious recombinant baculoviruses by site-specific transposition mediated insertion of foreign genes into a baculovirus genome propagated in *Escherichia coli*. *J. Virol.* **67:**4566–4579.

Maeda, S. (1989). Increased insecticidal effect by a recombinant baculovirus carrying a synthetic diuretic hormone gene. *Biochem. Biophys. Res. Commun.* **165:**1177–1183.

Maeda, S., Kawai, T., Obinata, M., Chika, T., Horiuchi, T., Maekawa, K., Nakasuji, K., Saeki, Y., Sato, Y., Yamada, K., and Furusawa, M. (1984). Characteristics of human interferon-α produced by a gene transferred by a baculovirus vector in the silkworm, *Bombyx mori*. *Proc. Jpn Acad.* **60**(Ser. B):423–426.

Maeda, S., Kawai, T., Obinata, M., Fujiwara, H., Horiuchi, T., Saeki, Y., Sato, Y., and Furusawa, M. (1985). Production of human α-interferon in silkworm using a baculovirus vector. *Nature* **315:**592–594.

Maeda, S., Volrath, S. L., Hanzlik, T. N., Harper, S. A., Majima, K., Maddox, D. W., Hammock, B. D., and Fowler, E. (1991). Insecticidal effects of an insect specific neurotoxin expressed by a recombinant baculovirus. *Virology* **184:**777.

Maestri, A. (1856). Frammenti anatomici, fisiologici e patologici sul baco da seta (*Bombyx mori* Linn.). Fratelli Fusi, Pavia. 172 p.

Martens, J. (1994). Development of a baculovirus insecticide exploring the *Bacillus thuringiensis* insecticidal crystal protein. Thesis Wageningen, 135 p.

Martignoni, M. E., and Scallion, R. J. (1961). Establishment of strains of cells from insect tissue cultured *in vitro*. *Nature (London)* **190:**1133–1134.

Maruniak, J. E., Summers, M. D., Falcon, L. A., and Smith, G. E. (1979). *Autographa californica* nuclear polyhedrosis virus structural proteins compared from *in vivo* and *in vitro* sources. *Intervirology* **11:**82–88.

Matsuura, Y., Possee, R. D., and Bishop, D. H. L. (1986). Expression of the S-coded genes of *Lymphocytic choriomengitis* arena virus using a baculovirus vector. *J. Gen. Virol.* **67:**1515–1529.

Matsuura, Y., Possee, R. D., Overton, H. A., and Bishop, D. H. (1987). Baculovirus expression vectors: The requirements for high level expression of proteins, including glycoproteins. *J. Gen. Virol.* **68:**1233–1250.

McCutchen, B. F., Chandary, V., Crenshaw, R., Maddox, D., Kamita, S. G., Palekar, N., Volrath, S., Fowler, E., Hammock, B. D., and Maeda, S. (1991). Development of a recombinant baculovirus expressing an insect-selective neurotoxin: Potential for pest control. *Biotechnology* **9:**848.

Merryweather, R. A., Weyer, U., Harris, M. P., Hirst, M., Booth, T, and Possee, R. D. (1990). Construction of genetically engineered baculovirus insecticides containing the *Bacillus thuringiensis* subsp. *kurstaki* HD-73 delta endotoxin. *J. Gen. Virol.* **71:** 1535–1544.

Miller, D. W., and Miller, L. K. (1982). A virus mutant with an insertion of a Copia-like transposable element. *Nature* **299:**562–564.

Miller, L. K. (1981). Construction of a genetic map of the baculovirus *Autographa californica* nuclear polyhedrosis virus by marker rescue of temperature-sensitive mutants. *J. Virol.* **39:**973–976.

Miller, L. K., and Dawes, K. D. (1978). Restriction endonuclease analysis for the identification of baculovirus pesticides. *Appl. Environ. Microbiol.* **35:**411.

Miller, L. K., and Dawes, K. P. (1979). Physical map of the DNA genome of *Autographa californica* nuclear polyhedrosis virus. *J. Virol.* **29:**1044–1055.

Miller, L. K., Lingy, A. J., and Bulla, L. A. (1983). Bacterial, viral and fungal insecticides. *Science* **219:**715–721.

Monsma, S. A., Oomens, A. G. P., and Blissard, G. W. (1996). The GP64 envelope fusion protein is an essential baculovirus protein required for cell to cell transmission of infection. *J. Virol.* **70:**4607–4616.

Morgan, C., Bergold, G. H., Moore, D. H., and Rose, H. M. (1955). The macromolecular paracrystalline lattice of insect viral polyhedral bodies demonstrated in ultrathin sections examined in the electron microscope. *J. Biophys. Biochem. Cyto.* **1:**187–190.

Mori, H., Yamao, M., Nakazawa, H., Sugahara, Y., Shirai, N., Matsubara, F., Sumida, M., and Imamura, T. (1995). Transovarian transmission of a foreign gene in the silkworm, *Bombyx mori*, by *Autographa californica* nuclear polyhedrosis virus. *Nat. Biotechnol.* **13:**1005–1007.

Murges, D., Kremer, A., and Knebel-Moensdorf, D. (1997). Baculovirus transactivator IE1 is functional in mammalian cells. *J. Gen. Virol.* **78:**1507–1510.

Orentas, R. J., Heldreth, J. E. K., Obah, B., Polydefkis, M., Smith, G. E., Clements, M. L., and Siliciano, R. F. (1990). Induction of $CD4^+$ human cytolytic T cells specific for HIV-infected cells by a gp160 subunit vaccine. *Science* **248:**1234–1236.

Paillot, A. (1926). Sur une nouvelle maladie du noyau ou grasserie des chenilles de *P. brassicae* et un nouveau groupe de microorganismes parasites. *Compt. Rend. Acad. Sci.* **182:**180–182.

Paillot, A. (1930). *Traité des maladies du ver à soie*. G. Doin et Cie, Paris, 279 p.

Paillot, A. (1933). *L'Infection chez les insectes*. G. Patissier, Trévoux, 535 p.

Paillot, A., and Gratia, A. (1939). Essai d'isolement du virus de la grasserie des vers à soie par l'ultracentrifugation. *Arch Gesell. Virusforschung* **1:**120–129.

Pearson, M., Bjornson, R., Pearson, G., and Rohrmann, G. F. (1992). The *Autographa californica* baculovirus genome: Evidence for multiple replication origins. *Science* **257:**1382–1384.

Pearson, M. N., Quant-Russell, R. L., Rohrmann, G. F., and Beaudreau, G. S. (1988). P39, a major baculovirus structural protein: Immunocytochemical characterization and genetic location. *Virology* **167:**407–413.

Pennock, G. D., Shoemaker, C., and Miller, L. K. (1984). Strong and regulated expression of *Escherichia coli* β-galactosidase in insect cells with a baculovirus vector. *Mol. Cell. Biol.* **4:**399–406.

Potter, J. N., Faulkner, P., and MacKinnon, E. A. (1976). Strain selection during serial passage of *Trichoplusia ni* nuclear polyhedrosis virus. *J. Virol.* **18:**1040–1050.

Powers, D. C., Smith, G. E., Anderson, E. L., Kenney, D. J., Hanchett, C. S., Wilkinson, B. E., Volvovitz, F., Belshe, R. B., and Treanor, J. J. (1995). Influenza A virus vaccines containing purified recombinant H3 hemagglutinin are well tolerated and induce protective immune responses in healthy adults. *J. Infect. Dis.* **171:**1595–1599.

Ramoska, W. A., and Hink, W. F. (1974). Electron microscope examination of two plaque variants from a nuclear polyhedrosis virus of the alfalfa looper, *Autographa californica*. *J. Invertebr. Pathol.* **23:**197–201.

Rohel, D. Z., Cochran, M. A., and Faulkner, P. (1983). Characterization of two abundant mRNA's of *Autographa californica* nuclear polyhedrons virus present late in infection. *Virology* **124:**357–365.

Rohrmann, G. F. (1986). Review Article: Polyhedrin structure. *J. Gen. Virol.* **67:** 1499–1513.

Rohrmann, G. F., and Beaudreau, G. S. (1977). Characterization of DNA from polyhedral inclusion bodies of the nucleopolyhedrosis single-rod virus pathogenic for *Orgyia pseudotsugata*. *Virology* **83:**474–478.

Rohrmann, G. F., McParland, R. H., Martignoni, M. E., and Beadreau, G. S. (1978). Genetic relatedness of two nucleopolyhedrosis viruses pathogenic for *Orgyia pseudotsugata*. *Virology* **84:**213.

Rohrmann, G. F., Bailey, T. J., Brimhall, B., Becker, R. R., and Beaudreau, G. S. (1979). Tryptic peptide analysis and NH_2-terminal amino acid sequences of polyhedrins of two baculoviruses from *Orgyia pseudotsugata*. *Proc. Natl. Acad. Sci. USA* **76:** 4976–4980.

Rohrmann, G. F., Leisy, D. J., Chou, K-C., Pearson, G. D., and Beaudreau, G. S. (1982). Identification, cloning, and R-loop mapping of the polyhedrin gene from the multicapsid nuclear polyhedrosis virus of *Orgyia pseudotsugata*. *Virology* **121:**51–60.

Roy, P., Mikhailow, M., and Bishop, D. H. L. (1997). Baculovirus multigene expression vectors and their use for understanding the assembly process of architecturally complex virus particles. *Gene* **190:**119–129.

Ruzicka, J. (1925). Einige Bemerkungen über die Nonnenbekämpfung auf biologischen Wege. *Forstwiss. Zentr.* **47:**537–538.

Serebryani, S. B., Levitina, T. L., Kautsman, M. L., Radavski, Y. L., Gusak, N. M., Ovander, M. N., Sucharenko, N. V., and Kozlov, E. A. (1977). The primary structure of the polyhedrin protein of nuclear polyhedrosis virus (NPV) of *Bombyx mori*. *J. Invertebr. Pathol.* **30:**442–443.

Shvedchikova, N. G., Ulanov, V. P., and Tanasevich, L. M. (1969). Structure of the granulosis virus of Siberian silkworm *Dendrolinus sibiricus* Tschetw. *Molekulyarwaya Biologiya* **3:**361–365.

Smith, G. E., and Summers, M. D. (1978). Analysis of baculovirus genomes with restriction endonucleases. *Virology* **89:**517–527.

Smith, G. E., and Summers, M. D. (1979). Restriction maps of five *Autographa californica* MNPV variants, *Trichoplusia ni* MNPV, and *Galleria mellonella* MNPV DNAs with endonucleases *Sma*I, *Kpn*I, *Bam*HI, *Sac*I, *Xho*I, and *Eco*RI. *J. Virol.* **30:** 828–838.

Smith, G. E., and Summers, M. D. (1981). Application of a novel radioimmunoassay to identify baculovirus structural proteins that share interspecies antigenic determinants. *J. Virol.* **39:**125–137.

Smith, G. E., Vlak, J. M., and Summers, M. D. (1982). *In vitro* translation of *Autographa californica* nuclear polyhedrosis virus early and late mRNAs. *J. Virol.* **44:**199–208.

Smith, G. E., Vlak, J. M., and Summers, M. D. (1983a). Physical analysis of *Autographa californica* nuclear polyhedrosis virus transcripts for polyhedron and 10,000-molecular-weight protein. *J. Virol.* **45:**215–225.
Smith, G. E., Fraser, M. J., and Summers, M. D. (1983b). Molecular engineering of the *Autographa californica* nuclear polyhedrosis virus genome: Deletion mutations within the polyhedrin gene. *J. Virol.* **46:**584–593.
Smith, G. E., Summers, M. D., and Fraser, M. J. (1983c). Production of human beta interferon in insect cells infested with a baculovirus expression vector. *Mol. Cell. Biol.* **3:**2156–2165.
Steinhaus, E. A. (1949). "Principles of Insect Pathology," p. 757. McGraw-Hill, New York.
Steinhaus, E. A. (ed.) (1963). "Insect Pathology, an Advanced Treatise," Vol 1, p. 661; Vol 2, p. 689. Academic Press, New York and London.
Steinhaus, E. A. (1967). "Insect Microbiology," p. 763. Hafner Publishing Co., New York and London.
Steinhaus, E. A. (1975). "Disease in a Minor Chord," p. 488. Ohio State University Press. Edited collection of leading expert opinions on principal developments in insect pathology.
Stewart, L. M. D., Hirst, M., Ferber, M. L., Merryweather, A. T., Cayley, J., and Possee, R. D. (1991). Construction of an improved baculovirus insecticide containing an insect-specific toxin gene. *Nature* **352:**85.
Stoltz, D. B., Pavan, C., and da Cunha, A. B. (1973). Nuclear polyhedrosis virus: A possible example of *de novo* intranuclear membrane morphogenesis. *J. Gen. Virol.* **19:**145–150.
Summers, M. D. (1969). Apparent *in vivo* pathway of granulosis virus invasion and infection. *J. Virol.* **2:**188–190.
Summers, M. D. (1971). Detailed EM analysis of nuclear pore interaction. *J. Ultrastruct. Res.* **35:**606–625.
Summers, M. D., and Anderson, D. L. (1972). Granulosis virus deoxyribonucleic acid: A closed, double-stranded molecule. *J. Virol.* **9:**710–713.
Summers, M. D., and Egawa, K. (1973). Physical and chemical properties of *Trichoplusia ni* granulosis virus granulin. *J. Virol.* **12:**1092–1103.
Summers, M. D., and Kawanishi, C. Y. (1978). *EPA-600/9-78-026.* September, 1978, 311 p.
Summers, M. D., and Paschke, J. D. (1970). Alkali-liberated granulosis virus of *Trichoplusia ni* I. density gradient purification of virus components and some of their *in vitro* chemical and physical properties. *J. Invertebr. Pathol.* **16:**227–240.
Summers, M. D., and Smith, G. E. (1976). Comparative studies of baculovirus granulins and polyhedrins. *Intervirology* **6:**168–180.
Summers, M. D., and Smith, G. E. (1987). A manual of methods for baculovirus vectors and insect cell culture procedures. Texas Agric. Exp. Station Bulletin No. 1555. 57 p.
Summers, M. D., and Volkman, L. D. (1976). Comparison of biophysical and morphological properties of occluded and extracellular nonoccluded baculovirus from *in vivo* and *in vitro* host systems. *J. Virol.* **17:**962–972.
Summers, M. D., Engler, R., Falcon, L. A., and Vail, P. (1975). Baculoviruses for insect pest control: Safety considerations. *Am. Soc. Microbiol.* 1913 I Street, N.W. Washington, D.C. 20006. 186 p.
Summers, M. D., Volkman, L. E., and Hsieh, C-H. (1978). Immunoperoxidase detection of baculovirus antigens in insect cells. *J. Gen. Virol.* **40:**545–557.
Summers, M. D., Smith, G. E., Knell, J. D., and Burand, J. P. (1980). Physical maps of *Autographa californica* and *Rachoplusia ou* nuclear polyhedrosis virus recombinants. *J. Virol.* **34:**693–703.
Tessier, D. C., Thomas, D. Y., Khouri, H. E., Laliberts, F., and Vernet, T. (1991). Enhanced secretion from insect cells of a foreign protein fused to the honeybee mellittin signal peptide. *Gene* **98:**177–183.

Thiem, S., and Miller, L. K. (1989). Identification, sequence, and transcription mapping of the major capsid protein gene of the baculovirus *Autgrapha californica* nuclear polyhedrosis virus. *J. Virol.* **63:**2008–2018.

Tinsley, T. W. (1977). Viruses and the biological control of insect pests. *Bioscience* **27:**659–661.

Tinsley, T. W. (1979). The potential of insect pathogenic viruses as pesticidal agents. *Ann. Rep. Entomol. Soc. Ont.* **24:**63–87.

Tinsley, T. W., and Melnick, J. L. (1973/74). Potential ecological hazards of pesticidal viruses. *Intervirology* **2:**206–208.

Tjia, S. T., Carstens, E. B., and Doerfler, W. (1979). Infection of *Spodoptera frugiperda* cells with *Autographa californica* nuclear polyhedrosis virus. II. The viral DNA and the kinetics of its replication. *Virology* **99:**399–409.

Tomalski, M. D., and Miller, L. K. (1991). Insect paralysis by baculovirus-mediated expression of a mite neurotoxin gene. *Nature* **352:**82–85.

Tomalski, M. D., and Miller, L. K. (1992). Expression of a paralytic neurotoxin gene to improve insect baculoviruses as biopesticides. *Biotechnology* **10:**545.

Toshiki, T., Chantal, T., Corinne, R., Toshio, K., Eappen. A., Kamba,M., Natus, K., Jean-Luc, T., Manchamp, B., Gerard, C., Shirk, P., Fraser, M. N. *et al.* (2000). Germline transformation of the silkworm *Bombyx mori* L. using a *piggybac* transposon-derived vector. *Nat. Biotechnol.* **18:**81–84.

Trager, W. (1935). Cultivation of the virus of grasserie in silkworm tissue cultures. *J. Exp. Med.* **61:**501–513.

Tramper, J., Williams, J. B., Joustra, D., and Vlak, J. M. (1986). Shear sensitivity of insect cells in suspension. *Enzyme Microb. Technol.* **8:**33–36.

Urakawa, T., Ferguson, M., Minor, P. D., Cooper, J., Sullivan, M., Almond, J. W., and Bishop, D. H. L. (1989). Synthesis of immunogenic, but non-infectious, poliovirus particles in insect cells by a baculovirus expression vector. *J. Gen. Virol.* **70:** 1453–1463.

Vail, P. V., Sutter, G., Jay, D. L., and Gough, D. (1971). Reciprocal infectivity of nuclear polyhedrosis viruses of the cabbage looper and alfalfa looper. *J. Invertebr. Pathol.* **17:**383–388.

Vail, P. V., Jay, D. L., and Hink, W. F. (1973). Replication and infectivity of the nuclear polyhedrosis virus of the alfalfa looper, *Autographa californica*, produced in cells grown *in vivo*. *J. Invertebr. Pathol.* **22:**231–237.

van der Beek, C. P., Saaijer-Riep, J. D., and Vlak, J. M. (der Beek 1980). On the origin of the polyhedral protein of *Autographa californica* nuclear polyhedrosis virus. *Virology* **100:**326–333.

Vaughn, J. L., and Faulkner, P (1963). Susceptibility of an insect tissue culture to infection by virus preparations of the nuclear polyhedrosis of the silkworm (*Bombyx mori*). *Virology* **20:**484–489.

Vlak, J. M., and Smith, G. E. (1982). Orientation of the genome of *Autographa californica* nuclear polyhedrosis virus: A proposal. *J. Virol.* **41:**1118–1121.

Vlak, J. M., Smith, G. E., and Summers, M. D. (1981). Hybridization selection and *in vitro* translation of *Autographa californica* nuclear polyhedrosis virus mRNA. *J. Virol.* **40:**762–771.

Vlak, J. M., Klinkenberg, F. A., Zaal, K. J., Usmany, M., Klinge-Roode, E. C., Geervliet, J. B., Roosien, J., and van Lent, J. W. M. (1988). Functional studies on the p10 gene of *Autographa californica* nuclear polyhedrosis virus using a recombinant expressing a p10-β-galactosidase fusion gene. *J. Gen. Virol.* **69:**765–776.

Volkman, L. E., and Goldsmith, P. A. (1982). Generalized immunoassay for *Ac*NPV infectivity *in vitro*. *Appl. Environ. Microbiol.* **44:**227–233.

Volkman, L. E., and Summers, M. D. (1975). Nuclear polyhedrosis virus detection: Relative capabilities of clones developed from *Trichoplusia ni* ovarian cell line TN-368 to serve as indicator cells in a plaque assay. *J. Virol.* **16:**1630–1637.

Volkman, L. E., and Summers, M. D. (1977). *Autographa californica* nuclear polyhedrosis virus: Comparative infectivity of the occluded, alkali-liberated and non-occluded forms. *J. Invertebr. Pathol.* **30:**102–103.

Volkman, L. E., Summers, M. D., and Hsieh, C-H. (1976). Occluded and nonoccluded nuclear polyhedrosis virus grown in *Trichoplusia ni*: Comparative neutralizations, comparative infectivity and *in vitro* growth studies. *J. Virol.* **19:**820–832.

Volkman, L. E., Goldsmith, P. A., Hess, R. T., and Faulkner, P. (1984). Neutralization of budded *Autographa californica* NPV by a monoclonal antibody: Identification of the target antigen. *Virology* **133:**354–362.

Wang, P., and Granados, R. R. (1997). An intestinal mucin is the target substrate for a baculovirus enhancin. *Proc. Natl. Acad. Sci. USA* **94:**6977–6982.

Wang, X., Ooi, B. G., and Miller, L. K. (1991). Baculovirus vectors for multiple gene expression and for occluded virus production. *Gene* **100:**131–137.

Whitford, M., Stewart, S., Kuzio, J., and Faulkner, P. (1989). Identification and sequence analysis of a gene encoding *gp67*, an abundant envelope glycoprotein of the baculovirus *Autographa californica* nuclear polyhedrosis virus. *J. Virol.* **63:**1393–1399.

Wilkie, G. E., Stockdale, H., and Pirt, S. V. (1980). Chemically-defined media for production of insect cells and viruses *in vitro*. *Dev. Biol. Stand.* **46:**29–37.

Wood, H. A. (1977). An agar overlay plaque assay method for *Autographa californica* nuclear-polyhedrosis virus. *J. Invertebr. Pathol.* **29:**304–307.

Wood, H. A., Hughes, P. R., and Shelton, A. (1994). Field studies of the co-occlusion strategy with a genetically altered isolate of the *Autographa californica* nuclear polyhedrosis virus. *Environ. Entomol.* **23:**211.

Yamafuji, K., Yoshinara, Y., and Hirayama, K. (1958). Protease and desoxyribonuclease in viral polyhedral crystal. *Enzymol. Biol. Clin.* **19:**53–58.

Yamao, M., Katayama, N., Nakazawa, H., Yamakawa, M., Hayashi, Y., Hara, S., Kamei, K., and Hajime, M. (1999). Gene targeting in the silk worm by use of a baculovirus. *Genes & Devel.* **13:**511–516.

Zhao, Y., Chapman, D. A., and Jones, I. M. (2003). Improving baculovirus recombination. *Nucleic Acid Res.* **31:**e6.

Zlotkin, E., Rochart, H., Kopeyan, C., Miranda, F., and Lissitzky, S. (1971a). Purification and properties of the insect toxin from the venom of the scorpion, *Androctonus australis* Hector. *Biochimie* **53:**1073–1078.

Zlotkin, E., Rochart, H., Kopeyan, C., Miranda, F., and Lissitzky, S. (1971b). The effect of scorpion venom on blowfly larvae—a new method for the evaluation of scorpion venoms potency. *Toxicon* **9:**1–8.

Zlotkin, E., Rochart, H., Kopeyan, C., Miranda, F., and Lissitzky, S. (1971c). A new toxic protein in the venom of the scorpion *Androctonus australis* Hector. *Toxicon* **9:**9–13.

Zuidema, D., Klinge Roode, E. C., van Lent, J. W. M., and Vlak, J. M. (1989). Construction and analysis of an *Autographa californica* nuclear polyhedrosis virus mutant lacking the polyhedral envelope. *Virology* **173:**98–108.

Zummer, M., and Faulkner, P. (1979). Absence of protease in baculovirus polyhedral bodies propagated. *in vitro*. *J. Invertebr. Pathol.* **33:**382–384.

POLYDNAVIRUS GENES THAT ENHANCE THE BACULOVIRUS EXPRESSION VECTOR SYSTEM

Angelika Fath-Goodin,[*,†] Jeremy Kroemer,[*,†] Stacy Martin,[†] Krista Reeves,[*] and Bruce A. Webb[*]

[*]Department of Entomology, S-225 Agricultural Science Building North
University of Kentucky, Lexington, Kentucky 40546
[†]ParaTechs Corp., 105c KTRDC Building, University and Cooper Drs.
Lexington, Kentucky 40546

Abstract

The baculovirus expression vector system (BEVS) is a powerful and versatile system for protein expression, which has many advantages. However, a limitation of any lytic viral expression system, including BEVS, is that death and lysis of infected insect cells terminates protein production. This results in interruption of protein production and higher production costs due to the need to set up new infections, maintain uninfected cells, and produce pure viral stocks. Genetic methods to slow or prevent cell death while maintaining high-level, virus-driven protein production could dramatically increase protein yields.

Several approaches have been used to improve the BEVS and increase the synthesis of functional proteins. Successful enhancement of the BEVS was obtained when various gene elements were added to the virus, secretion and posttranslational processing were modified, or protein integrity was improved. A gene family from the insect virus *Campoletis*

0065-3527/06 $35.00
DOI: 10.1016/S0065-3527(06)68002-0

sonorensis ichnovirus (CsIV) was discovered that delays lysis of baculovirus-infected cells, thereby significantly enhancing recombinant protein production in the BEVS system. By using the CsIV *vankyrin* gene family, protein production in the vankyrin-enhanced BEVS (VE-BEVS) was increased by a factor of 4- to 15-fold by either coexpressing the vankyrin protein from a dual BEVS or by providing its activity *in trans* by expressing the vankyrin protein from a stably transformed cell line. In sum, VE-BEVS is an enhancement of the existing BEVS technology that markedly improves protein expression levels while reducing the cost of labor and materials.

I. The BEVS: Advantages and Limitations

The baculovirus expression vector system (BEVS) is universally recognized as a powerful and versatile tool for producing recombinant proteins. The BEVS is a safe, easy, and effective eukaryotic expression system. There are many advantages of using the BEVS, including high levels of protein expression, expression of large proteins, efficient cleavage of signal peptides and processing of the protein, posttranslational modifications, simultaneous expression of multiple genes, and the system is readily amenable to scale-up. In addition to these advantages, expressed proteins are usually correctly folded and biologically active. The BEVS is used to design and synthesize recombinant pharmaceuticals, to develop faster acting biological insecticides, and as a protein expression system for a multitude of research projects (see Summers, this volume, pp. 3–73; van Oers, this volume, pp. 193–253; Inceoglu *et al.*, this volume, pp. 323–360).

The strong polyhedrin promoter in the BEVS is advantageous compared to other eukaryotic expression vector systems, as it promotes hypertranscription of the foreign gene of interest, with the protein product accumulating in large quantities during infection of lepidopteran cells (Luckow and Summers, 1988; Miller, 1988). However, since the polyhedrin promoter requires other viral gene products for its activity, foreign gene expression occurs only during the very late phase of the viral infection. This means that recombinant gene expression in BEVS is limited because the recombinant baculovirus will eventually kill the host cells. Furthermore, there is evidence that the host cell secretory pathways are compromised during the later phase of baculovirus infection (Jarvis and Summers, 1989). Due to the limitations in the posttranslational processing machinery, heterologous secreted and membrane proteins are often poorly processed, insoluble, or contain improper modifications (Ailor and Betenbaugh, 1999). Taken together, death and lysis of the BEVS-infected cells results in decreased productivity levels and higher production costs due to the need to set up newly infected cells.

II. Enhancement of the BEVS in Insect Cells: General

Several approaches that have been used to overcome limitations of the BEVS and increase yields of functional proteins are considered later (Kost *et al.*, 2005).

A. Addition of Various Gene Elements to the Virus

Dramatic increases in foreign gene expression have been reported by the addition of various DNA elements to the virus. Introduction of an additional copy of the homologous region (hr1) sequence downstream of the polyhedrin promoter locus of *Autographa californica* multiple nucleopolyhedrovirus (AcMNPV) resulted in an ~95-fold enhancement of luciferase expression (Venkaiah *et al.*, 2004). Introduction of a 21-bp sequence element derived from a 5′ untranslated leader sequence of a lobster tropomyosin cDNA (L21) containing both the Kozak sequence and an A-rich sequence into a baculovirus transfer vector increased the expression levels of exogenous genes by 7- to 20-fold (Sano *et al.*, 2002). As mentioned earlier, the use of the polyhedrin promoter often leads to incomplete posttranslational modifications because most cellular functions are inhibited in the very late phase of viral infection whereas the early *ie-1* promoter can be used without the help of viral gene products and function well throughout infection (Jarvis *et al.*, 1996). However, expression of foreign genes from the *ie-1* promoter is 10- to 50-fold lower than expression from the polyhedrin promoter. Chen *et al.* (2004) reported that the homologous region 3 of *Bombyx mori* NPV can greatly improve the transcription of the *ie-1* promoter to levels up to threefold higher than those obtained with the polyhedrin promoter alone, which will make the *ie-1* promoter more attractive to the end-user.

B. Modification of Secretion and Posttranslational Processing

To overcome the problem of improper secretory processing in insect cells during the late stages of baculovirus infection, researchers have engineered the secretory pathway by supplementing secretory processing proteins in the host insect cell (Ailor and Betenbaugh, 1999). Protein production was enhanced when insect cells were cotransfected with baculoviruses expressing chaperone proteins, such as calnexin, calreticulin, and Hsp70, which are known to facilitate folding and modification of newly synthesized proteins (reviewed by Ailor and Betenbaugh, 1999; Kost *et al.*, 2005). Furthermore, stably transformed insect cells expressing human calnexin and human calreticulin have been constructed and can be used as modified hosts for conventional baculovirus expression vectors to provide the required chaperones *in trans* (Kato *et al.*, 2005). Another limitation of the BEVS is that insect cells lack galactosyltransferase and sialyltransferase, two enzymes necessary to convert most N-linked side chains to complex N-linked oligosaccharides often found in mammalian glycoproteins. The modification of lepidopteran *N*-glycosylation pathways for improved processing and function of glycoproteins produced in lepidopteran cells is covered by Harrison and Jarvis (this volume, pp. 159–191).

C. Improvement of Protein Integrity

A difficulty of the BEVS associated with being a lytic expression system is that death of cells 3–5 days after baculovirus infection may lead to increased proteolytic activity that can result in the degradation of the recombinant protein. Ho *et al.* (2004) have isolated a nonlytic baculovirus by random mutagenesis of viral genomes. At 5 days postinfection, the nonlytic baculovirus showed only 7% lysis of Sf21 cells, whereas the conventional BEVS showed 60% lysis of the cells. The authors used a novel fluorescence resonance energy transfer (FRET)-based assay to analyze the integrity of proteins expressed in the nonlytic BEVS compared to conventional BEVS. They demonstrated that the recombinant protein produced in the nonlytic BEVS was compactly folded with less degradation than that found in the parental virus. Another strategy to enhance the integrity of secreted recombinant proteins is the development of baculovirus expression vectors that lack the *chitinase* and *v-cathepsin* genes (Kaba *et al.*, 2004). Chitinase in conjunction with v-cathepsin promotes liquefaction of the host in the late stage of baculovirus infection (Hawtin *et al.*, 1997). In addition, the protease v-cathepsin was shown to be responsible for the proteolysis of recombinant proteins (Hom and Volkman, 1998). When the unstable, secreted form of *Theileria parva* sporozoite surface protein p67 was expressed by the chitinase and v-cathepsin-negative AcMNPV, the recombinant protein was protected from degradation (Kaba *et al.*, 2004).

III. Enhancement of the BEVS in Insect Cells by *Campoletis sonorensis* Ichnovirus Vankyrin Proteins (Vankyrin-Enhanced BEVS)

Polydnaviruses (PDVs) are obligate symbionts of some parasitic hymenopteran wasps. PDVs are responsible for modifying the physiology of the host lepidopteran larva and overcoming host immunity to the benefit of the developing endoparasitoid (Webb, 1998). A family of polydnavirus genes, the viral ankyrins, or *vankyrin* genes, disrupts insect host cellular immunity (Kroemer and Webb, 2004). Two *Microplitis demoliter* bracovirus vankyrin proteins (H4 and H5) have homology to inhibitor κB proteins from insects and mammals. Activation of NF-κB transcription factors, which play a central role in activation of antimicrobial peptides and other genes of the insect immune system, are inhibited in the presence of H4 or H5 such that the insect immune response is suppressed (Thoetkiattikul *et al.*, 2005). These data suggest that these viral proteins function as IκBs and hence are potent inhibitors of the insect immune system. These genes may reduce antiviral immune responses in polydnavirus-infected cells, thereby enabling persistent polydnavirus gene expression in infected fat body cells.

Unexpectedly, the expression of some *Campoletis sonorensis* ichnovirus (CsIV) vankyrin proteins in BEVS altered viability of the baculovirus-infected cells and this discovery led to the development of the vankyrin-enhanced BEVS or VE-BEVS, which will be described in more detail later.

A. *CsIV* Vankyrin *Gene Family*

The *vankyrin* gene family is composed of seven genes on CsIV segments P and I^2 (Kroemer and Webb, 2004). Each *vankyrin* gene encodes a 500-bp open reading frame possessing four ankyrin repeat protein motifs (Fig. 1). The vankyrin protein motifs show significant identities to ankyrin motifs in cactus, the *Drosophila* IκB protein (Meng *et al.*, 1999). Typical IκB structure consists of (1) an N-terminal domain for signal-induced degradation of IκB; (2) an internal ankyrin repeat domain for nuclear localization signal shielding and inhibitory interactions with NF-κB DNA-binding domains; and (3) a C-terminal PEST domain involved in basal regulation of IκB proteolysis (Ghosh *et al.*, 1998; Huxford *et al.*, 1998). N-terminal serine motifs are phosphorylated in response to stimulation of the immunodeficiency (IMD) or Toll signal transduction pathways and target IκBs for polyubiquitination and destruction. The activity of IκB proteins are also regulated in their

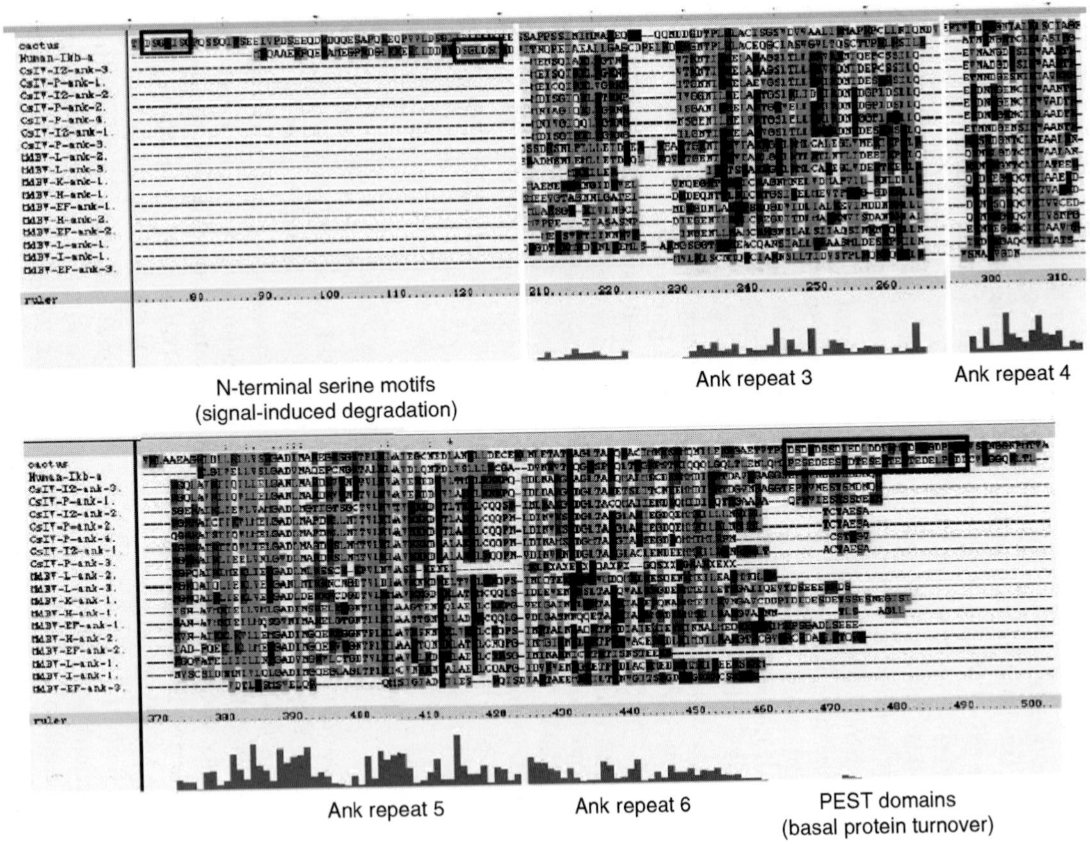
N-terminal serine motifs
(signal-induced degradation)
Ank repeat 3
Ank repeat 4
Ank repeat 5
Ank repeat 6
PEST domains
(basal protein turnover)

C-terminal PEST domains where acidic OH groups undergo phosphorylation at basal levels resulting in IκB degradation.

CsIV *vankyrin* genes, like those identified from four other polydnaviruses (PDVs), align with the four C-terminal inhibitory ankyrin repeat domains of IκBs, but lack N- and C-terminal regulatory domains important for signal-induced and basal degradation of typical NF-κB inhibitors (Fig. 1). The data suggest that there has been independent acquisition of wasp *IκB* genes by PDVs, followed by loss of the IκB protein domains that were nonessential or deleterious to the function of the proteins in PDVs (Kroemer and Webb, 2004; Turnbull and Webb, 2002). The loss of the IκB protease-sensitive domains would likely increase stability of the vankyrin proteins, while loss of ankyrin repeats 1–2 may expose nuclear import signals on bound NF-κB dimers. Thus, the structure of the vankyrin proteins suggests that they may disrupt insect immunity and/or development by irreversibly inhibiting host NF-κB-signaling pathways. In response to pathogens, the Toll and/or IMD pathways activate transcription and release of effector molecules. If NF-κB is irreversibly bound to a PDV IκB this pathway would be blocked (Santoro *et al.*, 2003; Silverman and Maniatis, 2001).

After the CsIV *vankyrin* genes were identified through genome sequence analyses their transcription was studied. Northern blot and rapid amplification of cDNA 3′ ends (3′RACE) analyses were used to detect expression of all seven CsIV *vankyrin* genes (*I^2-vank-1*, *I^2-vank-2*, *I^2-vank-3*, *P-vank-1*, *P-vank-2*, *P-vank-3*, *P-vank-4*) in parasitized larvae. Expression levels were highest at 4 h and detectable through at least 5 days in parasitized hosts (Kroemer and Webb, 2005). Relative Quantitative Real-time PCR data show tissue-specific expression of individual CsIV *vankyrin* genes. *I^2-vank-1*, *P-vank-2*, *P-vank-3*, and *P-vank-4* genes exhibited highest levels of expression in 3-day postparasitization (pp) hemocytes relative to other infected tissues. *I^2-vank-2*, *I^2-vank-3*, and *P-vank-1* genes were preferentially expressed in the 3-day pp fat body. Thus *vankyrin* gene expression is tissue specific and temporally variable.

To study the function(s) of this gene family, seven BEVS were constructed each of which expressed an individual *vankyrin* gene. The unanticipated results of *vankyrin* gene expression in BEVS clearly establish that the two *vankyrin* genes (*P-vank-1* and *I^2-vank-3*) that

Fig 1. *Microplitis demolitor* bracovirus and CsIV *vankyrin* genes align in similar regions spanning the ankyrin repeat domains (bold type) of typical IκB gene family members. *Vankyrin* genes lack N- and C-terminal destruction domains (black boxes) involved in the regulation of typical IκB activity. (See Color Insert.)

are expressed in fat body, but not in hemocytes, stabilized BEVS-infected cells by altering viability of the baculovirus-infected cells (Fig. 2). Death and lysis of baculovirus-infected cells was delayed with some cells surviving twice as long as normal (Fig. 2). This intriguing observation was pursued by evaluating expression of recombinant vankyrin proteins by protein blotting and determined that BEVS constructs with delayed host cell lysis have higher levels of protein production (Fig. 3). Delayed detection of proteins from P-vank-1 and I^2-vank-3 viruses until day 3 postinfection (d p.i.) was due to enhanced longevity of Sf9 cells infected by these viruses (consistent with Fig. 2). Conservatively, it has been estimated that a 4- to 10-fold increase in recombinant protein production occurs in VE-BEVS-infected cells (Fig. 3). The *vankyrin* BEVS constructs were expressed under the same promoter and were identical in design and size differing only in the identity of the expressed gene. Thus, the effects on cell viability and protein expression must result from differential functions of vankyrin proteins. Preserving fat body functions may be important for the survival and development of parasitized larvae while simultaneously being advantageous to the parasitoid due to elimination of hemocytes through apoptosis.

B. Enhancement of BEVS by Coexpressing a Vankyrin Protein from a Dual Expression Vector

It is assumed that the increased longevity of VE-BEVS-infected insect cells will increase the efficacy of expression of other heterologous proteins. To test this hypothesis, the production of yellow fluorescent protein (YFP) expressed in the VE-BEVS was quantified and compared to the yields obtained when expressed in conventional BEVS (Figs. 4 and 5). Therefore, dual expression vectors were constructed with the *vankyrin* gene under the control of the polyhedrin promoter and the gene of interest, in this case YFP, under the control of the p10 promoter. Fluorescence microscopic analyses revealed that YFP expression is enhanced when the *vankyrin* gene *P-vank-1* is coexpressed from a dual expression vector compared to YFP expression from a conventional BEVS (Fig. 4). Quantification of YFP production by measuring fluorescence intensity using a fluorometer indicated a 16-fold increase of YFP synthesis when expressed from VE-BEVS (Fig. 5).

The VE-BEVS has proven successful for enhanced expression of the intracellular protein YFP. To determine the general applicability of VE-BEVS for enhancing protein yields, evaluation of additional proteins expressed from dual expression vectors are under investigation.

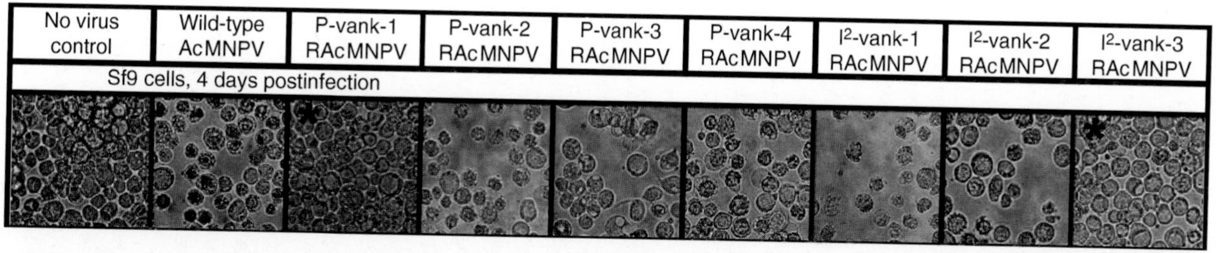

FIG 2. Effect of recombinant CsIV vankyrin proteins on Sf9 cells. Cells infected with recombinant baculoviruses (RAcMNPV) expressing fat body-specific P-vank-1 and I^2-vank-3 proteins (asterisks) exhibit enhanced longevity and resemble noninfected control cells at 4d p.i. Cells exposed to recombinant viruses expressing other vankyrin proteins undergo lysis by 4d p.i. and resemble cells infected with wild-type AcMNPV. 40× magnification.

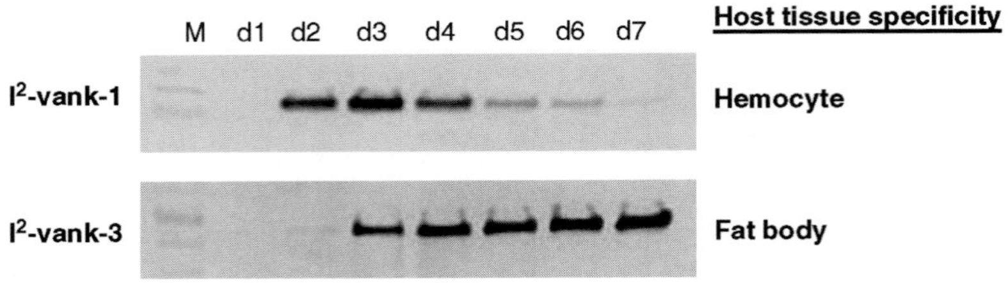

FIG 3. Protein expression is enhanced in Sf9 cells infected with recombinant AcMNPV expressing fat body-specific CsIV vankyrin protein I^2-vank-3 when compared to the expression of the hemocyte-specific I^2-vank-1. Western blots show proteins released into culture media each day after infection. The CsIV vankyrin proteins are intracellular proteins and lack secretory signals, thus protein detected in the medium results from cell lysis induced by infection.

We anticipate that VE-BEVS will effectively enhance expression of secreted and membrane-bound proteins, which would be of great interest.

C. Expression of Conventional BEVS Is Enhanced in a Transformed Cell Line Expressing Vankyrin *Genes*

A transformed cell line expressing vankyrin protein is likely to improve survivorship and protein expression of cells infected with conventional BEVS. To test the hypothesis that I^2-vank-3 and P-vank-1 proteins could indeed enhance protein expression, Sf9 cells were stably transformed with either *P-vank-1* or *I^2-vank-3* to provide the vankyrin proteins *in trans*. The transformed cell lines pIB-P-vank-1 and pIB-I^2-vank-3 were infected with conventional recombinant baculoviruses expressing YFP. As determined by fluorescence microscopy, YFP expression levels were enhanced and a prolonged heterologous protein synthesis was detected in the pIB-I^2-vank-3 cell line (Fig. 6) and to a lower extent in the pIB-P-vank-1 line compared to control Sf9 cells (not shown). Expression levels of recombinant protein were quantified by fluorometric measurements and YFP levels were up to eight-fold higher when expressed from the cell line harboring I^2-vank-3 compared to infection with control Sf9 cells (not shown). Interestingly, the number of fluorescent cells was directly correlated to the number of living cells, confirming that the presence of I^2-vank-3 enhanced cell integrity (Fig. 7).

The prolonged longevity of the BEVS-infected vankyrin cell line may support more efficient protein secretion. To test this hypothesis,

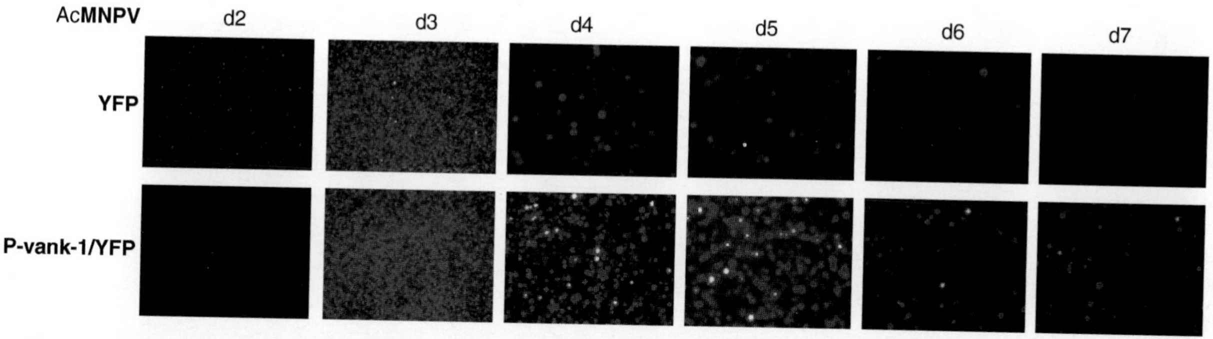

FIG 4. YFP expression is enhanced when coexpressed with P-vank-1 from a baculovirus dual expression vector. Fluorescence microscopic analyses of Sf9 cells infected with YFP baculovirus (YFP) or P-vank-1/YFP baculovirus (P-vank-1/YFP) at a MOI of 10 are shown. Pictures were taken at a 20× magnification 2–7d p.i. (See Color Insert.)

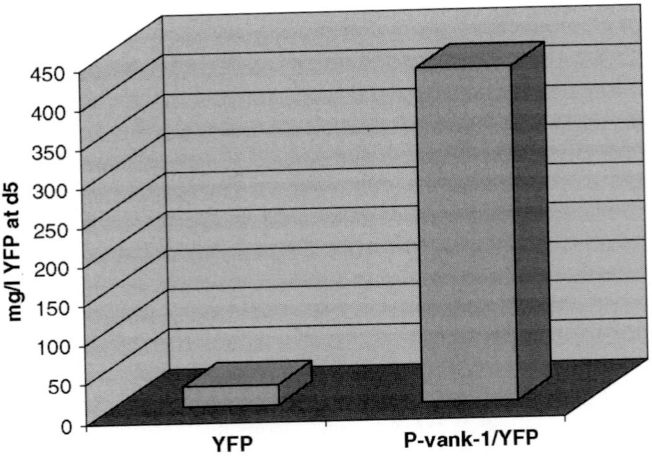

FIG 5. YFP yield is increased up to 16-fold when expressed in VE-BEVS. Quantification of YFP expression in Sf9 cells infected with YFP baculovirus (YFP) or P-vank-1/YFP baculovirus (P1-vank-1/YFP) was performed by fluorometry (excitation: 485 nm; emission: 520 nm). Total amount of YFP expressed at 5d p.i. was calculated using a standard curve for YFP.

a pIB-P-vank-1 cell line was infected with recombinant AcMNPV expressing the secreted CsIV Cys-motif protein VHv1.1. Western blotting revealed that release of VHv1.1 into culture medium is extended by at least 3 days when expressed in a pIB-P-vank-1 cell line compared to regular Sf9 cells infected with recombinant AcMNPV-expressing VHv1.1 (Fig. 8).

These experiments demonstrated that vankyrin protein function can be provided by transformation of cells with *vankyrin* genes and it is not essential for this gene to be expressed by the recombinant virus. This result indicated that recombinant baculoviruses can take advantage of VE-BEVS technology without redesign of the expression constructs.

D. Summary and Further Application

Taken together, we demonstrated that VE-BEVS is a powerful method to enhance conventional BEVS. Based on its unique mode of action, we anticipate that *vankyrin* genes can by used alongside other methods (described in Section II) to further improve the performance of recombinant baculoviruses for various biotechnological applications. *Vankyrin* genes have also been identified from four other PDVs,

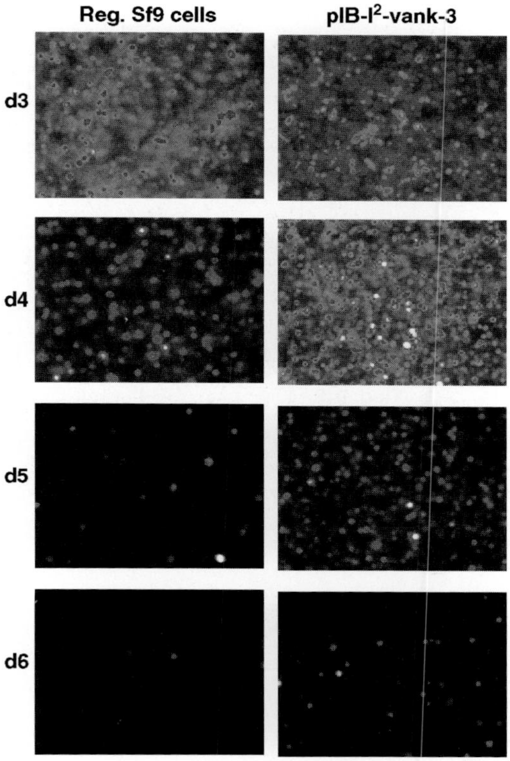

FIG 6. YFP expression is enhanced when vankyrin proteins are provided *in trans* from a stably transformed cell line. Fluorescence microscopic analysis of the blasticidin-resistant Sf9 pIB-I^2-vank-3 cell line expressing I^2-vank-3 protein and regular Sf9 cells infected with YFP baculovirus at an MOI of 10 are shown. Pictures were taken at a 20× magnification 3–6d p.i. (See Color Insert.)

Microplitis demolitor bracovirus (12 *vankyrin* genes); *Hyposoter fugitivus* ichnovirus (10 *vankyrin* genes); *Toxoneuron nigriceps* bracovirus (at least two *vankyrin* genes); *Glypta fumiferana* ichnovirus (at least four *vankyrin* genes), so it is quite possible that evaluation of these *vankyrin* genes may identify vankyrin variants that further enhance survivorship of cells after baculovirus infection.

The working hypothesis for the mechanism by which *vankyrin* genes stabilize cells is that apoptotic signal transduction pathways that are activated by NF-κB signaling are blocked to enable prolonged cell

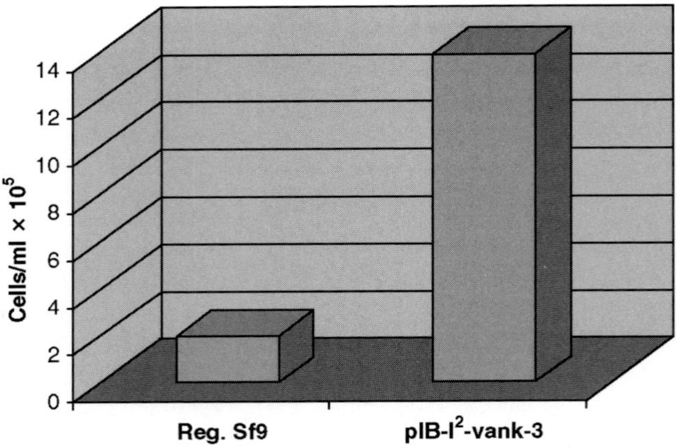

FIG 7. Viability of YFP baculovirus-infected cells is increased in the stably transformed Sf9 cell line expressing the vankyrin protein I^2-vank-3 (pIB-I^2-vank-3). Number of cells/ml was determined by counting living cells with a hemocytometer 4d p.i.

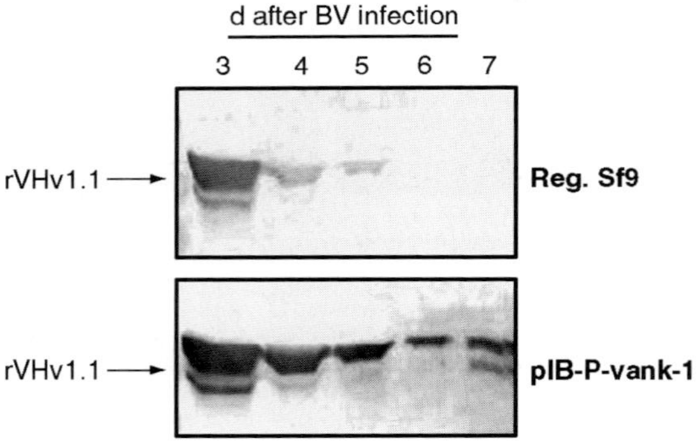

FIG 8. Secretion of the CsIV Cys-motif protein VHv1.1 is prolonged in a P-vank-1 expressing stably transformed cell line (pIB-P-vank-1) infected with recombinant AcMNPV expressing VHv1.1. Western blots show proteins released into culture medium each day starting at 3d p.i.

survival. As apoptotic pathways are highly conserved it is possible that *vankyrin* genes may have similar utility in other lytic virus expression vector systems, such as the adenovirus expression system, or in basic cell biology research.

Acknowledgments

This is publication # 06-08-033 of the University of Kentucky Agricultural Experiment Station.

References

Ailor, E., and Betenbaugh, M. J. (1999). Modifying secretion and post-translational processing in insect cells. *Curr. Opin. Biotechnol.* **10:**142–145.

Chen, Y., Yao, B., Zhu, Z., Yi, Y., Lin, X., Zhang, Z., and Shen, G. (2004). A constitutive super-enhancer: Homologous region 3 of *Bombyx mori* nucleopolyhedrovirus. *Biochem. Biophys. Res. Commun.* **318:**1039–1044.

Ghosh, S., May, M. J., and Kopp, E. B. (1998). NF-κB and rel proteins: Evolutionarily conserved mediators of immune responses. *Annu. Rev. Immunol.* **16:**225–260.

Hawtin, R. E., Zarkowska, T., Arnold, K., Thomas, C. J., Gooday, G. W., King, L. A., Kuzio, J. A., and Possee, R. D. (1997). Liquefaction of *Autographa californica* nucleopolyhedrovirus-infected insects is dependent on the integrity of virus-encoded chitinase and cathepsin genes. *Virology* **238:**243–253.

Hom, L. G., and Volkman, L. E. (1998). Preventing proteolytic artifacts in the baculovirus expression system. *Biotechniques* **25:**18–20.

Ho, Y., Lo, H.-R., Lee, T.-C., Wu, C. P. Y., and Chao, Y.-C. (2004). Enhancement of correct protein folding *in vivo* by a non-lytic baculovirus. *Biochem. J.* **382:**695–702.

Huxford, T., Huang, D.-B., Malek, S., and Ghosh, G. (1998). The crystal structure of the IκBα/NF-κB complex reveals mechanisms of NF-κB inactivation. *Cell* **95:**759–770.

Jarvis, D. L., and Summers, M. D. (1989). Glycosylation and secretion of human tissue plasminogen activator in recombinant baculovirus-infected insect cells. *Mol. Cell. Biol.* **9:**214–223.

Jarvis, D. L., Weinkauf, C., and Guarino, L. A. (1996). Immediate-early baculovirus vectors for foreign gene expression in transformed or infected insect cells. *Protein Expr. Purif.* **8:**191–203.

Kaba, S. A., Salcedo, A. M., Wafula, P. O., Vlak, J. M., and van Oers, M. M. (2004). Development of a chitinase and v-cathepsin negative bacmid for improved integrity of secreted recombinant proteins. *J. Virol. Methods* **122:**113–118.

Kato, T., Murata, T., Usui, T., and Park, E. Y. (2005). Improvement of the production of GFP_{uv}-β1,3-N-acetylglucosaminyltransferase 2 fusion protein using a molecular chaperone-assisted insect-cell based expression system. *Biotechnol. Bioeng.* **89:**424–433.

Kost, T. A., Condreay, J. P., and Jarvis, D. L. (2005). Baculovirus as versatile vectors for protein expression in insect and mammalian cells. *Nat. Biotechnol.* **23:**567–575.

Kroemer, J. A., and Webb, B. A. (2004). Polydnavirus genes and genomes: Emerging gene families and new insights into polydnavirus replication. *Annu. Rev. Entomol.* **49:** 431–456.

Kroemer, J. A., and Webb, B. A. (2005). Ikappabeta-related *vankyrin* genes in the *Campoletis sonorensis* ichnovirus: Temporal and tissue-specific patterns of expression in parasitized *Heliothis virescens* lepidopteran hosts. *J. Virol.* **79:**7617–7628.

Luckow, V. A., and Summers, M. D. (1988). Trends in the development of baculovirus vectors. *Bio/Technology* **6:**47–55.

Meng, X., Khanuja, B. S., and Ip, Y. T. (1999). Toll receptor-mediated *Drosophila* immune response requires Dif, an NF-κB factor. *Genes Dev.* **13:**792–797.

Miller, L. K. (1988). Baculoviruses as gene expression vectors. *Ann. Rev. Microbiol.* **42:** 177–199.

Sano, K.-I., Maeda, K., Oki, M., and Maéda, Y. (2002). Enhancement of protein expression in insect cells by a lobster tropomyosin cDNA leader sequence. *FEBS Lett.* **532:**143–146.

Santoro, M. G., Rossi, A., and Amici, C. (2003). NF-κB and virus infection: Who controls whom. *EMBO J.* **22:**2552–2560.

Silverman, N., and Maniatis, T. (2001). NF-κB signaling pathways in mammalian and insect innate immunity. *Genes Dev.* **15:**2321–2342.

Thoetkiattikul, H., Beck, M. H., and Strand, M. R. (2005). Inhibitor kappaB-like proteins from a polydnavirus inhibit NF-kappaB activation and suppress the insect immune response. *PNAS* **102:**11426–11431.

Turnbull, M. W., and Webb, B. A. (2002). Perspectives on polydnavirus origins and evolution. *Annu. Rev. Virol.* **58:**203–254.

Venkaiah, B., Viswanathan, P., Habib, S., and Hasnain, S. E. (2004). An additional copy of the homologous region (hr1) sequence in the *Autographica californica* Mutinucleocapsid polyhedrosis virus genome promotes hyperexpression of foreign genes. *Biochemistry* **43:**8143–8151.

Webb, B. A. (1998). Polydnavirus biology, genome structure, and evolution. *In* "The Insect Viruses" (L. K. Miller and L. A. Ball, eds.), pp. 105–139. Plenum Publishing Corporation, New York.

ADVANCES IN VIRUS RESEARCH, VOL 68

BACULOVIRUS DISPLAY: A MULTIFUNCTIONAL TECHNOLOGY FOR GENE DELIVERY AND EUKARYOTIC LIBRARY DEVELOPMENT

Anna R. Mäkelä and Christian Oker-Blom

Department of Biological and Environmental Science, NanoScience Center
University of Jyväskylä, FIN-40014, Finland

Abstract

For over a decade, phage display has proven to be of immense value, allowing selection of a large variety of genes with novel functions from diverse libraries. However, the folding and modification requirements of complex proteins place a severe constraint on the type of protein that can be successfully displayed using this strategy, a restriction that could be resolved by similarly engineering a eukaryotic virus for display purposes. The quite recently established eukaryotic molecular biology tool, the baculovirus display vector system (BDVS), allows combination of genotype with phenotype and thereby enables presentation of eukaryotic proteins on the viral envelope or capsid. Data have shown that the baculovirus, *Autographa californica* multiple nucleopolyhedrovirus (AcMNPV), is a versatile tool for eukaryotic virus display. Insertion of heterologous peptides and/or proteins into the viral surface by utilizing the major envelope glycoprotein gp64, or foreign membrane-derived counterparts, allows incorporation of the sequence of interest onto the surface of infected cells and virus particles. A number of strategies are being investigated in order to further develop the display capabilities of AcMNPV and improve the complexity of a library that may be accommodated. Numerous expression vectors for various approaches of surface display have already been developed. Further improvement of both insertion and selection strategies toward development of a refined tool for

0065-3527/06 $35.00
DOI: 10.1016/S0065-3527(06)68003-2

use in the creation of useful eukaryotic libraries is, however, needed. Here, the status of baculovirus display with respect to alteration of virus tropism, antigen presentation, transgene expression in mammalian cells, and development of eukaryotic libraries will be reviewed.

I. Introduction

In the era of genomics and proteomics, direct coupling of proteins to their DNA-coding sequence is extremely valuable as it holds potential to derive functional information from unknown open reading frames. Proteins can be identified by virtue of their unique functional properties and their encoding gene subsequently isolated. Replicating nanoparticles, such as bacteriophages, have proven ideal for this type of application as they can be designed to display peptides or proteins of interest on their surface while encapsidating the gene of interest. Due to the exceptional titers that phages can achieve, the diversity of the resultant prokaryote-based libraries is very high. Successful examples of such a display technology include isolation of antibodies from large combinatorial libraries displayed on the surface of bacteriophages. Phage display, however, has notable limitations due to the simple posttranslational machinery provided by the prokaryotic host.

The eukaryote-based baculovirus expression vector system (BEVS), primarily based on the use of *Autographa californica* multiple nucleopolyhedrovirus (AcMNPV), was developed during the 1980s (Luckow and Summers, 1988; Miller, 1988a,b, 1989; Sherman and McIntosh, 1979; Smith *et al.*, 1983). Complex animal, human, and viral proteins, requiring folding, subunit assembly, and/or extensive posttranslational modification, can be successfully expressed using this system (Kost *et al.*, 2005). The successful and wide adoption of BEVS benefits the choice of AcMNPV as a candidate for the development of a safe eukaryotic display system aimed at proper presentation of antigens, gene delivery to mammalian cells as well as development of eukaryotic libraries.

II. Targeting of Baculoviral Vectors by Surface Display

Surface glycoproteins of enveloped viruses are attractive candidates for control and manipulation of cellular recognition. The limitations of prokaryotic display systems regarding posttranslational modifications and folding of the displayed proteins has led to the development of alternative eukaryotic display systems. During the last decade, the expression of foreign peptides and proteins on the baculoviral surface has been quite extensively studied and display has already been employed in a number of applications. Although an insect virus, the tropism and transduction efficiency of AcMNPV with respect to mammalian cells or tissues can be manipulated by a variety of techniques including mutation of the major viral envelope protein (gp64), incorporation of targeting peptides or antibodies into virions, and vector pseudotyping (Fig. 1).

The extensive diversity of mammalian cells that can be transduced by baculovirus vectors implies that the entry and uptake mechanisms of this insect virus by mammalian cells are universal (Kost and Condreay, 2002). Therefore, several strategies have been developed to restrict viral transduction to desired cell types. Baculovirus transduction has generally been considered as safe and nontoxic to mammalian cells, and cell growth has not been stalled even at notably high MOIs (Ho *et al.*, 2004). Baculovirus enters insect cells by endocytosis followed by a low-pH–induced fusion of the viral envelope with the endosomal membrane, consequently permitting viral entrance into the cytoplasm and nucleus (Dee and Shuler, 1997). Correspondingly, baculovirus is considered to enter mammalian cells via the same route (Fig. 1), as gene expression is inhibited by lysosomotropic agents that inhibit endosomal maturation (Boyce and Bucher, 1996; Hofmann *et al.*, 1995; van Loo *et al.*, 2001). In addition to clathrin-mediated endocytosis, baculovirus is presumably internalized by macropinocytosis (Matilainen *et al.*, 2005). Although the receptor molecule(s) of AcMNPV are unknown, the cell surface molecules for the attachment and entry of the virus have been suggested to involve common constituents of the cell membrane including phospholipids or heparan sulfate proteoglycans (Duisit *et al.*, 1999; Tani *et al.*, 2001). Despite the somewhat limited knowledge regarding the molecular mechanism involved in baculovirus entry, functional alteration of baculovirus tropism has been achieved.

Gp64 is the major baculoviral envelope (phospho)glycoprotein (Whitford *et al.*, 1989) that is present on the surface of infected insect cells and on budded virions as homotrimers, forming typical peplomer structures at the pole of the virion (Markovic *et al.*, 1998; Oomens *et al.*, 1995). For the budded form of AcMNPV, it has been shown that gp64 determines the viral receptor preference in inhibition studies with a monoclonal anti-gp64 antibody and, therefore, defines both the host range and the infection efficiency of the host (Hohmann and Faulkner, 1983; Volkman and Goldsmith, 1985). Gp64 is necessary for the low-pH–triggered membrane fusion activity (Blissard and Wenz, 1992; Jarvis and Garcia, 1994; Markovic *et al.*, 1998; Oomens *et al.*, 1995;

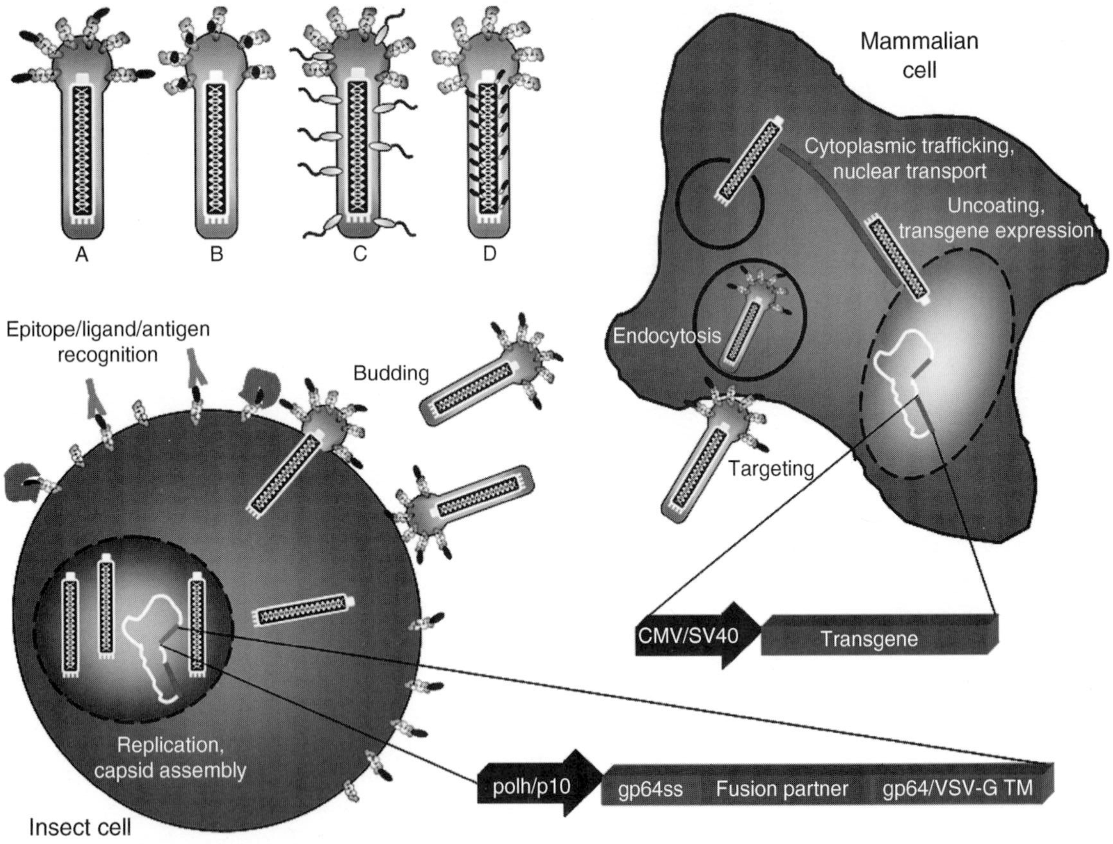
A
B
C
D
Mammalian cell
Cytoplasmic trafficking, nuclear transport
Uncoating, transgene expression
Endocytosis
Targeting
Epitope/ligand/antigen recognition
Budding
CMV/SV40
Transgene
Replication, capsid assembly
Insect cell
polh/p10
gp64ss
Fusion partner
gp64/VSV-G TM

Plonsky *et al.*, 1999) and is essential for viral budding from insect cells (Oomens and Blissard, 1999) as well as spreading the infection through cell-to-cell transmission (Monsma *et al.*, 1996). Permissive epitope insertions into the gp64 have been achieved without altering or disturbing viral infectivity. One example of this approach is a study where Ernst *et al.* (2000) took advantage of the naturally occurring *Not*I restriction site of gp64 at amino acid position 278 and nondestructively inserted two short peptides, that is, ELDKVA of the human immunodeficiency virus type 1 (HIV-1) gp41 and an eight amino acid streptavidin binding streptagII into the coding region of gp64. In a subsequent study, the ELDKVA peptide was also inserted into 17 different positions of gp64, where viral propagation was retained in as many as 13 cases, indicating that insertions of the affinity tags did not considerably affect the expression or function of gp64 (Spenger *et al.*, 2002). Thus, small peptides with specific and high affinities to receptors on mammalian cells could be introduced into gp64 for targeting to desired cell types.

While direct modification of native gp64 may be advantageous (Ernst *et al.*, 1998, 2000; Spenger *et al.*, 2002), fusion of heterologous proteins and ligand-binding moieties to an extra copy of the *gp64* gene

FIG 1. A generalized schematic outline of the multifunctional baculovirus display technology used for eukaryotic library development and mammalian gene delivery. Entry of baculovirus into mammalian cells is thought to be similar to that for insect cells. The receptor molecule(s) for baculovirus binding and entry are unknown, but they have been suggested to involve common cell surface components. The virus enters mammalian cells by endocytosis followed by low-pH–induced membrane fusion with endosomes and capsid release into the cytoplasm. The capsid is then transported toward the nucleus along actin filaments and enters the nucleus presumably through nuclear pores followed by uncoating and release of the genome. Through incorporation of ligands with high and specific affinities into the virus envelope, it is possible to target baculoviral transduction to desired cell types expressing the receptor molecule for the displayed ligand. Foreign peptides or proteins can be displayed on the baculovirus envelope as N-terminal (A) or internal (B) fusions to gp64, using a heterologous membrane anchor derived from VSV-G (C) for example, or by fusion to the major capsid protein, vp39 (D). The displayed proteins are directed to the surface of the recombinant baculovirus-infected insect cells using a signal sequence derived from the major AcMNPV envelope protein gp64 (gp64ss) for example. Genes encoding the fusion proteins are then coupled with a strong baculoviral promoter (e.g., polh or p10) for strong expression in insect cells. The vectors can be further equipped with an expression cassette encoding a reporter or a suicide gene under a mammalian promoter (e.g., CMV or SV40), enabling transduction monitoring in mammalian cells. Cell surface display can be applied in library screening for studying ligand–receptor interactions and antigen recognition. Display on the viral envelope provides possibilities for transductional targeting, whereas capsid display rather facilitates studies on intracellular trafficking as well as nuclear targeting of the virus. (See Color Insert.)

has generally been the method of choice for altering the baculovirus tropism. Boublik *et al.* (1995) were the first to demonstrate display of foreign proteins on the surface of baculovirus analogous to the established prokaryotic phage display systems. Glutathione-S-transferase was used to construct several fusion variants with the *gp64* gene. In addition, the HIV-1 major surface glycoprotein, gp120, was successfully incorporated to the amino terminus of gp64, illustrating functional ligand-binding activity when displayed on the viral surface (Boublik *et al.*, 1995). This report was followed by a study where another HIV-1–derived protein, the ectodomain of gp41, was coupled with both the native and truncated (membrane anchor) forms of gp64 (Grabherr *et al.*, 1997). To diversify the growing collection of heterologous gp64 fusion constructs, Mottershead *et al.* (1997) published a novel report where the green fluorescent protein (GFP) of *Aequorea victoria* or *Rubella virus* spike proteins E1 and E2 were displayed for the first time on the surface of an enveloped virus as N-terminal fusions to gp64. Together, these initial reports illustrated that gp64 is a suitable fusion partner for functional display of foreign polypeptides on the viral surface, providing growing possibilities for viral vector targeting. Next, the gp64 display strategy was expanded to include presentation of targeting moieties, that is, murine and human single chain antibody fragments (scFv) for the hapten 2-phenyloxazolone and carcinoembryonic antigen (CEA), respectively, and synthetic IgG-binding Z/ZZ-domains of protein A (Mottershead *et al.*, 2000). The viruses exhibited strong binding ability to the corresponding antigens and intact antibodies *in vitro*, demonstrating that the characteristics of the displayed polypeptides were preserved when presented on the viral surface. To achieve targeted baculovirus transduction, the viruses displaying ZZ-domains or scFv specific for the CEA were further modified to include a dual expression cassette containing GFP and enhanced green fluorescent protein (EGFP) under the transcriptional regulation of the polyhedrin and CMV promoters, respectively, allowing monitoring of transgene expression in both insect and mammalian cells (Ojala *et al.*, 2001). Despite improved binding, enhanced transgene expression was not observed. A baculovirus displaying an integrin-specific motif, RKK, as a part of two different loops of GFP fused with the gp64 was shown to bind a peptide representing the receptor-binding site of an $\alpha 2$ integrin, the $\alpha 2$I domain, by ELISA. Again, this interaction was not strong enough to overcome binding of wild-type gp64 to unknown cellular receptor(s) on the surface of $\alpha 2$ integrin-expressing Chinese hamster ovary (CHO) cells or to improve virus uptake (Riikonen *et al.*, 2005).

Thus, although enhanced viral binding to desired targets has been achieved, this interaction has generally not led to improved internalization and gene transduction of the vectors. In a study, however, the avidin-biotin technology was used for baculovirus targeting, resulting in both enhanced and targeted transduction (Raty *et al.*, 2004). Due to its positive charge at physiological pH, avidin itself was demonstrated to enhance viral transduction of rat malignant glioma cells (BT4C) and rabbit aortic smooth muscle (RAASMC) cells by 5- and 26-fold, respectively. Moreover, chimeric avidin–gp64 enabled viral targeting and efficient gene transfer to biotinylated cells, possibly providing a versatile tool for gene delivery. The display of $\alpha V\beta 3$-integrin–specific RGD motifs, derived from the C-terminus of coxsackievirus A9 or human parechovirus 1 VP protein, on the viral surface, resulted in both improved binding and, thus, enhanced transduction of human lung carcinoma cells expressing $\alpha V\beta 3$-integrins (Ernst *et al.*, 2006; Matilainen *et al.*, 2006).

Although foreign protein sequences have mainly been displayed on the viral surface after fusion to gp64, heterologous viral glycoproteins are capable of serving in the same context. As a model system, a truncated form of the vesicular stomatitis virus glycoprotein (VSV-G) was discovered to enhance display and enabled scattered distribution of the EGFP fusion proteins on the viral envelope (Chapple and Jones, 2002), whereas gp64 fusions normally accumulate only at the pole of the virion. To apply this fusion strategy to baculovirus targeting, the VSV-G transmembrane anchor, comprising 29 amino acids of the cytoplasmic domain, the 20-amino acid membrane spanning region in addition to the 21-amino acid truncated ectodomain, was fused with the IgG-binding ZZ domains of protein A (Ojala *et al.*, 2004) and displayed on the surface of baculovirus vectors. The ZZ-displaying viruses showed improved binding to IgG and, in principle, these vectors could be targeted to any desired cell type when a suitable IgG antibody is available, eliminating the need of preparing distinct vectors for each application. Improved transduction was not observed, however, when GFP was used as a reporter. To gain cancer cell-selective tropism of baculovirus, the LyP-1 (Laakkonen *et al.*, 2002), F3 (Porkka *et al.*, 2002), and CGKRK (Hoffman *et al.*, 2003) tumor-homing peptides were displayed on the surface of baculovirus by fusion to the membrane-anchoring signal of VSV-G (Mäkelä *et al.*, 2006). To increase the specificity, the VSV-G fusion strategy was further modified by excluding the 21-amino acid VSV-G ectodomain, known to mediate nonspecific binding and transduction of the baculovirus vectors (Kaikkonen *et al.*, 2005; Ojala *et al.*, 2004). These vectors exhibited significantly improved binding and transgene delivery to both human breast carcinoma and hepatocarcinoma cells, highlighting the potential of targeted baculovirus vectors in cancer gene therapy. In addition to VSV-G, a class I transmembrane protein, the membrane-spanning region of a class II membrane protein, neuraminidase A of influenza virus, is capable of serving as an N-terminal anchor domain for efficient display of EGFP on the viral surface (Borg *et al.*, 2004), providing an alternative display strategy to gp64 or VSV-G membrane anchors.

Analogous to surface display, a novel baculovirus capsid display method has been developed (Kukkonen *et al.*, 2003; Oker-Blom *et al.*, 2003). This technique is based on the display of foreign proteins or peptides on the surface of the viral capsid as amino or carboxy terminal fusions to the major AcMNPV capsid protein vp39 (Thiem and Miller, 1989a,b). In the first capsid display report, vp39 was demonstrated to be compatible for incorporation of a foreign protein molecule, EGFP, in large quantities. EGFP was successfully fused either to the N- or C-terminus of vp39 without compromising the viral titer or functionality (Kukkonen *et al.*, 2003). In addition, it was proposed that the block in transduction of mammalian cells by baculovirus lies in the cytoplasmic trafficking or nuclear import instead of viral escape from the endosomes, as had previously been suggested. Thus, this new tool provides possibilities for specific intracellular and nuclear targeting of the viral capsids, and facilitates baculovirus entry and nuclear import studies in both insect and mammalian cells.

It is now evident that heterologous targeting moieties with specific and high avidities can be functionally displayed in large quantities on both the baculovirus envelope and capsid. Although targeting appears to be an attractive concept to enhance baculoviral transduction of specific cell types, its applicability in human disease could be partly limited by the fact that mammalian host cell nonpermissiveness cannot always be reversed simply by making baculoviral binding and entry possible. Therefore, tissue targeting and nuclear localization signals could be displayed in different combinations on the viral envelope and nucleocapsid, respectively, enabling both cellular and nuclear targeting. In addition to appropriate targeting molecules, introduction of tissue-specific promoters and complement resistance (Huser *et al.*, 2001) into the baculovirus vectors could enable targeting of this insect virus to desired cells and tissues *in vivo* and therefore provide potential applications in gene therapy.

III. Altering the Tropism of Baculoviral Vectors Through Pseudotyping

Pseudotyping, that is, phenotypic mixing, is a process in which the natural envelope glycoproteins of the virus are modified, replaced, or expressed with surface (glyco)proteins from a donor virus. In this way, the host range of virus vectors can be expanded or altered. If successful, such particles possess the tropism of the virus from which the protein was derived. The vesicular stomatitis virus (VSV) G-protein (VSV-G) is among the first and still most widely used glycoprotein for pseudotyping viral vectors due to the very extensive tropism and stability of the resulting pseudotypes. Generation of VSV-G pseudotypes from a number of viruses has been described earlier. Both native and modified VSV-G have been extensively used for pseudotyping retroviruses (Croyle *et al.*, 2004; Emi *et al.*, 1991; Guibinga *et al.*, 2004; Schnitzer *et al.*, 1977), adenoviruses (Yun *et al.*, 2003), and herpesviruses (Anderson *et al.*, 2000; Tang *et al.*, 2001), for example. In the case of VSV-G–pseudotyped viral vectors where tissue targeting through ligand incorporation into VSV-G has been endeavored, several factors including the lack of three dimensional crystal structure of the glycoprotein, has rendered the tropism modification challenging. Regardless, permissive epitope/ligand insertion sites have been identified within native VSV-G that allow modification of the protein without compromising folding or oligomerization (Guibinga *et al.*, 2004). In addition, the use of different recombinant VSV vectors for vaccine production has been broadly studied (McKenna *et al.*, 2003; Schlehuber and Rose, 2004) and different truncated forms of VSV-G have served as partners for constructing chimeric fusion proteins to facilitate the study of the biological properties of viral or cellular membrane glycoproteins (Basu *et al.*, 2004; Buonocore *et al.*, 2002; Lagging *et al.*, 1998; Schnell *et al.*, 1996). While the VSV-G–pseudotyped vectors are valuable for many diverse studies and even for some preliminary clinical applications, their promiscuous susceptibility for target cells and tissues may contribute to toxicity and serious adverse effects through transduction of nontarget cells (Burns *et al.*, 1993; Naldini, 1999; VandenDriessche *et al.*, 2002).

Pseudotyped baculovirus vectors engineered to date represent viral particles bearing heterologous glycoproteins on their envelope, similar to other virus vectors, mainly the VSV-G, expressed either alone or with the endogenous baculovirus surface glycoprotein, gp64

(Barsoum *et al.*, 1997; Facciabene *et al.*, 2004; Kitagawa *et al.*, 2005; Mangor *et al.*, 2001; Park *et al.*, 2001; Pieroni *et al.*, 2001; Tani *et al.*, 2001, 2003). The primary objective of these studies was to engineer vectors possessing a wider tropism and improved transduction capacity of the target cells as compared to wild-type baculovirus. Secondarily, the VSV-G is expected to provide protection for baculovirus vectors against complement inactivation in potential *in vivo* gene therapy applications as has previously been demonstrated for VSV-G–pseudotyped retroviral vectors (Ory *et al.*, 1996).

Barsoum *et al.* (1997) demonstrated that AcMNPV can be pseudotyped with an envelope glycoprotein derived from another virus. The gene encoding VSV-G was placed under the transcriptional control of the polyhedrin promoter, providing abundant expression in infected insect cells and subsequent incorporation into budded virions, which exhibited atypical oval-shaped morphology and occasionally tail-like structures. However, no further studies have been published where the effect of VSV-G on the morphology of the budded form of AcMNPV has been described. These pseudotyped viruses improved transduction of HepG2 cells tenfold and also augmented transgene delivery to certain established as well as primary cell lines that are weakly or not susceptible to transduction by wild-type baculovirus, thus broadening the tropism. In addition, it was speculated that the VSV-G may augment the escape of the virus from intracellular vesicles via its membrane fusion activity rather than improve viral binding or entry into target cells, hence escalating transport of the viral genome into the nucleus (Barsoum *et al.*, 1997). Later, VSV-G and mouse hepatitis virus S protein (MHV-S)-pseudotyped baculovirus vectors were employed as a control system in a study where cell surface components involved in baculovirus infection of insect cells and entry into mammalian cells was explored using baculovirus displaying two copies of gp64 on the viral envelope (Tani *et al.*, 2001). It was demonstrated that the virus overexpressing gp64, in addition to its endogenous copy of gp64, can incorporate ~1.5- to 2-fold the normal quantity of gp64 on the budded virion. These modified viruses mediated transduction resulting in 10- to 100-fold increased reporter gene expression in a variety of cell lines as compared to the virus carrying an ordinary amount of gp64. It was also proposed that cell surface phospholipids provide a docking point for gp64, hence assisting viral entrance to mammalian cells (Tani *et al.*, 2001). Park *et al.* (2001) combined tropism modification of baculovirus with transcriptional targeting, designed to be limited to cells of hepatic origin. Accordingly, a VSV-G–pseudotyped virus, harboring an expression reporter (luciferase) gene placed under the control of a hepatocyte-specific AFP (α-fetoprotein) promoter/enhancer, was generated. The virus was able to transduce human hepatoma cells at an efficiency of approximately fivefold greater than the control virus lacking VSV-G and transgene expression was restricted to cells of hepatic origin expressing AFP, of which concentration is elevated in hepatocellular carcinomas.

The VSV-G is capable of complementing the function of gp64 by restoring the ability of a gp64-null virus to assemble and produce infectious budded virions, although the kinetics of infection is somewhat delayed and viral titers reduced by 1 to 2 logs as compared to wild-type AcMNPV (Mangor *et al.*, 2001). However, these gp64-null VSV-G–pseudotyped virions were not tested for transduction of vertebrate cells, thus, whether they could enhance transduction analogous to recombinant vectors coexpressing gp64 and VSV-G remains unanswered. In addition to VSV-G, the function of a *gp64*-deleted AcMNPV has been partially restored by inserting the recently identified F-proteins from two group II nucleopolyhedroviruses (NPVs), *Lymantria dispar* MNPV and *Spodoptera exigua* MNPV, into the gp64 locus, demonstrating that F-proteins derived from heterologous NPVs are functional analogs of gp64 (Lung *et al.*, 2002). Parallel to the VSV-G/gp64-null virus (Mangor *et al.*, 2001), infectious viral titers of the F-protein pseudotypes were somewhat compromised as compared to the wild-type counterpart, suggesting that the level of compatibility between the F-proteins and other AcMNPV proteins may not be optimal. Further, the capacity of these F-protein–pseudotyped vectors for gene transduction of mammalian cells remains to be explored.

The efficiency of gene delivery *in vivo* has also been explored using VSV-G–pseudotyped baculovirus vectors. The modified virus enhanced transgene delivery by five- to tenfold when mouse myoblasts and myotubes were transduced *in vitro* (Pieroni *et al.*, 2001). Similarly, the same increase in reporter gene (β-galactosidase) expression was detected *in vivo* after injection of the VSV-G–pseudotyped vector in the quadriceps of BALB/c and C57BL/6 mice. Moreover, expression of the transgene, mouse erythropoietin, was monitored to last for 35 and 178 days in the skeletal muscle of BALB/c or C57BL/6, and DBA/2J mice, respectively (Pieroni *et al.*, 2001). The VSV-G–coated baculovirus also exhibited improved resistance to inactivation by human, rabbit, guinea pig, hamster, and mouse, but not rat sera (Tani *et al.*, 2003). This modified virus could also be used for transduction of the cerebral cortex and testis of mice by direct inoculation *in vivo*. No comparisons were conducted, however, with the unmodified virus to evaluate putative enhancement in transduction efficiency. A truncated form of VSV-G (VSV-GED), composed of the cytoplasmic and membrane-spanning domains in addition to the 21-amino acid ectodomain, was shown to enhance transduction by the VSV-GED–pseudotyped baculovirus both *in vitro* and *in vivo* (Kaikkonen *et al.*, 2005). Thus, the enhancement of virus transduction, which is characteristic of full-length VSV-G, was retained by the truncated form. It was speculated that the improved gene delivery was due to possible augmentation of gp64-mediated release from endosomes during viral entry into the target cells. Moreover, induction of humoral and cell-mediated immune response has been studied with a recombinant baculovirus vector displaying VSV-G on the viral surface and expressing hepatitis C virus glycoprotein, E2, under the CMV promoter. The results demonstrated that cell-mediated immunity to the E2 antigen can be elicited in mice by injecting recombinant baculovirus vectors expressing the target antigen and that the display of VSV-G on the viral surface increases the immunogenic efficiency tenfold leading to greater induction of E2 antigen-specific $CD8^+$ T cells (Facciabene *et al.*, 2004). Ligand-directed gene delivery was achieved by pseudotyped *gp64*-deleted baculovirus vectors carrying measles virus receptors, CD46 and SLAM, on their surface (Kitagawa *et al.*, 2005). The viruses were able to replicate and spread infection in gp64-complementing *Sf*9 cells, whereas virus propagation was strongly reduced in cells not expressing gp64. However, after three rounds of passage of the pseudotyped viruses, the gp64-coding gene was integrated into the baculovirus genome probably through nonhomologous recombination. The corresponding viruses were able to target gene delivery to BHK cells expressing the measles virus H and F envelope glycoproteins and the transduction could be inhibited by pretreatment with specific monoclonal antibodies for the displayed ligands. A short hairpin RNA (shRNA) delivery system mediated by a VSV-G–displaying baculovirus vector was generated, resulting in knock down of an endogenous reporter gene, *EGFP*, and suppression of porcine reproductive and respiratory syndrome virus replication in tissue culture (Lu *et al.*, 2006), highlighting the potential of recombinant baculovirus as an alternative vehicle for antivirus shRNA delivery.

Overall, the AcMNPV-pseudotyping system provides an efficient and powerful method for examining the functions and compatibilities of heterologous viral or cellular membrane proteins as well as enabling diversification or constraint of the viral tropism. The selection of cell surface components during virus assembly in infected insect cells is flexible enough to allow incorporation of unrelated membrane proteins into baculovirus particles, yet specific enough to exclude the bulk of host proteins. The first proofs of the principle were the VSV-G–pseudotyped (gp64-null) baculovirus vectors, which retained their ability to replicate in insect cells and transduce a large collection of mammalian cells. VSV-G may use common cell surface determinants as putative receptor molecules, rendering the VSV-G–pseudotyped baculovirus vectors inappropriate for cell-specific gene delivery, but ideal in applications where a limited tropism is not required. Thus, such pseudotyped vectors would be particularly suitable for *ex vivo* gene therapeutic applications where there is no risk of transducing nontarget cell populations. The introduction of ligands with high and specific avidity into the viral envelope, as demonstrated by Kitagawa *et al.* (2005), could also enable baculovirus targeting *in vivo*.

IV. Baculovirus Display of Immunogens

For generation of antibodies by traditional procedures, the protein or peptide is produced in a system of choice and subsequently purified before immunization. This is often cumbersome, and more importantly, the final product may not be correctly folded—an essential requirement for an adequate immune response in the host, and thereby, for generation of functional antibodies. In addition to recombinant proteins, several other systems including phage display, DNA-based immunization, as well as recombinant viral infections and/or fusions to viral proteins are available for generation of antibodies. Here, examples from the literature are presented where baculovirus surface display has been employed for generation of functional monoclonal antibodies against proteins of different origin. Several reports also show clear evidence that display of the immunogen on the viral surface can elicit protective immune responses against viral or parasite infections by using animal models.

To produce monoclonal antibodies against the human nuclear receptors LXRβ and FXR, the N-terminal domains of these antigens were displayed on the baculoviral surface by inserting the corresponding coding sequences between the signal sequence and the mature domain of gp64 of AcMNPV (Lindley *et al.*, 2000). This study illustrated that baculovirus display is a versatile tool applicable for antigen presentation and for rapid production of functional monoclonal antibodies once the antigen-coding sequence is available (Lindley *et al.*, 2000). Similarly, monoclonal antibodies against human peroxisome proliferator-activated receptors (PPARs) using baculovirus display have been generated. The amino terminal sequences of human PPARd and PPARg2 were placed at the N-terminus of gp64 and antibodies were raised by immunization with whole virus without prior purification of the immunogens (Tanaka *et al.*, 2002). The antibodies generated by this method were functional in a variety of techniques such as immunohistochemistry, immunoblotting, and electrophoretic mobility shift assays. Antigenic epitopes of *Theileria parva*, an intracellular protozoan parasite that causes East Coast fever, a severe lymphoproliferative disease in cattle, were also successfully presented on the surface of AcMNPV by adopting the gp64 N-terminal fusion strategy (Kaba *et al.*, 2003). This approach was applied because previous attempts to produce recombinant sporozoite surface antigen (p67) in bacterial or insect cells for vaccine purposes had not resulted in correctly folded protein molecules. Further, a small, immunodominant antigenic site (site A) and the large polyprotein (P1) coding for the four structural proteins of foot-and-mouth disease virus (FMDV) have been displayed on the membrane of infected insect cells and consequently on the baculoviral surface by fusion to the N-terminus of gp64 (Tami *et al.*, 2000). Later, the investigators have shown that these FMVD antigens were able to elicit a specific immune response against FMVD in mice (Tami *et al.*, 2004). Similarly, Yoshida *et al.* (2003) have shown that the rodent malaria *Plasmodium berghei* circumsporozoite protein (PbCSP) displayed on the surface of baculovirus as a fusion to gp64 protects mice against a malaria sporozoite infection.

Urano *et al.* (2003) used an alternative approach of exploiting the baculovirus for monoclonal antibody production by displaying an integral ER membrane protein SCAP on the extracellular, budded form of the virus. Thus, SCAP was not displayed as a fusion to baculovirus specific proteins. SCAP is known to be involved in cleavage of sterol element-binding protein-2, hence its function is tightly coupled to cholesterol regulation (Urano *et al.*, 2003). Other membrane receptors, such as the β-adrenergic receptor (Loisel *et al.*, 1997) and the leukotriene B4 receptor (BLT1) (Masuda *et al.*, 2003) residing on the plasma membrane have also been functionally displayed in the same context.

In addition to AcMNPV, *Bombyx mori* NPV (BmNPV) has been modified to display immunogens with similar aims as described earlier. Rahman *et al.* (2003) displayed the immunodominant ectodomains of the fusion glycoprotein (F) of peste-des-petitis-ruminants virus (PPRV) and the hemagglutinin protein (H) of rinderpest virus (RPV), on budded virus particles. The strategy was identical, in that the antigens were fused to gp64 of BmNPV and expressed under transcriptional regulation of the polyhedrin promoter. The investigators showed that the antigenic epitopes were properly displayed and that the recombinant virions were able to induce an immune response in mice against both PPRV and RPV. Finally, Chang *et al.* (2004) aimed to produce a recombinant baculovirus that mimics severe acute respiratory syndrome corona virus (SARS-CoV) in its host range and infection mechanism. A baculovirus displaying a 688-amino acid fragment of SARS-CoV S glycoprotein as a gp64 fusion was generated and then used to examine the effect on the IL-8 release in A549, NCI-H520, HFL-1, and MRC-5 cells (Chang *et al.*, 2004).

Together, these reports provide convincing evidence that baculovirus can be used for the functional display of heterologous proteins on its surface through budding from the infected insect cell. Consequently, the baculovirus-displayed immunogens have been used to elicit immune responses needed for production of monoclonal antibodies and/ or to protect the animal host against a viral or parasite infection.

V. Generation of Display Libraries

Display on the surface of bacteriophage is currently the most widespread method for display and selection of large collections of antibodies. This approach is robust, simple to use and, in addition, highly versatile. The selection procedures can be adapted to many specific conditions including selections on whole cells, tissues, and even animals. Originally, generation of eukaryotic cDNA libraries was based on plasmid vectors capable of replicating in particular eukaryotic cell types. During the last decade, however, a variety of display methods and other library-screening techniques have been under study for isolating monoclonal antibodies from collections of recombinant antibody fragments. The development of virus-based cDNA expression libraries has offered several advantages over nonviral vectors regarding host cell tropism, transduction efficiency, stability of transgene expression, and production of the vector in high quantities.

In 1997, Granziero *et al.* (1997) aimed to develop a rapid method for generating baculovirus-based cDNA expression libraries for screening cell surface molecules, for which antibodies are available beforehand and whose expression pattern is restricted to particular cell types. The first proof of principle was gained by cloning a cDNA pool, reverse transcribed from human placenta, into the baculovirus genome and sorting the virus-infected insect cells by flow cytometry using monoclonal antibodies of an unknown specificity as probes. By this method, single positive cells could be sorted and viruses carrying the cDNAs encoding the cell surface epitopes isolated. The first demonstration of using baculovirus display for generation and screening of expression was described by Ernst *et al.* (1998). An HIV-1 gp41 epitope (ELDKWA), specific for the neutralizing human mAb 2F5, was inserted into the antigenic site B of influenza virus A hemagglutinin, and expressed on the surface of baculovirus-infected insect cells. The epitope was displayed in a library form, such that each clone contained different amino acids adjacent to the epitope. Thus, the purpose of the experiment was to alter the structural environment so that the corresponding epitope would be presented in the most accessible way, leading to an increased binding capacity of the mAb. The library consisted of 8000 variants out of which one clone showed an increased specific binding capacity when screened by fluorescence activated cell sorting (Ernst *et al.*, 1998). Later, the group also described a system where the same epitope as well as the biotin mimic streptag II were inserted at position 278 of gp64 (Ernst *et al.*, 2000). The fact that the insertions into the coding sequence of the major envelope protein of the virus did not alter virus propagation may be of value in further development of display libraries.

Crawford *et al.* (2004) have described the use of baculovirus-infected insect cells as a display platform for class II major histocompatibility complex (MHCII) molecules covalently bound to a library of potential peptide mimotopes. The sequence encoding the peptide was embedded within the genes for the MHC molecule in the viral genome. Thereby, each insect cell infected with a virus particle from a library coding for different peptides, displayed a unique peptide–MHC complex on its cellular membrane. Crawford *et al.* (2004) were able to identify such peptide mimotope–MHC complexes that bound to the soluble receptors and stimulating T cells bearing the same receptors by "fishing" with fluorescent, soluble T cell receptors. These findings should, therefore, have implications for the relative importance of peptide and MHC in T cell receptor-ligand recognition. Later, the same group used this baculovirus-based display system for identification of antigen mimotopes for MHC class I-specific T cells (Wang *et al.*, 2005). Here, a mouse MHC class I molecule was displayed on the surface of baculovirus-infected insect cells with a 9- to 10-mer peptide library tethered to the N-terminus of beta2 microglobulin via a flexible linker. Although there are relatively few studies on libraries generated by using baculovirus/insect cell technology, the present examples clearly show that this technology has potential and interest in further development and utilization of this technology will likely increase.

VI. Summary

In this chapter, we have given a "state of the art" overview of strategies and technologies developed for display of foreign peptides and/or proteins on the surface of baculovirus-infected insect cells and budded baculovirus particles. Data on virus targeting and transgene expression in mammalian cells, and on the generation of libraries for studying molecular recognition and protein–protein interactions using these techniques were summarized. Production of monoclonal antibodies by utilization of these techniques and the benefits of using baculovirus display to elicit protective immune responses in animal models were reviewed. Together, these studies show the potential for baculovirus within these areas of research and illustrate that further development and broadening of the interdisciplinary applications of this versatile and unique insect virus are justified.

References

Anderson, D. B., Laquerre, S., Ghosh, K., Ghosh, H. P., Goins, W. F., Cohen, J. B., and Glorioso, J. C. (2000). Pseudotyping of glycoprotein D-deficient herpes simplex virus type 1 with vesicular stomatitis virus glycoprotein G enables mutant virus attachment and entry. *J. Virol.* **74:**2481–2487.

Barsoum, J., Brown, R., McKee, M., and Boyce, F. M. (1997). Efficient transduction of mammalian cells by a recombinant baculovirus having the vesicular stomatitis virus G glycoprotein. *Hum. Gene Ther.* **8:**2011–2018.

Basu, A., Beyene, A., Meyer, K., and Ray, R. (2004). The hypervariable region 1 of the E2 glycoprotein of hepatitis C virus binds to glycosaminoglycans, but this binding does not lead to infection in a pseudotype system. *J. Virol.* **78:**4478–4486.

Blissard, G. W., and Wenz, J. R. (1992). Baculovirus gp64 envelope glycoprotein is sufficient to mediate pH-dependent membrane fusion. *J. Virol.* **66:**6829–6835.

Borg, J., Nevsten, P., Wallenberg, R., Stenstrom, M., Cardell, S., Falkenberg, C., and Holm, C. (2004). Amino-terminal anchored surface display in insect cells and budded baculovirus using the amino-terminal end of neuraminidase. *J. Biotechnol.* **114:**21–30.

Boublik, Y., Di Bonito, P., and Jones, I. M. (1995). Eukaryotic virus display: Engineering the major surface glycoprotein of the *Autographa californica* nuclear polyhedrosis virus (AcNPV) for the presentation of foreign proteins on the virus surface. *Biotechnology (NY)* **13:**1079–1084.

Boyce, F. M., and Bucher, N. L. (1996). Baculovirus-mediated gene transfer into mammalian cells. *Proc. Natl. Acad. Sci. USA* **93:**2348–2352.

Buonocore, L., Blight, K. J., Rice, C. M., and Rose, J. K. (2002). Characterization of vesicular stomatitis virus recombinants that express and incorporate high levels of hepatitis C virus glycoproteins. *J. Virol.* **76:**6865–6872.

Burns, J. C., Friedmann, T., Driever, W., Burrascano, M., and Yee, J. K. (1993). Vesicular stomatitis virus G glycoprotein pseudotyped retroviral vectors: Concentration to very high titer and efficient gene transfer into mammalian and nonmammalian cells. *Proc. Natl. Acad. Sci. USA* **90:**8033–8037.

Chang, Y. J., Liu, C. Y., Chiang, B. L., Chao, Y. C., and Chen, C. C. (2004). Induction of IL-8 release in lung cells via activator protein-1 by recombinant baculovirus displaying severe acute respiratory syndrome-coronavirus spike proteins: Identification of two functional regions. *J. Immunol.* **173:**7602–7614.

Chapple, S. D., and Jones, I. M. (2002). Non-polar distribution of green fluorescent protein on the surface of *Autographa californica* nucleopolyhedrovirus using a heterologous membrane anchor. *J. Biotechnol.* **95:**269–275.

Crawford, F., Huseby, E., White, J., Marrack, P., and Kappler, J. W. (2004). Mimotopes for alloreactive and conventional T cells in a peptide-MHC display library. *PLoS Biol.* **2:**E90.

Croyle, M. A., Callahan, S. M., Auricchio, A., Schumer, G., Linse, K. D., Wilson, J. M., Brunner, L. J., and Kobinger, G. P. (2004). PEGylation of a vesicular stomatitis virus G pseudotyped lentivirus vector prevents inactivation in serum. *J. Virol.* **78:**912–921.

Dee, K. U., and Shuler, M. L. (1997). Optimization of an assay for baculovirus titer and design of regimens for the synchronous infection of insect cells. *Biotechnol. Prog.* **13:**14–24.

Duisit, G., Saleun, S., Douthe, S., Barsoum, J., Chadeuf, G., and Moullier, P. (1999). Baculovirus vector requires electrostatic interactions including heparan sulfate for efficient gene transfer in mammalian cells. *J. Gene Med.* **1:**93–102.

Emi, N., Friedmann, T., and Yee, J. K. (1991). Pseudotype formation of murine leukemia virus with the G protein of vesicular stomatitis virus. *J. Virol.* **65:**1202–1207.

Ernst, W., Grabherr, R., Wegner, D., Borth, N., Grassauer, A., and Katinger, H. (1998). Baculovirus surface display: Construction and screening of a eukaryotic epitope library. *Nucleic Acids Res.* **26:**1718–1723.

Ernst, W., Schinko, T., Spenger, A., Oker-Blom, C., and Grabherr, R. (2006). Improving baculovirus transduction of mammalian cells by surface display of a RGD-motif. *J. Biotechnol.* (in press).

Ernst, W. J., Spenger, A., Toellner, L., Katinger, H., and Grabherr, R. M. (2000). Expanding baculovirus surface display. Modification of the native coat protein gp64 of *Autographa californica* NPV. *Eur. J. Biochem.* **267:**4033–4039.

Facciabene, A., Aurisicchio, L., and La Monica, N. (2004). Baculovirus vectors elicit antigen-specific immune responses in mice. *J. Virol.* **78:**8663–8672.

Grabherr, R., Ernst, W., Doblhoff-Dier, O., Sara, M., and Katinger, H. (1997). Expression of foreign proteins on the surface of Autographa californica nuclear polyhedrosis virus. *Biotechniques* **22:**730–735.

Granziero, L., Nelboeck, P., Bedoucha, M., Lanzavecchia, A., and Reid, H. H. (1997). Baculovirus cDNA libraries for expression cloning of genes encoding cell-surface antigens. *J. Immunol. Methods* **203:**131–139.

Guibinga, G. H., Hall, F. L., Gordon, E. M., Ruoslahti, E., and Friedmann, T. (2004). Ligand-modified vesicular stomatitis virus glycoprotein displays a temperature-sensitive intracellular trafficking and virus assembly phenotype. *Mol. Ther.* **9:**76–84.

Ho, Y. C., Chen, H. C., Wang, K. C., and Hu, Y. C. (2004). Highly efficient baculovirus-mediated gene transfer into rat chondrocytes. *Biotechnol. Bioeng.* **88:**643–651.

Hofmann, C., Sandig, V., Jennings, G., Rudolph, M., Schlag, P., and Strauss, M. (1995). Efficient gene transfer into human hepatocytes by baculovirus vectors. *Proc. Natl. Acad. Sci. USA* **92:**10099–10103.

Hoffman, J. A., Giraudo, E., Singh, M., Zhang, L., Inoue, M., Porkka, K., Hanahan, D., and Ruoslahti, E. (2003). Progressive vascular changes in a transgenic mouse model of squamous cell carcinoma. *Cancer Cell* **4:**383–391.

Hohmann, A. W., and Faulkner, P. (1983). Monoclonal antibodies to baculovirus structural proteins: Determination of specificities by Western blot analysis. *Virology* **125:**432–444.

Huser, A., Rudolph, M., and Hofmann, C. (2001). Incorporation of decay-accelerating factor into the baculovirus envelope generates complement-resistant gene transfer vectors. *Nat. Biotechnol.* **19:**451–455.

Jarvis, D. L., and Garcia, A., Jr. (1994). Biosynthesis and processing of the *Autographa californica* nuclear polyhedrosis virus gp64 protein. *Virology* **205:**300–313.

Kaba, S. A., Hemmes, J. C., van Lent, J. W., Vlak, J. M., Nene, V., Musoke, A. J., and van Oers, M. M. (2003). Baculovirus surface display of Theileria parva p67 antigen preserves the conformation of sporozoite-neutralizing epitopes. *Protein Eng.* **16:**73–78.

Kaikkonen, M. U., Raty, J. K., Airenne, K. J., Wirth, T., Heikura, T., and Yla-Herttuala, S. (2005). Truncated vesicular stomatitis virus G protein improves baculovirus transduction efficiency *in vitro* and *in vivo*. *Gene Ther.* Online publication November 3.

Kitagawa, Y., Tani, H., Limn, C. K., Matsunaga, T. M., Moriishi, K., and Matsuura, Y. (2005). Ligand-directed gene targeting to mammalian cells by pseudotype baculoviruses. *J. Virol.* **79:**3639–3652.

Kost, T. A., and Condreay, J. P. (2002). Recombinant baculoviruses as mammalian cell gene-delivery vectors. *Trends Biotechnol.* **20:**173–180.

Kost, T. A., Condreay, J. P., and Jarvis, D. L. (2005). Baculovirus as versatile vectors for protein expression in insect and mammalian cells. *Nat. Biotechnol.* **23**(5):567–575.

Kukkonen, S. P., Airenne, K. J., Marjomaki, V., Laitinen, O. H., Lehtolainen, P., Kankaanpaa, P., Mahonen, A. J., Raty, J. K., Nordlund, H. R., Oker-Blom, C., Kulomaa, M. S., and Yla-Herttuala, S. (2003). Baculovirus capsid display: A novel tool for transduction imaging. *Mol. Ther.* **8:**853–862.

Laakkonen, P., Porkka, K., Hoffman, J. A., and Ruoslahti, E. (2002). A tumor-homing peptide with a targeting specificity related to lymphatic vessels. *Nat. Med.* **8:**751–755.

Lagging, L. M., Meyer, K., Owens, R. J., and Ray, R. (1998). Functional role of hepatitis C virus chimeric glycoproteins in the infectivity of pseudotyped virus. *J. Virol.* **72:**3539–3546.

Lindley, K. M., Su, J. L., Hodges, P. K., Wisely, G. B., Bledsoe, R. K., Condreay, J. P., Winegar, D. A., Hutchins, J. T., and Kost, T. A. (2000). Production of monoclonal antibodies using recombinant baculovirus displaying gp64-fusion proteins. *J. Immunol. Methods* **234:**123–135.

Loisel, T. P., Ansanay, H., St-Onge, S., Gay, B., Boulanger, P., Strosberg, A. D., Marullo, S., and Bouvier, M. (1997). Recovery of homogeneous and functional beta 2-adrenergic receptors from extracellular baculovirus particles. *Nat. Biotechnol.* **15:**1300–1304.

Lu, L., Ho, Y., and Kwang, J. (2006). Suppression of porcine arterivirus replication by baculovirus-delivered shRNA targeting nucleoprotein. *Biochem. Biophys. Res. Commun.* **340:**1178–1183.

Luckow, V. A., and Summers, M. A. (1988). Trends in the development of baculovirus expression vectors. *Bio/Technology* **6:**47–55.

Lung, O., Westenberg, M., Vlak, J. M., Zuidema, D., and Blissard, G. W. (2002). Pseudotyping *Autographa californica* multicapsid nucleopolyhedrovirus (AcMNPV): F proteins from group II NPVs are functionally analogous to AcMNPV GP64. *J. Virol.* **76:** 5729–5736.

Mäkelä, A. R., Matilainen, H., White, D. J., Rouslahti, E., and Oker-Blom, C. (2006). Enhanced baculovirus-mediated transduction of human cancer cells by tumor-homing peptides. *J. Virol.* **80:** 6603–6611.

Mangor, J. T., Monsma, S. A., Johnson, M. C., and Blissard, G. W. (2001). A GP64-null baculovirus pseudotyped with vesicular stomatitis virus G protein. *J. Virol.* **75:**2544–2556.

Markovic, I., Pulyaeva, H., Sokoloff, A., and Chernomordik, L. V. (1998). Membrane fusion mediated by baculovirus gp64 involves assembly of stable gp64 trimers into multiprotein aggregates. *J. Cell Biol.* **143:**1155–1166.

Masuda, K., Itoh, H., Sakihama, T., Akiyama, C., Takahashi, K., Fukuda, R., Yokomizo, T., Shimizu, T., Kodama, T., and Hamakubo, T. (2003). A combinatorial G protein-coupled receptor reconstitution system on budded baculovirus. Evidence for Galpha and Galphao coupling to a human leukotriene B4 receptor. *J. Biol. Chem.* **278:**24552–24562.

Matilainen, H., Rinne, J., Gilbert, L., Marjomaki, V., Reunanen, H., and Oker-Blom, C. (2005). Baculovirus entry into human hepatoma cells. *J. Virol.* **79:**15452–15459.

Matilainen, H., Mäkelä, A. R., Riikonen, R., Saloniemi, T., Korhonen, E., Hyypiä, T., Heino, J., Grabherr, R., and Oker-Blom, C. (2006). RGD motifs on the surface of baculovirus enhance transduction of human lung carcinoma cells. *J. Biotechnol.* **125**(1):114–126.

McKenna, P. M., McGettigan, J. P., Pomerantz, R. J., Dietzschold, B., and Schnell, M. J. (2003). Recombinant rhabdoviruses as potential vaccines for HIV-1 and other diseases. *Curr. HIV Res.* **1:**229–237.

Miller, L. K. (1988a). Baculoviruses as gene expression vectors. *Annu. Rev. Microbiol.* **42:**177–199.

Miller, L. K. (1988b). Baculoviruses for foreign gene expression in insect cells. *Biotechnology* **10:**457–465.

Miller, L. K. (1989). Insect baculoviruses: Powerful gene expression vectors. *Bioessays* **11:**91–95.

Monsma, S. A., Oomens, A. G., and Blissard, G. W. (1996). The GP64 envelope fusion protein is an essential baculovirus protein required for cell-to-cell transmission of infection. *J. Virol.* **70:**4607–4616.

Mottershead, D., van der Linden, I., von Bonsdorff, C. H., Keinanen, K., and Oker-Blom, C. (1997). Baculoviral display of the green fluorescent protein and rubella virus envelope proteins. *Biochem. Biophys. Res. Commun.* **238:**717–722.

Mottershead, D. G., Alfthan, K., Ojala, K., Takkinen, K., and Oker-Blom, C. (2000). Baculoviral display of functional scFv and synthetic IgG-binding domains. *Biochem. Biophys. Res. Commun.* **275:**84–90.

Naldini, L. (1999). *In vivo* gene delivery by lentiviral vectors. *Thromb. Haemost.* **82:**552–554.

Ojala, K., Mottershead, D. G., Suokko, A., and Oker-Blom, C. (2001). Specific binding of baculoviruses displaying gp64 fusion proteins to mammalian cells. *Biochem. Biophys. Res. Commun.* **284:**777–784.

Ojala, K., Koski, J., Ernst, W., Grabherr, R., Jones, I., and Oker-Blom, C. (2004). Improved display of synthetic IgG-binding domains on the baculovirus surface. *Technol. Cancer Res. Treat.* **3:**77–84.

Oker-Blom, C., Airenne, K. J., and Grabherr, R. (2003). Baculovirus display strategies: Emerging tools for eukaryotic libraries and gene delivery. *Brief. Funct. Genomic. Proteomic.* **2:**244–253.

Oomens, A. G., and Blissard, G. W. (1999). Requirement for GP64 to drive efficient budding of Autographa californica multicapsid nucleopolyhedrovirus. *Virology* **254:**297–314.

Oomens, A. G., Monsma, S. A., and Blissard, G. W. (1995). The baculovirus GP64 envelope fusion protein: Synthesis, oligomerization, and processing. *Virology* **209:** 592–603.

Ory, D. S., Neugeboren, B. A., and Mulligan, R. C. (1996). A stable human-derived packaging cell line for production of high titer retrovirus/vesicular stomatitis virus G pseudotypes. *Proc. Natl. Acad. Sci. USA* **93:**11400–11406.

Park, S. W., Lee, H. K., Kim, T. G., Yoon, S. K., and Paik, S. Y. (2001). Hepatocyte-specific gene expression by baculovirus pseudotyped with vesicular stomatitis virus envelope glycoprotein. *Biochem. Biophys. Res. Commun.* **289:**444–450.

Pieroni, L., Maione, D., and La Monica, N. (2001). *In vivo* gene transfer in mouse skeletal muscle mediated by baculovirus vectors. *Hum. Gene Ther.* **12:**871–881.

Plonsky, I., Cho, M. S., Oomens, A. G., Blissard, G., and Zimmerberg, J. (1999). An analysis of the role of the target membrane on the Gp64-induced fusion pore. *Virology* **253:**65–76.

Porkka, K., Laakkonen, P., Hoffman, J. A., Bernasconi, M., and Ruoslahti, E. (2002). A fragment of the HMGN2 protein homes to the nuclei of tumor cells and tumor endothelial cells *in vivo*. *Proc. Natl. Acad. Sci. USA* **99:**7444–7449.

Rahman, M. M., Shaila, M. S., and Gopinathan, K. P. (2003). Baculovirus display of fusion protein of Peste des petits ruminants virus and hemagglutination protein of Rinderpest virus and immunogenicity of the displayed proteins in mouse model. *Virology* **317:**36–49.

Raty, J. K., Airenne, K. J., Marttila, A. T., Marjomaki, V., Hytonen, V. P., Lehtolainen, P., Laitinen, O. H., Mahonen, A. J., Kulomaa, M. S., and Yla-Herttuala, S. (2004). Enhanced gene delivery by avidin-displaying baculovirus. *Mol. Ther.* **9:**282–291.

Riikonen, R., Matilainen, H., Rajala, N., Pentikainen, O., Johnson, M., Heino, J., and Oker-Blom, C. (2005). Functional display of an alpha2 integrin-specific motif (RKK) on the surface of baculovirus particles. *Technol. Cancer Res. Treat* **4:**437–445.

Schlehuber, L. D., and Rose, J. K. (2004). Prediction and identification of a permissive epitope insertion site in the vesicular stomatitis virus glycoprotein. *J. Virol.* **78:** 5079–5087.

Schnell, M. J., Buonocore, L., Kretzschmar, E., Johnson, E., and Rose, J. K. (1996). Foreign glycoproteins expressed from recombinant vesicular stomatitis viruses are incorporated efficiently into virus particles. *Proc. Natl. Acad. Sci. USA* **93:**11359–11365.

Schnitzer, T. J., Weiss, R. A., and Zavada, J. (1977). Pseudotypes of vesicular stomatitis virus with the envelope properties of mammalian and primate retroviruses. *J. Virol.* **23:**449–454.

Sherman, K. E., and McIntosh, A. H. (1979). Baculovirus replication in a mosquito (dipteran) cell line. *Infect. Immun.* **26:**232–234.

Smith, G. E., Summers, M. D., and Fraser, M. J. (1983). Production of human beta interferon in insect cells infected with a baculovirus expression vector. *Mol. Cell. Biol.* **3:**2156–2165.

Spenger, A., Grabherr, R., Tollner, L., Katinger, H., and Ernst, W. (2002). Altering the surface properties of baculovirus Autographa californica NPV by insertional mutagenesis of the envelope protein gp64. *Eur. J. Biochem.* **269:**4458–4467.

Tami, C., Farber, M., Palma, E. L., and Taboga, O. (2000). Presentation of antigenic sites from foot-and-mouth disease virus on the surface of baculovirus and in the membrane of infected cells. *Arch. Virol.* **145:**1815–1828.

Tami, C., Peralta, A., Barbieri, R., Berinstein, A., Carrillo, E., and Taboga, O. (2004). Immunological properties of FMDV-gP64 fusion proteins expressed on SF9 cell and baculovirus surfaces. *Vaccine* **23:**840–845.

Tanaka, T., Takeno, T., Watanabe, Y., Uchiyama, Y., Murakami, T., Yamashita, H., Suzuki, A., Aoi, R., Iwanari, H., Jiang, S. Y., Naito, M., Tachibana, K., *et al.* (2002). The generation of monoclonal antibodies against human peroxisome proliferator-activated receptors (PPARs). *J. Atheroscler. Thromb.* **9:**233–242.

Tang, J., Yang, T., Ghosh, H. P., and Geller, A. I. (2001). Helper virus-free HSV-1 vectors packaged both in the presence of VSV G protein and in the absence of HSV-1 glycoprotein B support gene transfer into neurons in the rat striatum. *J. Neurovirol.* **7:**548–555.

Tani, H., Nishijima, M., Ushijima, H., Miyamura, T., and Matsuura, Y. (2001). Characterization of cell-surface determinants important for baculovirus infection. *Virology* **279:**343–353.

Tani, H., Limn, C. K., Yap, C. C., Onishi, M., Nozaki, M., Nishimune, Y., Okahashi, N., Kitagawa, Y., Watanabe, R., Mochizuki, R., Moriishi, K., and Matsuura, Y. (2003). *In vitro* and *in vivo* gene delivery by recombinant baculoviruses. *J. Virol.* **77:**9799–9808.

Thiem, S. M., and Miller, L. K. (1989a). A baculovirus gene with a novel transcription pattern encodes a polypeptide with a zinc finger and a leucine zipper. *J. Virol.* **63:**4489–4497.

Thiem, S. M., and Miller, L. K. (1989b). Identification, sequence, and transcriptional mapping of the major capsid protein gene of the baculovirus Autographa californica nuclear polyhedrosis virus. *J. Virol.* **63:**2008–2018.

Urano, Y., Yamaguchi, M., Fukuda, R., Masuda, K., Takahashi, K., Uchiyama, Y., Iwanari, H., Jiang, S. Y., Naito, M., Kodama, T., and Hamakubo, T. (2003). A novel method for viral display of ER membrane proteins on budded baculovirus. *Biochem. Biophys. Res. Commun.* **308:**191–196.

van Loo, N. D., Fortunati, E., Ehlert, E., Rabelink, M., Grosveld, F., and Scholte, B. J. (2001). Baculovirus infection of nondividing mammalian cells: Mechanisms of entry and nuclear transport of capsids. *J. Virol.* **75:**961–970.

VandenDriessche, T., Naldini, L., Collen, D., and Chuah, M. K. (2002). Oncoretroviral and lentiviral vector-mediated gene therapy. *Methods Enzymol.* **346:**573–589.

Wang, Y., Rubtsov, A., Heiser, R., White, J., Crawford, F., Marrack, P., and Kappler, J. W. (2005). Using a baculovirus display library to identify MHC class I mimotopes. *Proc. Natl. Acad. Sci. USA* **102:**2476–2481.

Whitford, M., Stewart, S., Kuzio, J., and Faulkner, P. (1989). Identification and sequence analysis of a gene encoding gp67, an abundant envelope glycoprotein of the baculovirus Autographa californica nuclear polyhedrosis virus. *J. Virol.* **63:**1393–1399.

Volkman, L. E., and Goldsmith, P. A. (1985). Mechanism of neutralization of budded *Autographa californica* nuclear polyhedrosis virus by a monoclonal antibody: Inhibition of entry by adsorptive endocytosis. *Virology* **143:**185–195.

Yoshida, S., Kondoh, D., Arai, E., Matsuoka, H., Seki, C., Tanaka, T., Okada, M., and Ishii, A. (2003). Baculovirus virions displaying Plasmodium berghei circumsporozoite protein protect mice against malaria sporozoite infection. *Virology* **316:**161–170.

Yun, C. O., Cho, E. A., Song, J. J., Kang, D. B., Kim, E., Sohn, J. H., and Kim, J. H. (2003). dl-VSVG-LacZ, a vesicular stomatitis virus glycoprotein epitope-incorporated adenovirus, exhibits marked enhancement in gene transduction efficiency. *Hum. Gene Ther.* **14:**1643–1652.

STABLY TRANSFORMED INSECT CELL LINES: TOOLS FOR EXPRESSION OF SECRETED AND MEMBRANE-ANCHORED PROTEINS AND HIGH-THROUGHPUT SCREENING PLATFORMS FOR DRUG AND INSECTICIDE DISCOVERY

Vassilis Douris,* Luc Swevers,* Vassiliki Labropoulou,* Evi Andronopoulou,* Zafiroula Georgoussi,† and Kostas Iatrou*

*Insect Molecular Genetics and Biotechnology Group, Institute of Biology National Centre for Scientific Research "Demokritos", GR 153 10 Aghia Paraskevi Attikis (Athens), Greece
†Laboratory of Cellular Signaling and Molecular Pharmacology, Institute of Biology National Centre for Scientific Research "Demokritos", GR 153 10 Aghia Paraskevi Attikis (Athens), Greece

Abstract

Insect cell-based expression systems are prominent amongst current expression platforms for their ability to express virtually all types of heterologous recombinant proteins. Stably transformed insect cell lines represent an attractive alternative to the baculovirus expression system, particularly for the production of secreted and membrane-anchored proteins. For this reason, transformed insect cell systems are receiving

0065-3527/06 $35.00
DOI: 10.1016/S0065-3527(06)68004-4

increased attention from the research community and the biotechnology industry. In this article, we review recent developments in the field of insect cell-based expression from two main perspectives, the production of secreted and membrane-anchored proteins and the establishment of novel methodological tools for the identification of bioactive compounds that can be used as research reagents and leads for new pharmaceuticals and insecticides.

I. Introduction

High-throughput protein expression is an essential tool for the development of multiple research and biotechnological applications in the postgenomic era. For example, characterization of novel genes may require overexpression of the encoded proteins in heterologous protein expression systems in order to proceed with functional or structural characterization. Production of therapeutic proteins largely relies on the use of genetically engineered host organisms that allow for protein production at high levels. Furthermore, the properties of a wide variety of compounds of natural or synthetic origin are explored through the use of high-throughput screens based on genetically engineered organisms or cell culture systems.

Expression tools derived from several biological systems are used by research and industrial laboratories to achieve efficient recombinant protein expression. The organisms from which the tools have been derived include bacteria, yeast and other fungi, plants, mammals, and insects, as well as cell lines derived from various mammalian and insect species. Each expression system has unique features and limitations that make it appropriate for certain applications but not as suitable for others.

Bacterial expression systems can generally direct very high levels of protein expression, but in many cases correct folding and biological activity of the proteins produced in bacteria are not achieved (Baneyx and Mujacic, 2004). Recombinant proteins produced in bacteria also lack the complex posttranslational modifications that take place in eukaryotic cells. As a result, eukaryotic proteins that require modifications cannot be functionally expressed in bacteria. Furthermore, bacteria lack eukaryotic-type secretion systems; secreted heterologous proteins localize in the periplasmic space from where they can be isolated using specific, but usually cumbersome and inefficient, protocols. Highly hydrophobic, unfolded, misfolded, and denatured proteins accumulate in the bacterial cytoplasm as insoluble "inclusion bodies." Thus, purification of functional proteins may become extremely difficult.

Expression systems based on yeast and filamentous fungi can be used to produce high levels of recombinant proteins (Gerngross, 2004). However, for secreted or membrane-anchored proteins, the levels of expression achieved by these systems are generally low compared to expression levels of intracellular proteins. Although posttranslational modifications including glycosylation take place in fungi, their pattern is more limited than in higher eukaryotes. This limitation may affect the biological activity of the recombinant proteins.

Plant expression systems for recombinant protein production have been developed. However, this type of expression system is still in its infancy and lacks several benefits of other established expression platforms (reviewed in Hellwig *et al.*, 2004). Plant cell cultures may prove useful in the future, when their properties and feasibility are fully investigated.

Cultured mammalian cells are generally considered an excellent means for expression of recombinant membrane-anchored and secreted proteins (reviewed in Wurm, 2004). However, mammalian cell cultures need to be maintained under carefully controlled conditions, in fairly expensive media and supplemented with CO_2. Serum that is usually added as supplement to these media represents a potential source of harmful pathogens, which can limit the utility of transformed mammalian cells for production of therapeutic proteins. Last, but not the least, several months of subculture are required for the generation of clones that stably overexpress recombinant proteins, making mammalian systems rather labor intensive.

Insect cell-based expression systems are prominent among expression platforms for their ability to express virtually all types of heterologous recombinant proteins. These systems exhibit a number of advantages that make them suitable not only for insect-related applications but also for therapeutic protein production and development of bioactivity assays for drug discovery purposes. Two types of insect cell-based systems are in use: the baculovirus expression system and stably transformed insect cell lines.

Since the development of the baculovirus expression system, insect cells have been extensively used in a wide range of applications (discussed in detail in several other chapters of this issue; reviewed in Farrell *et al.*, 2005; Kost *et al.*, 2005), including production of recombinant proteins, with excellent results. In general, baculovirus vector–mediated protein expression in insect cells is superior to other systems in terms of capacity to produce higher levels of soluble recombinant proteins with correct folding and extensive posttranslational modifications. Furthermore, insect cell-based expression systems are appropriate for therapeutic protein production because insect cells can be grown in media free of protein or potential pathogens. However, the baculovirus expression system exhibits certain inherent limitations that are dictated by its very nature; the production of proteins is transient because host cells are lysed and killed during each infection cycle. Furthermore, the cell breakdown associated with late stages of baculovirus infection may prevent efficient secretion as well as completion of the extensive posttranslational modifications at the stage of maximal production of the expressed proteins.

Stably transformed insect cell lines represent the most attractive alternative to the baculovirus expression system and are especially suited for the production of secreted and membrane-anchored proteins. A series of expression vectors for lepidopteran and dipteran cell lines has been developed that enable high-level protein production without the disadvantages associated with baculovirus infection. In certain cases reviewed in this chapter, the incorporation of baculovirus genetic elements in plasmid expression vectors has allowed the generation of extremely powerful expression systems that have been used successfully for the production of large quantities of recombinant proteins with yields superior to those achieved by other eukaryotic expression systems employed to date.

Insect cell-based expression systems for continuous production of recombinant proteins show certain distinct features that make them suitable for a growing number of applications. These features (summarized in Farrell *et al.*, 2005) include fast process from cDNA cloning to protein production (stable cell lines are developed within 1–2 months), correct intron splicing for expression from genomic DNA, full capacity of posttranslational modifications (since integrity of the cells is maintained throughout the production process), continuity of protein production (the transgene is stably integrated into the genome and proteins are produced continuously), limited proteolysis due to absence of cell lysis (because no virus infection occurs), stable physiological environment for membrane protein expression, high yields of secreted recombinant proteins, and easy purification of secreted proteins from serum-free media. On the other hand, possible limitations of this system may lie in the correct folding, and the extent and type of posttranslational modifications (especially glycosylation; see chapter by Harrison and Jarvis, this volume, pp. 159–191) taking place in insect cells, in order to obtain production of bioactive proteins (i.e., with regard to the need for "humanized" glycoprotein production for the pharmaceutical industry). In parallel to the popular baculovirus expression system, transformed insect cell lines have therefore received increasing attention from the research community and the biotechnology industry.

The scope of this chapter is to review developments in insect cell-based expression systems as tools for production of secreted and membrane-anchored proteins as well as novel applications like the development of high-throughput screening (HTS) tools for the identification of bioactive compounds, pharmaceuticals and insecticides.

II. Generation of Stably Transformed Cell Lines

A. *General Strategy*

Recombinant protein expression in insect cell lines typically employs a plasmid expression cassette that harbors the gene of interest under the control of a promoter that drives constitutive or inducible expression in insect cells. The plasmid is introduced into the host cells using some type of transfection technique and the recombinant protein is expressed transiently for a few days posttransfection. Generation of stable cell lines traditionally involves application of an antibiotic resistance-selection scheme, although more sophisticated genetic approaches have been developed. The stably transformed cell population is either used directly for protein production or highly expressing clones are selected from it. The culture is amplified and the cells or their media are harvested. A schematic overview of the whole strategy is depicted in Fig. 1.

B. *Cell Types*

The plasmid-based expression systems developed so far employ cell lines derived from two different insect orders, Diptera and Lepidoptera. Dipteran cell line expression concerns almost exclusively Schneider 2 (S2) cells (Schneider, 1972) derived from *Drosophila melanogaster*. Among the lepidopteran cell lines, the most frequently used are IPLB-Sf21AE (Sf21; Vaughn *et al.*, 1977) established from pupal ovaries of the fall armyworm *Spodoptera frugiperda*, its subclone Sf9, BTI-Tn-5B1–4 (or High Five^{TM}) cells established from embryos of the cabbage looper *Trichoplusia ni* (Granados *et al.*, 1994), and Bm5 (Grace, 1967) established from *Bombyx mori* ovarian tissue cells. Since most of these cell types are also associated with baculovirus propagation, their properties and growth characteristics have been analyzed in detail (see among others Agathos, 1991; Ikonomou *et al.*, 2003; Keith *et al.*, 1999; Rhiel *et al.*, 1997; Stavroulakis *et al.*, 1991a,b; Vlak *et al.*, 1996; Zhang *et al.*, 1992, 1994).

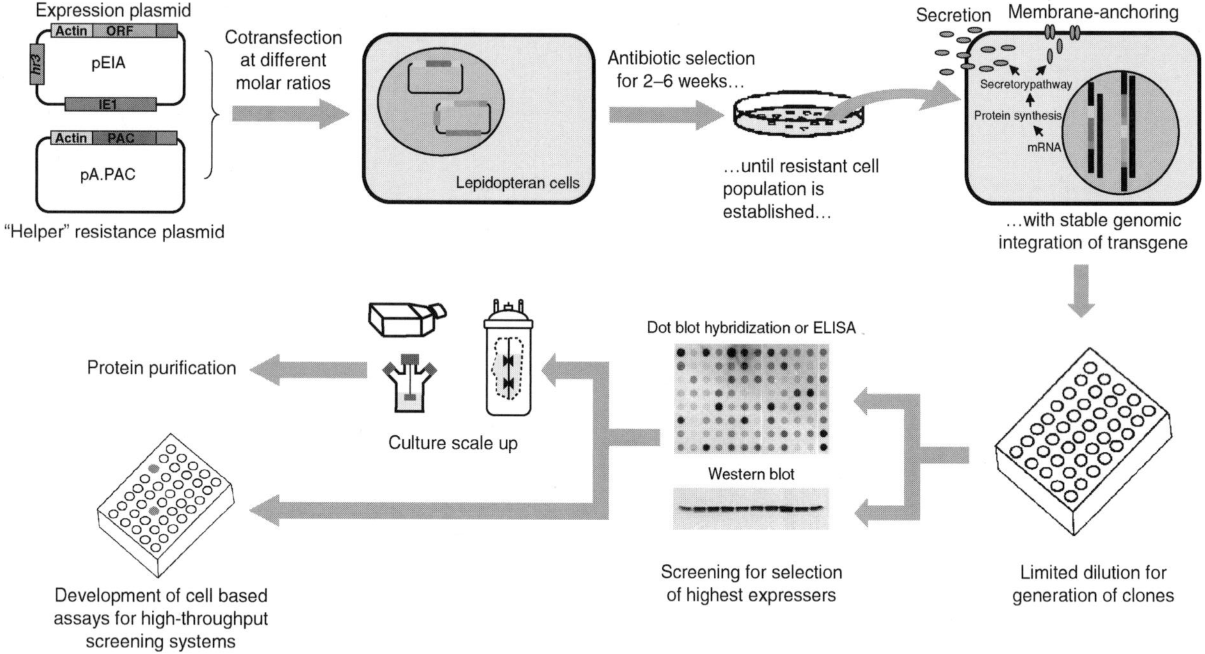

FIG 1. Overview of a general strategy for expression of secreted or membrane-anchored proteins using an insect cell-based expression system: The open reading frame (ORF) of interest is subcloned into an expression vector (the pEIA vector is shown as an example) and the expression plasmid is cotransfected into insect cells along with a "helper" plasmid conferring resistance to an antibiotic (pA.PAC for puromycin resistance is shown). Different expression versus helper plasmid molar ratios can be tested for optimal expression levels. Antibiotic selection is applied until a resistant polyclonal population is established; transformed cells with stable genomic integration of the transgene express the protein of interest. Clonal lines can be generated and screened for optimal expression levels by several methods (DNA or RNA hybridization, Western blot, ELISA, enzymatic, ligand-binding, or other functional assays). The selected clonal cultures can be scaled up for preparative protein purification or used directly for the development of cell-based functional assays. (See Color Insert.)

C. *Genetic Elements Used in the Expression Cassettes*

1. *Promoters and Polyadenylation Signals*

Different types of promoters are used for constitutive or inducible expression in dipteran and lepidopteran expression vectors; some of them are order-specific, while others are functional in cell lines derived from both orders. Dipteran expression vectors currently in use utilize primarily either the strong constitutive actin 5C promoter of *D. melanogaster* (Angelichio *et al.*, 1991) or the inducible *Drosophila* metallothionein promoter (Hegedus *et al.*, 1998; Johansen *et al.*, 1989; Kovach *et al.*, 1992; Millar *et al.*, 1995; Zhang *et al.*, 2001). These promoters are used in a series of vectors included in the *Drosophila* Expression System (DES®), commercially available through Invitrogen Corporation. These vectors make use of a polyadenylation signal sequence derived from the *Simian virus 40* (SV40). A growing number of proteins of different types (including enzymes, membrane receptors, ion channels, viral antigens, and monoclonal antibodies) have been successfully produced in S2 cells using this system (see Tables I and II for relevant references).

An interesting alternative for inducible gene expression in *Drosophila* might arise from the incorporation of the UAS/GAL4 system (Brand and Perrimon, 1993; Duffy, 2002) into S2 cells. Although the proposed expression strategies primarily aim for temporal regulation of gene expression for functional studies in transient assays (Klueg *et al.*, 2002; Roman *et al.*, 2001), generation of stable cell lines expressing several proteins under UAS/GAL4 control has been reported (Makridou *et al.*, 2003).

Inducible gene expression has also been achieved in certain cases via the *Drosophila* heat shock protein 70 (hsp70) promoter (Thummel and Pirrotta, 1992), which has a low level of basal expression and is coupled to a high degree of inducibility (>100-fold at 42°C; Huynh and Zieler, 1999). This promoter has limited use for high-level protein production, however, because the high temperature required for full induction is generally detrimental to insect cell cultures. Furthermore, the translational induction to heat shock is reported to be much lower than the transcriptional induction (Cherbas *et al.*, 1994).

In contrast to the *Drosophila* metallothionein promoter, which drives high-level expression only in dipteran cells (Hegedus *et al.*, 1998; V. D., L. S, and K. I., unpublished data), the hsp70 promoter is functional not only in *Drosophila* cells but also in other dipteran and lepidopteran cell lines (Crouch and Passarelli, 2005; Helgen and Fallon, 1990; Lan and Riddiford, 1997; Zhao and Eggleston, 1999). Gene expression directed by the hsp70 promoter in the context of the baculovirus genome in infected cells was found to be constitutive rather than heat inducible

TABLE I
SOME SECRETED PROTEINS EXPRESSED IN TRANSFORMED INSECT CELL-BASED EXPRESSION SYSTEMS

Protein	Cells	Level (mg/l)	Culture conditions	References
	D. melanogaster metallotheionein gene promoter			
Modified HIV gp120	S2	5–35	n.r.	Ivey-Hoyle *et al.* (1991)
Hu IgG_1 (dimers)	S2	>1	n.r.	Kirkpatrick *et al.* (1995)
Hu IL-5	S2	22	Spinner flasks	Johanson *et al.* (1995)
Hu dopamine β-hydroxylase	S2	>16	Shake flasks	Li *et al.* (1996)
Hu SPC1	S2	3	n.r.	Denault *et al.* (2000)
Mu scFv anti-ACMV	S2	20	Static	Reavy *et al.* (2000)
Hu IL-12 (dimer)	S2	10	n.r.	Lehr *et al.* (2000)
Hu EPO receptor	S2	5	n.r.	Lehr *et al.* (2000)
Hu EPO	S2	2	Spinner flasks	Shin and Cha (2003)
Hu IL-2/GFP fusion	S2	2.3	Spinner flasks	Shin *et al.* (2003)
Hu transferrin	S2	40.8	Spinner flasks	Lim *et al.* (2004)
Hu pro-CAT	S2	20	n.r.	Prosise *et al.* (2004)
Rat NCAM/mu L1	S2	1–3	n.r.	Kulahin *et al.* (2004)
Hu uPAR mutants	S2	0.13–12.2	Shake flasks	Gårdsvoll *et al.* (2004)
Hu XXI (trimers) with P4h	S2	3	Shake flasks	Li *et al.* (2005)
	D. melanogaster actin 5C gene promoter			
Rat pro-CCK	S2	0.5	Static–shake flasks	Kleditzsch *et al.* (2003)
Hu pro-CAT	S2	1	n.r.	Prosise *et al.* (2004)
	AcNPV *ie-1* gene promoter			
Hu tPA	Sf9	1.0	Static	Jarvis *et al.* (1990)
Hu anti-G2	High Five	0.06	Static	Guttieri *et al.* (2000)
Hu IgG_1 fragments	Sf9	0.8–2	Shake/spinner flasks	White *et al.* (2001)

		OpMNPV *ie-2* gene promoter		
Modified Hu p97	Sf9	10	Spinner flasks	Hegedus *et al.* (1999)
Hu plasminogen	S2	10–15	Spinner flasks	Nilsen and Castellino (1999)
Modified hu factor X	Sf9	18	Shake flasks	Pfeifer *et al.* (2001)
Mu IgG_1	Sf9	0.5–1	Spinner flasks	Li *et al.* (2001)
Hu α3/4 fucosyltransferase III	Sf9	13.4	Shake flasks	Morais and Costa (2003)
Hu xylosyltransferase I	High Five	5	Roller flasks	Kuhn *et al.* (2003)
Hu GM2AP	Sf21	0.1	Shaker flasks	Wendeler *et al.* (2003)
Hu tumstatin	High Five	4	Static	Chang *et al.* (2004)
Modified hu α1,3 fucosyltransferase V	High Five	3.92	Static	Münster *et al.* (2006)
		B. mori cytoplasmic actin gene promoter, pEIA vector		
In JHE	Bm5	130–190	Static–spinner flask	Farrell *et al.* (1998)
Hu tPA	Bm5	135–160	Static–spinner flask	Farrell *et al.* (1999)
Hu GM-CSF	High Five	27–46	Spinner–static flask	Keith *et al.* (1999)
Mu L1/F3 (soluble form)	High Five	10–25	Static	V. D. and K. I., unpublished data
B. mori promoting protein	Bm5 and High Five	6–10	Static	Iatrou and Swevers (2005), L. S. and K. I., unpublished data
Agam OBPs	High Five	10–100	Static	Andronopoulou *et al.* (2006)
Ms Hemolin	High Five	~30	Static	V. L., V. D., and K. I., unpublished data

n.r., not reported; Hu, human; Mu, murine; In, insect; Agam, *Anopheles gambiae*; Ms, *Manduca sexta*; HIV gp120, human immunodeficiency virus glycoprotein 120; IgG_1, immunoglobulin G_1; IL, interleukin; SPC1, subtilisin-like proprotein convertase 1; EPO, erythropoietin; tPA, tissue plasminogen activator; p97, melanotransferrin; JHE, *Heliothis virescens* juvenile hormone esterase; GM-CSF, granulocyte-macrophage colony stimulating factor; anti-G2, neutralizing monoclonal antibody specific to *Puumala virus* G2 protein; XXI, homotrimeric type XXI minicollagen; P4h, Prolyl 4-hydroxylase; pro-CCK, cholecystokinin precursor; pro-CAT, ADAM33 zymogen; NCAM/L1/F3, neural adhesion molecules; uPAR, urokinase-type plasminogen activator receptor; GM2AP, GM2-activator protein; scFv, single-chain variable fragment; ACMV, *African cassava mosaic virus*; OBPs, odorant-binding proteins; GFP, green fluorescent protein.

TABLE II
SOME MEMBRANE-BOUND PROTEINS FUNCTIONALLY EXPRESSED IN TRANSFORMED INSECT CELL-BASED EXPRESSION SYSTEMS

Protein	Cells	Level	References
		D. melanogaster metallotheionein gene promoter	
Dmel GPI-linked fasciclin I	S2	0.5 mg/l	Wang *et al.* (1993)
Dmel $GABA_A$ receptor	S2	2.7 pmoles/mg membrane protein ~35,000 sites/cell	Millar *et al.* (1994)
Hu IL-5 receptor α chain (membrane bound and soluble forms)	S2	17 and 10 mg/l, respectively, 1×10^6 sites/cell	Johanson *et al.* (1995)
Hu glucagon receptor	S2	250 pmoles/mg membrane protein	Tota *et al.* (1995)
MHC class II I-Ed molecules	S2	0.1–0.4 mg/l	Wallny *et al.* (1995)
Dmel muscarinic AchR	S2	2.4 pmoles/mg membrane protein	Millar *et al.* (1995)
Hu μ-opioid receptor	S2	20,000–30,000 receptors/cells	Perret *et al.* (2003)
		D. melanogaster hsp70 gene promoter	
Dmel GPI-linked chaoptin	S2	~1 μg/10^6 cells	Krantz and Zipursky (1990)
Ms GPI-linked APN	Sf21	n.r.	Luo *et al.* (1999)
		AcNPV *ie-1* gene promoter	
Hu b_2-adrenergic receptor	Sf9	350,000 receptors/cell	Kleymann *et al.* (1993)
In $GABA_A$ receptors	Sf9	n.r.	Joyce *et al.* (1993); Smith *et al.* (1995)
		OpMNPV *ie-2* gene promoter	
Hu μ-opioid receptor	Sf9	11,000–15,000 sites/cell	Kempf *et al.* (2002)
In APN	S2	n.r.	Banks *et al.* (2003)
		B. mori cytoplasmic actin gene promoter, pEA or pEIA vectors	
Mammalian/C.e. NCKX exchangers	High Five	n.r.	Szerencsei *et al.* (2000)
Hu δ-opioid receptor	Bm5	30,000 active sites/cell	Swevers *et al.* (2005)

n.r., not reported; Hu, human; Dmel, *Drosophila melanogaster*; In, insect; Ms, *Manduca sexta*; C.e., *Caenorhabditis elegans*; IL, interleukin; AchR, acetylcholine receptor; MHC, major histocompatibility complex; APN, aminopeptidase N; GPI, glycosylphosphatidylinositol; GABA, γ-aminobutyric acid; NCKX, Na/Ca-K exchanger.

(Lee *et al.*, 2000; Moto *et al.*, 2003). The characterization of a *B. mori* heat shock promoter (Lee *et al.*, 2003) whose activity can be enhanced by baculovirus elements (Tang *et al.*, 2005) may allow a broader utilization of heat-induced promoters in lepidopteran cell lines in the future.

A number of lepidopteran expression vectors make use of baculovirus immediate early promoters. The first lepidopteran cell-based expression system (Jarvis *et al.*, 1990) utilized *Autographa californica* multiple nucleopolyhedrovirus (AcMNPV) immediate early 1 (*ie-1*) gene promoter and a region containing mRNA polyadenylation signals (Guarino and Summers, 1987). The *ie-1* gene promoter has been shown to function in several dipteran cell lines as well (Gray and Coates, 2004; Vanden Broeck *et al.*, 1995). A series of vectors using enhanced versions of this promoter (see Section II.C.2) are available from EMD Biosciences (Novagen brand), as the InsectDirect™ system (Jarvis *et al.*, 1996; Loomis *et al.*, 2005). A similar set of expression vectors has been developed based on the *Orgyia pseudotsugata* multiple nucleopolyhedrovirus (OpMNPV) *ie-2* promoter (Hegedus *et al.*, 1998; Pfeifer *et al.*, 1997; Theilmann and Stewart, 1992a) and the relevant OpMNPV *ie-2* polyadenylation signal. These vectors are available from Invitrogen as the InsectSelect™ vector set.

Another series of lepidopteran expression vectors developed in the laboratory of Dr. Kostas Iatrou utilizes the silkmoth (*B. mori*) A3 cytoplasmic actin gene promoter (Johnson *et al.*, 1992; Mounier and Prudhomme, 1986), a strong constitutive promoter that is active in a variety of lepidopteran cell lines. Enhanced and double-enhanced versions of this promoter (see later and Fig. 2) have been used for the expression of a large number of proteins in Bm5 and High Five™ cells (Tables I and II). Terminator sequences with polyadenylation signals deriving from the cytoplasmic actin gene of *B. mori* as well as SV40 or the bovine growth hormone (BGH) gene (V. D., L. S., and K. I., unpublished data) have been successfully used with this promoter in a number of lepidopteran cell lines.

A basal silkmoth actin promoter containing multiple repeats of an ecdysone response element (ERE) derived from the *Drosophila* hsp27 promoter (Riddihough and Pelham, 1987) was also developed (Swevers *et al.*, 2004). This promoter was induced 2000-fold by micromolar quantities of 20-hydroxyecdysone (20E), with total expression levels comparable to the ones obtained by enhanced versions of strong constitutive promoters (Fig. 2). Transformed cell lines incorporating ecdysteroid-inducible expression elements have been described for production of recombinant proteins (Tomita *et al.*, 2001) and HTS for potential 20E agonists and antagonists (Swevers *et al.*, 2004).

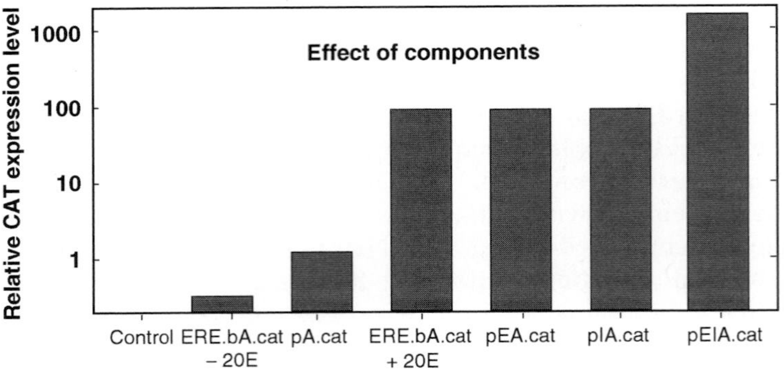

FIG 2. Expression levels achieved by a lepidopteran expression system using different expression modules. Stepwise increases in reporter gene expression (chloramphenicol acetyl transferase, CAT) are achieved through the use of three genetic elements: the silkmoth actin promoter, the baculoviral (BmNPV) *hr3* enhancer, and the baculoviral IE1 transactivator (Farrell *et al.*, 1998; Lu *et al.*, 1997). Expression levels varying over three orders of magnitude were achieved with different modules. An ecdysone-inducible vector construct has only basal expression levels but on induction with 20-hydroxyecdysone (20E), expression comparable to a single-enhanced constitutive promoter is achieved. pA: actin promoter alone, pEA: actin promoter enhanced by *hr3* enhancer, pIA: actin promoter transactivated by IE1 transactivator protein, pEIA: actin promoter double enhanced by both *hr3* and IE1, ERE.bA: basal actin promoter downstream of seven repeats of an ecdysone response element (Swevers *et al.*, 2004).

2. *Enhancers and Transactivators*

The activity of certain promoters used in insect cell-based expression systems can be enhanced by certain *cis*- or *trans*-acting elements of baculoviral origin. The AcMNPV homologous repeat (HR) 5 transcriptional enhancer element (Guarino *et al.*, 1986) was shown to act in *cis* to stimulate expression of reporter proteins from early baculovirus promoters, such as *ie-1* and *p35*, in transient expression assays (Pullen and Friesen, 1995; Rodems and Friesen, 1993). Thus, the *ie-1* gene promoter activity has been enhanced by the incorporation of HR5 upstream of the *ie-1* promoter (Jarvis *et al.*, 1996) in the relevant vector constructs. Similarly, HR sequence elements enhance the activity of the OpMNPV *ie-2* promoter (Theilmann and Stewart, 1992b). Linkage of *B. mori* nucleopolyhedrovirus (BmNPV) HR3 at various orientations to the cytoplasmic actin promoter of *B. mori* was also found to stimulate promoter activity by two orders of magnitude (Lu *et al.*, 1997). Stimulation by the HR3 element has been reported for other promoters of insect (Tang *et al.*, 2005) or even mammalian (Viswanathan *et al.*, 2003) origin.

Transcription from the cellular actin promoter of *B. mori* was also found to be stimulated by the protein IE1, the immediate early gene product of BmNPV. IE1 is a transcription factor capable of stimulating transcription from the actin promoter *in trans* by 100-fold (Lu *et al.*, 1996). The mode of action of the IE1 transactivator has been investigated only in the context of AcMNPV infection; IE1 activity is enhanced by binding to homologous regions *cis*-linked to the promoters to be transactivated (Kovacs *et al.*, 1991, 1992; Leisy and Rohrmann, 2000; Olson *et al.*, 2001, 2002, 2003; Rodems *et al.*, 1997). IE1 can also stimulate certain heterologous promoters linked to HR elements. Thus, the *Drosophila hsp70* promoter is stimulated ~40-fold by combined HR5 and IE1 enhancement in lepidopteran cells (Crouch and Passarelli, 2005), while the *Drosophila* actin5C and polyubiquitin Ubi-p63E promoters tested for BmNPV HR3/IE1 transactivation in mosquito cell lines are upregulated from 10- to 200-fold, based on different reporter constructs (Gray and Coates, 2004).

The most striking example of synergistic promoter stimulation though resulted from linkage of the BmNPV *ie-1* gene with the HR3 element and the *B. mori* A3 cytoplasmic actin promoter in the double-enhanced expression vector pEIA (formerly pIE1/153A), which resulted in stimulation of foreign gene expression directed by the actin promoter by 5000-fold (Fig. 2) in transient expression assays for two proteins (Lu *et al.*, 1997). This powerful expression tool was subsequently used for the generation of stable cell lines expressing several secreted proteins, with expression levels far exceeding those achieved for secreted proteins by the baculovirus expression system (Farrell *et al.*, 1998, 1999). The expression cassette was shown to function in all lepidopteran cell lines investigated (Keith *et al.*, 1999). The relevant expression system, which is marketed by CytoStore, Inc., Canada under the trade name of TripleXpress™ Insect Expression System, includes a variety of expression cassettes.

3. *Secretion Modules, Purification, and Epitope Tags*

A number of vectors harbor heterologous signal sequences derived from various sources such as honeybee mellitin (Tessier *et al.*, 1991) and immunoglobulin heavy chain binding protein (BiP) used in the DES® and InsectSelect™ systems of Invitrogen, the adipokinetic hormone (AKH) and a mouse IgM (Kim *et al.*, 2003) used in the Insect-Direct™ system of Novagen, and *B. mori* chorion proteins in certain pEIA derivatives (Farrell *et al.*, 2000).

The ultimate goal through development of these modified vectors is to facilitate secretion of heterologous proteins, especially intracellular ones (cytoplasmic or nuclear) to enable purification from culture media rather than cell extracts. However, it has been demonstrated that in contrast to proteins that are normally destined for secretion, in most cases, the fusion of a signal peptide to the N-terminus of normally intracellular (cytoplasmic or nuclear) proteins is not sufficient for their secretion (Farrell *et al.*, 2000). On the other hand, fusion of the complete coding sequence of a secreted protein, like the juvenile hormone esterase (JHE) of *Heliothis virescens* or the human granulocyte-macrophage colony stimulating factor (huGM-CSF), to intracellular proteins enables efficient secretion and purification of the fusion protein from cell culture supernatants (Farrell *et al.*, 2000). Thus, derivatives of pEIA with secretion modules with JHE or huGM-CSF open reading frames (ORFs) followed by a 6xHis tag and an enterokinase cleavage site (Fig. 3) allow for efficient secretion of recombinant intracellular proteins, as well as detection, purification, and release of the authentic protein after expression in lepidopteran cells. These derivatives are also marketed by CytoStore.

Several other fusion tags are frequently used for expression of secreted proteins, as C-terminal fusions. These tags may facilitate the detection of the protein and its purification from the culture medium. The most frequently employed tag in all available systems is polyhistidine (6xHis), which enables easy purification of tagged proteins from culture media via metal affinity chromatography, as well as antibody detection (Lindner *et al.*, 1997). Certain Novagen vectors make use of other affinity tags such as S-tag™ and Strep-tag® (Skerra and Schmidt, 2000) which facilitate binding to S-protein and streptavidin columns, respectively.

Many expression vectors also contain extra tag sequences, usually epitopes enabling easy detection by commercially available antibodies. Such epitopes are the V5 epitope (Southern *et al.*, 1991) used in the DES® and InsectSelect™ vectors of Invitrogen, and the HSV Tag® sequence in the InsectDirect™ vectors of Novagen, which is an epitope derived from herpes simplex glycoprotein D. Some derivatives of pEIA feature c-Myc (Alitalo *et al.*, 1983) or Glu–Glu (Grussenmeyer *et al.*, 1985) epitopes as C-terminal tags along with the 6xHis tag (Fig. 3).

Tagged expression of membrane-anchored proteins is also possible provided that the tags do not interfere with cellular localization and bioactivity. In addition to the C-terminal tags described earlier, appropriate N-terminal tags used for antibody detection may also be employed. Novagen provides InsectDirect™ vector permutations with N-terminal S-tag™ and Strep-tag® sequences, while pEIA derivatives with N-terminal c-Myc or FLAG tags are also available (Fig. 3). For cases in which removal of the tag is desired, most systems enable tag removal by incorporating a protease cleavage site between the native

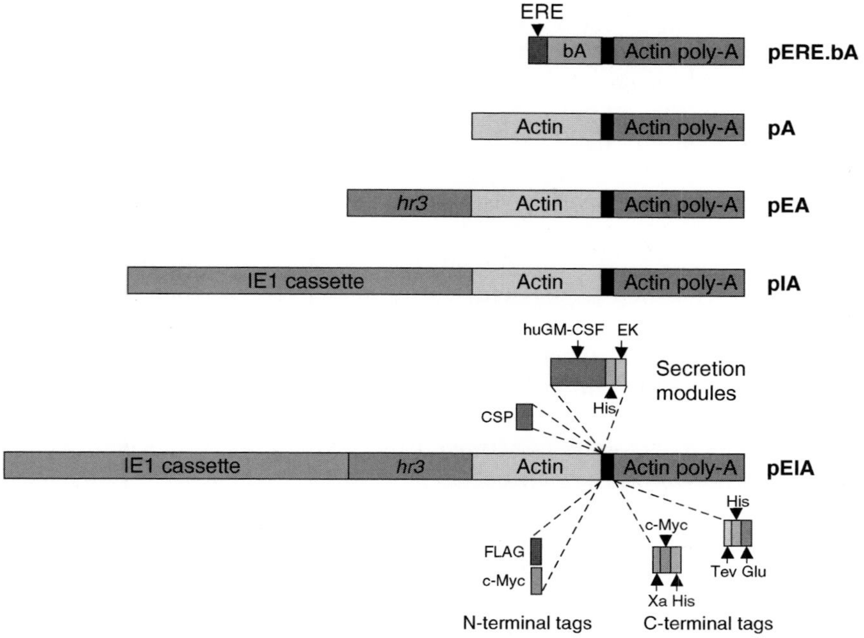

FIG 3. Overview of expression constructs available in a lepidopteran expression system. Versatility in the expression system is achieved by the use of constitutive promoters of different strength and inducible promoters. **pERE.bA**: inducible expression with 20E for toxic proteins and other applications. **pA**: constitutive expression of recombinant proteins in lepidopteran cells. **pEA**: **x100** enhancement of constitutive expression. **pIA**: **x100** enhancement of constitutive expression. **pEIA**: **x5000** enhancement of constitutive expression. Several types of expression tags were developed for this vector, located at the N- or C-terminus, and can be used for detection and purification as well as to direct secretion to the extracellular medium. Secretion can be facilitated either by *B. mori* chorion protein signal peptide or by a module containing an N-terminal huGM-CSF ORF, followed by a polyhistidine tag and an enterokinase cleavage site enabling release of the authentic protein (Farrell *et al.*, 2000). Two N-terminal tags (with either the c-Myc or FLAG epitope) are available for detection of cytoplasmic, nuclear, or membrane-anchored proteins, while two C-terminal double tags enable detection by c-Myc or Glu-Glu epitopes, purification by antibody or metal affinity chromatography and release of authentic protein after cleavage with a relevant protease. **ERE**: 7x ecdysone response element; **bA**: basal actin promoter of *B. mori*; **actin poly-A**: 3′ untranslated region of *B. mori* actin gene containing polyadenylation signals; **actin**: *B. mori* A3 cytoplasmic actin promoter; ***hr3***: baculoviral (BmNPV) homologous region 3 enhancer sequence; **IE1 cassette**: baculoviral (BmNPV) DNA fragment containing the *ie-1* transactivator gene under the control of its native viral promoter; **CSP**: *B. mori* chorion signal peptide; **huGM-CSF**: human granulocyte-macrophage colony stimulating factor coding sequence; **His**: 6x histidine tag; **EK**: enterokinase cleavage site; **c-Myc**: c-Myc epitope tag; **FLAG**: FLAG epitope tag; **Glu**: Glu–Glu epitope tag; **Xa**: factor Xa cleavage site; **Tev**: *Tobacco etch virus* protease cleavage site. The black box downstream of the promoter in all expression constructs indicates the position of the cloning sites. (See Color Insert.)

protein sequence and the tag. Thus, the fusion protein is immobilized on the relevant affinity matrix, and elution of the native protein is achieved by cleavage with the appropriate protease. Novagen Insect-Direct^{TM} vectors employ thrombin or enterokinase cleavage sites, while the different pEIA derivatives contain sites for enterokinase, factor Xa, and *Tobacco etch virus* (Tev) protease (Fig. 3).

D. Transformation Procedures

In cases in which a powerful expression cassette such as pEIA is used, moderate quantities of secreted proteins can be purified by using scaled-up transient expression protocols (Farrell and Iatrou, 2004). In most cases though, the generation of stably transformed cell lines is the optimal choice for protein production at high levels. Stable transformation of insect cells is achieved by using expression vectors that harbor an antibiotic resistance gene or by cotransfecting cells with expression plasmids and a selection plasmid containing a gene that confers antibiotic resistance to the cells. Then, antibiotic selection is applied for 2–6 weeks until a stably transformed cell population, which is resistant to the antibiotic, is established.

The molecular ratio of expression versus selection plasmids as well as the antibiotic concentration may be empirically optimized for different cell lines in order to obtain maximum expression levels without jeopardizing eventual recovery of a resistant cell population. In several series of experiments with Bm5 and High Five^{TM} cells, cotransfection of pEIA expression constructs with selection plasmids conferring resistance to hygromycin or puromycin at ratios ranging from 1:1 to 500:1, it was demonstrated that higher molecular ratios of expression versus selection plasmid lead to higher expression levels (Farrell, 1998; V. D. and K. I., unpublished data). This can primarily be attributed to the relatively high number of genome-integrated expression plasmids. Thus, Bm5 cell populations transformed at a 1:1 ratio of an expression versus selection plasmid had an average of 3 expression plasmid copies per haploid genome, while equivalent cell lines transformed at a 100:1 ratio had an average of 38 expression plasmid copies per haploid genome (Farrell, 1998). Given that the ploidy of Bm5 and High Five^{TM} cells is ~4 (Farrell, 1998; Farrell *et al.*, 1999; V. D. and K. I., unpublished data), this translates into ~100 transgene copies per cell.

Selection schemes conferring resistance to G418 (Jarvis *et al.*, 1990), hygromycin B (Johansen *et al.*, 1989), methotrexate (Shotkoski and Fallon, 1993), actinomycin D (McLachlin and Miller, 1997), puromycin (McLachlin and Miller, 1997), zeocin (Pfeifer *et al.*, 1997), and blasticidin

(Kimura *et al.*, 1994) have been used extensively with insect cell lines. In some cases, the antibiotic resistance gene is incorporated into the expression vector (pIB and pIZ plasmids in the InsectSelect™ vectors harbor blasticidin and Zeocin™ resistance, respectively), while in others there is a separate selection plasmid such as pCoBlast and pCoHygro for blasticidin and hygromycin selection in DES® (Invitrogen), pIE1-Neo for G418 selection in InsectDirect™ (EMD Biosciences), and pA.Hygro (Farrell *et al.*, 1998) as well as pA.PAC (CytoStore) and pIA.PAC (P. J. Farrell, V. D., and K. I., unpublished data) conferring hygromycin or puromycin resistance to lepidopteran cells. In several cases, resistance was still present in the absence of selection for several months, enabling stable heterologous gene expression without continuous selection (Farrell *et al.*, 1998; Hegedus *et al.*, 1998; Pfeifer *et al.*, 1997).

Antibiotic selection is the most widely used but not the only available method for stable cell transformation. Transposable elements regularly used for germ line transformation (reviewed in Handler and O'Brochta, 2005) have also been used for transformation of insect cells. These include *P* elements (Segal *et al.*, 1996), *Hermes* (Zhao and Eggleston, 1998), *Minos* (Catteruccia *et al.*, 2000; Klinakis *et al.*, 2000), *Mos1/mariner* (Wang *et al.*, 2000), and *piggyBac* (Grossman *et al.*, 2000; Mandrioli and Wimmer, 2003). Reporter transgenes are incorporated in vectors containing transposon-inverted repeats and cotransfected with a helper plasmid expressing the relevant transposase. Although the efficiency of transformation is not very high, transposon-mediated stable transformation of insect cells remains an option that should be considered for future developments related to the various expression platforms.

The major breakthrough in stable transformation of insect cells, however, has come with the development of a novel densovirus-based vector system that enables integration of multiple transgene copies into the host cell genome (Bossin *et al.*, 2003). This property was initially demonstrated in Sf9 cells and subsequently verified for Bm5 and High Five™ cells (C. Kenoutis, L. S., and K. I, unpublished data). However, despite the fact that densovirus vectors are functional in *Drosophila* embryos, larvae, and adults (Royer *et al.*, 2001), their activity could not be demonstrated in S2 cells (L. S. and K. I., unpublished data).

Establishment of a stably transformed cell population by any of the aforementioned procedures is not always sufficient for high-level protein production. The need to select for the highest expressers (or merely expressers) within a diverse polyclonal cell population is evident. This goal may be facilitated by expression of the gene of interest as a fusion with a fluorescent protein; transformed cells are fluorescent and can be separated by means of repeated fluorescence-activated cell sorting (FACS), before being scaled-up for protein production. The InsectSelect^{TM} vector set (Invitrogen) contains a vector (pIZT/V5-His) that expresses a fusion of the green fluorescent protein (GFP) with the Zeocin^{TM} resistance protein, permitting rapid selection of transfected cells. Some densoviral vectors also harbor fluorescent protein genes, enabling separation of transformed cells by FACS (Bossin *et al.*, 2003).

However, the most widely employed strategy for selection of highly expressing cells is the generation of clonal populations. As for mammalian cell cultures, the location of the integrated transgene may affect protein expression. It is advisable to isolate as many colonies as possible for expression testing. Clones can be picked with the help of special cloning rings or cylinders used in mammalian cell culture, but a serial dilution strategy is usually employed. For isolation of clones by serial dilution, cloning is achieved by repeatedly seeding small numbers of cells in a small culture volume (100–150 μl in 96-well plates). To assist growth, 50% conditioned medium is used to provide necessary growth factors. After 3–4 weeks of culture, cell populations resulting from one or a few cells are established. Supernatants or cell extracts from these populations are examined for optimal protein production, usually by Western blotting or some relevant activity assay, and the best performing pure (or semipure) clones are further expanded. Although somewhat time consuming, this strategy enables maximal performance of the expression system and may prove cost effective in the long run because it may provide protein yields that far exceed those obtained by the initial polyclonal cell lines.

III. Expression of Secreted, Intracellular, and Membrane-Anchored Proteins

A. *Secreted Proteins*

Numerous secreted proteins from multiple organisms have been expressed from stably transformed insect cell lines (Table I). The yields obtained differ for each protein and expression system. Yields ranging from 2 to 20 mg/liter are typical for secreted protein expression in the DES®, although yields of up to 35 mg/liter have been reported for the modified HIV glycoprotein 120 (Ivey-Hoyle *et al.*, 1991) and up to 40 mg/liter for human transferrin (Lim *et al.*, 2004). For the InsectSelect^{TM} system, reported expression levels typically range from 12 mg/liter for human IL-6 (Invitrogen) to 8–10 mg/liter for human melanotransferrin (Hegedus *et al.*, 1999), although levels of 13.4 mg/liter have been reported for human $\alpha3/4$ fucosyltransferase III (Morais and Costa, 2003). Much higher levels have been achieved with pEIA, including 46 mg/liter for huGM-CSF (Keith *et al.*, 1999) and 130–190 mg/liter for insect JHE (Farrell *et al.*, 1998).

These expression levels are comparable or higher to those obtained for secreted proteins using baculovirus expression vectors. Despite the fact that cellular or early phase baculovirus promoters used in the stably transformed insect cell systems are not as powerful as late and very late phase baculovirus promoters, such as polyhedrin and p10, in the absence of viral infection the secretory pathway remains intact, contributing to improved expression levels and protein quality when compared to expression with the baculovirus system.

Most proteins containing eukaryotic-type signal peptides are generally secreted from insect cells. The substitution of native mammalian signal peptide-encoding sequences with sequences encoding insect-specific signal peptides does not have any significant effect on protein expression levels in transfected insect cells (Farrell *et al.*, 2000), although in other studies an effect has been observed at least for some cell lines (Kock *et al.*, 2004).

Several vectors harboring heterologous signal peptides or other detection and purification tags (described in Section II.C.3) have been employed for expression of secreted proteins and purification from cell culture supernatants. Tagged vectors offer the opportunity for rapid functional expression and characterization in a high-throughput fashion; a relevant example is shown in Fig. 4 in which expression of multiple *Anopheles gambiae* odorant-binding proteins in pEIA derivatives with C-terminal tags enabled fast verification of the interactions among them without the need to generate specific antibodies against each one (Andronopoulou *et al.*, 2006).

Although tags are useful, they are not always necessary for purification of secreted protein. Proteins may be purified from culture media by conventional chromatography, affinity antibody columns, or by using specific substrates. For example, a *Cotesia congregata* bracovirus cystatin (Espagne *et al.*, 2005) expressed via pEIA in two stably transfected High Five™ cell lines grown in serum-free medium was detected in cell culture supernatants by an enzymatic (papain) inhibition assay (Fig. 5A) and later purified from the supernatants by affinity to immobilized carboxymethylated papain (Fig. 5B).

B. Intracellular (Cytoplasmic or Nuclear) Proteins

Although stably transformed insect cells are not frequently used for expression of intracellular (cytoplasmic or nuclear) proteins, there are

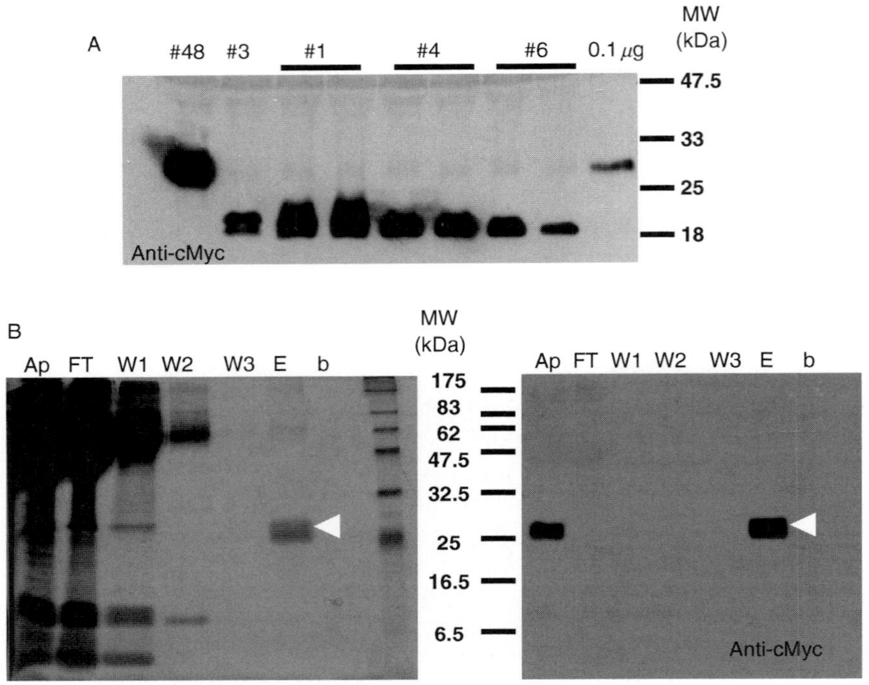

FIG 4. Expression, detection, and purification of secreted proteins with a pEIA derivative with C-terminal c-Myc and 6xHis tags. (A) Western analysis using anti-cMyc antibody of cell culture supernatants from lepidopteran cell lines stably transformed to express several *Anopheles gambiae* odorant-binding proteins (#1, 3, 4, 6, and 48). An aliquot (0.1 μg) of purified tagged OBP48 was used as positive control. (B) Silver stain (left) and Western analysis (right) of fractions collected throughout metal affinity purification from 2 ml of tagged OBP48 cell culture supernatant. Ap: applied supernatant; FT: flow-through fraction; W1–3: successive wash fractions; E: elution fraction, b: beads after elution. The arrow indicates the purified tagged OBP48 protein (V. D. and K. I., unpublished data).

some cases in which such systems have been successfully employed. A cytotoxic protein from the sea hare *Aplysia punctata*, cyplasin, that was inactive when expressed in bacteria and toxic when expressed in mammalian cells, was successfully expressed in Sf9 cells as a non-secreted enhanced green fluorescent protein (EGFP) fusion using the InsectSelect^{TM} system and amplification of stable cell lines expressing the fusion protein enabled further characterization (Petzelt *et al.*, 2002). The InsectDirect^{TM} system has been used for transient, small-scale expression of several protein kinases, phospholipases, and heat

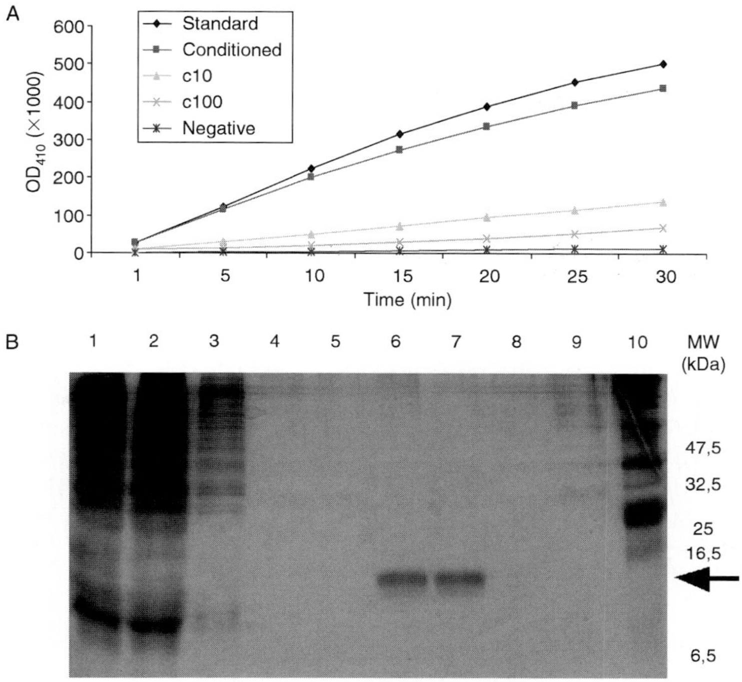

FIG 5. Functional expression and purification of a cysteine protease inhibitor (cystatin) expressed in stably transformed High Five™ cells using pEIA. (A) Cell culture supernatants from two cystatin-expressing lines (c10, c100) strongly inhibit papain (cysteine protease) activity compared to supernatants from nontransformed cell cultures. (B) Silver stained gel for evaluation of cystatin purification by affinity chromatography. 1: supernatant; 2: flow through, 3–4: wash fractions, 5–8: elution fractions, 9: beads after elution. The arrow indicates the unique band at the expected molecular mass in the elution fractions. (Reprinted with permission from Espagne, E., Douris, V., Lalmanach, G., Provost, B., Cattolico, L., Lesobre, J., Kurata, S., Iatrou, K., Drezen, J.-M., and Huguet, E. (2005). A virus essential for insect host–parasite interactions encodes cystatins. *J. Virol.* **79:**9765–9776; © the American Society for Microbiology.) (See Color Insert.)

shock proteins, with yields ranging from 0 to 48 mg/liter, while higher yields were obtained in certain medium-scale expression trials (Loomis *et al.*, 2005). Transient expression of GFP and chloramphenicol acetyl transferase (CAT) using the pEIA system resulted in a total yield of 6 and 14 mg/liter, respectively (Farrell and Iatrou, 2004).

Given that accumulation of the expressed protein in the cytoplasm may lead to cytotoxicity and restrict expression levels, an alternative approach is to drive secretion of intracellular proteins by expressing them as fusions with a secretion module of the type discussed under Section II.C.3. Thus, a

B. mori orphan nuclear receptor, BmCF1, was expressed in Bm5 and High Five^{TM} cells as a fusion with JHE, and the fusion protein was efficiently secreted into the culture medium at levels of 10 and 28 mg/liter, respectively, despite the presence of two nuclear localization signals within the BmCF1 amino acid sequence (Farrell, 1998; Farrell *et al.*, 2000). Similarly, CAT expression levels were significantly improved when CAT was expressed as a fusion with huGM-CSF using the secretion module described in Section II.C.3, resulting in an expression level of 30 mg/liter for the secreted fusion protein (Farrell and Iatrou, 2004).

The same pEIA secretion module with fusion to huGM-CSF was successfully used for expression of TnBV1, a protein that induces apoptosis-like programmed cell death in insect cells (Lapointe *et al.*, 2005). Expression of the fusion protein and efficient secretion rescued the cell population, while expression of native TnBV1 resulted in rapid cell death (V. D. and K. I., unpublished data). This strategy enabled construction of stably transformed insect cell lines efficiently expressing an otherwise toxic protein and facilitated purification of TnBV1 from cell culture supernatants.

C. Membrane-Anchored Proteins

As already noted for secreted proteins, the absence of viral infection and cell lysis may provide an appropriate cellular environment for production of membrane proteins that also enter the secretory pathway. Several functional membrane-anchored proteins have been successfully expressed in stable insect cell lines (Table II). These include ion exchangers, transmitter-gated ion channels, and receptors of different classes. Expression levels in the case of membrane-anchored proteins are probably better represented in numbers of active protein molecules per cell. Protein quantity is not so important per se; correct folding, posttranslational modifications, and localization are much more critical for membrane protein expression, particularly when the expressing cells are destined for use as functional expression platforms for various applications.

Depending on the membrane protein expressed, different types of assays have been employed to monitor bioactivity. When heterologous proteins are expressed, all required components for a specific bioassay may not be present in the expression system. For example, as was found with the baculovirus expression system (Bouvier *et al.*, 1998; Wehmeyer and Schulz, 1997), coupling of heterologous G-protein–coupled receptors (GPCRs) following ligand binding to endogenous insect G-proteins may be inefficient and this may hold even with transformed insect cells (Farrell *et al.*, 2005; Kempf *et al.*, 2002; Torfs *et al.*, 2002; Vanden Broeck, 1996). However, coexpression of mammalian G-proteins has proven useful for functional expression of heterologous GPCRs in lepidopteran cell lines (Francken *et al.*, 2000; Knight *et al.*, 2003), allowing for the development of screening platforms for specific ligand mimetics (see Section IV).

IV. Screening Platforms for Drug and Insecticide Discovery

In the postgenomic era, the need for high-throughput production of recombinant proteins for functional and structural analysis is, in many cases, coupled to that for the development of screening platforms that allow for mass-detection of bioactive compounds using HTS formats. This use of transformed insect cell lines has somewhat lagged behind their more traditional use as protein expression systems. As illustrated later, however, developments have confirmed the potential of insect cell lines for development of HTS systems for fast identification of bioactive substances with defined specificities. Thus, for receptors and other functional regulators of mammalian (human) origin, the availability of HTS systems is predicted to lead to the identification of new drugs for improvement of human health. Screening systems that target equivalent regulators of insect origin, on the other hand, will aid in the development of new strategies for efficient and environmentally safe insect pest control.

Cell-based screening systems generally rely on the presence of two elements: (1) an expression element that produces the target regulator against which bioactive substances need to be selected, and (2) a detection element that allows for rapid and easy observation of the activation or suppression of the activity of the regulator by particular compounds (Swevers *et al.*, 2003). Thus far, screening systems using transformed insect cell lines have targeted primarily two major classes of regulators, nuclear receptors and GPCRs. By and large, the employment of transformed insect cell-based screening systems for identification of activators and suppressors of other classes of cellular regulators has yet to be tested (but see Section IV.C).

A. Nuclear Receptors

Ligand-bound nuclear receptors activate responsive (target) genes in the genome after binding of ligand to specific target sites in promoter and enhancer regions. Detection systems for nuclear receptors are therefore usually based on the activation of gene reporter (GFP, luciferase, CAT, β-galactosidase) cassettes that are engineered to contain multiple copies of the binding sites for the nuclear receptors upstream of a basal promoter (Gustafsson, 1999; Kliewer *et al.*, 1999). Activation of the nuclear receptors is recorded by induction of reporter gene activity in the engineered cell lines.

Of the ~20 nuclear receptors identified in insects, only 1 has a clearly identified ligand, the ecdysone (molting hormone) receptor (EcR; King-Jones and Thummel, 2005). Because the two nuclear proteins that constitute the ecdysone receptor heterodimer, EcR and ultraspiracle (USP), are endogenously expressed in a variety of insect cell lines (Chen *et al.*, 2002; Sohi *et al.*, 1995; Swevers *et al.*, 2003), development of an ecdysone-responsive HTS system necessitates only the insertion of appropriate ecdysone-responsive reporter cassettes into the insect cell genomes. Such an HTS system for ecdysone mimetics has been developed using silkmoth-derived Bm5 cells that have ecdysone-responsive GFP reporter cassettes incorporated into their genomes (Swevers *et al.*, 2004). The ecdysone-responsive GFP reporter cassette is stimulated more than 1000-fold by 20E at a concentration range of 10 nM to 1 μM. The intense fluorescence induced in the cells following administration of 20E can be easily quantified in individual wells of a 96-well plate by using a fluorescence microplate reader, thus making the system amenable to a high-throughput format (Fig. 6). The system has been used successfully to screen for ecdysone agonists and antagonists in plant extracts and in chemical libraries of dibenzoyl hydrazine compounds (Swevers *et al.*, 2004). Because the system is extremely rapid and robust, the EC_{50} values for a large number of dibenzoyl hydrazines could be determined and this allowed for generation of improved quantitative structure–activity relationship (QSAR) models of molting hormone activity in lepidopteran insects (Wheelock *et al.*, 2006). Finally, the ecdysone-inducible reporter cassette is also active in dipteran cell lines, such as *Drosophila* S2 cells (T. Soin, G. Smagghe, L. S., and K. I., unpublished data), and could therefore be employed for the generation of cell-based ecdysone-responsive screening systems that are specific to dipteran insects.

For other (orphan) insect nuclear receptors, such as the HNF-4 receptor (Kapitskaya *et al.*, 1998), the FTZ-F1 receptor (Suzuki *et al.*, 2001), and the HR3 and E75 receptors (Swevers *et al.*, 2002), DNA-binding target sites have been identified and it should therefore be feasible to develop reporter-based screening systems for these receptors. Which class of ligands activates (or inhibits) these receptors remains speculative. However, it was shown that the E75 receptor contains a heme group in its ligand-binding domain that may function

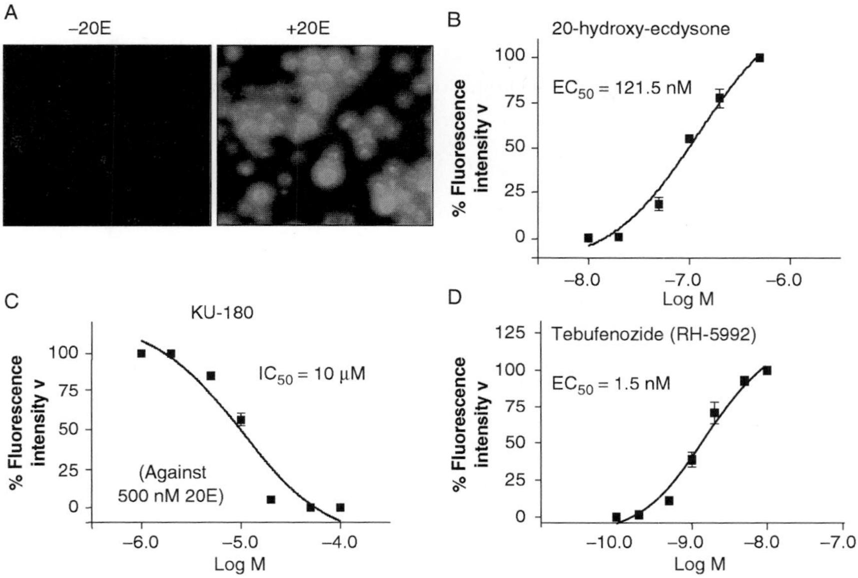

FIG 6. HTS system for ecdysone mimetics based on transformed silkmoth-derived Bm5 cells. (A) Fluorescence photographs of transformed cells before (left) and after challenge with 1 μM 20E (right). (B) and (D) Dose–response curves of the natural insect molting hormone 20E and the synthetic ecdysone agonist tebufenozide (RH-5992) as determined by measurements using a fluorescence microplate reader. (C) Identification of an ecdysone antagonist (KU-180) after screening of a library of dibenzoyl hydrazine compounds (provided by Dr. Y. Nakagawa, University of Kyoto, Japan). Shown is the inhibition of the response by 500 nM 20E using different concentrations of KU-180. The median effective concentration (EC_{50}) of the agonist compounds (Panels B and D) and the median inhibitory concentration (IC_{50}) of the antagonist compound (Panel B) are indicated. (See Color Insert.)

as a redox sensor (Reinking *et al.*, 2005). For this particular receptor, intracellular messengers, such as nitric oxide (NO) or carbon monoxide (CO), are predicted to modulate the function of the receptor. Thus, compounds that increase NO or CO production are good candidates to act as regulators of E75 receptor function.

Whether insect cell lines can be engineered to act as screening systems for mammalian nuclear receptors has yet to be determined. Because it was observed that the authentic ecdysone receptor is only marginally functional in mammalian cell lines (Christopherson *et al.*, 1992), a similar situation may exist for the function of mammalian receptors in insect cell lines. Thus, the functional expression of mammalian nuclear receptors in insect cell lines may require appropriate engineering of promoter-reporter cassettes and careful assessment of the pharmaceutical profiles of the receptors in insect cells to deduce whether they match those reported for mammalian cells.

B. G-Protein–Coupled Receptors

GPCRs constitute a large superfamily of transmembrane proteins that mediate cellular responses to diverse extracellular stimuli that include light, odorants, phospholipids, neurotransmitters, and hormones. GPCRs are subdivided into several subclasses according to their coupling specificity to different members of G-protein complexes which, in turn, modulate the activity of various effector molecules, such as adenylyl cyclase and phopsholipase C, to generate a variety of second messengers (Hamm, 1998; Lefkowitz, 2000; McCudden *et al.*, 2005; Neves *et al.*, 2002).

For some GPCRs, for example, those that couple to Gαs protein complexes and stimulate adenylyl cyclase to produce elevated cAMP levels, gene reporter assay systems have been developed, mostly for use in mammalian cells, that are based on the presence of appropriate DNA target sites upstream of a basal reporter cassette (in this example, cAMP-responsive elements or CREs; Gonzalez and Montminy, 1989; Williams, 2004). Other detection systems have been developed, which rely on easy detection of induced intracellular messengers by fluorescent or luminescent methods (Rudolf *et al.*, 2003; Williams, 2004) or increased interaction of activated GPCRs to G-proteins by fluorescence or bioluminescence resonance energy transfer (FRET or BRET) technologies (Angers *et al.*, 2000; Janetopoulos *et al.*, 2001). One particular detection system that has found wide application is based on the detection of calcium release on GPCR activation using fluorescent dyes or the aequorin (luminescence) technology (Grynkiewicz *et al.*, 1985; Knight *et al.*, 1991; Milligan, 2003; Milligan *et al.*, 1996).

Lepidopteran cell lines (High Five^{TM}, Bm5) express an array of different heterotrimeric G-proteins such as Gαs, Gαi, Gαo, Gαq as well as Gβ subunits of G-proteins (Knight and Grigliatti, 2004a; Swevers *et al.*, 2005; Z. G., L. S., and K. I., unpublished data; see also Fig. 7). Treatment of lepidopteran Bm5 cells with forskolin, an activator of adenylate cyclase, also results in the accumulation of cAMP in these cells (L. S., Z. G., and K. I., unpublished data). However, CRE-linked reporter constructs are not activated by forskolin in lepidopteran or dipteran cell lines (L. S., Z. G., and K. I., unpublished data; Poels *et al.*, 2004), suggesting that certain events downstream of the cAMP release are abrogated. In *Drosophila* S2 cells, several different isoforms of

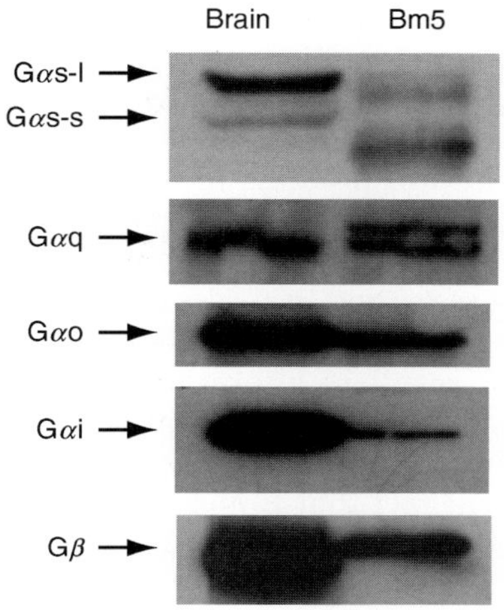

FIG 7. Immunological detection of G-proteins in silkmoth Bm5 cells. Membranes from rat brain and Bm5 cells were immunoblotted for Gαs with antiserum purchased from Chemicon, Temecula, California, for Gαo with antiserum OC1 kindly provided by Prof. G. Milligan, University of Glasgow, United Kingdom, and for Gαq/11, Gαi and Gβ with antisera E17, C10, and T20 (Santa Cruz Biotechnology), respectively.

cyclic-AMP response element-binding protein (CREB) have been identified that are inhibitors, rather than activators, of the cAMP response (Poels *et al.*, 2004). Whether a similar situation exists in lepidopteran cells and results in the abrogation of the cAMP response at the transcriptional level remains to be investigated. These observations suggest that the development of screening systems for Gαs-coupled receptors through the detection of CRE-dependent reporter gene activation probably requires major engineering of these cell lines. However, it should also be noted that cAMP detection systems amenable to HTS format and not requiring reporter gene activation have become available (Williams, 2004). Thus, the development of high-throughput systems for GPCRs that signal via cAMP stimulation or inhibition can occur using these alternative techniques, which detect the levels of cellular cAMP by use of fluorescent or luminescent probes.

GPCR functional assays in mammalian cell lines were revolutionized by the discovery of the promiscuous Gα15/16 proteins that allow coupling of almost any GPCR to phospholipase Cβ (PLCβ) to generate both diacylglycerol and inositol (1,4,5)-trisphosphate (IP3) with subsequent activation of protein kinase C and elevation of intracellular levels of Ca^{2+} (Kostenis, 2001; Milligan *et al.*, 1996; Offermanns and Simon, 1995). Thus, coexpression of $G\alpha15$ or $G\alpha16$ in lepidopteran cell lines will redirect GPCRs that normally couple to $G\alpha s$ or $G\alpha i$ to the PLCβ/Ca^{2+} pathway (Knight *et al.*, 2003). It was also found that expression of chimeric $G\alpha q$ proteins in Sf9 cells is more effective than expression of $G\alpha15$ or $G\alpha16$ for the purpose of redirecting $G\alpha i$-coupled receptors to the Ca^{2+} pathway (Knight and Grigliatti, 2004b). On the other hand, coexpression of $G\alpha16$ in Bm5 cells stably expressing the δ-opioid receptor (a $G\alpha i$/$G\alpha o$-coupled receptor; Georgoussi *et al.*, 1995, 1997) did not result in alteration of the levels of intracellular Ca^{2+} observed on opioid agonist stimulation relative to cells that were expressing the receptor alone (Swevers *et al.,* 2005). This was presumably due to the fact that Bm5 cells contain sufficient quantities of endogenous $G\alpha o$ and $G\alpha i$ (and $G\beta\gamma$ subunits; Fig. 7) to direct activation of the Ca^{2+} release pathway.

GPCRs that could be functionally expressed in transformed lepidopteran cell lines (High Five^{TM}, Sf9, and Bm5) by coupling to the Ca^{2+} release pathway (directed by a coexpressed $G\alpha16$ or $G\alpha q$) include $G\alpha s$-coupled receptors (dopamine D1, adrenergic $\beta2$, histamine H2, and serotonin 4A receptors; Knight *et al.*, 2003; L. S., Z. G., and K. I., unpublished data), $G\alpha q$-coupled receptors (thromboxane A2, muscarinic acetylcholine M1, and histamine H1 receptors; Knight *et al.*, 2003), and $G\alpha i$-coupled receptors (serotonin 1A, serotonin 1D, and dopamine D2 receptors; Knight and Grigliatti, 2004b; Knight *et al.*, 2003). In all cases, the pharmacological properties of the receptors expressed in lepidopteran cells were similar to those in mammalian cell lines. In the case of the δ-opioid receptor, its expression in Bm5 cells allowed coupling to the Ca^{2+} release pathway at concentrations (EC50s) similar to those for transformed human embryonic kidney (HEK) 293 cells (Swevers *et al.*, 2005; Fig. 8). The magnitude of the response in the transformed Bm5 cells was also similar to that of the HEK293 cells, indicating that the two systems have similar sensitivities as screening systems. Furthermore, our studies have shown that transformed Bm5 cells can be used successfully to detect both opioid agonists and antagonists by measuring alterations of Ca^{2+}-induced fluorescence, confirming their applicability for fast detection of δ-opioid receptor ligand mimetics in an HTS format (Swevers *et al.*, 2005). Because lepidopteran cell lines have some beneficial features compared to their mammalian counterparts, such as low maintenance costs, they provide a valuable alternative as screening systems for drug ligands that target GPCRs.

Drosophila S2 cells have also been used successfully for functional expression of insect as well as mammalian GPCRs (Cordova *et al.*, 2003;

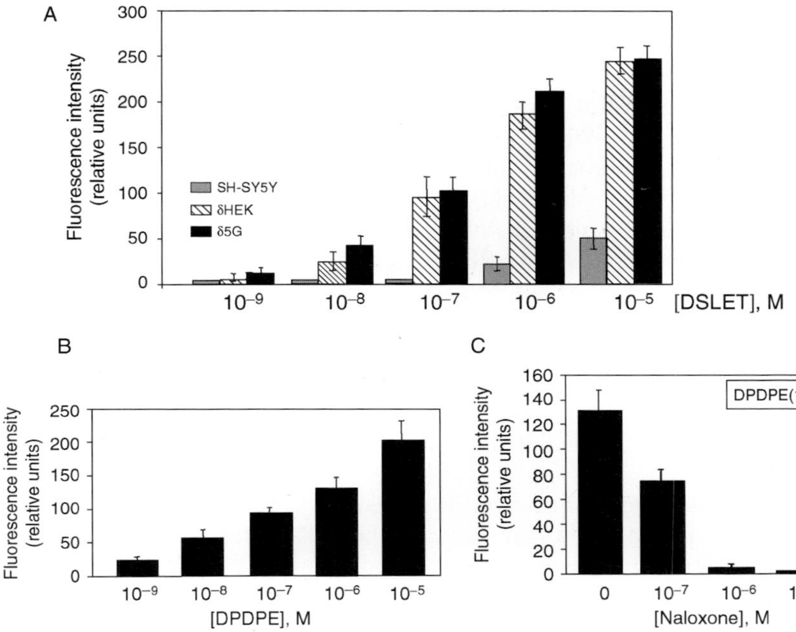

FIG 8. Functional expression of the mouse δ1-opioid receptor in transformed silkmoth-derived Bm5 cell lines. (A) Comparison of the response to the opioid agonist DSLET between the mammalian cell lines SY-SY5Y (a neural line that endogenously expresses the δ1-opioid receptor) and δHEK (a transformed HEK293 cell line) and the transformed Bm5 cell line δ5G. (B) Calcium response assay for the agonist DPDPE that is specific for the δ1-opioid receptor. (C) Blocking of the calcium response induced by 1 μM DPDPE in the presence of different concentrations of the opioid antagonist naloxone. In all cases, receptor activation was detected by measurement of calcium-induced fluorescence using the calcium-sensitive marker Fluo-3. (Modified from Swevers, L., Morou, E., Balatsos, N., Iatrou, K., and Georgoussi, Z. (2005). Functional expression of mammalian opioid receptors in insect cells and high-throughput screening platforms for receptor ligand mimetics. *Cell. Mol. Life Sci.* **62:** 919–930; © Birkhäuser Verlag, Basel.)

Perret *et al.*, 2003; Radford *et al.*, 2002, 2004; Torfs *et al.*, 2002). In most cases, GPCR activation was monitored by the detection of Ca^{2+} release and/or aequorin/coelenterazine-induced luminescence.

C. Other Cellular Regulators

Although most HTS systems developed to date target receptors that respond to extracellular signals, a few instances of transformed cell lines that were engineered to detect changes of other cellular processes in an HTS format also exist. Thus, a transformed S2 cell-based assay for the inhibition of human β-secretase, the enzyme that generates amyloid β-peptide and is implicated in Alzheimer's disease, has been reported (Oh *et al.*, 2003). The use of S2 cells is beneficial relative to use of mammalian cell lines because of the lack of endogenous β-secretase activity. Similarly, Sf9 cells have been engineered for a rapid and quantitative cell-to-cell-fusion assay that is suitable for HTS (Slack and Blissard, 2001). These examples illustrate that insect cell lines are flexible tools that can be engineered to function as detectors for a wide variety of cellular processes.

V. Host Cell Engineering

Stably transformed insect cell lines can be viewed as an "engineered host" environment for a growing number of applications. As already noted (Section II), expression of mammalian glycoproteins in insect cells, either by the baculovirus expression system or by stably transformed cell lines, is occasionally compromised by the lack of complex glycosylation patterns (reviewed in Jarvis, 2003; Harrison and Jarvis, 2006). To overcome this limitation, stably transformed insect cell lines that express several mammalian glycotransferases have been engineered (Hollister *et al.*, 1998; Jarvis *et al.*, 1998). These lines have been used for efficient expression of properly *N*-glycosylated and terminally sialylated glycoproteins in the context of both the baculovirus and transformed insect cell expression systems (see chapter by Harrison and Jarvis, this volume, pp. 159–191 for more detailed information and relevant references). One of these lines is commercially available through Invitrogen Corporation as Mimic™ cells. These are transformed Sf9 cells identical to the SfSWT-1 cells developed by the Jarvis *et al.* (1998). Further engineering of these cells was also reported later (Aumiller *et al.*, 2003). Other studies have reported similar engineering of *Drosophila* S2 cells, resulting in enhanced activity of heterologous recombinant enzymes (Chang *et al.*, 2005).

A complementary approach, originally developed for the baculovirus expression system, involves coexpression of molecular chaperones and folding factors. Coexpression of molecular chaperones, like calnexin and calreticulin, in stably transformed High Five™ cells (Deo and Park, 2006; Kato *et al.*, 2004) resulted in improved activity relative to lines expressing only the recombinant glycoprotein.

Other applications that make use of an engineered host insect cell environment include cell lines that are engineered for rescue of baculovirus mutants (Farrell *et al.*, 2005; Iatrou *et al.*, 2000). In this application, baculoviruses that are incapacitated by disruption or deletion of a gene vital for baculovirus replication or transcription can be propagated in cell lines providing the relevant gene product *in trans*. Several Bm5-based cell lines have been engineered in order to stably express gene products of BmNPV genes such as *lef-8* and *ie-1* (C. Kenoutis, P. J. Farrell, V. D., and K. I., unpublished data).

Finally, the generation of Bm5 cell lines transformed with the *B. mori* promoting protein (PP) gene has also been reported (Iatrou and Swevers, 2005). This protein increases budded virus production (Kanaya and Kobayashi, 2000). Bm5 cell lines engineered for overexpression of PP exhibit up to a 1000-fold enhancement of viral infectivity in serum-free media (Iatrou and Swevers, 2005). Thus, PP allows for production of high titers of baculoviruses and high quantities of recombinant proteins obtained through the baculovirus expression system in serum-free media in the absence of potential pathogenic factors such as viruses, mycoplasmas, and prions. In turn, PP allows for the development of new baculovirus tools for safe therapeutic protein production and mammalian cell transduction for gene therapy applications.

VI. Conclusions—Future Perspectives

A variety of insect cell-based expression systems with several modifications tailored for specific needs is available. Stably transformed insect cell lines and systems that employ baculovirus-derived genetic elements and direct continuous high-level expression of recombinant proteins in particular have drawn significant attention as means for expression of proteins that require an intact cellular environment, that is, secreted and membrane-anchored proteins. Furthermore, such systems can be employed as HTS tools for the identification of bioactive substances of natural or synthetic origin.

The development of plasmid-based expression systems is relatively recent but is advancing at a rapid pace, with important improvements and new applications likely. New generation, multipurpose vectors may include features from systems that are currently available. Thus, secretion modules can be further developed to facilitate efficient secretion of intracellular proteins with alternative affinity tags and proteolytic sites. Double-tagged proteins can be generated and purified by a "tandem affinity purification" strategy enabling purification of multiprotein complexes and functional characterization in a high-throughput fashion (for an application in S2 cells, see Forler *et al.*, 2003). Certain pEIA derivatives (Fig. 3) have already been used for generation of proteins with double C-terminal tags and their application for multiprotein complex purification is under way. Densovirus-based vectors can be combined with available expression cassettes to form powerful systems for fast and easy transformation and detection as well as high-level protein expression. Novel HTS systems may also be designed in the future for orphan nuclear receptors whose ligands remain unknown. In addition, insect cell lines are expected to provide better host systems for insect GPCRs and can be employed for more effective screening for compounds that mediate their function. For example, functional expression of insect odorant receptors in stably transformed insect cell lines may allow for generation of reliable assays to investigate mechanisms of olfaction and development of screening platforms for the identification of natural sources of and/or new synthetic repellants and attractants. Similarly, additional receptors and other cellular and developmental regulators may be assayed as targets for interference by "endocrine disruptors," which may be developed into environmentally friendly insecticides. In conclusion, insect cell-based expression systems are expected to play key roles in many research efforts. Most probably, the peak contribution of insect cell-based expression systems to biotechnology is yet to come.

Acknowledgments

Recent work of the "Insect Molecular Genetics and Biotechnology" Group has been supported in part by grants from the General Secretariat of Research and Technology, Greek Ministry of Development (contract numbers 02/PRAXE/181, EPAN YB/11, and USA 052), and the European Community (contract No. QLK3-CT-2001–01586). V. D. was the recipient of a NCSR "Demokritos" research fellowship.

References

Agathos, S. N. (1991). Production scale insect cell culture. *Biotechnol. Adv.* **9:**51–68.

Alitalo, K., Schwab, M., Lin, C. C., Vatmus, H. E., and Bishop, M. (1983). Homogeneously staining chromosomal regions containing amplified copies of an abundantly expressed cellular oncogene (c-myc) in malignant neuroendocrine cells from a human colon carcinoma. *Proc. Natl. Acad. Sci. USA* **80:**1707–1711.

Andronopoulou, E., Labropoulou, V., Douris, V., Woods, D. F., Biessmann, H., and Iatrou, K. (2006). Specific interactions amongst odorant binding proteins of the African malaria vector *Anopheles gambiae. Insect Mol. Biol.* (in press).

Angelichio, M. L., Beck, J. A., Johansen, H., and Ivey-Hoyle, M. (1991). Comparison of several promoters and polyadenylation signals for use in heterologous gene expression in cultured *Drosophila* cells. *Nucleic Acids Res.* **19:**5037–5043.

Angers, S., Salahpour, A., Joly, E., Hilairet, S., Chelsky, D., Dennis, M., and Bouvier, M. (2000). Detection of β2-adrenergic receptor dimerization in living cells using bioluminescence resonance energy transfer (BRET). *Proc. Natl. Acad. Sci. USA* **97:**3684–3689.

Aumiller, J. J., Hollister, J. R., and Jarvis, D. L. (2003). A transgenic insect cell line engineered to produce CMP-sialic acid and sialylated glycoproteins. *Glycobiology* **13:**497–507.

Baneyx, F., and Mujacic, M. (2004). Recombinant protein folding and misfolding in *Escherichia coli*. *Nat. Biotechnol.* **22:**1399–1408.

Banks, D. J., Hua, G., and Adang, M. J. (2003). Cloning of a *Heliothis virescens* 110 kDa aminopeptodase N and expression in *Drosophila* S2 cells. *Insect Biochem. Mol. Biol.* **33:**499–508.

Bossin, H., Fournier, P., Royer, C., Barry, P., Cerutti, P., Gimenez, S., Couble, P., and Bergoin, M. (2003). *Junonia coenia* densovirus-based vectors for stable transgene expression in Sf9 cells: Influence of the densovirus sequences on genomic integration. *J. Virol.* **77:**11060–11071.

Bouvier, M., Menard, L., Dennis, M., and Marullo, S. (1998). Expression and recovery of functional G-protein coupled receptors using baculovirus expression systems. *Curr. Opin. Biotechnol.* **9:**522–527.

Brand, A. H., and Perrimon, N. (1993). Targeted gene expression as a means of altering cell fates and generating dominant phenotypes. *Development* **118:**401–415.

Catteruccia, F., Nolan, T., Blass, C., Muller, H. M., Crisanti, A., Kafatos, F. C., and Loukeris, T. G. (2000). Toward *Anopheles* transformation: *Minos* element activity in anopheline cells and embryos. *Proc. Natl. Acad. Sci. USA* **97:**2157–2162.

Chang, K. H., Lee, J. M., Jeon, H. K., and Chung, I. S. (2004). Improved production of recombinant tumstatin in stably transformed *Trichoplusia ni* BTI Tn 5B1–4 cells. *Protein Expr. Purif.* **35:**69–75.

Chang, K. H., Yang, J. M., Chun, H. O. K., and Chung, I. S. (2005). Enhanced activity of recombinant β-secretase from Drosophila melanogaster S2 cells transformed with cDNAs encoding human β1,4-galactosyltransferase and Galβ1,4-GlcNAc α2,6-sialyltransferase. *J. Biotechnol.* **116:**359–367.

Chen, J. H., Turner, P. C., and Rees, H. H. (2002). Molecular cloning and induction of nuclear receptors from insect cell lines. *Insect Biochem. Mol. Biol.* **32:**657–667.

Cherbas, L., Moss, R., and Cherbas, P. (1994). Transformation techniques for *Drosophila* cell lines. *Methods Cell Biol.* **44:**161–179.

Christopherson, K. S., Mark, M. R., Bajaj, V., and Godowski, P. J. (1992). Ecdysteroid-dependent regulation of genes in mammalian cells by a *Drosophila* ecdysone receptor and chimeric transactivators. *Proc. Natl. Acad. Sci. USA* **89:**6314–6318.

Cordova, D., Delpech, V. R., Sattelle, D. B., and Rauh, J. J. (2003). Spatiotemporal calcium signaling in a *Drosophila melanogaster* cell line stably expressing a *Drosophila* muscarinic acetylcholine receptor. *Invert. Neurosc.* **5:**19–28.

Crouch, E. A., and Passarelli, A. L. (2005). Effects of baculovirus transactivators IE-1 and IE-2 on the *Drosophila* heat shock 70 promoter in two insect cell lines. *Arch. Virol.* **150:**1563–1578.

Denault, J.-B., Lazure, C., Day, R., and Lrduc, R. (2000). Comparative characterization of two forms of recombinant human SPC1 secreted from Schneider 2 cells. *Protein Expr. Purif.* **19:**113–124.

Deo, V. K., and Park, E. Y. (2006). Multiple co-transfection and co-expression of human beta-1,3-N-acetylglucosaminyltransferase with human calreticulin chaperone cDNA in a single step in insect cells. *Biotechnol. Appl. Biochem.* **43:**129–135.
Duffy, J. B. (2002). The GAL4 system in *Drosophila*: A fly geneticist's Swiss army knife. *Genesis* **34:**1–15.
Espagne, E., Douris, V., Lalmanach, G., Provost, B., Cattolico, L., Lesobre, J., Kurata, S., Iatrou, K., Drezen, J.-M., and Huguet, E. (2005). A virus essential for insect host-parasite interactions encodes cystatins. *J. Virol.* **79:**9765–9776.
Farrell, P. J. (1998). The development of stably transformed lepidopteran insect cell technology for both the expression of recombinant proteins and the generation of baculovirus artificial chromosomes. Ph.D. thesis. The University of Calgary, Calgary, Alberta, Canada.
Farrell, P. J., and Iatrou, K. (2004). Transfected insect cells in suspension culture rapidly yield moderate quantities of recombinant proteins in protein-free culture medium. *Protein Expr. Purif.* **36:**177–185.
Farrell, P. J., Lu, M., Prevost, J., Brown, C., Behie, L., and Iatrou, K. (1998). High-level expression of secreted glycoproteins in transformed lepidopteran insect cells using a novel expression vector. *Biotechnol. Bioeng.* **60:**656–663.
Farrell, P. J., Behie, L., and Iatrou, K. (1999). Transformed lepidopteran insect cells: New sources of recombinant human tissue plasminogen activator. *Biotechnol. Bioeng.* **64:**426–433.
Farrell, P. J., Behie, L., and Iatrou, K. (2000). Secretion of cytoplasmic and nuclear proteins from insect cells using a novel secretion module. *Proteins: Struct. Function Genet.* **41:**144–153.
Farrell, P. J., Swevers, L., and Iatrou, K. (2005). Insect cell culture and recombinant protein expression systems. *In* "Comprehensive Molecular Insect Science" (L. Gilbert, K. Iatrou, and S. S. Gill, eds.), Vol. 4, pp. 475–507. Elsevier, San Diego.
Forler, D., Köcher, T., Rode, M., Gentzel, M., Izaurralde, E., and Wilm, M. (2003). An efficient protein complex purification method for functional proteomics in higher eukaryotes. *Nat. Biotechnol.* **21:**89–92.
Francken, B. J., Josson, K., Lijnen, P., Jurzak, M., Luyten, W. H., and Leysen, J. E. (2000). Human 5-hydroxytryptamine(5A) receptors activate coexpressed G(i) and G(o) proteins in *Spodoptera frugiperda* 9 cells. *Mol. Pharmacol.* **57:**1034–1044.
Gårdsvoll, H., Werner, F., Søndergaard, L., Danø, K., and Ploug, M. (2004). Characterization of low-glycosylated forms of soluble human urokinase receptor expressed in *Drosophila* Schneider 2 cells after deletion of glycosylation-sites. *Protein Expr. Purif.* **34:**284–295.
Georgoussi, Z., Milligan, G., and Zioudrou, C. (1995). Immunoprecipitation of opioid receptor-GTP-binding protein complexes using selective GTP-binding protein antisera. *Biochem. J.* **306:**71–75.
Georgoussi, Z., Merkouris, M., Mullaney, I., Megaritis, G., Carr, G., Zioudrou, C., and Milligan, G. (1997). Selective interactions of the μ-opioid receptors with pertussis toxin-sensitive G proteins: Involvement of the third intracellular loop. *Biochim. Biophys. Acta* **1359:**263–274.
Gerngross, T. U. (2004). Advances in the production of human therapeutic proteins in yeasts and filamentous fungi. *Nat. Biotechnol.* **22:**1409–1414.
Gonzalez, G. A., and Montminy, M. R. (1989). Cyclic AMP stimulates somatostatin gene transcription by phosphorylation of CREB at serine 133. *Cell* **59:**675–680.

Grace, T. D. C. (1967). Establishment of a line of cells from the silkworm *Bombyx mori*. *Nature* **216:**613.

Granados, R. R., Li, G., Derkensen, A. C. G., and McKenna, K. A. (1994). A new insect cell line from *Trichoplusia ni* (BTI-Tn-5B1–4) susceptible to *Trichoplusia ni* single enveloped nuclear polyhedrosis virus. *J. Inverteb. Pathol.* **64:**260–266.

Gray, C. E., and Coates, C. J. (2004). High-level gene expression in *Aedes albopictus* cells using a baculovirus Hr3 enhancer and IE1 *trans* activator. *BMC Mol. Biol.* **5:**8.

Grossman, G. L., Rafferty, C. S., Fraser, M. J., and Benedict, M. Q. (2000). The piggyBac element is capable of precise excision and transposition in cells and embryos of the mosquito, *Anopheles gambiae*. *Insect Biochem. Mol. Biol.* **10:**909–914.

Grussenmeyer, T., Schneidtmann, K. H., Hutchinson, M. A., Eckhart, W., and Walter, G. (1985). Complexes of polyoma virus medium T antigen and cellular proteins. *Proc. Natl. Acad. Sci. USA* **82:**7952–7954.

Grynkiewicz, G., Poenie, M., and Tsien, R. Y. (1985). A new generation of Ca^{2+} indicators with greatly improved fluorescence properties. *J. Biol. Chem.* **260:**3440–3450.

Guarino, L. A., and Summers, M. D. (1987). Nucleotide sequence and temporal expression of a baculovirus regulatory gene. *J. Virol.* **61:**2091–2099.

Guarino, L. A., Gonzalez, M. A., and Summers, M. D. (1986). Complete sequence and enhancer function of the homologous DNA regions of *Autographa californica* nuclear polyhedrosis virus. *J. Virol.* **60:**224–229.

Gustafsson, J. A. (1999). Seeking ligands for lonely orphan receptors. *Science* **284:** 1285–1286.

Guttieri, M. C., Bookwalter, C., and Schmaljohn, C. (2000). Expression of a human, neutralizing monoclonal antibody specific to Puumala virus G2-protein in stably-transformed insect cells. *J. Immunol. Methods* **246:**97–108.

Hamm, H. E. (1998). The many faces of G protein signaling. *J. Biol. Chem.* **273:**669–672.

Handler, A. M., and O'Brochta, D. A. (2005). Transposable elements for insect transformation. *In* "Comprehensive Molecular Insect Science" (L. Gilbert, K. Iatrou, and S. S. Gill, eds.), Vol. 4, pp. 437–474. Elsevier, San Diego.

Harrison, R., and Jarvis, D. (2006). Protein N-glycosylation in the baculovirus-insect cell expression system. *Adv. Vir. Res.* **68:**159–191.

Hegedus, D. D., Pfeifer, T. A., Hendry, J., Theilmann, D. A., and Grigliatti, T. A. (1998). A series of broad host range shuttle vectors for constitutive and inducible expression of heterologous proteins in insect cells. *Gene* **207:**241–249.

Hegedus, D. D., Pfeifer, T. A., Theilmann, D. A., Kennard, M. L., Gabathuler, R., Jefferies, W. A., and Grigliatti, T. A. (1999). Differences in the expression and localization of human melanotransferrin in lepidopteran and dipteran insect cell lines. *Prot. Expr. Purif.* **15:**296–307.

Helgen, J. C., and Fallon, A. M. (1990). Polybrene-mediated transfection of cultured lepidopteran cells: Induction of a *Drosophila* heat shock promoter. *In Vitro Cell Dev. Biol.* **26:**731–736.

Hellwig, S., Drossard, J., Twyman, R. M., and Fischer, R. (2004). Plant cell cultures for the production of recombinant proteins. *Nat. Biotechnol.* **22:**1415–1422.

Hollister, J. R., Shaper, J. H., and Jarvis, D. L. (1998). Stable expression of mammalian beta 1,4-galactosyltransferase extends the N-glycosylation pathway in insect cells. *Glycobiology* **8:**473–480.

Huynh, C. Q., and Zieler, H. (1999). Construction of modular and versatile plasmid vectors for the high-level expression of single or multiple genes in insects and insect cell lines. *J. Mol. Biol.* **288:**13–20.

Iatrou, K., and Swevers, L. (2005). Transformed lepidopteran cells expressing a protein of the silkmoth fat body display enhanced susceptibility to baculovirus infection and produce high titers of budded virus in serum-free media. *J. Biotechnol.* **120:**237–250.

Iatrou, K., Farrell, P., and Hashimoto, Y. (2000). Baculovirus artificial chromosomes and methods of use. US patent number 6,090,584.

Ikonomou, L., Schneider, Y. J., and Agathos, S. N. (2003). Insect cell culture for industrial production of recombinant proteins. *Appl. Microbiol. Biotechnol.* **62:**1–20.

Ivey-Hoyle, M., Culp, J. S., Chaikin, M. A., Hellmig, B. D., Matthews, T. J., Sweet, R. W., and Rosenberg, M. (1991). Envelope glycoproteins from biologically diverse isolates of immunodeficiency viruses have widely different affinities for CD4. *Proc. Natl. Acad. Sci. USA* **88:**512–516.

Janetopoulos, C., Jin, T., and Devreotes, P. (2001). Receptor-mediated activation of heterotrimeric G-proteins in living cells. *Science* **291:**1408–2411.

Jarvis, D. L. (2003). Developing baculovirus—insect cell expression systems for humanized recombinant glycoprotein production. *Virology* **310:**1–7.

Jarvis, D. L., Fleming, J.-A. G. W., Kovacs, G. R., Summers, M. D., and Guarino, L. A. (1990). Use of early baculovirus promoters for continuous expression and efficient processing of foreign gene products in stably transformed lepidopteran cells. *Biotechnology* **8:**950–955.

Jarvis, D. L., Weinkauf, C., and Guarino, L. A. (1996). Immediate-early baculovirus vectors for foreign gene expression in transformed or infected insect cells. *Protein Expr. Purif.* **8:**191–203.

Jarvis, D. L., Kawar, Z. S., and Hollister, J. R. (1998). Engineering N-glycosylation pathways in the baculovirus-insect cell system. *Curr. Opin. Biotechnol.* **9:**528–533.

Johansen, H., van der Straten, A., Sweet, R., Otto, E., Maroni, G., and Rosenberg, M. (1989). Regulated expression of high copy number allows production of a growth inhibitory oncogene product in *Drosophila* Schneider cells. *Genes Devel.* **3:**882–889.

Johanson, K., Appelbaum, E., Doyle, M., Hensley, P., Zhao, B., Abdel-Meguid, S. S., Young, P., Cook, R., Carr, S., Matico, R., Cusimano, D., Dul, E., *et al.* (1995). Binding interactions of human interleukin 5 with its receptor a subunit: Large scale production, structural, and functional studies of *Drosophil*a-expressed recombinant proteins. *J. Biol. Chem.* **270:**9459–9471.

Johnson, R., Meidinger, R. G., and Iatrou, K. (1992). A cellular promoter-based expression cassette for generating recombinant baculoviruses directing rapid expression of passenger genes in infected insects. *Virology* **190:**815–823.

Joyce, K. A., Atkinson, A. E., Bermudez, I., Beadle, D. J., and King, L. A. (1993). Synthesis of functional GABAA receptors in stable insect cell lines. *FEBS Lett.* **335:**61–64.

Kanaya, T., and Kobayashi, J. (2000). Purification and characterization of an insect haemolymph protein promoting *in vitro* replication of the *Bombyx mori* nucleoplyhedrovirus. *J. Gen. Virol.* **81:**1135–1141.

Kapitskaya, M. Z., Dittmer, N. T., Deitsch, K. W., Cho, W.-L., Taylor, D. G., Leff, T., and Raikhel, A. S. (1998). Three isoforms of a hepatocyte muclear factor-4 transcription factor with tissue- and stage-specific expression in the adult mosquito. *J. Biol. Chem.* **273:**29801–29810.

Kato, T., Murata, T., Usui, T., and Park, E. Y. (2004). Improvement of the production of GFP_{uv}-β1,3-*N*-acetylglucosaminyltransferase 2 fusion protein using a molecular chaperone-assisted insect-cell-based expression system. *Biotechnol. Bioeng.* **89:**424–433.

Keith, M. B. A., Farrell, P. J., Iatrou, K., and Behie, L. A. (1999). Screening of transformed insect cell lines for recombinant protein production. *Biotechnol. Prog.* **15:**1046–1052.

Kempf, J., Snook, L. A., Vonesch, J.-L., Dahms, T. E. S., Pattus, F., and Massotte, D. (2002). Expression of the human μ-opioid receptor in a stable Sf9 cell line. *J. Biotechnol.* **95:**181–187.

Kim, H. G., Yang, S. M., Lee, Y. C., Do, S. I., Chung, I. S., and Yang, J. M. (2003). High-level expression of human glycosyltransferases in insect cells as biochemically active form. *Biochem. Biophys. Res. Commun.* **305:**488–493.

Kimura, M., Takatsuki, A., and Yamaguchi, I. (1994). Blasticidin S deaminase gene from *Aspergillus terreus* (BSD): A new drug resistance gene for transfection of mammalian cells. *Biochim. Biophys. Acta* **1219:**653–659.

King-Jones, K., and Thummel, C. S. (2005). Nuclear receptors—a perspective from *Drosophila*. *Nat. Rev. Genet.* **6:**311–323.

Kirkpatrick, R. B., Ganguly, S., Angelichio, M., Griego, S., Shatzman, A., Silverman, C., and Rosenberg, M. (1995). Heavy chain dimers as well as complete antibodies are efficiently formed and secreted from *Drosophila* via a BiP-mediated pathway. *J. Biol. Chem.* **270:**19800–19805.

Kleditzsch, P., Pratt, J., Vishnuvardhan, D., Henklein, P., Schade, R., and Beinfeld, M. C. (2003). Production, purification and characterization of rat pro-CCK from serum-free adapted *Drosophila* cells. *Protein Expr. Purif.* **31:**56–63.

Kleymann, G., Boege, F., Hahn, M., Hampe, W., Vasudevan, S., and Reilander, H. (1993). Human beta 2-adrenergic receptor produced in stably transformed insect cells is functionally coupled via endogenous GTP-binding protein to adenylyl cyclase. *Eur. J. Biochem.* **213:**797–804.

Kliewer, S. A., Lehmann, J. M., and Wilson, T. M. (1999). Orphan nuclear receptors: Shifting endocrinology into reverse. *Science* **284:**757–760.

Klinakis, A. G., Loukeris, T. G., Pavlopoulos, A., and Savakis, C. (2000). Mobility assays confirm the broad host-range activity of the *Minos* transposable element and validate new transformation tools. *Insect Mol. Biol.* **9:**269–275.

Klueg, K. M., Avarado, D., Muskavitch, M. A., and Duffy, J. B. (2002). Creation of a GAL4/UAS coupled inducible gene expression system for use in *Drosophila* cultured cell lines. *Genesis* **34:**119–122.

Knight, M. R., Campbell, A. K., Smith, S. M., and Trewavas, A. J. (1991). Recombinant aequorin as a probe for cytosolic free Ca^{2+} in *Escherichia coli*. *FEBS Lett.* **282:**405–408.

Knight, P. J., and Grigliatti, T. A. (2004a). Diversity of G proteins in lepidopteran cell lines: Partial sequences of six G protein alpha subunits. *Arch. Insect Biochem. Physiol.* **57:**142–150.

Knight, P. J., and Grigliatti, T. A. (2004b). Chimeric G proteins extend the range of insect cell-based functional assays for human G protein-coupled receptors. *J. Recept. Signal Transduct. Res.* **24:**241–256.

Knight, P. J. K., Pfeifer, T. A., and Grigliatti, T. A. (2003). A functional assay for G-protein-coupled receptors using stably transformed insect tissue culture cell lines. *Anal. Biochem.* **320:**88–103.

Kock, M. A., Hew, B. E., Bammert, H., Fritzinger, D. C., and Vogel, C.-W. (2004). Structure and function of recombinant cobra venom factor. *J. Biol. Chem.* **279:** 30836–30843.

Kost, T. A., Condreay, J. P., and Jarvis, D. L. (2005). Baculovirus as versatile vectors for protein expression in insect and mammalian cells. *Nat. Biotechnol.* **23:**567–575.

Kostenis, E. (2001). Is Gα_{16} the optimal tool for fishing ligands of orphan G-protein-coupled receptors? *Trends Pharmac. Sci.* **22:**560–564.

Kovach, M. J., Carlson, J. O., and Beaty, B. J. (1992). A *Drosophila* metallothionein promoter is inducible in mosquito cells. *Insect Mol. Biol.* **1:**37–43.

Kovacs, G. R., Guarino, L. A., and Summers, M. D. (1991). Novel regulatory properties of the IE1 and IE0 transactivators encoded by the baculovirus *Autographa californica* nuclear polyhedrosis virus. *J. Virol.* **65:**5281–5288.

Kovacs, G. R., Choi, J., Guarino, L. A., and Summers, M. D. (1992). Functional dissection of the *Autographa californica* nuclear polyhedrosis virus immediate-early transcriptional regulatory protein. *J. Virol.* **66:**7429–7437.

Krantz, D. E., and Zipursky, S. L. (1990). *Drosophila* chaoptin, a member of the leucine-rich repeat family, is a photoreceptor cell-specific adhesion molecule. *EMBO J.* **9:** 1969–1977.

Kuhn, J., Müller, S., Schnölzer, M., Kempf, T., Schön, S., Brinkmann, T., Schöttler, M., Götting, C., and Kleesiek, K. (2003). High-level expression and purification of human xylosyltransferase I in High Five insect cells as biochemically active form. *Biochem. Biophys. Res. Commun.* **312:**537–544.

Kulahin, N., Kasper, C., Gajhede, M., Berezin, V., Bock, E., and Kastrup, J. S. (2004). Expression, crystallization and preliminary X-ray analysis of extracellular modules of the neural cell-adhesion molecules NCAM and L1. *Acta Crystallogr. D Biol. Crystallogr.* **60:**591–593.

Lan, Q., and Riddiford, L. M. (1997). DNA transfection in the ecdysteroid-responsive GV1 cell line from the tobacco hornworm, *Manduca sexta. In Vitro Cell Dev. Biol. Anim.* **33:**615–621.

Lapointe, R., Wilson, R., Vilaplana, L., O'Reilly, D. R., Falabella, P., Douris, V., Bernier-Cardou, M., Pennacchio, F., Iatrou, K., Malva, C., and Olszewski, J. A. (2005). Expression of a *Toxoneuron nigriceps* polydnavirus (TnBV) encoded protein, TnBV1, causes apoptosis-like programmed cell death in lepidopteran insect cells. *J. Gen. Vir.* **86:** 963–971.

Lee, D. F., Chen, C.-C., Hsu, T.-A., and Juang, J.-L. (2000). A baculovirus superinfection system: Efficient vehicle for gene transfer into *Drosophila* S2 cells. *J. Virol.* **74:** 11873–11880.

Lee, J. M., Kusakabe, T., Kawaguchi, Y., Yasunaga-Aiki, C., Nho, S., Nakajima, Y., and Koga, K. (2003). Molecular characterization of a heat shock cognate 70–4 promoter from the silkworm, *Bombyx mori*. *J. Insect Biotech. Sericol.* **72:**33–39.

Lefkowitz, R. J. (2000). The superfamily of hepthelical receptors. *Nat. Cell Biol.* **2:** E133–E136.

Lehr, R. V., Elefante, L. C., Kikly, K. K., O'Brien, S. P., and Kirkpatrick, R. B. (2000). A modified metal-ion affinity chromatography procedure for the purification of histidine-tagged recombinant proteins expressed in *Drosophila* S2 cells. *Protein Expr. Purif.* **19:**362–368.

Leisy, D. J., and Rohrmann, G. F. (2000). The *Autographa californica* nucleopolyhedrovirus IE-1 protein complex has two modes of specific DNA binding. *Virology* **274:** 196–202.

Li, B., Tsing, S., Kosaka, A. H., Nguyen, B., Osen, E. G., Bach, C., Chan, H., and Barnett, J. (1996). Expression of human dopamine b-hydroxylase in *Drosophila* Schneider 2 cells. *Biochem. J.* **313:**57–64.

Li, E., Brown, S. L., Dolman, C. S., Brown, G. B., and Nemerow, G. R. (2001). Production of functional antibodies generated in a nonlytic insect cell expression system. *Protein Expr. Purif.* **21:**121–128.

Li, H.-C., Huang, C.-C., Chen, S.-F., and Chou, M.-Y. (2005). Assembly of homotrimeric type XXI minicollagen by coexpression of prolyl 4-hydroxylase in stably transfected *Drosophila melanogaster* S2 cells. *Biochem. Biophys. Res. Commun.* **336:**375–385.

Lim, H. J., Kim, Y. K., Hwang, D. S., and Cha, H. J. (2004). Expression of functional human transferrin in stably transfected *Drosophila* S2 cells. *Biotechnol. Prog.* **20:**1192–1197.

Lindner, P., Bauer, K., Krebber, A., Nieba, L., Kremmer, E., Krebber, C., Honegger, A., Klinger, B., Mocikat, R., and Pluckthun, A. (1997). Specific detection of His-tagged proteins with recombinant anti-His tag scFv-phosphatase or scFv-phage fusions. *BioTechniques* **22:**140–149.

Loomis, K. H., Yaeger, K. W., Batenjany, M. M., Mehler, M. M., Grabski. A. C., Wong, S. C., and Novy, R. E. (2005). InsectDirect™ System: Rapid, high-level protein expression and purification from insect cells. *J. Struct. Funct. Genomics* **6:**189–194.

Lu, M., Johnson, R. R., and Iatrou, K. (1996). Trans-activation of a cell house-keeping gene promoter by the IE1 gene product of baculoviruses. *Virology* **218:**103–113.

Lu, M., Johnson, R. R., and Iatrou, K. (1997). A baculovirus (BmNPV) repeat element functions as a powerful constitutive enhancer in transfected insect cells. *J. Biol. Chem.* **272:**30724–30728.

Luo, K., McLachlin, J. R., Brown, M. R., and Adang, M. J. (1999). Expression of a glycosylphosphatidylinositol-linked *Manduca sexta* aminopeptidase N in insect cells. *Protein Expr. Purif.* **17:**113–122.

Makridou, P., Burnett, C., Landy, T., and Howard, K. (2003). Hygromycin B-selected cell lines from GAL4-regulated pUAST constructs. *Genesis* **36:**83–87.

Mandrioli, M., and Wimmer, E. A. (2003). Stable transformation of a *Mamestra brassicae* (lepidoptera) cell line with the lepidopteran-derived transposon piggyBac. *Insect Biochem. Mol. Biol.* **33:**1–5.

McCudden, C. R., Hains, M. D., Kimple, R. J., Siderovski, D. P., and Willard, F. S. (2005). G-protein signaling: Back to the future. *Cell. Mol. Life Sci.* **62:**551–577.

McLachlin, J. R., and Miller, L. K. (1997). Stable transformation of insect cells to coexpress a rapidly selectable marker gene and an inhibitor of apoptosis. *In Vitro Cell. Dev. Biol. Anim.* **33:**575–579.

Millar, N. S., Buckingham, S. D., and Sattelle, D. B. (1994). Stable expression of a functional homo-oligomeric *Drosophila* GABA receptor in a *Drosophila* cell line. *Proc. R. Soc. Lond. B* **258:**307–314.

Millar, N. S., Baylis, H. A., Reaper, C., Bunting, R., Mason, W. T., and Sattelle, D. B. (1995). Functional expression of a cloned *Drosophila* muscarinic acetylcholine receptor in a stable *Drosophila* cell line. *J. Exp. Biol.* **198:**1843–1850.

Milligan, G. (2003). High-content assays for ligand regulation of G-protein-coupled receptors. *Drug Discov. Today* **8:**579–585.

Milligan, G., Marshall, F., and Rees, S. (1996). G16 as a universal G protein adapter: Implications for agonist screening strategies. *Trends Pharmacol. Sci.* **17:**235–237.

Morais, V. A., and Costa, J. (2003). Stable expression of recombinant human α3/4 fucosyltransferase III in *Spodoptera frugiperda* Sf9 cells. *J. Biotechnol.* **106:**69–75.

Moto, K., Kojima, H., Kurihara, M., Iwami, M., and Matsumoto, S. (2003). Cell-specific expression of enhanced green fluorescence protein under the control of neuropeptide gene promoters in the brain of the silkworm, *Bombyx mori* using *Bombyx mori* nucleopolyhedrosis-derived vectors. *Insect Biochem. Molec. Biol.* **33:**7–12.

Mounier, N., and Prudhomme, J. C. (1986). Isolation of actin genes in *Bombyx mori*. The coding sequences of a cytoplasmic actin gene expressed in the silk gland is interrupted by a single intron in an unusual position. *Biochemie* **68:**1053–1061.

Münster, J., Ziegelmüller, P., Spillner, E., and Bredehorst, R. (2006). High level expression of monomeric and dimeric human α1,3-fucosyltransferase V. *J. Biotechnol.* **121:**448–457.

Neves, S. R., Ram, P. T., and Iyengar, R. (2002). G protein pathways. *Science* **296:**1636–1639.

Nilsen, S. L., and Castellino, F. J. (1999). Expression of human plasminogen in *Drosophila* Schneider S2 cells. *Protein Expr. Purif.* **16:**136–143.

Offermanns, S., and Simon, M. I. (1995). Gα_{15} and Gα_{16} couple a wide variety of receptors to phospholipase C. *J. Biol. Chem.* **270:**15175–15180.

Oh, M., Kim, S. Y., Oh, Y. S., Choi, D.-Y., Sin, H. J., Jung, I. M., and Park, W. J. (2003). Cell-based assay for β-secretase activity. *Anal. Biochem.* **323:**7–11.

Olson, V. A., Wetter, J. A., and Friesen, P. D. (2001). Oligomerization mediated by a helix-loop-helix-like domain of baculovirus IE1 is required for early promoter transactivation. *J. Virol.* **75:**6042–6051.

Olson, V. A., Wetter, J. A., and Friesen, P. D. (2002). Baculovirus transregulator IE1 requires a dimeric nuclear localization element for nuclear import and promoter activation. *J. Virol.* **76:**9505–9515.

Olson, V. A., Wetter, J. A., and Friesen, P. D. (2003). The highly conserved basic domain I of baculovirus IE1 is required for *hr* enhancer DNA binding and *hr*-dependent transactivation. *J. Virol.* **77:**5668–5677.

Perret, B. G., Wagner, R., Lecat, S., Brillet, K., Rabut, G., Bucher, B., and Pattus, F. (2003). Expression of EGFP-amino-tagged human mu opioid receptor in *Drosophila* Schneider 2 cells: A potential expression system for large-scale production of G-protein coupled receptors. *Protein Expr. Purif.* **31:**123–132.

Petzelt, C., Joswig, G., Stammer, H., and Werner, D. (2002). Cytotoxic cyplasin of the sea hare, *Aplysia punctata*, cDNA cloning, and expression of bioactive recombinants in insect cells. *Neoplasia* **4:**49–59.

Pfeifer, T. A., Hegedus, D. D., Grigliatti, T. A., and Theilmann, D. A. (1997). Baculovirus immediate-early promoter-mediated expression of the Zeocin resistance gene for use as a dominant selectable marker in dipteran and lepidopteran insect cell lines. *Gene* **188:**183–190.

Pfeifer, T. A., Guarna, M. M., Kwan, E. M., Lesnicki, G., Theilmann, D. A., Grigliatti, T. A., and Kilburn, D. G. (2001). Expression analysis of a modified factor X in stably transformed insect cell lines. *Protein Expr. Purif.* **23:**233–241.

Poels, J., Franssens, V., Van Loy, T., Martinez, A., Suner, M.-M., Dunbar, S. J., De Loof, A., and Vanden Broeck, J. (2004). Isoforms of cyclic AMP response element binding proteins in *Drosophila* S2 cells. *Biochem. Biophys. Res. Commun.* **320:** 318–324.

Prosise, W. W., Yarosh-Tomaine, T., Lozewski, Z., Ingram, R. N., Zou, J., Liu, J.-J., Zhu, F., Taremi, S. S., Le, H. V., and Wang, W. (2004). Protease domain of human ADAM33 produced by *Drosophila* S2 cells. *Protein Expr. Purif.* **38:**292–301.

Pullen, S. S., and Friesen, P. D. (1995). Early transcription of the ie-1 translegulator gene of *Autographa californica* nuclear polyhedrosis virus is regulated by DNA sequences within its 5′ noncoding leader sequence. *J. Virol.* **69:**156–165.

Radford, J. C., Davies, S. A., and Dow, J. A. T. (2002). Systematic G-protein-coupled receptor analysis in *Drosophila melanogaster* identifies a leukokinin receptor with novel roles. *J. Biol. Chem.* **277:**38810–38817.

Radford, J. C., Terhzaz, S., Cabrero, P., Davies, S. A., and Dow, J. A. T. (2004). Functional characterization of the *Anopheles* leukokinins and their cognate G-protein coupled receptor. *J. Exp. Biol.* **207:**4573–4586.

Reavy, B., Ziegler, A., Diplexcito, J., Macintosh, S. M., Torrance, L., and Mayo, M. (2000). Expression of functional recombinant antibody molecules in insect cell expression system. *Protein Expr. Purif.* **18:**221–228.

Reinking, J., Lam, M. M. S., Pardee, K., Sampson, H. M., Liu, S., Yang, P., Williams, S., White, W., Lajoie, G., Edwards, A., and Krause, H. M. (2005). The *Drosophila* nuclear receptor E75 contains heme and is gas responsive. *Cell* **122:**195–207.

Rhiel, M., Mitchell-Logean, C. M., and Murhammer, D. W. (1997). Comparison of *Trichoplusia ni* BTI-Tn-5B1–4 (High Five TM) and *Spodoptera frugiperda* Sf-9 insect cell line metabolism in suspension cultures. *Biotechnol. Bioeng.* **55:**909–920.

Riddihough, G., and Pelham, H. R. B. (1987). An ecdysone response element in the *Drosophila* hsp27 promoter. *EMBO J.* **6:**3729–3734.

Rodems, S. M., and Friesen, P. D. (1993). The *hr5* transcriptional enhancer stimulates early expression from the *Autographa californica* nuclear polyhedrosis virus genome but is not required for virus replication. *J. Virol.* **67:**5776–5785.

Rodems, S. M., Pullen, S. S., and Friesen, P. D. (1997). DNA-dependant transregulation by IE1 of *Autographa californica* nuclear polyhedrosis virus: IE1 domains required for transactivation and DNA binding. *J. Virol.* **71:**9270–9277.

Roman, G., Endo, K., Zong, L., and Davis, R. L. (2001). P(Switch), a system for spatial and temporal control of gene expression in *Drosophila melanogaster. Proc. Natl. Acad. Sci. USA* **98:**12602–12607.

Royer, C., Bossin, H., Romane, C., and Couble, P. (2001). High amplification of a densovirus-derived vector in larval and adult tissues of *Drosophila*. *Insect Mol. Biol.* **10:**275–280.

Rudolf, R., Mongillo, M., Rizzuto, R., and Pozzan, T. (2003). Looking forward to seeing calcium. *Nat. Rev. Mol. Cell Biol.* **4:**579–586.

Schneider, I. (1972). Cell lines derived from late embryonic stages of *Drosophila melanogaster. J. Embryol. Exp. Morphol.* **27:**353–356.

Segal, D., Cherbas, L., and Cherbas, P. (1996). Genetic transformation of *Drosophila* cells in culture by *P* element-mediated transposition. *Somat. Cell Mol. Genet.* **22:**159–165.

Shin, H. S., and Cha, H. J. (2003). Statistical optimization for immobilized metal affinity purification of secreted human erythropoietin from *Drosophila* S2 cells. *Protein Expr. Purif.* **28:**331–339.

Shin, H. S., Lim, H. J., and Cha, H. J. (2003). Quantitative monitoring for secreted production of human interleukin-2 in stable insect *Drosophila* S2 cells using a green fluorescent protein fusion partner. *Biotechnol. Prog.* **19:**152–157.

Shotkoski, E. A., and Fallon, A. M. (1993). The mosquito dihydrofolate reductase gene functions as a dominant selectable marker in transfected cells. *Insect Biochem. Mol. Biol.* **23:**883–893.

Skerra, A., and Schmidt, T. G. (2000). Use of the Strep-Tag and streptavidin for detection and purification of recombinant proteins. *Meth. Enzymol.* **326:**271–304.

Slack, J. M., and Blissard, G. W. (2001). Measurement of membrane fusion activity from viral membrane fusion proteins based on a fusion-dependent promoter induction system in insect cells. *J. Gen. Virol.* **82:**2519–2529.

Smith, L. A., Amar, M., Harvey, R. J., Darlison, M. G., Earley, F. G., Beadle, D. J., King, L. A., and Bermudez, I. (1995). The production of a stably transformed insect cell line expressing an invertebrate GABAA receptor beta-subunit. *J. Recept. Signal. Transduct. Res.* **15:**33–41.

Sohi, S. S., Palli, S. R., Cook, B. J., and Retnakaran, A. (1995). Forest insect cell lines responsive to 20-hydroxyecdysone and two nonsteroidal ecdysone agonists, RH-5849 and RH-5992. *J. Insect Physiol.* **41:**457–464.

Southern, J. A., Young, D. F., Heaney, F., Baumgartner, W., and Randall, R. E. (1991). Identification of an epitope on the P and V proteins of Simian virus 5 that distinguishes between two isolates with different biological characteristics. *J. Gen. Virol.* **72:**1551–1557.

Stavroulakis, D. A., Kalogerakis, N., Behie, L. A., and Iatrou, K. (1991a). Growth characteristics of a *Bombyx mori* insect cell line in stationary and suspension cultures. *Can. J. Chem. Engin.* **69:**457–464.

Stavroulakis, D. A., Kalogerakis, N., Behie, L. A., and Iatrou, K. (1991b). Kinetic data for the Bm-5 insect line in repeated-batch suspension cultures. *Biotechnol. Bioeng.* **38:**116–126.

Suzuki, T., Kawasaki, H., Yu, R. T., Ueda, H., and Umesono, K. (2001). Segmentation gene product Fushi tarazu is an LXXL motif-dependent coactivator for orphan receptor FTZ-F1. *Proc. Natl. Acad. Sci. USA* **98:**12403–12408.

Swevers, L., Ito, K., and Iatrou, K. (2002). The BmE75 nuclear receptors function as dominant repressors of the nuclear receptor BmHR3A. *J. Biol. Chem.* **277:**41637–41644.

Swevers, L., Farrell, P. J., Kravariti, L., Xenou-Kokoletsi, M., Sdralia, N., Lioupis, A., Morou, E., Balatsos, N. A. A., Douris, V., Georgoussi, Z., Mazomenos, B., and Iatrou, K. (2003). Transformed insect cells as high throughput screening tools for the discovery of new bioactive compounds. *Comm. Agric. Appl. Biol. Sci.* **68/2(b):**333–341.

Swevers, L., Kravariti, L., Ciolfi, S., Xenou-Kokoletsi, M., Ragoussis, N., Smagghe, G., Nakagawa, Y., Mazomenos, B., and Iatrou, K. (2004). A cell-based high-throughput screening system for detecting ecdysteroid agonists and antagonists in plant extracts and libraries of synthetic compounds. *FASEB J.* **18:**134–136.

Swevers, L., Morou, E., Balatsos, N., Iatrou, K., and Georgoussi, Z. (2005). Functional expression of mammalian opioid receptors in insect cells and high-throughput screening platforms for receptor ligand mimetics. *Cell. Mol. Life Sci.* **62:**919–930.

Szerencsei, R. T., Tucker, J. E., Cooper, C. B., Winkfein, R. J., Farrell, P. J., Iatrou, K., and Schnetkamp, P. P. (2000). Minimal domain requirement for cation transport by the potassium-dependent Na/Ca-K exchanger. Comparison with an NCKX paralog from *Caenorhabditis elegans*. *J. Biol. Chem.* **275:**669–676.

Tang, S., Zhao, Q., Yi, Y., Zhang, Z., and Li, Y. (2005). Homologous region 3 from *Bombyx mori* nucleopolyhedrovirus enhancing the transcriptional activity of heat shock cognate 70–4 promoter from *Bombyx mori* and *Bombyx mandarina in vitro* and *in vivo*. *Biosci. Biotechnol. Biochem.* **69:**1014–1017.

Tessier, D. C., Thomas, D. Y., Khouri, H. E., Laliberte, F., and Vernet, T. (1991). Enhanced secretion from insect cells of a foreign protein fused to the honeybee melittin signal peptide. *Gene* **98:**177–183.

Theilmann, D. A., and Stewart, S. (1992a). Molecular analysis of the trans-activating IE-2 gene of *Orgyia pseudotsugata* multicapsid nuclear polyhedrosis virus. *Virology* **187:** 84–96.

Theilmann, D. A., and Stewart, S. (1992b). Tandemly repeated sequence at the 3′ end of the IE2 gene of the baculovirus *Orgyia pseudotsugata* multicapsid nuclear polyhedrosis virus is an enhancer element. *Virology* **187:**97–106.

Thummel, C. S., and Pirrotta, V. (1992). New pCaSpeR P-element vectors. *Drosophila Inf. Serv.* **71:**150.

Tomita, S., Kawai, Y., Woo, S. D., Kamimura, M., Iwabuichi, K., and Imanishi, A. S. (2001). Ecdysone-inducible foreign gene expression in stably transformed lepidopteran insect cells. *In Vitro Cell Dev. Biol. Anim.* **37:**564–571.

Torfs, H., Poels, J., Detheux, M., Dupriez, V., Van Loy, T., Vercammen, L., Vassart, G., Parmentier, M., and Vanden Broeck, J. (2002). Recombinant aequorin as a reporter for receptor-mediated changes of intracellular Ca^{2+}-levels in *Drosophila* S2 cells. *Invert. Neurosci.* **4:**119–124.

Tota, M. R., Xu, L., Sirotina, A., Strader, C. D., and Graziano, M. P. (1995). Interaction of [flourescein-Trp25] glucagon with human glucagon receptor expressed in *Drosophila* Schneider 2 cells. *J. Biol. Chem.* **270:**26466–26472.

Vanden Broeck, J. (1996). G-protein-coupled receptors in insect cells. *Int. Rev. Cytol.* **164:**189–268.

Vanden Broeck, J., Vulsteke, V., Huybrechts, R., and De Loof, A. (1995). Characterization of a cloned locust tyramine receptor cDNA by functional expression in permanently transformed *Drosophila* S2 cells. *J. Neurochem.* **64:**2387–2395.

Vaughn, J. L., Goodwin, R. H., Tompkins, G. J., and McCawley, P. (1977). The establishment of two cell lines from the insect *Spodoptera frugiperda* (Lepidoptera: Noctuidae). *In Vitro* **13:**213–217.

Viswanathan, P., Venkaiah, B., Kumar, M. S., Rasheedi, S., Vrati, S., Bashyam, M. D., and Hasnain, S. E. (2003). The homologous region sequence (hr1) of *Autographa californica* multinucleocapsid polyhedrosis virus can enhance transcription from non-baculoviral promoters in mammalian cells. *J. Biol. Chem.* **278:**52564–52571.

Vlak, J. M., de Gooijer, C. D., Tramper, J., and Miltenburger, H. G. (1996). Insect cell cultures: Fundamental and applied aspects. Kluwer Academic, Dordrecht.

Wallny, H. J., Sollami, G., and Karjalainen, K. (1995). Soluble mouse major histocompatibility complex class II molecules produced in *Drosophila* cells. *Eur. J. Immunol.* **25:**1262–1266.

Wang, W., Swevers, L., and Iatrou, K. (2000). Mariner (Mos1) transposase and genomic integration of foreign gene sequences in *Bombyx mori* cells. *Insect Mol. Biol.* **9:**145–155.

Wang, W.-C., Zinn, K., and Bjorkman, P. J. (1993). Expression and structural studies of fasciclin I, an insect cell adhesion molecule. *J. Biol. Chem.* **268:**1448–1455.

Wehmeyer, A., and Schulz, R. (1997). Overexpression of δ-opioid receptors in recombinant baculovirus-infected *Trichoplusia ni* "High 5" insect cells *J. Neurochem.* **68:**1361–1371.

Wendeler, M., Lemm, T., Weisgerber, J., Hoernschemeyer, J., Bartelsen, O., Schepers, U., and Sandhoff, K. (2003). Expression of recombinant human GM2-activator protein in insect cells: Purification and characterization by mass spectrometry. *Protein Expr. Purif.* **27:**259–266.

Wheelock, C. E., Nakagawa, Y., Harada, T., Oikawa, N., Akamatsu, M., Smagghe, G., Stefanou, D., Iatrou, K., and Swevers, L. (2006). High-throughput screening of ecdysone agonists using a reporter gene assay followed by 3-D QSAR analysis of the molting hormonal activity. *Bioorg. Med. Chem.* **14:**1143–1159.

White, D. M., Jensen, M. A., Shi, X., Qu, Z.-X., and Arnason, B. G. W. (2001). Design and expression of polymeric immunoglobulin fusion proteins: A strategy for targeting low-affinity Fcγ receptors. *Protein Expr. Purif.* **21:**446–455.

Williams, C. (2004). cAMP detection methods in HTS: Selecting the best from the rest. *Nat. Rev. Drug Discov.* **3:**125–135.

Wurm, F. M. (2004). Production of recombinant protein therapeutics in cultivated mammalian cells. *Nat. Biotechnol.* **22:**1393–1398.

Zhang, B., Egli, D., Georgiev, O., and Schaffner, W. (2001). The *Drosophila* homolog of mammalian zinc finger factor MTF-1 activates transcription in response to heavy metals. *Mol. Cell. Biol.* **21:**4505–4514.

Zhang, J., Kalogerakis, N., Behie, L. A., and Iatrou, K. (1992). Investigation of reduced serum and serum-free media for the cultivation of insect cells (Bm5) and the production of baculovirus (BmNPV). *Biotechn. Bioeng.* **40:**1165–1172.

Zhang, J., Kalogerakis, N., Behie, L. A., and Iatrou, K. (1994). Optimization of the physiochemical parameters for the culture of the *Bombyx mori* insect cells used in recombinant protein production. *J. Biotechnol.* **33:**249–258.

Zhao, Y.-G., and Eggleston, P. (1998). Stable transformation of an *Anopheles gambiae* cell line mediated by the *Hermes* mobile genetic element. *Insect Biochem. Mol. Biol.* **28:**213–219.

Zhao, Y.-G., and Eggleston, P. (1999). Comparative analysis of promoters for transient gene expression in cultured mosquito cells. *Insect Mol. Biol.* **8:**31–38.

SECTION II

APPLICATIONS TO HUMAN AND ANIMAL HEALTH

PROTEIN *N*-GLYCOSYLATION IN THE BACULOVIRUS–INSECT CELL EXPRESSION SYSTEM AND ENGINEERING OF INSECT CELLS TO PRODUCE "MAMMALIANIZED" RECOMBINANT GLYCOPROTEINS

Robert L. Harrison* and Donald L. Jarvis†

*Insect Biocontrol Laboratory, USDA Agricultural Research Service, Plant Sciences Institute, 10300 Baltimore Avenue, Beltsville, Maryland 20705
†Department of Molecular Biology, University of Wyoming, 1000 East University Avenue, Laramie, Wyoming 82071

Abstract

Baculovirus expression vectors are frequently used to express glycoproteins, a subclass of proteins that includes many products with therapeutic value. The insect cells that serve as hosts for baculovirus vector infection are capable of transferring oligosaccharide side chains (glycans) to the same sites in recombinant proteins as those that are used for native protein *N*-glycosylation in mammalian cells. However, while mammalian cells produce compositionally more complex *N*-glycans containing terminal sialic acids, insect cells mostly produce simpler *N*-glycans with terminal mannose residues. This structural difference between insect and mammalian *N*-glycans compromises the *in vivo* bioactivity of glycoproteins and can potentially induce allergenic reactions in humans. These features obviously compromise

0065-3527/06 $35.00
DOI: 10.1016/S0065-3527(06)68005-6

the biomedical value of recombinant glycoproteins produced in the baculovirus expression vector system. Thus, much effort has been expended to characterize the potential and limits of *N*-glycosylation in insect cell systems. Discoveries from this research have led to the engineering of insect *N*-glycosylation pathways for assembly of mammalian-style glycans on baculovirus-expressed glycoproteins. This chapter summarizes our knowledge of insect *N*-glycosylation pathways and describes efforts to engineer baculovirus vectors and insect cell lines to overcome the limits of insect cell glycosylation. In addition, we consider other possible strategies for improving glycosylation in insect cells.

I. Insect Protein *N*-Glycosylation and Its Importance

A. *Significance of* N-*Glycosylation for Functional Glycoprotein Production*

One of the most common eukaryotic posttranslational protein modifications is *N*-glycosylation, which involves the addition of oligosaccharides on asparagine side chains and their subsequent processing. Many proteins expressed in baculovirus–insect cell and other expression systems are glycoproteins, which include many proteins of medical and veterinary therapeutic and diagnostic significance, such as hormones, receptors, and antibodies. The addition and processing of *N*-linked oligosaccharides (referred to as *N*-glycans) play an essential role in the folding and quality control of most membrane-associated and secreted glycoproteins (Helenius and Aebi, 2001). The presence of *N*-glycans can influence and stabilize protein structure and protect against proteolysis (Imperiali and O'Connor, 1999; Lis and Sharon, 1993; Wyss and Wagner, 1996). *N*-Glycans also can play an important role in determining the specificity of protein–protein interactions involved in a wide variety of processes, including clearance of glycoproteins from mammalian circulatory systems, intracellular trafficking of enzymes, cell–cell interactions, signal transduction, and antigen recognition (Lis and Sharon, 1993; Opdenakker *et al.*, 1993; Rudd *et al.*, 2001; Varki, 1993). In addition, enzyme activity can be influenced by glycosylation. For example, the activities of plasminogen and tissue plasminogen activator are altered by glycosylation at specific sites, while the activity of RNase A is inversely proportional to the size of its *N*-glycans (Rudd *et al.*, 1995). Another example is the loss of membrane fusion activity in paramyxovirus fusion (F) proteins, which occurs when specific *N*-glycosylation sites are obliterated by mutagenesis of the asparagine codon (McGinnes *et al.*, 2001; Segawa *et al.*, 2000).

Because of the role that *N*-glycosylation plays in protein folding, the efficiency of *N*-glycosylation in insect cells potentially can affect the yield of secreted, baculovirus-expressed glycoprotein. In addition, the structural characteristics of insect *N*-glycans can affect the utility of recombinant glycoproteins produced with baculovirus expression vectors. Glycosylation is a very important factor controlling the *in vivo* behavior and activity of candidate therapeutics. Thus, the value of baculovirus vector-derived glycoproteins for therapeutic use in humans depends on the ability of the host-insect cells to efficiently and accurately synthesize mammalian-like *N*-glycans. Hence, much effort has been dedicated to elucidating the characteristics of insect *N*-glycosylation, and to engineering both cell lines and baculoviral vectors to modify *N*-glycosylation pathways in the virus–host system. This chapter describes what we know about *N*-glycosylation of both endogenous proteins and recombinant, baculovirus-expressed proteins produced in insect cells. We also summarize the results of efforts to engineer the insect *N*-glycosylation pathway to produce mammalian-type glycans and discuss possible approaches that might be used to further improve protein glycosylation in insect cells. Readers are referred to other reviews focusing on insect *N*-glycosylation and the engineering of insect protein *N*-glycosylation pathways (Jarvis, 2003; Marchal *et al.*, 2001; Tomiya *et al.*, 2003a, 2004).

B. The Insect N-*Glycosylation Pathway*

The assembly and processing of *N*-glycans in mammalian cells have been extensively characterized, but we know much less about these same pathways in insects. Much of what is known about insect protein *N*-glycosylation was derived from studies conducted with material from insect cell lines. The sequence of the *Drosophila melanogaster* genome (Adams *et al.*, 2000) has further expanded our knowledge of insect *N*-glycosylation by facilitating the identification and characterization of additional mammalian *N*-glycosylation enzyme homologues. The recently reported genome sequences of other insects, such as the mosquito *Anopheles gambiae* (Holt *et al.*, 2002) and the silkworm *Bombyx mori* (Xia *et al.*, 2004), promise to lead to an even more complete and detailed portrayal of the *N*-glycan biosynthetic capacity of insects.

The initial events of protein *N*-glycosylation are conserved among vertebrates, invertebrates, plants, and fungi (Kornfeld and Kornfeld, 1985). The first step in the pathway is a cotranslational process that takes place in the endoplasmic reticulum (ER). A branched oligosaccharide is transferred *en bloc* from a lipid (dolichyl pyrophosphate)-linked precursor to the side chain amide of an Asn residue in the consensus sequence Asn-Xaa-Ser/Thr (where Xaa is any amino acid other than proline) (Burda and Aebi, 1999). The shorthand formula for this oligosaccharide is $Glc_3Man_9GlcNAc_2$, where Glc is glucose, Man is mannose, GlcNAc is *N*-acetylglucosamine, and the subscripts refer to the numbers of each residue in the oligosaccharide (Fig. 1). Lipid-linked oligosaccharides with the same structure have been identified in cell lines derived from the dipterans *Aedes albopictus* (Hsieh and Robbins, 1984) and *D. melanogaster* (Parker *et al.*, 1991; Sagami and Lennarz, 1987) and from the lepidopteran *Spodoptera frugiperda* (Marchal *et al.*, 1999).

After being transferred to asparagine, the core oligosaccharide is trimmed by a series of glycosidases in the ER and Golgi apparatus (Trombetta, 2003) (Fig. 1). The first trimming reaction removes the terminal glucose and is catalyzed by α-glucosidase I. α-Glucosidase II then removes the next two terminal glucose residues, yielding $Man_9GlcNAc_2$. Although genes demonstrated to encode processing ER α-glucosidases have not been identified in any insect system, studies using a glucosidase inhibitor indicate that they are present in lepidopteran cells (Davis *et al.*, 1993; Jarvis and Summers, 1989; Marchal *et al.*, 1999).

The removal of terminal glucose is followed by the removal of four mannose residues by class I α-mannosidases to produce $Man_5GlcNAc_2$ (Fig. 1). If perfectly analogous to higher eukaryotes, this process would begin in the ER and continue through the action of several additional mannosidases in the Golgi apparatus. A class I α-mannosidase gene, *mas-1*, has been identified in *D. melanogaster* and characterized (Kerscher *et al.*, 1995). The *mas-1* gene does not seem to be strictly required for *N*-glycan processing in the fruit fly, suggesting that a redundant function or alternative pathway for mannose trimming exists in *Drosophila* (Roberts *et al.*, 1998). A class I α-mannosidase enzyme and gene have been isolated from *S. frugiperda* cells and characterized. With the exception of a difference in the precise sequence of mannose trimming, this Golgi enzyme appears to be orthologous to one or more of the class I Golgi α-mannosidases of higher eukaryotes (Kawar and Jarvis, 2001; Kawar *et al.*, 1997, 2000; Ren *et al.*, 1995).

Following trimming of the α-1,2-linked mannose residues, *N*-acetylglucosamine is added to the lower (α-1,3) branch of the remaining structure by *N*-acetylglucosaminyltransferase I (GlcNAcT-I) to produce $GlcNAcMan_5GlcNAc_2$ (Fig. 1). A *D. melanogaster* gene encoding an active GlcNAcT-I enzyme has been identified and characterized (Sarkar and Schachter, 2001). GlcNAcT-I activity has also been

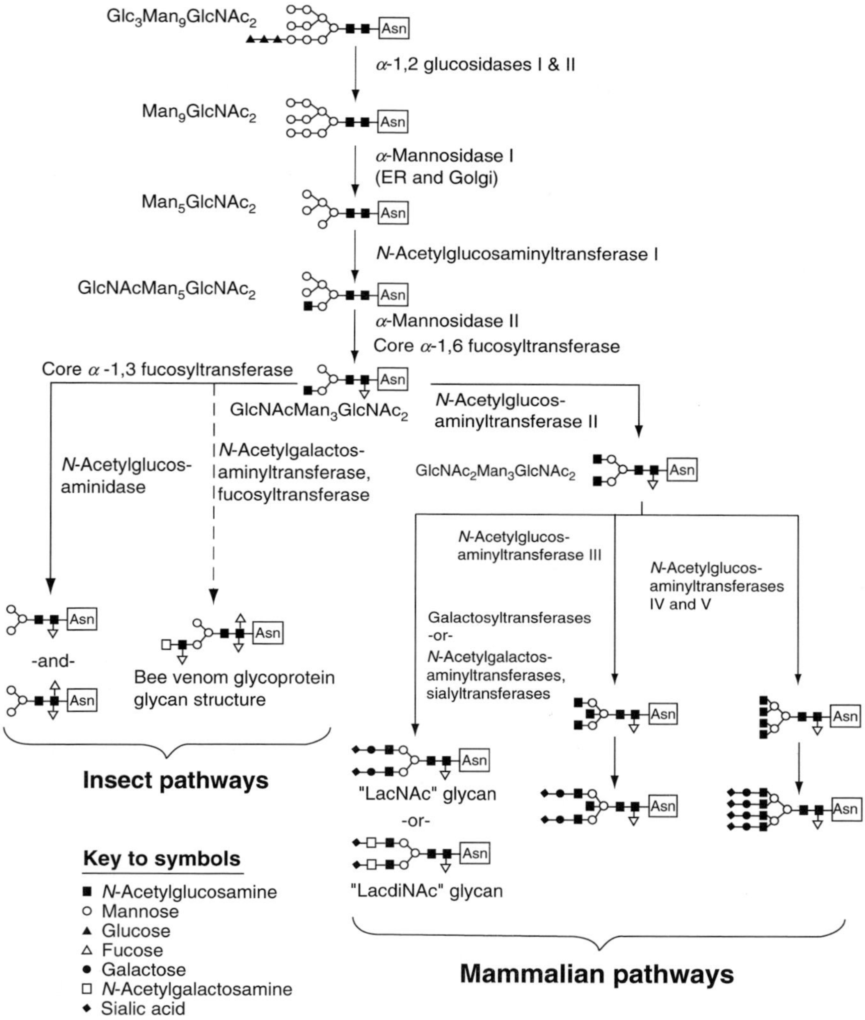

FIG 1. Protein *N*-glycosylation pathways in insects and mammals, showing structures and symbolic descriptions of the *N*-glycan–processing intermediates and indicating the enzymes involved at each step. Each monosaccharide is represented by its standard symbol, as defined in "Essentials of Glycobiology" (Varki *et al.*, 1999). The major product of the insect *N*-glycan–processing pathway, classified as a paucimannose structure, is presented alongside the most complex insect *N*-glycan confirmed by mass spectrometry. Mammalian disialylated, biantennary *N*-glycans are presented alongside biantennary structures bisected with *N*-acetylglucosamine and sialylated tri- and tetraantennary structures.

detected in cell lines derived from the lepidopterans *S. frugiperda, B. mori, Mamestra brassicae*, and *Estigmene acrea* (Altmann *et al.*, 1993; Velardo *et al.*, 1993; Wagner *et al.*, 1996a).

The class II Golgi α-1,2-mannosidase subsequently trims the α-1,3- and α-1,6-linked mannose residues from the upper (α-1,6) branch of the biantennary glycan intermediate, thereby converting $GlcNAcMan_5GlcNAc_2$ to $GlcNAcMan_3GlcNAc_2$ (Fig. 1). A gene encoding this enzyme has been identified in the genome of *D. melanogaster* (Foster *et al.*, 1995; Rabouille *et al.*, 1999) and an enzyme with this same hydrolytic activity and substrate specificity has been identified in *S. frugiperda* Sf9 lysates and cell membranes (Ren *et al.*, 1997; Wagner *et al.*, 1996a). It has also been shown that *S. frugiperda* encodes a separate class II α-mannosidase, which cannot remove mannose residues from $GlcNAcMan_5GlcNAc_2$ but removes terminal α-1,3- and α-1,6-linked mannose residues from *N*-glycans lacking a terminal *N*-acetylglucosamine residue instead (Jarvis *et al.*, 1997; Kawar *et al.*, 2001). To date, this is the only clear example of a cloned gene encoding the so-called α-mannosidase III activity that was described in knockout mice lacking the classic Golgi class II α-1,2-mannosidase gene (Chui *et al.*, 1997).

At this stage of the *N*-glycan processing pathway, mammalian cells can produce a wide variety of complex, branched oligosaccharide side chains by elongating both branches of $GlcNAcMan_3GlcNAc_2$. The first step in this process involves the transfer of an *N*-acetylglucosamine to the upper (α-1,6) branch by *N*-acetylglucosaminyltransferase II (GlcNAcT-II) to make $GlcNAc_2Man_3GlcNAc_2$ (Fig. 1). A low level of GlcNAcT-II activity has been detected in cell lines from three different lepidopteran species (Altmann *et al.*, 1993), and a *Drosophila* gene (*Mgat2)* with sequence similarity to GlcNAcT-II has been identified, although the activity of the encoded gene product was not characterized (Tsitilou and Grammenoudi, 2003).

The branches on mammalian *N*-glycans are often elongated by the transfer of a single galactose to the terminal GlcNAc residues by β-1,4-galactosyltransferase, producing a lactosamine (LacNAc) disaccharide unit (Fig. 1). Alternatively, *N*-acetylgalactosamine may be added to GlcNAc, producing a LacdiNAc disaccharide unit (Fig. 1). β-1,4-Galactosyltransferase activity has been detected at low levels in *Trichoplusia ni* Tn-5B1-4 (High Five^{TM}) cells and also in *Danaus plexippus* DpN1 cells but not in *S. frugiperda* Sf9 or *Pseudoletia unipuncta* A7S cells (Abdul-Rahman *et al.*, 2002; Palomares *et al.*, 2003; van Die *et al.*, 1996). Genes with significant sequence similarity to mammalian β-1,4-galactosyltransferases have been identified in *D. melanogaster* and in the lepidopteran *T. ni*, but in both cases the encoded enzymes preferentially transferred *N*-acetylgalactosamine (GalNAc) to GlcNAc (Haines and Irvine, 2005; Vadaie and Jarvis, 2004). This result is consistent with an earlier observation of *N*-acetylgalactosaminyltransferase activity in cell lines derived from *T. ni, S. frugiperda*, and *M. brassicae* (van Die *et al.*, 1996). An *N*-acetylgalactosaminyltransferase activity would produce the LacdiNAc disaccharide unit, which is found on honeybee venom phospholipase and hyaluronidase (Fig. 1).

Many mature mammalian *N*-glycans terminate in sialic acid residues, which are added to galactose residues by α-2,3- and α-2,6-sialyltransferases (Fig. 1). Sialyltransferase activity has not been detected by enzymatic assay in any lepidopteran cell line examined (Hollister and Jarvis, 2001; Hooker *et al.*, 1999; Joshi *et al.*, 2001), but an α-2,6-sialyltransferase homologue was identified in the *D. melanogaster* genome (Koles *et al.*, 2004). The encoded enzyme preferentially sialylated terminal LacdiNAc units in both free oligosaccharide and *N*-glycoprotein substrates (Koles *et al.*, 2004).

The studies cited above, with insects from two different orders, suggest that insects have the potential to form complex *N*-glycans similar in structure to those found on mammalian glycoproteins. However, analyses of native insect *N*-glycans from dipterans and lepidopterans have revealed that most *N*-glycans found in these organisms have simple oligomannose ($Man_{5–9}GlcNAc_2$) or paucimannose ($Man_{1–3}$ $GlcNAc_2$) structures containing no galactose or sialic acids (Fig. 1) (Butters and Hughes, 1981; Fabini *et al.*, 2001; Kim *et al.*, 2003; Kubelka *et al.*, 1994; Park *et al.*, 1999; Williams *et al.*, 1991). The most complex endogenous insect *N*-glycan unambiguously identified by direct structural analysis by mass spectrometry has a hybrid structure in which the upper branch terminates with mannose and the lower branch terminates with LacdiNAc (Fig. 1). This structure is a minor *N*-glycan, which occupies a subpopulation of the glycosylation sites in honeybee venom phospholipase A2 (Kubelka *et al.*, 1993) and hyaluronidase (Kubelka *et al.*, 1995). In addition, other, more unusual *N*-glycan structures have been described in insects. The *N*-glycans on apolipophorin isolated from the locust *Locusta migratoria* were found to terminate in GlcNAc on both branches and were modified on the lower-branch terminal GlcNAc and the upper-branch subterminal mannose residues with 2-aminoethylphosphonate (Hard *et al.*, 1993). A minority of the *N*-glycans on *Manduca sexta* aminopeptidase N consist of highly fucosylated structures with fucose attached to both GlcNAc residues in the chitobiose core and also to terminal GlcNAc residues in the antennae (Stephens *et al.*, 2004).

If insects have the potential to form complex *N*-glycans, why do oligomannose and paucimannose structures predominate? One possible reason is that the expression of key glycosyltransferases is restricted to specific tissues and developmental stages in insects. The *D. melanogaster* *N*-acetylgalactosaminyl- and *N*-acetylglucosaminyltransferase genes were found to be transcribed ubiquitously throughout development, but in embryos and larvae GlcNAcT-II transcripts were restricted primarily to neural tissue and eye imaginal discs (Haines and Irvine, 2005; Tsitilou and Grammenoudi, 2003). Expression of the *D. melanogaster* α-2,6-sialyltransferase also was restricted to a subset of central nervous system cells in embryos (Koles *et al.*, 2004). Thus, it is possible that many of the genes encoding enzymes involved in producing complex *N*-glycans are simply inactive in many insect tissues or in many of the insect cell lines that have been used to study insect protein *N*-glycosylation.

In addition, several studies have identified intracellular and extracellular sialidase, β-galactosidase, β-*N*-acetylgalactosaminidase, and β-*N*-acetylglucosaminidase activities (Joosten and Shuler, 2003a; Licari *et al.*, 1993; Sommer and Spindler, 1991; van Die *et al.*, 1996; Wagner *et al.*, 1996a), which could either prevent the formation of complex *N*-glycans or degrade them to paucimannose structures prior to analysis.

There is also a membrane-bound β-*N*-acetylglucosaminidase activity in *D. melanogaster* and lepidopteran insect cell lines (Altmann *et al.*, 1995). This appears to be a branch specific, processing enzyme activity, which removes the terminal *N*-acetylglucosamine residue from the lower (α-1,3) branch of *N*-glycans in the final processing step leading to the production of paucimannose end products (Fig. 1). The presence of terminal *N*-acetylglucosamine on influenza virus *N*-glycans produced by *E. acrea* EaA cells, but not *S. frugiperda* Sf9 cells, was attributed to the absence of this putative processing *N*-acetylglucosaminidase activity in the former (Wagner *et al.*, 1996a). Removal of the lower-branch *N*-acetylglucosamine not only blocks elongation of the lower branch but might also prevent elongation of the upper branch, as mammalian GlcNAcT-II requires terminal *N*-acetylglucosamine on the lower branch for activity (Bendiak and Schachter, 1987).

Finally, the failure to detect terminal sialic acids in endogenous insect *N*-glycans could also reflect the absence of a conventional donor substrate for sialylation, which is the nucleotide sugar, CMP-sialic acid. In vertebrates, the pathway for the production of CMP-sialic acid (Angata and Varki, 2002) begins with the conversion of UDP-GlcNAc to *N*-acetylmannosamine, or ManNAc, which is then phosphorylated to produce ManNAc-6-phosphate (Fig. 2). These reactions are catalyzed

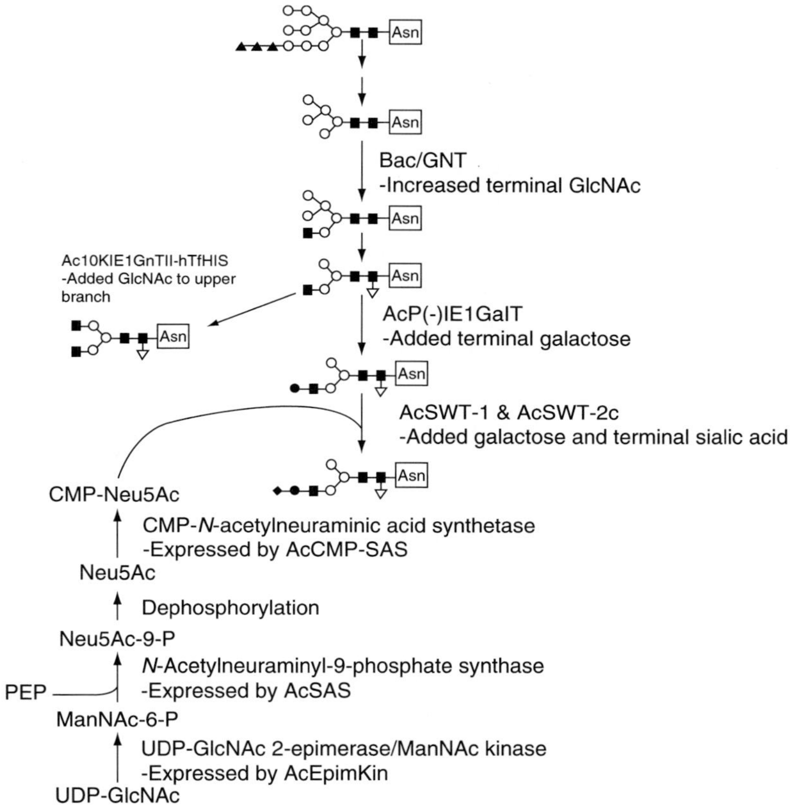

FIG 2. Structures of the *N*-glycans produced in insect cells infected with recombinant baculoviruses expressing mammalian glycosyltransferases. The pathway for synthesis of the sialylation donor substrate CMP-sialic acid (CMP-Neu5Ac) also is shown, and the names of the recombinant baculoviruses encoding mammalian enzymes that participate in this pathway are indicated. The standard monosaccharide symbols used in this figure are defined in the key shown in Fig. 1.

by the bifunctional enzyme, UDP-GlcNAc 2-epimerase/ManNAc kinase. ManNAc-6-phosphate is then condensed with phospoenol-pyruvate by *N*-acetylneuraminyl-9-phosphate synthase to produce *N*-acetylneuraminyl-9-phosphate, which is then dephosphorylated to produce *N*-acetylneuraminic acid (Neu5Ac), the form of sialic acid most commonly found in *N*-glycans. Finally, Neu5Ac is conjugated to cytosine monophosphate by CMP-*N*-acetylneuraminic acid synthetase to produce CMP-*N*-acetylneuraminic acid (CMP-Neu5Ac). Although free sialic acid has been unambiguously identified in *D. melanogaster,* in the cicada *Philaenus spumarius* (Malykh *et al.*, 2000; Roth *et al.*, 1992), and at low levels in Sf9 cells (Lawrence *et al.*, 2000), the levels of CMP-sialic acid were found to be negligible in lepidopteran cell lines (Hooker *et al.*, 1999; Tomiya *et al.*, 2001). A gene encoding a sialic acid phosphate synthase homologue has been identified in *D. melanogaster* (Kim *et al.*, 2002). This gene was transcribed at every developmental stage, and the encoded product was able to synthesize sialic acid phosphate from *N*-acetylmannosamine-6-phosphate and phosphoenolpyruvate. These same authors also performed a homology search and found a CMP-sialic acid synthetase homologue in the fly genome, suggesting that dipterans encode at least some of the enzymes involved in CMP-sialic acid biosynthesis.

C. N-*Glycosylation of Proteins Produced with Baculovirus Expression Vectors*

The preponderance of studies on the glycans of recombinant *N*-glycoproteins produced in lepidopteran cell lines with baculovirus expression vectors report mostly oligomannose or paucimannose structures (e.g., Ding *et al.*, 2003; Grabenhorst *et al.*, 1993; Hollister *et al.*, 2002; Kuroda *et al.*, 1990; Lopez *et al.*, 1997; Manneberg *et al.*, 1994; Takahashi *et al.*, 1999; Wendeler *et al.*, 2003), with a minority of *N*-glycans ($<10\%$) containing terminal GlcNAc on one branch in some studies (Ailor *et al.*, 2000; Choi *et al.*, 2003). This trend also has been observed with recombinant glycoproteins produced in baculovirus-infected lepidopteran larvae (Hogeland and Deinzer, 1994; Kulakosky *et al.*, 1998). One study reported that a majority of the *N*-glycans assembled in *E. acrea* EaA cells contained terminal GlcNAc (Wagner *et al.*, 1996a). Yet other studies have reported the assembly of complex *N*-glycans containing terminal galactose or sialic acid residues on recombinant glycoproteins expressed in a variety of lepidopteran cell lines (Davidson and Castellino, 1991a,b, 1993; Davidson *et al.*, 1990; Hsu *et al.*, 1997; Joosten and Shuler, 2003a,b; Joosten *et al.*, 2003; Joshi *et al.*, 2000, 2001; Ogonah *et al.*, 1996; Pajot-Augy *et al.*, 1999; Palomares *et al.*, 2003; Rudd *et al.*, 2000; Watanabe *et al.*, 2001). In two of these studies, the presence of galactose was confirmed by direct compositional analysis with mass spectrometry (Ogonah *et al.*, 1996; Rudd *et al.*, 2000). In both cases, the authors used matrix-assisted laser desorption/ionization-time of flight (MALDI-TOF) mass spectrometry to verify that a minority of the *N*-glycans on glycoproteins expressed in *E. acrea* Ea4 (Ogonah *et al.*, 1996) and *T. ni*

Tn-5B1–4 cells (Rudd *et al.*, 2000) contained terminal galactose on a single branch. To date, terminal sialylation of recombinant glycoproteins synthesized in insect cells has not been confirmed by mass spectrometry (Marchal *et al.*, 2001).

Although it is generally the case that recombinant proteins synthesized in insect cells bear *N*-glycans at the sites where they occur in the natural products, one might expect *N*-glycan assembly and processing to be less efficient in cells infected with a baculovirus expression vector. The steady state levels of transcripts derived from host cell genes decrease as viral infection progresses (Nobiron *et al.*, 2003; Ooi and Miller, 1988) and this general effect might extend to the expression of host genes encoding glycosylation pathway functions. In addition, the *N*-glycosylation capabilities of the host might simply be overwhelmed by the massive amounts of recombinant protein produced by a baculovirus expression vector, especially one that uses the highly active polyhedrin promoter to drive foreign gene expression. In their assessment of these issues, Kretzschmar *et al.* (1994) reported no discernible difference in the structures of the *N*-glycans of total glycoproteins isolated from mock- and baculovirus-infected *S. frugiperda* cells. Similarly, Joosten and Shuler (2003a) reported no significant differences among the profiles of secreted alkaline phosphatase *N*-glycans harvested at 48, 72, and 96 h postinfection. However, in a study of baculovirus-mediated expression of HIV envelope proteins, Murphy *et al.* (1990) found that the proportion of nonglycosylated protein increased as infection progressed from 24 to 48 h postinfection. Baculovirus-mediated expression of human thyrotropin receptor under the control of the *p6.9* promoter, which is activated earlier than the polyhedrin promoter, produced a more extensively glycosylated form of the receptor than conventional baculovirus-mediated expression under the control of the polyhedrin promoter (Chazenbalk and Rapoport, 1995). A similar effect on the extent of lutropin receptor glycosylation was observed with baculoviruses that expressed the porcine lutropin receptor under the control of either the polyhedrin or *p10* promoters, perhaps reflecting the fact that the latter is activated a few hours earlier and mediates a weaker level of expression than the polyhedrin promoter (Pajot-Augy *et al.*, 1999; Roelvink *et al.*, 1992). Thus, the results of these studies indicate that, for at least some recombinant products, the quality and extent of protein *N*-glycosylation can be affected by baculoviral infection, with glycosylation efficiency decreasing as the infection proceeds. This conclusion is consistent with the original observation that baculovirus infection has an adverse effect on secretory pathway function (Jarvis and Summers, 1989). It also is consistent with the general idea that baculovirus infection represses host gene expression (Nobiron *et al.*, 2003; Ooi and Miller, 1988) and with the pathway-specific observation that host *N*-acetylgalactosaminyltransferase activity decreases with time of baculovirus infection (van Die *et al.*, 1996). On the other hand, this conclusion is inconsistent with results indicating that *N*-glycan–processing activities, including both GlcNAc-TI (Velardo *et al.*, 1993) and α-mannosidase (Davidson *et al.*, 1991) activities, are induced at later times after baculovirus infection.

The general inability of established lepidopteran cell lines to produce terminally sialylated *N*-glycans is problematic for the production of efficacious therapeutic glycoproteins. Terminal sialic acids protect glycoproteins from being cleared from the bloodstream by various mammalian lectins (Ashwell and Harford, 1982). Thus, it was not surprising to find that glycoproteins produced with a baculovirus expression vector were rapidly cleared from the bloodstream after being injected into rodents (Kurschat *et al.*, 1995; Sareneva *et al.*, 1993). A comparison of the clearance rates of the glycopeptide hormone thyrotropin produced in either baculovirus-infected insect cells or DNA-transfected Chinese hamster ovary (CHO) cells revealed that only 0.1% of the former product, as compared to >10% of the latter, remained in the bloodstream 2 h after injection (Grossmann *et al.*, 1997). Because prolonged residence time in the plasma is necessary for therapeutic efficacy, the rapid clearance of nonsialylated products is a significant impediment to the commercialization of baculovirus vector-produced therapeutic glycoproteins.

Another impediment is the presence of a fucose residue linked to the asparagine-bound GlcNAc residue in the *N*-glycan chitobiose core produced by some lepidopteran insect cells. Two main types of core fucosylation actually occur in insects: (1) monofucosylation with an α-1,6 linkage, which is also found in mammalian *N*-glycans, and (2) difucosylation with α-1,6 and α-1,3 linkages, which is not found in mammalian *N*-glycans (Staudacher *et al.*, 1999) (Fig. 1). Core difucosylation requires two separate enzymes that transfer first the α-1,6- and then the α-1,3-linked fucose residues to the chitobiose core (Staudacher and Marz, 1998). The presence of core α-1,3-linked fucose is a potential problem for the use of baculovirus vector-generated glycoproteins for therapeutic purposes because it represents an allergenic carbohydrate epitope in mammalian species, one that plays a significant role in immunogenic responses to plant and invertebrate glycoproteins (Prenner *et al.*, 1992; Tretter *et al.*, 1993; Wilson *et al.*, 1998). However, it is important to recognize that the extent to which core difucosylation occurs varies among different insect cell lines and, in some cases, it is actually a relatively infrequent modification of endogenous insect *N*-glycans. The molar percentage of native insect *N*-glycans with core α-1,3-fucose ranges from 2.5% for a *B. mori* cell line (BmN) to 30% for an *M. brassicae* cell line (Mb-0503) (Kubelka *et al.*, 1994). The *S. frugiperda* Sf9 cell line has no detectable α-1,3-fucosyltransferase activity (Staudacher *et al.*, 1992), but a subsequent study revealed that these cells contain very small subpopulations of *N*-glycans with either difucosylated or α-1,3-monofucosylated cores (Kubelka *et al.*, 1994). The presence of core α-1,3 fucosylation, as assessed by staining with anti-horseradish peroxidase polyclonal antibodies, was restricted to neurons in grasshopper and *D. melanogaster* embryos (Fabini *et al.*, 2001; Snow *et al.*, 1987). A gene encoding an enzymatically active core α-1,3-fucosyltransferase has been identified in the *D. melanogaster* genome (Fabini *et al.*, 2001). However, in adult *D. melanogaster*, *N*-glycans with a difucosylated core GlcNAc represent only 0.8% of total, as assessed by MALDI-TOF (Fabini *et al.*, 2001). No core difucosylated *N*-glycans were detected in hemolymph glycoproteins isolated from larvae of *Antheraea pernyi*, the Chinese oak silkworm (Kim *et al.*, 2003). Core difucosylation of recombinant glycoproteins is also highly variable. Approximately 12–23% of the *N*-glycans on baculovirus-expressed recombinant glycoproteins isolated from *T. ni* Tn-5B1-4 cells carried core α-1,3-fucose (Hsu *et al.*, 1997; Takahashi *et al.*, 1999). On the other hand, there was no detectable core α-1,3 fucosylation of latent TGF-β–binding protein-1 expressed in Sf9 cells (Rudd *et al.*, 2000), human interferon gamma expressed in *E. acrea* Ea4 cells (Ogonah *et al.*, 1996), or human transferrin expressed in *Lymantria dispar* Ld652Y cells (Choi *et al.*, 2003). Thus, the extent to which core difucosylation will be a problem associated with the use of baculovirus–insect cell expression systems will depend strongly on the nature of the host and the nature of the recombinant glycoprotein being produced.

II. Modification of Lepidopteran *N*-Glycosylation Pathways for Improved Processing and Function of Glycoproteins Produced with Baculovirus Expression Vectors

A. Baculovirus Expression of Mammalian N-*Glycosylation Pathway Enzymes*

Much effort has been directed toward modifying protein *N*-glycosylation pathways in lepidopteran cells in order to obtain baculovirus vector-expressed glycoproteins bearing mammalian-style, complex, terminally sialylated *N*-glycans. This effort has involved introducing genes encoding functions that are either missing or present at suboptimal levels in lepidopteran host cells but required for complex *N*-glycan assembly.

One approach to providing these missing processing activities has entailed inserting the desired *N*-glycan–processing genes into a baculovirus vector so that they can be expressed during infection (Fig. 2). Coexpression of human GlcNAcT-I and fowl plaque virus hemagglutinin in *S. frugiperda* cells using two separate baculoviruses carrying the genes for these proteins resulted in a fourfold increase in the proportion of hemagglutinin *N*-glycans bearing terminal GlcNAc (Wagner *et al.*, 1996b). Expression of a bovine β-1,4-galactosyltransferase gene under the control of *ie1* promoter and *hr5* enhancer sequences with the recombinant baculovirus AcP(−)IE1GalT resulted in the addition of galactose to the *N*-glycans of the viral envelope fusion protein, GP64 (Jarvis and Finn, 1996). This virus also added terminal galactose residues to the *N*-glycans of human transferrin when the latter was coexpressed with a separate baculovirus (Ailor *et al.*, 2000). In this case, 12.6% of the *N*-glycans were terminally galactosylated on the lower (α-1,3) branch. The recombinant baculoviruses AcSWT-1 and AcSWT-2c were designed to express both bovine β-1,4-galactosyltransferase and rat α-2,6-sialyltransferase genes from a bidirectional *ie-1* promoter/*hr5* enhancer element, with the processing genes inserted into different locations in the two recombinant viral genomes (Jarvis *et al.*, 2001). Infections with either of these viruses resulted in the addition of both galactose and sialic acid to the *N*-glycans of the GP64 viral envelope protein. Similar results were obtained when human α1-antitrypsin was expressed in Ea4 cells infected with a single recombinant baculovirus (BacATgng26) encoding this product together with human GlcNAcT-II, β-1,4-galactosyltransferase, and α-2,6-sialyltransferase under the control of *p10* and polyhedrin promoters (Chang *et al.*, 2003).

Other recombinant baculoviruses have been designed and constructed to express enzymes of the CMP-sialic acid biosynthetic pathway (Fig. 2). Infection of Sf9 cells with the recombinant virus AcSAS, which expresses human *N*-acetylneuraminyl-9-phosphate synthase under the control of the polyhedrin promoter, resulted in the production of high levels of Neu5Ac when the cells were incubated in serum-free medium containing the sialic acid precursor, ManNAc (Lawrence *et al.*, 2000). This result indicated that endogenous Sf9 cell enzymes could phosphorylate ManNAc and dephosphorylate *N*-acetylneuraminyl-9-phosphate. Coinfection of Sf9 cells with AcSAS and AcCMP-SAS, which encodes the human CMP-*N*-acetylneuraminic acid synthetase under control of the polyhedrin promoter, resulted in the production of CMP-Neu5Ac in Sf9 cells fed with serum-free medium supplemented with ManNAc (Lawrence *et al.*, 2001). The recombinant virus AcEpimKin, which encodes the UDP-GlcNAc 2-epimerase/ManNAc kinase gene under control of the *polh* promoter (Effertz *et al.*, 1999), induced Neu5Ac production in Sf9 cells coinfected with AcSAS and cultured without ManNAc supplementation (Viswanathan *et al.*, 2003). Infecting Sf9 cells simultaneously with AcEpimKin, AcSAS, and AcCMP-SAS resulted in the production of CMP-Neu5Ac without ManNAc supplementation, indicating that Sf9 cells can produce CMP-Neu5Ac from endogenous UDP-GlcNAc when mammalian sialic acid synthesis pathway functions are provided in *trans* (Viswanathan *et al.*, 2005).

B. Stable Transformation of Lepidopteran Host Cells with Genes for Mammalian Pathway Enzymes

Another approach that has been used to provide host-insect cells with missing or suboptimal *N*-glycan–processing activities involves stable transformation of cells with processing genes designed to be expressed under the control of a constitutively active promoter. This "metabolic engineering" approach yields new cell lines that are pre-equipped with a mammalian-like glycosylation pathway prior to the time of baculovirus vector infection. It also permits one to use standard transfer vector/recombinant baculovirus production systems and/or preexisting recombinant baculovirus vectors for the expression of glycoproteins with "humanized" *N*-glycans.

This approach was first used to create an Sf9 cell clone that was stably transformed with a bovine β-1,4-galactosyltransferase cDNA positioned under the control of an *ie1* promoter-*hr5* enhancer element (Hollister and Jarvis, 2001). This cell line, called Sfβ4GalT (Fig. 3), had high levels of β-1,4-galactosyltransferase activity. Hence, when these cells were infected with wild-type *Autographa californica* multiple nucleopolyhedrovirus (AcMNPV) or a conventional baculovirus vector expressing human tissue plasminogen activator (t-PA) under the control of the *polh* promoter, galactose was detected in the *N*-glycans of both GP64 and t-PA by lectin blotting (Hollister and Jarvis, 2001). Further, when a virus expressing rat α-2,6-sialyltransferase under the control of the *ie1* promoter-*hr5* enhancer element (AcP(+)IE1α26ST) was used to infect Sfβ4GalT cells, the GP64 produced during infection was found to be both galactosylated and sialylated (Seo *et al.*, 2001).

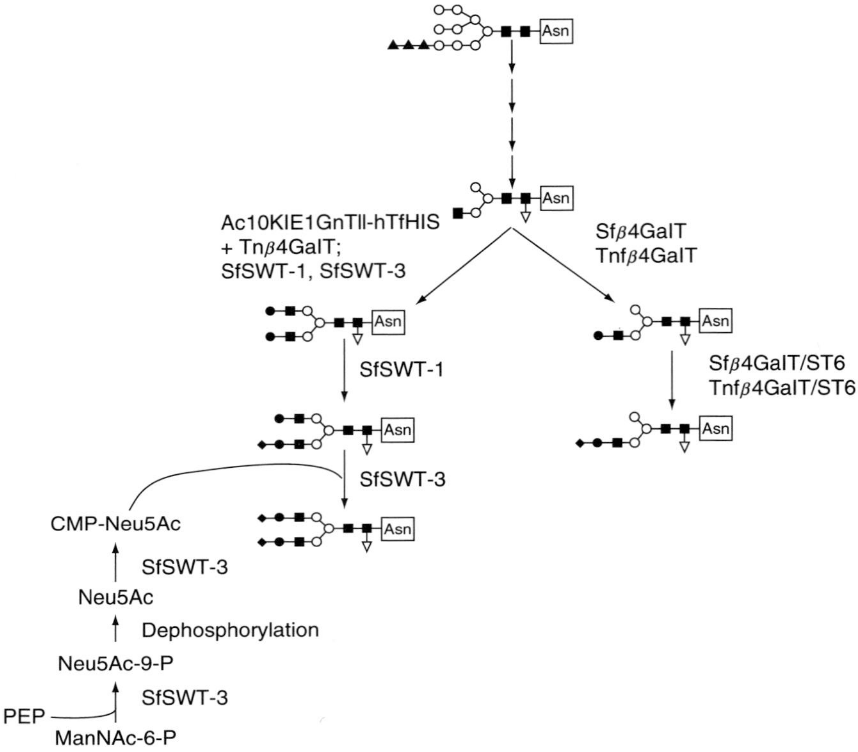

FIG 3. Structures of the *N*-glycans produced by insect cells stably transformed to encode and express mammalian *N*-glycosylation functions. The cell lines capable of producing each intermediate are indicated, as is the SfSWT-3 cell line engineered to express enzymes required for CMP-sialic acid synthesis. The standard monosaccharide symbols used in this figure are defined in the key shown in Fig. 1.

The Sfβ4GalT cell line subsequently was transformed with a rat α-2,6-sialyltransferase cDNA positioned under the control of the *ie1* promoter-*hr5* enhancer element (Hollister and Jarvis, 2001). This cell line, called Sfβ4GalT/ST6 (Fig. 3), had both galactosyltransferase and sialyltransferase activities, and produced sialylated *N*-glycans on both GP64 and an additional model glycoprotein, the *S. frugiperda* class I Golgi α-mannosidase fused to glutathione S-transferase (GST-SfManI).

The *T. ni* cell line Tn-5B1–4 was also transformed with the same constructs to produce cell lines designated Tn5β4GalT, which encoded β-1,4-galactosyltransferase, and Tn5β4GalT/ST6, which encoded both galactosyltransferase and sialyltransferase under the control of the bidirectional *ie-1* promoter/*hr5* enhancer (Fig. 3; Breitbach and Jarvis, 2001). These cell lines contained the expected glycosyltransferase activities and produced GP64 with galactosylated (Tn5β4GalT) or galactosylated and sialylated (Tn5β4GalT/ST6) *N*-glycans during infection with wild-type AcMNPV.

Structural analyses of the *N*-glycans produced by insect cells that were either infected or transformed to express the above-mentioned glycosyltransferase genes revealed that they were monoantennary structures in which only the lower (α-1,3) branch was elongated (Ailor *et al.*, 2000; Hollister *et al.*, 2002). Subsequently, it was found that infection of Tn5β4GalT cells with a virus (Ac10KIEGnTII-hTfHIS) encoding human GlcNAcT-II resulted in the production of human transferrin with biantennary *N*-glycans that contained terminal galactose on both branches (Tomiya *et al.*, 2003b). This result indicated that both Sf9 and Tn-5B1-4 cells engineered to express galactosyl- and sialyltransferase genes produced monoantennary *N*-glycans because they lacked sufficient levels of GlcNAcT-II, which is needed to initiate elongation of the upper (α-1,6) branch and produce biantennary *N*-glycans. This observation was consistent with the previous finding that GlcNAcT-II activity was present, but only at extremely low levels in three lepidopteran cell lines (Altmann *et al.*, 1993). To produce a cell line that could produce biantennary, complex *N*-glycans, Sfβ4GalT cells were transformed with constructs encoding *N*-acetylglucosaminyltransferase I, *N*-acetylglucosaminyltransferase II, α-2,6-sialyltransferase (ST6GalI), and α-2,3-sialyltransferase (ST3GalIV) cDNAs, all under the control of the *ie1* promoter-*hr5* enhancer element. This cell line, named SfSWT-1 (Fig. 3), was, in fact, able to produce complex, biantennary *N*-glycans on GST-SfManI (Hollister *et al.*, 2002). However, only the α-1,3 branch was sialylated, and the sialic acid was attached with an α-2,6 linkage. This result suggested that ST3GalIV, which transfers sialic acid to galactose in an α-2,3 linkage, was not active in the SWT-1 cell line.

Since Sf9 cells contain very little sialic acid (Lawrence *et al.*, 2000) and no detectable CMP-sialic acid (Hooker *et al.*, 1999; Tomiya *et al.*, 2001), the source of a donor substrate for the sialylation of *N*-glycans assembled in wild-type cells infected with the AcSWT viruses and in the Sfβ4GalT/ST6, Tn5β4GalT/ST6, and SfSWT-1 cell lines was unknown. Subsequently, it was found that the *N*-glycans on GST-SfManI were sialylated only when Sfβ4GalT/ST6 cells were cultured in the presence of serum or a purified, sialylated glycoprotein (Hollister *et al.*, 2003). No sialylation was detected when the cells were cultured in serum-free medium. Serum that had been dialyzed with a 50,000-Da cutoff membrane still supported sialylation, suggesting that the molecules in serum providing the donor substrate source were relatively large. Sfβ4GalT/ST6 cells also produced a very small quantity of sialylated *N*-glycans when cultured in serum-free medium supplemented with Neu5Ac or ManNAc. These results suggested that Sf9 cells have a mechanism for salvaging sialic acids from external sources and, together with the discovery of a gene encoding an active *N*-acetylneuraminic acid synthase and a potential homologue for CMP-Neu5Ac synthetase in *D. melanogaster* (Kim *et al.*, 2002), these results are consistent with the idea that insects have the ability to produce CMP-Neu5Ac.

To produce a cell line that could sialylate *N*-glycoproteins when cultured in serum-free media, SfSWT-1 cells were transformed with mammalian genes encoding Neu5Ac synthase and CMP-Neu5Ac synthetase under the control of the *hr5-ie1* enhancer/promoter element (Aumiller *et al.*, 2003). The resulting cell line, designated SfSWT-3 (Fig. 3), contained CMP-Neu5Ac and was able to produce complex, terminally sialylated *N*-glycans on GST-SfManI when cultured in a serum-free medium supplemented with ManNAc. Like SfSWT-1, SfSWT-3 produced mainly monosialylated, biantennary *N*-glycans, but they also appeared to produce a very minor population of disialylated, biantennary *N*-glycans.

Assuming that the endogenous membrane glycoproteins of Tnβ4GalT/ST6, Sfβ4GalT/ST6, SfSWT-1, and SfSWT-3 cells are terminally sialylated, the surfaces of these cells should be dramatically altered relative to the surfaces of the untransformed parental cell lines (Sf9 and Tn-5B1-4). These structural changes to cell surface glycoconjugates might be expected to inhibit baculovirus virion attachment, penetration, assembly, and/or release. However, all of these transformed cell lines supported wild-type and recombinant AcMNPV infection and replication at levels comparable to those of the untransformed parental cell lines (Hollister and Jarvis, 2001; Hollister *et al.*, 1998, 2002). The transformed cell lines described above also had similar growth properties and morphologies, relative to their untransformed progenitors (Breitbach and Jarvis, 2001). However, SfSWT-1 cells were noticeably smaller than Sf9 and achieved higher final densities in suspension culture than Sf-9 and SfSWT-3 cell lines after an initial period of slower growth (Aumiller *et al.*, 2003). Finally, the GST-S-fManI model glycoprotein was found to be expressed at approximately equal levels in Sf9, SfSWT-1, and SfSWT-3 cells after baculovirus vector infection, although there were some differences in the expression kinetics (Aumiller *et al.*, 2003).

III. Other Considerations and Future Improvements

A. *Manipulation of Cell Culture Conditions*

Studies with mammalian cells have shown that culture conditions can be manipulated to alter protein *N*-glycosylation profiles (reviewed in Jenkins *et al.*, 1996). For example, supplementation of CHO cell medium with ManNAc increased the extent of *N*-glycan sialylation (Gu and Wang, 1998). These types of environmental manipulations can be used to alter the *N*-glycosylation profiles of baculovirus-expressed recombinant proteins as well. The addition of mannosamine to lepidopteran cell culture medium increased the proportion of *N*-glycans with terminal *N*-acetylglucosamine (Donaldson *et al.*, 1999; Estrada-Mondaca *et al.*, 2005). ManNAc supplementation increased *N*-glycan sialylation in *T. ni* Tn-4h cells (Joshi *et al.*, 2001). In the same study, culturing Tn-4h cells in a bioreactor that simulated weightlessness and free fall had a similar effect (Joshi *et al.*, 2001). Supplementation of culture medium with *B. mori* (silkworm) hemolymph increased *N*-glycan sialylation by Tn-4s cells (Joosten *et al.*, 2003). In contrast to reports on the effects of fetal bovine serum on glycosylation in mammalian cell lines (Jenkins *et al.*, 1996), the presence or absence of fetal bovine serum had little to no effect on the *N*-glycan profiles of human transferrin expressed in *L. dispar* Ld652Y cells (Choi *et al.*, 2003).

It is also possible to add exoglycosidase inhibitors to influence the structures of the *N*-glycans found on insect cell-produced glycoproteins. The addition of swainsonine, an α-mannosidase II inhibitor, increased the quantity of terminal *N*-acetylglucosamine on fowl plague hemagglutinin *N*-glycans produced in Sf9 cells (Wagner *et al.*, 1996a). This effect was explained by the fact that this inhibitor would block the conversion of $GlcNAcMan_5GlcNAc_2$ to $GlcNAcMan_3GlcNAc_2$, thereby blocking production of the substrate for the *N*-acetylglucosaminidase activity in these cells (Wagner *et al.*, 1996a). Supplementation of insect cell medium with an *N*-acetylglucosaminidase inhibitor, 2-acetamido-1,2,5-trideoxy-1,5-imino-D-glucitol, increased the proportion of *N*-glycans with terminal GlcNAc in Sf9 cells (Wagner *et al.*, 1996a) but failed to significantly alter the *N*-glycan profiles of secreted alkaline phosphatase produced in *T. ni* Tn-4s cells (Joosten and Shuler, 2003a). The addition of a different *N*-acetylglucosaminidase inhibitor, 2-acetamido-1,2-dideoxynojirimycin, resulted in the appearance of sialylated *N*-glycans in *T. ni* Tn-5B1-4 cells, as assessed by lectin blotting analysis (Watanabe *et al.*, 2001). The addition of lactose, a competitive inhibitor of β-galactosidase, to infected

Tn-4s cell culture medium did not alter the *N*-glycan profiles of secreted alkaline phosphatase (Joosten and Shuler, 2003a).

B. Eliminating Unwanted Enzymatic Activities

As discussed earlier, the potential for at least some hosts to produce oligomannose or paucimannose *N*-glycans with an allergenic core fucose residue seriously impedes their use to produce recombinant glycoproteins for therapeutic applications. One potential way to address this problem would be to selectively downregulate or eliminate the relevant enzymatic activities. The processing *N*-acetylglucosaminidase and core α-1,3-fucosyltransferase are two potential targets.

Antisense RNA and RNA interference (RNAi) technologies can be used to reduce or eliminate unwanted enzymatic activities in cells (Lee and Roth, 2003). For example, an antisense RNA approach was used successfully to reduce sialidase activity in CHO cell culture supernatants (Ferrari *et al.*, 1998). A sialidase secreted from CHO cells causes a progressive reduction in sialic acid content of glycoproteins produced by CHO cells over time. Thus, Ferrari *et al.* (1998) stably transformed CHO cells with constructs that constitutively produced sialidase antisense RNAs. The transformed CHO clones exhibited reduced sialidase transcript levels, a 60% reduction in extracellular sialidase activity, and a 20–37% increase in sialic acid content of DNase (Ferrari *et al.*, 1998).

While the antisense RNA method rarely leads to complete shutdown of target gene expression, a higher degree of downregulation can be achieved using the RNAi approach (Hannon, 2002). Generally, this approach involves the introduction of double-stranded RNA molecules composed of sense and antisense transcripts derived from the target gene. These transcripts may be introduced by transfection or produced endogenously after stable transformation of cells with a DNA construct that induces the production of the desired double-stranded RNAs. The double-stranded RNA is then processed via established RNAi pathways, which ultimately target the desired mRNAs for selective cleavage and degradation. RNAi has been applied successfully to lepidopterans, both to inhibit expression of an endogenous gene (Rajagopal *et al.*, 2002) and to inhibit expression of a baculovirus gene during infection (Means *et al.*, 2003). In both cases, target gene transcripts were nearly undetectable after inoculation with the double-stranded RNA trigger.

C. *Nucleotide Sugar Transporters*

The addition of monosaccharides to *N*-glycan antennae occurs in the Golgi compartments. The nucleotide-sugar donor substrates utilized by glycosyltranferases are synthesized in the cytoplasm, except for CMP-Neu5Ac, which is produced in the nucleus, and each must be imported into the Golgi by specific nucleotide-sugar transporters (Gerardy-Schahn *et al.*, 2001). A gene encoding a putative CMP-sialic acid/UDP-galactose transporter was identified in the *D. melanogaster* genome, but the gene product actually transported UDP-galactose, not CMP-sialic acid (Aumiller and Jarvis, 2002; Segawa *et al.*, 2002). These studies did not reveal whether this fly transporter is expressed in tissue- or developmental stage-specific fashion, nor did they examine its expression levels.

The ability of Sf9 cells infected with the recombinant baculoviruses, AcSWT-1 and AcSWT-2c, and the transformed cell lines, SfSWT-1 and SfSWT-3, to produce sialylated *N*-glycoproteins indicates that these cells must have nucleotide-sugar transporters capable of importing CMP-Neu5Ac into the Golgi apparatus. However, the nature of these transporters remains unknown. Meanwhile, engineering insect cells to express an exogenous CMP-Neu5Ac transporter could increase the level of CMP-Neu5Ac import into the Golgi and possibly improve the efficiency with which *N*-glycans are sialylated during baculovirus infection.

D. *Bisected, Tri- and Tetraantennary* N-*Glycans*

In addition to producing glycoproteins with complex, terminally sialylated *N*-glycans, glycoprotein functionality can be extended by supplying additional *N*-glycan–processing activities. For example, the antibody-dependent cellular cytotoxicity of an antineuroblastoma IgG1 monoclonal antibody being developed to treat some forms of cancer correlates with the presence of a bisecting GlcNAc in its *N*-glycans (Lifely *et al.*, 1995). The enzyme GlcNAcT-III catalyzes the transfer of this bisecting GlcNAc to the core mannose residue of biantennary *N*-glycans (Fig. 1). Doxycyclin induction of CHO cells transformed with the GlcNAcT-III gene under the control of a tetracycline-repressible promoter increased the proportion of *N*-glycans with bisecting GlcNAc and concomitantly increased the antibody-dependent cellular toxicity of the IgG1 produced by this transformed cell line (Umana *et al.*, 1999).

The enzymes GlcNAcT-IV and GlcNAcT-V transfer additional *N*-acetylglucosamine residues to the mannose residues on the α-1,3 and α-1,6 branches, leading to the formation of tri- and tetraantennary *N*-glycans (Fig. 1). The glycoprotein hormone erythropoietin (EPO) had more activity *in vivo* if it had a tetraantennary rather than a biantennary *N*-glycan (Takeuchi *et al.*, 1989) and it also had a longer *in vivo* half-life (Misaizu *et al.*, 1995). The degree of sialylation of the tetraantennary form of EPO correlated positively with its *in vivo* bioactivity, presumably due to reduced clearance rates (Yuen *et al.*, 2003). Hence, engineering insect cells to produce *N*-glycans with more than two sialylated antennae could similarly improve the *in vivo* bioactivity of baculovirus vector-produced therapeutic glycoproteins. CHO cells engineered to express higher levels of GlcNAcT-IV and GlcNAcT-V produced glycoproteins with larger proportions of tri- and tetraantennary *N*-glycans (Fukuta *et al.*, 2000), indicating that branching can be increased by introducing genes encoding the appropriate glycosyltransferases.

E. Protein Specificity of N-*Glycan Processing by Engineered Insect Cell Lines?*

Lepidopteran insect cell lines engineered to express mammalian *N*-glycan–processing activities have been shown to sialylate a variety of different recombinant glycoproteins, as detailed earlier. However, one study has challenged the breadth of this capability. Legardinier *et al.* (2005) used the commercial version of SfSWT-1 cells (MIMIC^{TM}, Invitrogen) as a host for baculovirus-mediated expression of equine luteinizing hormone/chorionic gonadotropin (eLH/CG) and were unable to detect sialylation of this product using lectin-based assays. This result could indicate that these cells cannot universally sialylate any recombinant *N*-glycoprotein that one might want to produce using a baculovirus expression vector. Further work is clearly needed to examine this possibility and assess the breadth of the recombinant glycoprotein sialylation capabilities of SfSWT-1 and other transgenic lepidopteran insect cell lines. Another possibility raised by the results of Legardinier and coworkers is that SfSWT-1/MIMIC^{TM} cells might be genetically unstable under some conditions. With respect to this latter possibility, it is noteworthy that the eLH/CG preparations produced by Legardinier and coworkers were galactosylated, as this finding indicated that their cells had retained at least one of the five transgenes that were originally introduced. To our knowledge, the stability of the transgenes in SfSWT-1/MIMIC^{TM} cells has not been rigorously monitored. However, a previous study showed that another transformed Sf9 cell line retained its transgene for at least 55 serial passages in culture (Jarvis *et al.*, 1990). Similarly, we have observed GST-SfManI sialylation using SfSWT-1 cells that have been routinely subcultured three times a week and subjected to well over 100 serial passages in the laboratory (D. L. Jarvis, unpublished data). Thus, while it remains a distinct possibility, it seems relatively unlikely to us that the results of Legardinier *et al.* reflect genetic instability of the MIMIC^{TM} cells used in their study.

IV. Concluding Remarks

Baculovirus expression vectors are a popular and safe means of producing large quantities of recombinant protein. However, the use of baculovirus vectors for the production of therapeutic glycoproteins is limited by the nature of insect cell glycosylation. Several studies have now shown that insects have the potential to produce complex, sialylated *N*-glycan. However, the *N*-glycans produced by the insect cell lines that serve as hosts for baculovirus vectors frequently consist of simple paucimannose structures. To overcome the deficiencies of insect glycosylation, baculovirus expression vectors and/or their insect-cell hosts have been supplemented with genes encoding enzymes required for complex, sialylated *N*-glycan biosynthesis. These approaches have been used successfully to produce glycoproteins with mammalianized *N*-glycans in the baculovirus–insect cell system. These results, along with efforts to address other problems with insect glycosylation, such as the presence of allergenic core α-1,3-fucose, will facilitate the adaptation of baculovirus expression vectors for the production of glycoproteins for biomedical purposes.

Acknowledgments

Figures 1, 2, and 3 are adapted and reprinted from *Virology* **310**, D. L. Jarvis, "Developing baculovirus-insect cell expression systems for humanized recombinant protein production," pp. 1–7, 2003, with permission from Elsevier. Mention of trade names or commercial products in this chapter is solely for the purpose of providing specific information and does not imply recommendation or endorsement by the US Department of Agriculture. Insect glycobiology in the Jarvis lab was supported by the grants NIH (GM49734), NSF (BES 9814157 and BES 9818001), and NIST (70NANB3H3042).

References

Abdul-Rahman, B., Ailor, E., Jarvis, D., Betenbaugh, M., and Lee, Y. C. (2002). Beta-(1 → 4)-galactosyltransferase activity in native and engineered insect cells measured with time-resolved europium fluorescence. *Carbohydr. Res.* **337**:2181–2186.

Adams, M. D., Celniker, S. E., Holt, R. A., Evans, C. A., Gocayne, J. D., Amanatides, P. G., Scherer, S. E., Li, S. E., Hoskins, R. A., Galle, R. F., George, R. A., Lewis, S. E. *et al.* (2000). The genome sequence of *Drosophila melanogaster*. *Science* **287:**2185–2195.

Ailor, E., Takahashi, N., Tsukamoto, Y., Masuda, K., Rahman, B. A., Jarvis, D. L., Lee, Y. C., and Betenbaugh, M. J. (2000). N-Glycan patterns of human transferrin produced in *Trichoplusia ni* insect cells: Effects of mammalian galactosyltransferase. *Glycobiology* **10:**837–847.

Altmann, F., Kornfeld, G., Dalik, T., Staudacher, E., and Glossl, J. (1993). Processing of asparagine-linked oligosaccharides in insect cells. N-Acetylglucosaminyltransferase I and II activities in cultured lepidopteran cells. *Glycobiology* **3:**619–625.

Altmann, F., Schwihla, H., Staudacher, E., Glossl, J., and Marz, L. (1995). Insect cells contain an unusual, membrane-bound β-N-acetylglucosaminidase probably involved in the processing of protein N-glycans. *J. Biol. Chem.* **270:**17344–17349.

Angata, T., and Varki, A. (2002). Chemical diversity in the sialic acids and related alpha-keto acids: An evolutionary perspective. *Chem. Rev.* **102:**439–469.

Ashwell, G., and Harford, J. (1982). Carbohydrate-specific receptors of the liver. *Annu. Rev. Biochem.* **51:**531–554.

Aumiller, J. J., and Jarvis, D. L. (2002). Expression and functional characterization of a nucleotide sugar transporter from *Drosophila melanogaster*: Relevance to protein glycosylation in insect cell expression systems. *Protein Expr. Purif.* **26:**438–448.

Aumiller, J. J., Hollister, J. R., and Jarvis, D. L. (2003). A transgenic lepidopteran insect cell line engineered to produce CMP-sialic acid and sialoglycoproteins. *Glycobiology* **13:**497–507.

Bendiak, B., and Schachter, H. (1987). Control of glycoprotein synthesis. Kinetic mechanism, substrate specificity, and inhibition characteristics of UDP-N-acetylglucosamine: Alpha-D-mannoside beta 1-2 N-acetylglucosaminyltransferase II from rat liver. *J. Biol. Chem.* **262:**5784–5790.

Breitbach, K., and Jarvis, D. L. (2001). Improved glycosylation of a foreign protein by Tn-5B1-4 cells engineered to express mammalian glycosyltransferases. *Biotech. Bioengr.* **74:**230–239.

Burda, P., and Aebi, M. (1999). The dolichol pathway of N-linked glycosylation. *Biochim. Biophys. Acta* **1426:**239–257.

Butters, T. D., and Hughes, R. C. (1981). Isolation and characterization of mosquito cell membrane glycoproteins. *Biochim. Biophys. Acta* **640:**655–671.

Chang, G. D., Chen, C. J., Lin, C. Y., Chen, H. C., and Chen, H. (2003). Improvement of glycosylation in insect cells with mammalian glycosyltransferases. *J. Biotechnol.* **102:**61–71.

Chazenbalk, G. D., and Rapoport, B. (1995). Expression of the extracellular domain of the thyrotropin receptor in the baculovirus system using a promoter active earlier than the polyhedrin promoter. Implications for the expression of functional highly glycosylated proteins. *J. Biol. Chem.* **270:**1543–1549.

Choi, O., Tomiya, N., Kim, J. H., Slavicek, J. M., Betenbaugh, M. J., and Lee, Y. C. (2003). *N*-Glycan structures of human transferrin produced by *Lymantria dispar* (gypsy moth) cells using the LdMNPV expression system. *Glycobiology* **13:**539–548.

Chui, D., Oh-Eda, M., Liao, Y. F., Panneerselvam, K., Lal, A., Marek, K. W., Freeze, H. H., Moremen, K. W., Fukuda, M. N., and Marth, J. D. (1997). Alpha-mannosidase-II deficiency results in dyserythropoiesis and unveils an alternate pathway in oligosaccharide biosynthesis. *Cell* **90:**157–167.

Davidson, D. J., and Castellino, F. J. (1991a). Structures of the asparagine-289-linked oligosaccharides assembled on recombinant human plasminogen expressed in a *Mamestra brassicae* cell line (IZD-MBO503). *Biochemistry* **30:**6689–6696.

Davidson, D. J., and Castellino, F. J. (1991b). Asparagine-linked oligosaccharide processing in lepidopteran insect cells. Temporal dependence of the nature of the oligosaccharides assembled on asparagine-289 of recombinant human plasminogen produced in baculovirus vector infected *Spodoptera frugiperda* (IPLB-SF-21AE) cells. *Biochemistry* **30:**6167–6174.

Davidson, D. J., and Castellino, F. J. (1993). The influence of the nature of the asparagine 289-linked oligosaccharide on the activation by urokinase and lysine binding properties of natural and recombinant human plasminogens. *J. Clin. Invest.* **92:**249–254.

Davidson, D. J., Fraser, M. J., and Castellino, F. J. (1990). Oligosaccharide processing in the expression of human plasminogen cDNA by lepidopteran insect (*Spodoptera frugiperda*) cells. *Biochemistry* **29:**5584–5590.

Davidson, D. J., Bretthauer, R. K., and Castellino, F. J. (1991). Alpha-mannosidase-catalyzed trimming of high-mannose glycans in noninfected and baculovirus-infected *Spodoptera frugiperda* cells (IPLB-SF-21AE). A possible contributing regulatory mechanism for assembly of complex-type oligosaccharides in infected cells. *Biochemistry* **30:**9811–9815.

Davis, T. R., Schuler, M. L., Granados, R. R., and Wood, H. A. (1993). Comparison of oligosaccharide processing among various insect cell lines expressing a secreted glycoprotein. *In Vitro Cell. Dev. Biol. Anim.* **29A:**842–846.

Ding, H., Griesel, C., Nimtz, M., Conradt, H. S., Weich, H. A., and Jager, V. (2003). Molecular cloning, expression, purification, and characterization of soluble full-length, human interleukin-3 with a baculovirus-insect cell expression system. *Protein Expr. Purif.* **31:**34–41.

Donaldson, M., Wood, H. A., Kulakosky, P. C., and Shuler, M. L. (1999). Use of mannosamine for inducing the addition of outer arm N-acetylglucosamine onto N-linked oligosaccharides of recombinant proteins in insect cells. *Biotechnol. Prog.* **15:**168–173.

Effertz, K., Hinderlich, S., and Reutter, W. (1999). Selective loss of either the epimerase or kinase activity of UDP-N-acetylglucosamine 2-epimerase/N-acetylmannosamine kinase due to site-directed mutagenesis based on sequence alignments. *J. Biol. Chem.* **274:**28771–28778.

Estrada-Mondaca, S., Delgado-Bustos, L. A., and Ramirez, O. T. (2005). Mannosamine supplementation extends the *N*-acetylglucosaminylation of recombinant human secreted alkaline phosphatase produced in *Trichoplusia ni* (cabbage looper) insect cell cultures. *Biotechnol. Appl. Biochem.* **42:**25–34.

Fabini, G., Freilinger, A., Altmann, F., and Wilson, I. B. (2001). Identification of core alpha 1,3-fucosylated glycans and cloning of the requisite fucosyltransferase cDNA from *Drosophila melanogaster.* Potential basis of the neural anti-horseadish peroxidase epitope. *J. Biol. Chem.* **276:**28058–28067.

Ferrari, J., Gunson, J., Lofgren, J., Krummen, L., and Warner, T. G. (1998). Chinese hamster ovary cells with constitutively expressed sialidase antisense RNA produce recombinant DNase in batch culture with increased sialic acid. *Biotechnol. Bioeng.* **60:**589–595.

Foster, J. M., Yudkin, B., Lockyer, A. E., and Roberts, D. B. (1995). Cloning and sequence analysis of GmII, a *Drosophila melanogaster* homologue of the cDNA encoding murine Golgi alpha-mannosidase II. *Gene* **154:**183–186.

Fukuta, K., Abe, R., Yokomatsu, T., Kono, N., Asanagi, M., Omae, F., Minowa, M. T., Takeuchi, M., and Makino, T. (2000). Remodeling of sugar chain structures of human interferon-gamma. *Glycobiology* **10:**421–430.

Gerardy-Schahn, R., Oelmann, S., and Bakker, H. (2001). Nucleotide sugar transporters: Biological and functional aspects. *Biochimie* **83:**775–782.

Grabenhorst, E., Hofer, B., Nimtz, M., Jager, V., and Conradt, H. S. (1993). Biosynthesis and secretion of human interleukin 2 glycoprotein variants from baculovirus-infected Sf21 cells. Characterization of polypeptides and posttranslational modifications. *Eur. J. Biochem.* **215:**189–197.

Grossmann, M., Wong, R., Teh, N. G., Tropea, J. E., East-Palmer, J., Weintraub, B. D., and Szkudlinski, M. W. (1997). Expression of biologically active human thyrotropin (hTSH) in a baculovirus system: Effect of insect cell glycosylation on hTSH activity *in vitro* and *in vivo*. *Endocrinology* **138:**92–100.

Gu, X., and Wang, D. I. (1998). Improvement of interferon-gamma sialylation in Chinese hamster ovary cell culture by feeding of N-acetylmannosamine. *Biotechnol. Bioeng.* **58:**642–648.

Haines, N., and Irvine, K. D. (2005). Functional analysis of *Drosophila* beta1,4-N-acetlygalactosaminyltransferases. *Glycobiology* **15:**335–346.

Hannon, G. J. (2002). RNA interference. *Nature* **418:**244–251.

Hard, K., Van Doorn, J. M., Thomas-Oates, J. E., Kamerling, J. P., and Van der Horst, D. J. (1993). Structure of the asn-linked oligosaccharides of apolipophorin III from the insect *Locusta migratoria*. Carbohydrate-linked 2-aminoethylphosphonate as a constituent of a glycoprotein. *Biochemistry* **32:**766–775.

Helenius, A., and Aebi, M. (2001). Intracellular functions of N-linked glycans. *Science* **291:**2364–2369.

Hogeland, K. E., Jr., and Deinzer, M. L. (1994). Mass spectrometric studies on the N-linked oligosaccharides of baculovirus-expressed mouse interleukin-3. *Biol. Mass Spec.* **23:**218–224.

Hollister, J., and Jarvis, D. L. (2001). Engineering lepidopteran insect cells for sialoglycoprotein production by genetic transformation with mammalian β1,4-galactosyltransferase and α2,6-sialyltransferase genes. *Glycobiology* **11:**1–9.

Hollister, J. R., Shaper, J. H., and Jarvis, D. L. (1998). Stable expression of mammalian beta 1,4-galactosyltransferase extends the N-glycosylation pathway in insect cells. *Glycobiology* **8:**473–480.

Hollister, J. R., Grabenhorst, E., Nimtz, M., Conradt, H. O., and Jarvis, D. L. (2002). Engineering the protein N-glycosylation pathway in insect cells for production of biantennary, complex N-glycans. *Biochemistry* **41:**15093–15104.

Hollister, J. R., Conradt, H. O., and Jarvis, D. L. (2003). Evidence for a sialic acid salvaging pathway in lepidopteran insect cell lines. *Glycobiology* **13:**487–495.

Holt, R. A., Subramanian, G. M., Halpern, A., Sutton, G. G., Charlab, R., Nusskern, D. R., Wincker, P., Clark, A. G., Ribeiro, J. M., Wides, R., Salzberg, S. L., Loftus, B. *et al.* (2002). The genome sequence of the malaria mosquito *Anopheles gambiae*. *Science* **298:** 129–149.

Hooker, A. D., Green, N. H., Baines, A. J., Bull, A. T., Jenkins, N., Strange, P. G., and James, D. C. (1999). Constraints on the transport and glycosylation of recombinant IFN-gamma in Chinese hamster ovary and insect cells. *Biotechnol. Bioeng.* **63:**559–572.

Hsieh, P., and Robbins, P. W. (1984). Regulation of asparagine-linked oligosaccharide processing. Oligosaccharide processing in *Aedes albopictus* mosquito cells. *J. Biol. Chem.* **259:**2375–2382.

Hsu, T. A., Takahashi, N., Tsukamoto, Y., Kato, K., Shimada, I., Masuda, K., Whiteley, E. M., Fan, J. Q., Lee, Y. C., and Betenbaugh, M. J. (1997). Differential *N*-glycan patterns of secreted and intracellular IgG produced in *Trichoplusia ni* cells. *J. Biol. Chem.* **272:**9062–9070.

Imperiali, B., and O'Connor, S. E. (1999). Effect of N-linked glycosylation on glycopeptide and glycoprotein structure. *Curr. Opin. Chem. Biol.* **3:**643–649.

Jarvis, D. L. (2003). Humanizing recombinant glycoprotein production in the baculovirus-insect cell expression system. *Virology* **310:**1–7.

Jarvis, D. L., and Finn, E. E. (1996). Modifying the insect cell *N*-glycosylation pathway with immediate early baculovirus expression vectors. *Nat. Biotechnol.* **14:** 1288–1292.

Jarvis, D. L., and Summers, M. D. (1989). Glycosylation and secretion of human tissue plasminogen activator in recombinant baculovirus-infected insect cells. *Mol. Cell. Biol.* **9:**214–223.

Jarvis, D. L., Fleming, J. A., Kovacs, G. R., Summers, M. D., and Guarino, L. A. (1990). Use of early baculovirus promoters for continuous expression and efficient processing of foreign gene products in stably transformed lepidopteran cells. *Bio/technology* **8:**950–955.

Jarvis, D. L., Bohlmeyer, D. A., Liao, Y. F., Lomax, K. K., Merkle, R. K., Weinkauf, C., and Moremen, K. W. (1997). Isolation and characterization of a class II alpha-mannosidase cDNA from lepidopteran insect cells. *Glycobiology* **7:**113–127.

Jarvis, D. L., Howe, D., and Aumiller, J. J. (2001). Novel baculovirus expression vectors that provide sialylation of recombinant glycoproteins in lepidopteran insect cells. *J. Virol.* **75:**6223–6227.

Jenkins, N., Parekh, R. B., and James, D. C. (1996). Getting the glycosylation right: Implications for the biotechnology industry. *Nat. Biotechnol.* **14:**975–981.

Joosten, C. E., and Shuler, M. L. (2003a). Production of a sialylated *N*-linked glycoprotein in insect cells: Role of glycosidases and effect of harvest time on glycosylation. *Biotechnol. Prog.* **19:**193–201.

Joosten, C. E., and Shuler, M. L. (2003b). Effect of culture conditions on the degree of sialylation of a recombinant glycoprotein expressed in insect cells. *Biotechnol. Prog.* **19:**739–749.

Joosten, C. E., Park, T. H., and Shuler, M. L. (2003). Effect of silkworm hemolymph on N-linked glycosylation in two *Trichoplusia ni* insect cell lines. *Biotechnol. Bioeng.* **83:**695–705.

Joshi, L., Davis, T. R., Mattu, T. S., Rudd, P. M., Dwek, R. A., Shuler, M. L., and Wood, H. A. (2000). Influence of baculovirus-host cell interactions on complex N-linked glycosylation of a recombinant human protein. *Biotechnol. Prog.* **16:**650–656.

Joshi, L., Shuler, M. L., and Wood, H. A. (2001). Production of a sialylated N-linked glycoprotein in insect cells. *Biotechnol. Prog.* **17:**822–827.

Kawar, Z., and Jarvis, D. L. (2001). Biosynthesis and intracellular localization of a lepidopteran insect alpha 1,2-mannosidase. *Insect Biochem. Mol. Biol.* **31:**289–297.

Kawar, Z., Herscovics, A., and Jarvis, D. L. (1997). Isolation and characterization of an alpha 1,2-mannosidase cDNA from the lepidopteran insect cell line Sf9. *Glycobiology* **7:**433–443.

Kawar, Z., Romero, P. A., Herscovics, A., and Jarvis, D. L. (2000). N-Glycan processing by a lepidopteran insect alpha1,2-mannosidase. *Glycobiology* **10:**347–355.

Kawar, Z., Moremen, K. W., and Jarvis, D. L. (2001). Insect cells encode a class II alpha-mannosidase with unique properties. *J. Biol. Chem.* **276:**16335–16340.

Kerscher, S., Albert, S., Wucherpfennig, D., Heisenberg, M., and Schneuwly, S. (1995). Molecular and genetic analysis of the *Drosophila* mas-1 (mannosidase-1) gene which encodes a glycoprotein processing alpha 1,2-mannosidase. *Dev. Biol.* **168:**613–626.

Kim, K., Lawrence, S. M., Park, J., Pitts, L., Vann, W. F., Betenbaugh, M. J., and Palter, K. B. (2002). Expression of a functional *Drosophila melanogaster* N-acetylneuraminic acid (Neu5Ac) phosphate synthase gene: Evidence for endogenous sialic acid biosynthetic ability in insects. *Glycobiology* **12:**73–83.

Kim, S., Hwang, S. K., Dwek, R. A., Rudd, P. M., Ahn, Y. H., Kim, E. H., Cheong, C., Kim, S. I., Park, N. S., and Lee, S. M. (2003). Structural determination of the N-glycans of a lepidopteran arylphorin reveals the presence of a monoglucosylated oligosaccharide in the storage protein. *Glycobiology* **13:**147–157.

Koles, K., Irvine, K. D., and Panin, V. M. (2004). Functional characterization of a *Drosophila* sialyltransferase. *J. Biol. Chem.* **279:**4346–4357.

Kornfeld, R., and Kornfeld, S. (1985). Assembly of asparagine-linked oligosaccharides. *Ann. Rev. Biochem.* **54:**631–664.

Kretzschmar, E., Geyer, R., and Klenk, H. D. (1994). Baculovirus infection does not alter N-glycosylation in Spodoptera frugiperda cells. *Biol. Chem. Hoppe-Seyler* **375:**23–27.

Kubelka, V., Altmann, F., Staudacher, E., Tretter, V., Marz, L., Hard, K., Kamerling, J. P., and Vliegenthart, J. F. (1993). Primary structures of the N-linked carbohydrate chains from honeybee venom phospholipase A2. *Eur. J. Biochem.* **213:**1193–1204.

Kubelka, V., Altmann, F., Kornfeld, G., and Marz, L. (1994). Structures of the N-linked oligosaccharides of the membrane glycoproteins from three lepidopteran cell lines (Sf-21, IZD-Mb-0503, Bm-N). *Arch. Biochem. Biophys.* **308:**148–157.

Kubelka, V., Altmann, F., and Marz, L. (1995). The asparagine-linked carbohydrate of honeybee venom hyaluronidase. *Glycoconj. J.* **12:**77–83.

Kulakosky, P. C., Hughes, P. R., and Wood, H. A. (1998). N-Linked glycosylation of a baculovirus-expressed recombinant glycoprotein in insect larvae and tissue culture cells. *Glycobiology* **8:**741–745.

Kuroda, K., Geyer, H., Geyer, R., Doerfler, W., and Klenk, H. D. (1990). The oligosaccharides of influenza virus hemagglutinin expressed in insect cells by a baculovirus vector. *Virology* **174:**418–429.

Kurschat, P., Graeve, L., Erren, A., Gatsios, P., Rose-John, S., Roeb, E., Tschesche, H., Koj, A., and Heinrich, P. C. (1995). Expression of a biologically active murine tissue inhibitor of metalloproteinases-1 (TIMP-1) in baculovirus-infected insect cells. Purification and tissue distribution in the rat. *Eur. J. Biochem.* **234:**485–491.

Lawrence, S. M., Huddleston, K. A., Pitts, L. R., Nguyen, N., Lee, Y. C., Vann, W. F., Coleman, T. A., and Betenbaugh, M. J. (2000). Cloning and expression of the human N-acetylneuraminic acid phosphate synthase gene with 2-keto-3-deoxy-D-glycero-D-galacto-nononic acid biosynthetic ability. *J. Biol. Chem.* **275:**17869–17877.

Lawrence, S. M., Huddleston, K. A., Tomiya, N., Nguyen, N., Lee, Y. C., Vann, W. F., Coleman, T. A., and Betenbaugh, M. J. (2001). Cloning and expression of human sialic acid pathway genes to generate CMP-sialic acids in insect cells. *Glycoconj. J.* **18:**205–213.

Lee, L. K., and Roth, C. M. (2003). Antisense technology in molecular and cellular bioengineering. *Curr. Opin. Biotechnol.* **14:**505–511.

Legardinier, S., Klett, D., Poirier, J. C., Combarnous, Y., and Cahoreau, C. (2005). Mammalian-like nonsialyl complex-type N-glycosylation of equine gonadotropins in Mimic insect cells. *Glycobiology* **15:**776–790.

Licari, P. J., Jarvis, D. L., and Bailey, J. E. (1993). Insect cell hosts for baculovirus expression vectors contain endogenous exoglycosidase activity. *Biotechnol. Prog* **9:**146–152.

Lifely, M. R., Hale, C., Boyce, S., Keen, M. J., and Phillips, J. (1995). Glycosylation and biological activity of CAMPATH-1H expressed in different cell lines and grown under different culture conditions. *Glycobiology* **5:**813–822.

Lis, H., and Sharon, N. (1993). Protein glycosylation. Structural and functional aspects. *Eur. J. Biochem.* **218:**1–27.

Lopez, M., Coddeville, B., Langridge, J., Plancke, Y., Sautiere, P., Chaabihi, H., Chirat, F., Harduin-Lepers, A., Cerutti, M., Verbert, A., and Delannoy, P. (1997). Microheterogeneity of the oligosaccharides carried by the recombinant bovine lactoferrin expressed in *Mamestra brassicae* cells. *Glycobiology* **7:**635–651.

Malykh, Y. N., Krisch, B., Gerardy-Schahn, R., Lapina, E. B., Shaw, L., and Schauer, R. (2000). The presence of N-acetylneuraminic acid in Malpighian tubules of larvae of the cicada *Philaenus spumarius*. *Glycoconj. J.* **16:**731–739.

Manneberg, M., Friedlein, A., Kurth, H., Lahm, H. W., and Fountoulakis, M. (1994). Structural analysis and localization of the carbohydrate moieties of a soluble human interferon gamma receptor produced in baculovirus-infected insect cells. *Protein Sci.* **3:**30–38.

Marchal, I., Mir, A. M., Kmiecik, D., Verbert, A., and Cacan, R. (1999). Use of inhibitors to characterize intermediates in the processing of N-glycans synthesized by insect cells: A metabolic study with Sf9 cell line. *Glycobiology* **9:**645–654.

Marchal, I., Jarvis, D. L., Cacan, R., and Verbert, A. (2001). Glycoproteins from insect cells: Sialylated or not? *Biol. Chem.* **382:**151–159.

McGinnes, L., Sergel, T., Reitter, J., and Morrison, T. (2001). Carbohydrate modifications of the NDV fusion protein heptad repeat domains influence maturation and fusion activity. *Virology* **283:**332–342.

Means, J. C., Muro, I., and Clem, R. J. (2003). Silencing of the baculovirus Op-*iap3* gene by RNA interference reveals that it is required for prevention of apoptosis during *Orgyia pseudotsugata M* nucleopolyhedrovirus infection of Ld652Y cells. *J. Virol.* **77:**4481–4488.

Misaizu, T., Matsuki, S., Strickland, T. W., Takeuchi, M., Kobata, A., and Takasaki, S. (1995). Role of antennary structure of N-linked sugar chains in renal handling of recombinant human erythropoietin. *Blood* **86:**4097–4104.

Murphy, C. I., Lennick, M., Lehar, S. M., Beltz, G. A., and Young, E. (1990). Temporal expression of HIV-1 envelope proteins in baculovirus-infected insect cells: Implications for glycosylation and CD4 binding. *Genet. Anal. Tech. Appl.* **7:**160–171.

Nobiron, I., O'Reilly, D. R., and Olszewski, J. A. (2003). Autographa californica nucleopolyhedrovirus infection of Spodoptera frugiperda cells: A global analysis of host gene regulation during infection, using a differential display approach. *J. Gen. Virol.* **84:**3029–3039.

Ogonah, O. W., Freedman, R. B., Jenkins, N., Patel, K., and Rooney, B. (1996). Isolation and characterization of an insect cell line able to perform complex N-linked glycosylation on recombinant proteins. *Bio/technology* **14:**197–202.

Ooi, B. G., and Miller, L. K. (1988). Regulation of host RNA levels during baculovirus infection. *Virology* **166:**515–523.

Opdenakker, G., Rudd, P. M., Ponting, C. P., and Dwek, R. A. (1993). Concepts and principles of glycobiology. *FASEB J.* **7:**1330–1337.

Pajot-Augy, E., Bozon, V., Remy, J. J., Couture, L., and Salesse, R. (1999). Critical relationship between glycosylation of recombinant lutropin receptor ectodomain and its secretion from baculovirus-infected insect cells. *Eur. J. Biochem.* **260:**635–648.

Palomares, L., Joosten, C. E., Hughes, P. R., Granados, R. R., and Shuler, M. L. (2003). Novel insect cell line capable of complex N-glycosylation and sialylation of recombinant proteins. *Biotechnol. Prog.* **19:**185–192.

Park, Y. I., Wood, H. A., and Lee, Y. C. (1999). Monosaccharide compositions of Danaus plexippus (monarch butterfly) and Trichoplusia ni (cabbage looper) egg glycoproteins. *Glycoconj. J.* **16:**629–638.

Parker, G. F., Williams, P. J., Butters, T. D., and Roberts, D. B. (1991). Detection of the lipid-linked precursor oligosaccharide of N-linked protein glycosylation in *Drosophila melanogaster*. *FEBS Lett.* **290:**58–60.

Prenner, C., Mach, L., Glossl, J., and Marz, L. (1992). The antigenicity of the carbohydrate moiety of an insect glycoprotein, honey-bee (Apis mellifera) venom phospholipase A2. The role of alpha 1,3-fucosylation of the asparagine-bound N-acetylglucosamine. *Biochem. J.* **284**(Pt. 2):377–380.

Rabouille, C., Kuntz, D. A., Lockyer, A., Watson, R., Signorelli, T., Rose, D. R., van den Heuvel, M., and Roberts, D. B. (1999). The *Drosophila* GMII gene encodes a Golgi alpha-mannosidase II. *J. Cell Sci.* **112:**3319–3330.

Rajagopal, R., Sivakumar, S., Agrawal, N., Malhotra, P., and Bhatnagar, R. K. (2002). Silencing of midgut aminopeptidase N of *Spodoptera litura* by double-stranded RNA establishes its role as *Bacillus thuringiensis* toxin receptor. *J. Biol. Chem.* **277:** 46849–46851.

Ren, J., Bretthauer, R. K., and Castellino, F. J. (1995). Purification and properties of a Golgi-derived (alpha 1,2)-mannosidase-I from baculovirus-infected lepidopteran insect cells (IPLB-SF21AE) with preferential activity toward mannose6-N-acetylglucosamine2. *Biochemistry* **34:**2489–2495.

Ren, J., Castellino, F. J., and Bretthauer, R. K. (1997). Purification and properties of alpha-mannosidase II from Golgi-like membranes of baculovirus-infected Spodoptera frugiperda (IPLB-SF-21 AE) cells. *Biochem. J.* **15:**951–956.

Roberts, D. B., Mulvany, W. J., Dwek, R. A., and Rudd, P. M. (1998). Mutant analysis reveals an alternative pathway for N-linked glycosylation in *Drosophila melanogaster. Eur. J. Biochem.* **253:**494–498.

Roelvink, P. W., van Meer, M. M., de Kort, C. A., Possee, R. D., Hammock, B. D., and Vlak, J. M. (1992). Dissimilar expression of *Autographa californica* multiple nucleocapsid nuclear polyhedrosis virus polyhedrin and p10 genes. *J. Gen. Virol.* **73:** 1481–1489.

Roth, J., Kempf, A., Reuter, G., Schauer, R., and Gehring, W. (1992). Occurrence of sialic acids in *Drosophila melanogaster. Science* **256:**673–675.

Rudd, P. M., Woods, R. J., Wormald, M. R., Opdenakker, G., Downing, A. K., Campbell, I. D., and Dwek, R. A. (1995). The effects of variable glycosylation on the functional activities of ribonuclease, plasminogen and tissue plasminogen activator. *Biochim. Biophys. Acta* **1248:**1–10.

Rudd, P. M., Downing, A. K., Cadene, M., Harvey, D. J., Wormald, M. R., Weir, I., Dwek, R. A., Rifkin, D. B., and Gleizes, P. E. (2000). Hybrid and complex glycans are linked to the conserved N-glycosylation site of the third eight-cysteine domain of LTBP-1 in insect cells. *Biochemistry* **39:**1596–1603.

Rudd, P. M., Elliott, T., Cresswell, P., Wilson, I. A., and Dwek, R. A. (2001). Glycosylation and the immune system. *Science* **291:**2370–2376.

Sagami, H., and Lennarz, W. J. (1987). Glycoprotein synthesis in *Drosophila* Kc cells. Biosynthesis of dolichol-linked saccharides. *J. Biol. Chem.* **262:**15610–15617.

Sareneva, T., Cantell, K., Pyhala, L., Pirhonen, J., and Julkunen, I. (1993). Effect of carbohydrates on the pharmacokinetics of human interferon-gamma. *J. Interferon Res.* **13:**267–269.

Sarkar, M., and Schachter, H. (2001). Cloning and expression of *Drosophila melanogaster* UDP-GlcNAc: Alpha-3-D-mannoside beta 1,2-N-acetylglucosaminyltransferase I. *Biol. Chem.* **382:**209–217.

Segawa, H., Yamashita, T., Kawakita, M., and Taira, H. (2000). Functional analysis of the individual oligosaccharide chains of sendai virus fusion protein. *J. Biochem. (Tokyo)* **128:**65–72.

Segawa, H., Kawakita, M., and Ishida, N. (2002). Human and *Drosophila* UDP-galactose transporters transport UDP-N-acetylgalactosamine in addition to UDP-galactose. *Eur. J. Biochem.* **269:**128–138.

Seo, N. S., Hollister, J. R., and Jarvis, D. L. (2001). Mammalian glycosyltransferase expression allows sialoglycoprotein production by baculovirus-infected insect cells. *Protein Expr. Purif.* **22:**234–241.

Snow, P. M., Patel, N. H., Harrelson, A. L., and Goodman, C. S. (1987). Neural-specific carbohydrate moiety shared by many surface glycoproteins in *Drosophila* and grasshopper embryos. *J. Neurosci.* **7:**4137–4144.

Sommer, U., and Spindler, K. D. (1991). Demonstration of beta-N-acetyl-D-glucosaminidase and beta-N-acetyl-D-hexosaminidase in *Drosophila* Kc-cells. *Arch. Insect Biochem. Physiol.* **17:**3–13.

Staudacher, E., and Marz, L. (1998). Strict order of (Fuc to Asn-linked GlcNAc) fucosyltransferases forming core-difucosylated structures. *Glycoconj. J.* **15:**355–360.

Staudacher, E., Kubelka, V., and Marz, L. (1992). Distinct N-glycan fucosylation potentials of three lepidopteran cell lines. *Eur. J. Biochem.* **207:**987–993.

Staudacher, E., Altmann, F., Wilson, I. B., and Marz, L. (1999). Fucose in N-glycans: From plant to man. *Biochim. Biophys. Acta* **1473:**216–236.

Stephens, E., Sugars, J., Maslen, S. L., Williams, D. H., Packman, L. C., and Ellar, D. J. (2004). The N-linked oligosaccharides of aminopeptidase N from Manduca sexta: Site localization and identification of novel N-glycan structures. *Eur. J. Biochem.* **271:**4241–4258.

Takahashi, N., Tsukamoto, Y., Shiosaka, S., Kishi, T., Hakoshima, T., Arata, Y., Yamaguchi, Y., Kato, K., and Shimada, I. (1999). N-Glycan structures of murine hippocampus serine protease, neuropsin, produced in Trichoplusia ni cells. *Glycoconj. J.* **16:**405–414.

Takeuchi, M., Inoue, N., Strickland, T. W., Kubota, M., Wada, M., Shimizu, R., Hoshi, S., Kozutsumi, H., Takasaki, S., and Kobata, A. (1989). Relationship between sugar chain structure and biological activity of recombinant human erythropoietin produced in Chinese hamster ovary cells. *Proc. Natl. Acad. Sci. USA* **86:**7819–7822.

Tomiya, N., Ailor, E., Lawrence, S. M., Betenbaugh, M. J., and Lee, Y. C. (2001). Determination of nucleotides and sugar nucleotides involved in protein glycosylation by high-performance anion-exchange chromatography: Sugar nucleotide contents in cultured insect cells and mammalian cells. *Analyt. Biochem.* **293:**129–137.

Tomiya, N., Betenbaugh, M. J., and Lee, Y. C. (2003a). Humanization of lepidopteran insect-cell-produced glycoproteins. *Acc. Chem. Res.* **36:**613–620.

Tomiya, N., Howe, D., Aumiller, J. J., Pathak, M., Park, J., Palter, K., Jarvis, D. L., Betenbaugh, M. J., and Lee, Y. C. (2003b). Complex-type biantennary *N*-glycans of recombinant human transferrin from *Trichoplusia ni* insect cells expressing mammalian β1,4-galactosyltransferase and β1,2-*N*-acetylglucosaminyltransferase II. *Glycobiology* **13:**23–34.

Tomiya, N., Narang, S., Lee, Y. C., and Betenbaugh, M. J. (2004). Comparing N-glycan processing in mammalian cell lines to native and engineered lepidopteran insect cell lines. *Glycoconj. J.* **21:**343–360.

Tretter, V., Altmann, F., Kubelka, V., Marz, L., and Becker, W. M. (1993). Fucose alpha 1,3-linked to the core region of glycoprotein N-glycans creates an important epitope for IgE from honeybee venom allergic individuals. *Int. Arch. Allergy Immunol.* **102:** 259–266.

Trombetta, E. S. (2003). The contribution of N-glycans and their processing in the endoplasmic reticulum to glycoprotein biosynthesis. *Glycobiology* **13:**77R–91R.

Tsitilou, S. G., and Grammenoudi, S. (2003). Evidence for alternative splicing and developmental regulation of the *Drosophila melanogaster* Mgat2 (N-acetylglucosaminyltransferase II) gene. *Biochem. Biophys. Res. Commun.* **312:**1372–1376.

Umana, P., Jean-Mairet, J., Moudry, R., Amstutz, H., and Bailey, J. E. (1999). Engineered glycoforms of an antineuroblastoma IgG1 with optimized antibody-dependent cellular cytotoxic activity. *Nat. Biotechnol.* **17:**176–180.

Vadaie, N., and Jarvis, D. L. (2004). Molecular cloning and functional characterization of a lepidopteran insect beta4-N-acetylgalactosaminyltransferase with broad substrate specificity, a functional role in glycoprotein biosynthesis, and a potential functional role in glycolipid biosynthesis. *J. Biol. Chem.* **279:**33501–33518.

van Die, I., van Tetering, A., Bakker, H., van den Eijnden, D. H., and Joziasse, D. H. (1996). Glycosylation in lepidopteran insect cells: Identification of a β1,4-N-acetylgalactosaminyltransferase involved in the synthesis of complex-type oligosaccharide chains. *Glycobiology* **6:**157–164.

Varki, A. (1993). Biological roles of oligosaccharides: All of the theories are correct. *Glycobiology* **3:**97–130.

Varki, A., Cummings, R., Esko, J., Freeze, H., Hart, G., and Marth, J. (1999). "Essentials of Glycobiology." Cold Spring Harbor Press, Cold Spring Harbor, New York.

Velardo, M. A., Bretthauer, R. K., Boutaud, A., Reinhold, B., Reinhold, V. N., and Castellino, F. J. (1993). The presence of UDP-N-acetylglucosamine: Alpha-3-D-mannoside beta 1,2-N-acetylglucosaminyltransferase I activity in Spodoptera frugiperda cells (IPLB-SF-21AE) and its enhancement as a result of baculovirus infection. *J. Biol. Chem.* **268:**17902–17907.

Viswanathan, K., Lawrence, S., Hinderlich, S., Yarema, K. J., Lee, Y. C., and Betenbaugh, M. J. (2003). Engineering sialic acid synthetic ability into insect cells: Identifying metabolic bottlenecks and devising strategies to overcome them. *Biochemistry* **42:**15215–15225.

Viswanathan, K., Narang, S., Hinderlich, S., Lee, Y. C., and Betenbaugh, M. J. (2005). Engineering intracellular CMP-sialic acid metabolism into insect cells and methods to enhance its generation. *Biochemistry* **44:**7526–7534.

Wagner, R., Geyer, H., Geyer, R., and Klenk, H. D. (1996a). N-Acetyl-beta-glucosaminidase accounts for differences in glycosylation of influenza virus hemagglutinin expressed in insect cells from a baculovirus vector. *J. Virol.* **70:**4103–4109.

Wagner, R., Liedtke, S., Kretzschmar, E., Geyer, H., Geyer, R., and Klenk, H. D. (1996b). Elongation of the N-glycans of fowl plague virus hemagglutinin expressed in Spodoptera frugiperda (Sf9) cells by coexpression of human β1,2-N-acetylglucosaminyltransferase I. *Glycobiology* **6:**165–175.

Watanabe, S., Kokuho, T., Takahashi, H., Takahashi, M., Kubota, T., and Inumaru, S. (2001). Sialylation of N-glycans on the recombinant proteins expressed by a baculovirus-insect cell system under β-N-acetylglucosaminidase inhibition. *J. Biol. Chem.* **277:**5090–5093.

Wendeler, M., Lemm, T., Weisgerber, J., Hoernschemeyer, J., Bartelsen, O., Schepers, U., and Sandhoff, K. (2003). Expression of recombinant human GM2-activator protein in insect cells: Purification and characterization by mass spectrometry. *Protein Expr. Purif.* **27:**259–266.

Williams, P. J., Wormald, M. R., Dwek, R. A., Rademacher, T. W., Parker, G. F., and Roberts, D. R. (1991). Characterization of oligosaccharides from *Drosophila melanogaster* glycoproteins. *Biochim. Biophys. Acta* **1075:**146–153.

Wilson, I. B., Harthill, J. E., Mullin, N. P., Ashford, D. A., and Altmann, F. (1998). Core alpha1,3-fucose is a key part of the epitope recognized by antibodies reacting against plant N-linked oligosaccharides and is present in a wide variety of plant extracts. *Glycobiology* **8:**651–661.

Wyss, D. F., and Wagner, G. (1996). The structural role of sugars in glycoproteins. *Curr. Opin. Biotechnol.* **7:**409–416.

Xia, Q., Zhou, Z., Lu, C., Cheng, D., Dai, F., Li, B., Zhao, P., Zha, X., Cheng, T., Chai, C., Pan, G., Xu, J. *et al.* (2004). A draft sequence for the genome of the domesticated silkworm (*Bombyx mori*). *Science* **306:**1937–1940.

Yuen, C. T., Storring, P. L., Tiplady, R. J., Izquierdo, M., Wait, R., Gee, C. K., Gerson, P., Lloyd, P., and Cremata, J. A. (2003). Relationships between the N-glycan structures and biological activities of recombinant human erythropoietins produced using different culture conditions and purification procedures. *Br. J. Haematol.* **121:**511–526.

ADVANCES IN VIRUS RESEARCH, VOL 68

VACCINES FOR VIRAL AND PARASITIC DISEASES PRODUCED WITH BACULOVIRUS VECTORS

Monique M. van Oers

Laboratory of Virology, Wageningen University, Binnenhaven 11
6709 PD, Wageningen, The Netherlands

Abstract

The baculovirus–insect cell expression system is an approved system for the production of viral antigens with vaccine potential for humans and animals and has been used for production of subunit vaccines against parasitic diseases as well. Many candidate subunit vaccines have been expressed in this system and immunization commonly led to protective immunity against pathogen challenge. The first vaccines produced in insect cells for animal use are now on the market. This chapter deals with the tailoring of the baculovirus–insect cell expression system for vaccine production in terms of expression levels, integrity and immunogenicity of recombinant proteins, and baculovirus genome stability. Various expression strategies are discussed including chimeric, virus-like particles, baculovirus display of foreign antigens on budded virions or

0065-3527/06 $35.00
DOI: 10.1016/S0065-3527(06)68006-8

in occlusion bodies, and specialized baculovirus vectors with mammalian promoters that express the antigen in the immunized individual. A historical overview shows the wide variety of viral (glyco)proteins that have successfully been expressed in this system for vaccine purposes. The potential of this expression system for antiparasite vaccines is illustrated. The combination of subunit vaccines and marker tests, both based on antigens expressed in insect cells, provides a powerful tool to combat disease and to monitor infectious agents.

I. Introduction to Recombinant Subunit Vaccines

Historically, vaccines have been one of the most cost-effective and easily administered means of controlling infectious diseases in humans and animals. Vaccine development has its roots in the work of Edward Jenner (1749–1823) who discovered that man could be protected from smallpox by inoculation with cowpox (Fenner, 2000) and the work of Louis Pasteur (1822–1895) who developed the first rabies vaccine (Fu, 1997). These pioneering efforts led to vaccines against diseases that had once claimed millions of lives worldwide (Andre, 2003). Childhood vaccination programs are now common practice and elaborate vaccination programs have been set up by the World Health Organization (WHO), leading to the official eradication of smallpox in 1979 (Fenner, 2000). Today large parts of the world are also declared poliomyelitis free, and measles is the next target for eradication. Vaccines have controlled major bacterial and viral diseases in humans, and effective vaccines are available against many more (Andre, 2003; Hansson *et al.*, 2000b). Vaccination also protects our livestock and pet animals (Pastoret *et al.*, 1997). For some diseases, however, such as malaria and acquired immunodeficiency syndrome (AIDS), vaccines are desparately sought.

Most human and animal vaccines are based on killed or live-attenuated pathogens. Killed vaccines require the production of large amounts of often highly virulent pathogens and these types of vaccines are therefore risky to produce. Another risk lies in the potential for incomplete inactivation of the pathogens. Inactivation on the other hand affects the immunogenic properties of the pathogen, and hence the efficacy as a vaccine, and it is often difficult to find the balance between efficient inactivation and conservation of immunogenicity. Live-attenuated vaccines consist of pathogens that are reduced in virulence or have been attenuated either by growing them in alternative hosts or under unfavorable growing conditions, or by recombinant DNA technology. These live-attenuated vaccines can potentially replicate in their host, but are typically attenuated in their pathogenicity to avoid the development of severe disease. Live vaccines elicit humoral and cellular immunity, and may provide lifelong protection with a single or a few doses (Dertzbaugh, 1998; Hansson *et al.*, 2000b; Schijns, 2003). Such long-term protection is advantageous in developing countries where individuals are often only immunized once. A drawback of live-attenuated vaccines is that they can cause side effects, which may be dangerous when used for prophylaxis in immunocompromised persons such as the elderly or individuals with genetic or acquired diseases of the immune system (e.g., AIDS or severe combined immunodeficiency; SCID). Live-attenuated vaccines may also convert to virulent strains and spread to nonimmunized persons as observed during recent poliomyelitis outbreaks (Kew *et al.*, 2004). Adverse effects with both killed and live-attenuated vaccines can also be due to allergic reactions to components of the vaccine such as residual egg proteins in the case of influenza vaccines (Kelso and Yunginger, 2003) or gelatin in the measles-mumps-rubella (MMR) vaccine (Patja *et al.*, 2001).

The development of vaccines is not easy for all infectious diseases and the medical and veterinary world is challenged frequently by the emergence of novel diseases such as AIDS, severe acute respiratory syndrome (SARS), and *West Nile virus* infection. The vaccine industry is under constant pressure for rapidly changing pathogens, for which large amounts of vaccines are needed annually, such as influenza viruses (Palese, 2004), and flexible vaccine production techniques are required. For several infectious diseases vaccines cannot be developed using conventional approaches, for instance due to a lack of appropriate animal production systems or the high-mutation frequency of the pathogen (*Human immunodeficiency virus* (HIV), malaria). A vaccine against the H5N1 influenza strain that is currently epidemic in Asian poultry could not be produced the classical way, by using embryonized chicken eggs without reducing the virulence of the virus by reverse genetics, due to high-mortality rates of the chicken embryos (Horimoto *et al.*, 2006).

Recombinant protein production systems may provide good alternatives for the development of vaccines that are more difficult to produce *in vivo* for manufacture of so-called subunit vaccines. A pathogen consists of many proteins, frequently with carbohydrate moieties, but these are not all equally important for generation of an adequate immunological response. Subunit vaccines contain the immunodominant components of a pathogen and in the case of viral vaccines these are often (glyco)proteins of the viral coat or envelope such as the hepatitis B surface antigen (Valenzuela *et al.*, 1982) or the classical swine fever virus (CSFV) E2 glycoprotein (Bouma *et al.*, 1999). Viral coat proteins sometimes form virus-like particles (VLPs) when expressed in heterologous systems (Brown *et al.*, 1991), which are often immunogenic and may induce both humoral and cellular responses. The subunit vaccine against hepatitis B produced in yeast is highly succesful. An extreme example of subunit vaccines are peptide-based vaccines which consist of small amino acid chains harboring the part of the antigenic protein that is recognized by antibodies. Typically, subunit vaccines do not contain the genetic material of the pathogen or only a small part thereof. Therefore, these vaccines cannot cause disease and do not introduce pathogens into nonendemic regions. An additional advantage of subunit vaccines is that they can be used in combination with specific marker tests, which make it possible to *d*ifferentiate *i*nfected from *v*accinated *a*nimals, the so-called DIVA vaccines (Capua *et al.*, 2003; van Oirschot, 1999); an important issue in monitoring virus prevalence and virus-free export of animals and their products.

Immunogenic subunits can be isolated chemically from the pathogen, such as the purified capsular polysaccharides present in the *Streptococcus pneumonia* vaccine (Pneumovax23; Merck). This process still requires the production of virulent pathogens, which is not without risk. An alternative is the use of recombinant DNA technology to produce protein subunits in a heterologous system, and a variety of expression systems are available (Clark and Cassidy-Hanley, 2005; Hansson *et al.*, 2000b). The yeast system *Saccharomyces cerevisiae* for instance is used to produce the hepatitis B subunit vaccine (Valenzuela *et al.*, 1982), which is currently the only licensed recombinant subunit vaccine for human use. The yeast *Pichia pastoris* is used for production of the antitick vaccine Gavac™ (Canales *et al.*, 1997), which protects cattle against the tick *Boophilus microplus*, the transmitter of *Babesia* and *Anaplasma* parasite species. Insect cells are used to produce vaccines against classical swine fever or hog cholera (Depner *et al.*, 2001; van Aarle, 2003). For the production of recombinant proteins in higher eukaryotics, mammalian, insect, and plant expression systems are available that either use trangenes or viral vectors for protein expression. Plants have been recognized for the production of so-called edible subunit vaccines to be administered by ingestion of vegetable foods (Ma *et al.*, 2005; Streatfield and Howard, 2003).

This chapter concentrates on the use of cultured insect cells or larvae in combination with baculovirus expression vectors for the production of subunit vaccines. The baculovirus expression system is an accepted and well-developed system for the production of viral antigens with vaccine potential (Dertzbaugh, 1998; Hansson *et al.*, 2000a; Vlak and Keus, 1990). This system has also been explored for development of vaccines against protozoan parasites (Kaba *et al.*, 2005) and for therapeutic vaccines against tumors. A vaccine against prostate-cancer (Provenge) is in phase II/III clinical trials and is based on combining recombinant prostatic acid phosphatase (characteristic of 95% of prostate cancers) with the patient's own dendritic cells before immunization (Beinart *et al.*, 2005; Rini, 2002). Trials have also been initiated for a prophylactic vaccine using VLPs produced in insect cells against cervical cancer caused by *Human papillomavirus* (HPV) *16* (Mao *et al.*, 2006).

Each expression system has advantages and drawbacks (Table I) and the system of choice depends very much on the specific requirements for a particular vaccine and is often based, at least partly, on trial and error. Before a definitive choice can be made, the expression levels achieved, the adequacy of posttranslational modifications, the immunological performance, the possibilites for scale-up, the costs, the risk of contamination, the method of administration, and legal aspects must all be taken into account.

TABLE I

POTENTIAL OF VARIOUS EXPRESSION SYSTEMS FOR RECOMBINANT SUBUNIT VACCINE PRODUCTION[a]

Processing/ feature	*E. coli*	Yeast	Mammalian cells	Insect cells	Plants
Glycosylation	−	+	+ + +	+ +	+
Phosphorylation	−	+	+ +	+ +	+
Acylation	−	+	+	+	+
Amidation	−	−	+	+	−
Proteolysis	+/−	+/−	+	+	+
Folding	+/−	+/−	+ + +	+ +	+
Secretion	+/−	+	+ +	+ +	+/−
Serum free	Not relevant	Not relevant	+	+	Not relevant
Yield (%dry mass)	1–5	1	<1	Up to 30	<5
Scale-up	+ + +	+ + +	+	+	+ + +
Downstream processing	+	+	+ +	++	− −
Costs	Low	Low	High	Intermediate	Low
Safety	+ +	+ +	+	+ +	+ +
Versatility	+	+	+ +	+ + +	+

[a] Adapted from Vlak and Keus, Baculovirus Expression Vector System for Production of Viral Vaccines, Advances in Biotechnological Processes 14, pp. 19–28. Copyright © (1990, John Wiley & Sons, Inc.). Reprinted with permission of John Wiley & Sons, Inc.

II. The Baculovirus–Insect Cell Expression System for Vaccine Production

A. *Characteristics*

The baculovirus–insect cell expression system (Smith *et al.*, 1983) has been developed for the production of biologically active (glyco) proteins in a well-established and safe eukaryotic environment (Kost *et al.*, 2005). The family *Baculoviridae* contains rod-shaped, invertebrate-infecting viruses, which have large double-stranded, covalently closed circular DNA genomes (Table II). The members of this large virus family are taxonomically divided into the genera Nucleopolyhedrovirus (NPV) and Granulovirus (GV), based on occlusion body morphology (Theilmann *et al.*, 2005). NPVs express two genes, *polyhedrin* and *p10*, at very high levels in the very late phase of infection. The polyhedrin protein forms the viral occlusion bodies or polyhedra and p10 is present in fibrillar structures, which function in polyhedron morphology and in breakdown of infected cell-nuclei to release the polyhedra (Okano *et al.*, 2006; Van Oers and Vlak, 1997). These two genes are not essential for virus replication in cell culture and, therefore, their promoters are exploited to drive foreign gene expression, which forms the basis for the baculovirus–insect cell expression system. Since baculoviruses are rod-shaped, large amounts of foreign DNA can be accommodated within the virus particle, in contrast to vaccinia and especially adenovirus expression vectors (Table II).

The type member of the NPVs is *Autographa californica* multiple nucleopolyhedrovirus (AcMNPV), a virus with a genome of 133 kilobase pairs (Ayres *et al.*, 1994). This baculovirus is routinely used for foreign gene expression. The baculovirus *Bombyx mori* NPV is being used for vaccine purposes to a much lesser extent (Choi *et al.*, 2000; Mori *et al.*, 1994). Baculovirus expression vectors replicate in cultured insect cells or larvae and high yields of heterologous protein are generally obtained when the strong viral *polyhedrin* and *p10* promoters are exploited (King and Possee, 1992; O'Reilly *et al.*, 1992). The insect cell lines used in the baculovirus expression system are derived from lepidopteran insects (moths) and are most often *Spodoptera frugiperda* lines (Sf9 or Sf21) and *Trichoplusia ni* (High Five™) cells, which can be used in combination with AcMNPV-based vectors. *B. mori* cells (e.g., Bm5) are used for BmNPV. Insect cell lines vary in their characteristics in terms of growth rate, protein production, secretion efficiency and glycosylation pattern, and interference with viral genome stability (Pijlman *et al.*, 2003b; Vlak *et al.*, 1996). These insect cells are

TABLE II

CHARACTERISTICS OF BACULOVIRUS VECTORS VERSUS VACCINIA AND ADENOVIRUS VECTORS[a]

Feature	Baculovirus[b]	Adenovirus	Vaccinia
Virus morphology	Enveloped, rod shaped	Nonenveloped, icosahedral	Brick shaped
Genome structure	Circular dsDNA	Linear dsDNA	Linear dsDNA
Genome size	130 kbp	±35 kbp	190 kbp
Expandability	Large	Low	Intermediate
Particle dimensions	30–60 × 250 –300 nm	80–110 nm	250 × 250 × 200 nm
Replication site	Nucleus	Nucleus	Cytoplasm
Replication in humans	None	Replication competent or defective	Yes
Progeny virus	Budding BVs/lysis ODVs	Accumulation in the nucleus	Exocytosis/lysis
Pathogenicity for mammals including humans	Nonpathogenic	Low due to host defense and attenuation	Reduced with modified strains
Immunological complications	Complement inactivation	Strong protective responses of the host	–
Immunological history	–	Preexisting immunity due to natural infections	Preexisting immunity due to smallpox vaccination
Protein production system in cell lines	Yes	Less frequently	Yes
Applications:			
Antigen display vector	Surface display vectors	No	No
Carrier DNA vaccine vector	Yes	Yes	Yes
Gene therapy	+	++	–
Vaccine examples	Therapeutic prostate cancer vaccine (see text for further information)	Immunomodulators, therapeutic cancer vaccines	Mucosal immunity against tuberculosis and HIV

[a] Gherardi and Esteban, 2005; Russell, 2000; Young *et al.*, 2006; Universal data base of International Committee on Virus Taxonomy (http://www.ncbi.nlm.nih.gov/ICTVdb/index.htm; January 2006).

[b] AcMNPV, *Autographa californica* multiple nucleopolyhedrovirus.

relatively easy to maintain and many grow equally well in suspension in large volumes (up to 2000 L reactions) and at high densities as on solid supports, and can be cultivated in serum-free media which facilitates purification of recombinant proteins. Unlike mammalian cells, they do not require CO_2 and can easily withstand temperature fluctuations. An extra advantage is that the chance of contamination with human or mammalian viruses, especially in serum-free cultures, is small compared to mammalian production systems because these vertebrate viruses do not replicate in lepidopteran cells. These cells do not support the growth of mammalian mycoplasmas either. Instead of insect cells, whole insect larvae may be used as live bioreactors for vaccine production. The use of whole insect larvae has the advantage that the simple insect-rearing technology and downstream processing can be exploited. Such *in vivo* production could be performed by small-scale local industries, especially if the larvae can be fed directly to animals such as for an experimental Newcastle disease vaccine for chickens (Mori *et al.*, 1994). Such vaccines are less well defined however and quality control may therefore be more difficult to achieve.

Expression of proteins in insect cells allows for appropriate folding, posttranslational modification, and oligomerization and therefore, biological activity is normally preserved. Protein glycosylation in insects and mammals is not identical though: the *N*-glycan–processing pathway in insects results in glycoproteins with paucimannose glycan groups, in contrast to mammalian glycoproteins which contain complex sialylated glycans (see also Harrison, this volume, pp. 159–191). The exact glycan composition varies between different insect cell lines (Kost *et al.*, 2005; Tomiya *et al.*, 2004). In general, glycan groups are not very immunogenic and therefore this does not seem to be a major disadvantage for subunit vaccines. In situations where more authentic glycosylation is required, for instance for preserving functional activity, transformed "humanized" insect cell lines expressing mammalian glycosylation enzymes are available (Jarvis, 2003; Kost *et al.*, 2005; Tomiya *et al.*, 2004). For some insect cell lines it has been reported that fucose groups are added to *N*-glycans. The impact of this remains to be determined, but since fucans may cause allergic reactions, it may be a point for consideration when choosing an insect cell line for vaccine production (Long *et al.*, 2006; Tomiya *et al.*, 2004).

B. Baculovirus Vectors

Originally, the baculovirus expression system was based on the allelic exchange of the baculovirus polyhedrin gene for a heterologous gene by recombination in insect cells (Smith *et al.,* 1983). In a similar way, baculovirus vectors have since been developed which exploit the nonessential very late baculovirus *p10* promoter (Vlak *et al.*, 1990; Weyer and Possee, 1991). Vectors that leave the polyhedrin gene intact can be used for the production of recombinant proteins in insect larvae (Fig. 1). The *in vivo* recombination protocol was improved by using linearized viral DNA in the allelic replacement, which resulted in dominant selection and much higher percentages of recombinant viruses (Kitts *et al.*, 1990; Martens *et al.*, 1995). In vectors of this type (BacPAK™ vectors, BaculoGold™, Bac-N-Blue™) the linearized viral DNA carries a lethal deletion (ORF1609) and becomes replication competent only after recombination with a transfer plasmid carrying the foreign gene, thereby restoring the deletion (Kitts and Possee, 1993). Baculovirus vectors based on Gateway technology (BaculoDirect™) are linear baculovirus vectors in which foreign genes are introduced through site-specific *in vitro* recombination.

At about the same time, another efficient and rapid method for generation of recombinant baculoviruses was developed (Luckow *et al.*, 1993) that employed transposition of a foreign gene expression cassette from a donor plasmid into a bacterial artificial chromosome (BAC) which contains the entire AcMNPV genome (bacmid). In this system (Bac-to-Bac™) recombinant baculovirus genomes are generated in *Escherichia coli* and then used to transfect insect cells to obtain recombinant baculovirus particles. After generating high-titer virus stocks, insect cells are infected to produce recombinant proteins. With the bacmid-based methodology the time to generate recombinant viruses is reduced considerably. Another advantage is that the recombinant bacmid can be stored in *E. coli* and recovered when needed. A disadvantage is that the bacterial gene cassette present in bacmid-derived viruses may easily be lost during virus passaging (Pijlman *et al.*, 2003a). In addition to AcMNPV, bacmids have also been constructed for *Spodoptera exigua* MNPV and *Helicoverpa armigera* SNPV (Pijlman *et al.*, 2002; Wang *et al.*, 2003). The most recent method combines bacmid technologies with allelic replacement (FlashBac™; Oxford Expression Technologies) and thereby removes the BAC sequences from the viral genome. This latter system is especially suitable for high-throughput screening.

Over the years, novel baculovirus vectors have been developed with special features and for more specific applications: transfer vectors have been modified to express polyhistidine-tagged proteins for easy purification (pFastBac-His™). Transfer vectors with dual, triple, or quadruple promoters usually *p10* and *polyhedrin*, have been developed

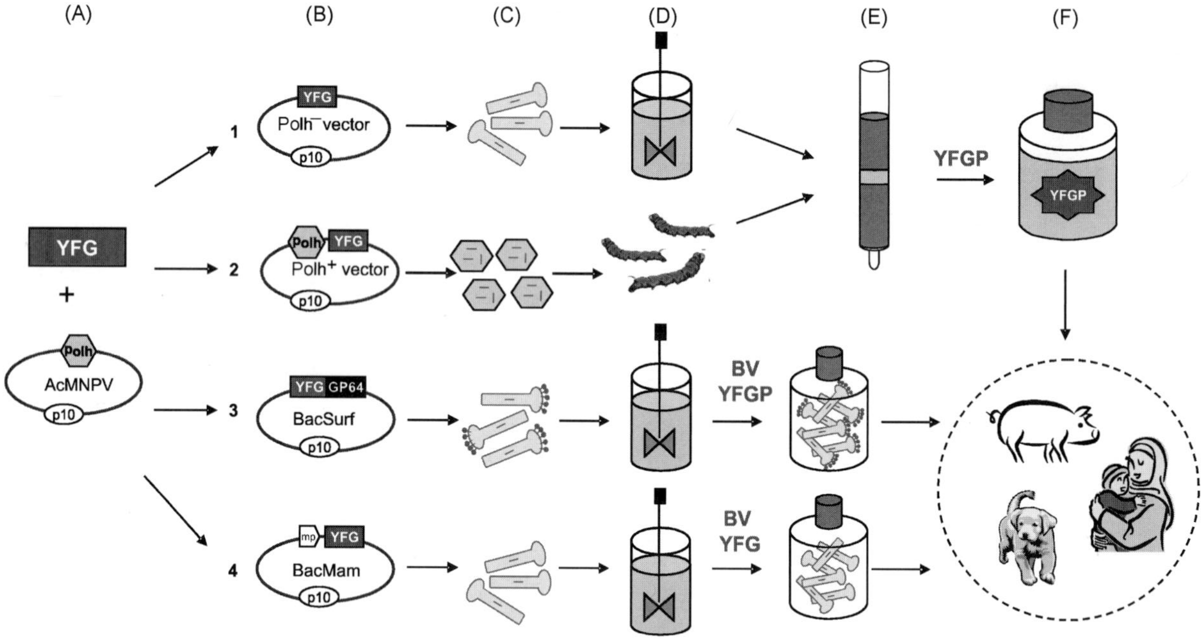

FIG 1. Flow chart showing four different methods to make a vaccine based on your favorite gene (*YFG*) in the baculovirus expression system: (1) protein expression in insect cell bioreactors using the *polyhedrin* locus for expression, (2) protein expression in insect larvae leaving the *polyhedrin* gene intact; expression is driven either by a duplicated *p10* promoter or by the original *p10* promoter, (3) baculovirus surface display methods where YFG is fused to GP64, and (4) DNA vectors with a mammalian promoter (*mp*) for synthesis of your favorite gene product (YFGP) in the target species. Subsequent steps in the process are: (A) selection and PCR amplification of *YFG*, (B) cloning into the appropriate baculovirus vector, (C) generation of recombinant BV particles or occlusion bodies, (D) production in insect cells in bioreactors or larvae, (E) purification of recombinant YFGP or collection of BVs loaded with either YFGP or YFG, and (F) delivery of prophylactic or therapeutic vaccines. (See Color Insert.)

for allelic replacement (Belyaev and Roy, 1993; Weyer and Possee, 1991); dual (pFastBacDual™) and quadruple vectors (Tareilus *et al.*, 2003) have also been developed for bacmid technology. Such multiple vectors can be used to express various proteins simultaneously, and hence are useful for producing multimeric complexes, including viral capsids consisting of more than one viral protein (Belyaev and Roy, 1993). Balancing expression levels is sometimes a problem in these vectors and may require coinfection with a vector expressing only the dominant protein, or a modification of the promoters. One of the promoters in multiple promoter vectors may be used to express a reporter gene, such as green fluorescent protein (GFP), which makes it easy to follow the infection process in cells, perform virus titrations, and track baculovirus infection in the insect (Cha *et al.*, 1997; Kaba *et al.*, 2003). The recently developed vector system (UltraBac) uses the baculovirus late basic protein (P6.9) promoter to express GFP together with the foreign gene to allow earlier monitoring of infection (Philipps *et al.*, 2005).

Baculovirus surface display vectors (Grabherr *et al.*, 2001) expose the antigen on the surface of budded baculovirus particles. This is achieved by fusing the foreign antigen to the baculovirus envelope glycoprotein GP64 (Monsma *et al.*, 1996). The chimeric protein is transported to the cell membrane and is taken up in the viral envelope during budding. This system has also been combined with bacmid technology (Kaba *et al.*, 2003). The recombinant budded virus (BV) particles and lysates of cells infected with a display vector have been shown to evoke protective immune responses (Kaba *et al.*, 2005; Tami *et al.*, 2004; Yoshida *et al.*, 2003). Baculovirus vectors that express foreign genes in fusion with polyhedrin along with wild-type polyhedrin allow for incorporation of antigens into baculovirus occlusion bodies (Je *et al.*, 2003). These occlusion bodies are stable and easy to purify and can be used directly for immunization (Wilson *et al.*, 2005).

C. Adaptations for Secreted Proteins

Expression of surface (glyco)proteins that go through the export pathway is in general more difficult than expression of soluble cytoplasmic proteins and results in much lower yields (van Oers *et al.*, 2001). To increase the production level, surface proteins are often expressed as secreted proteins by removing hydrophobic transmembrane regions (TMR) that serve to anchor the protein to cell membranes. Removing these domains by recombinant DNA technology leads to secreted proteins which can then be purified from the culture medium. However, some caution is needed because this approach may affect folding and vaccine efficacy.

Not all proteins present at the surface under native conditions are automatically transported to the cell surface when expressed in insect cells, such as the p67 surface protein of the bovine parasite *Theileria parva* (Nene *et al.*, 1995). When the original signal peptide was replaced with an insect analogue, such as the honeybee mellitin signal peptide (Tessier *et al.*, 1991), p67 was properly routed to the cell surface (Kaba *et al.*, 2004a). A similar routing of p67 to the export pathway could be obtained by fusion to GP64 in a surface display vector (Kaba *et al.*, 2002), where the GP64 signal peptide directed the protein to the cell surface.

Membrane and secreted proteins pass through the endoplasmic reticulum (ER) and the Golgi apparatus on their way to the cell surface and may become glycosylated during this process. The abundant baculovirus protein chitinase is also transported to the ER and accumulates there due to a KDEL retention sequence (Saville *et al.*, 2004; Thomas *et al.*, 1998). Chitinase is expressed in the late phase of baculovirus infection and is involved in the dissolution of the insect chitinous cuticle to enhance the spread of viral occlusion bodies (Hawtin *et al.*, 1997). Deletion of chitinase from the baculovirus vector resulted in higher levels of secreted recombinant protein (Possee *et al.*, 1999) possibly because chitinase "cloggs up" the protein translocation machinery and competes with recombinant secretory proteins. The FlashBac system described earlier lacks this chitinase gene. Another baculovirus protein, v-cathepsin also accumulates in the ER and is activated on cell death by proteolytic cleavage (Hom *et al.*, 2002). Processing of pro-v-cathepsin into active cathepsin is also triggered by chaotropic agents, such as sodium dodecyl sulfate, and this may result in proteolysis of recombinant proteins during extraction and purification (Hom and Volkman, 1998). A bacmid vector that lacked both chitinase and v-cathepsin (AcBacΔCC) improved the stability of a secreted recombinant protein, thereby increasing the yield of full-length protein molecules (Kaba *et al.*, 2004a).

Folding of complicated transmembrane glycoproteins can be improved by coexpression of molecular chaperones. The serotonin transporter (SERT) protein is a brain glycoprotein with 12 predicted transmembrane domains. Coexpression of the chaperones calnexin and, to a lesser extent, of immunoglobulin heavy chain-binding protein (BiP) or calreticulin increased the yield of functional SERT threefold. The foldase ERp57 did not have this effect (Tate *et al.*, 1999). Calreticulin and calnexin were also shown to increase the level of active lipoprotein lipase when coexpressed in insect cells, and to stimulate dimerization of the recombinant protein (Zhang *et al.*, 2003). Expression of calnexin and calreticulin in a stable transgenic insect cell line, which was than infected with a recombinant baculovirus, resulted in a lower ratio of secreted versus intracellular recombinant protein than when cells were coinfected with two baculoviruses, one carrying the gene of interest and the other a chaperone (Kato *et al.*, 2005). This result suggests that chaperone expression levels should be of the same order as recombinant protein levels.

D. Baculovirus Vectors with Mammalian Promoters

Another special adaptation is the incorporation of mammalian promoters in baculovirus vectors to drive foreign gene expression. Baculovirus vectors with mammalian promoters (BacMam™ viruses) have the potential to serve as gene delivery vectors in gene therapy (Huser and Hofmann, 2003; Kost and Condreay, 2002) and have also been tested for vaccination purposes (Abe *et al.*, 2003; Aoki *et al.*, 1999; Facciabene *et al.*, 2004; Poomputsa *et al.*, 2003). In this case, a mammalian promoter or a viral promoter active in mammalian cells, such as the human cytomegalovirus (HCMV) IE1 promoter, drives intracellular expression of the antigen. Exposure on the cell surface via the major histocompatibility complex (MHC) activates the cellular immune system and in this respect, these types of vaccines resemble DNA vaccines. BacMam™ vectors are produced in insect cells and are replication incompetent in mammalian cells (Table II). A further advantage is that multiple genes can be inserted simultaneously into the baculovirus genome allowing for multivalent vaccines. Expression of multiple proteins is an advantage of the baculovirus expression system over other systems, especially adenovirus vectors, where the maximal increase in genome size is more limited due to packaging restrictions.

E. Adaptations for Vector Genome Stability

For manufacturing subunit vaccines, large-scale production units will be needed, for instance for the production of malaria or the annual influenza subunit vaccines. Baculovirus–insect cell systems have been scaled-up to large-scale cultures in either fermentors (bioreactors) or cellbag devices (WAVE reactors). Insect cell bioreactors up to 2000 L have been reported. The bioprocess technology behind this large scale production has been reviewed by others (Hunt, 2005; Ikonomou *et al.*, 2003; Vlak *et al.*, 1996). A problem repeatedly encountered when expressing recombinant proteins with baculovirus vectors is a drop in expression levels with increasing virus passage (reviewed in Krell, 1996). This so-called "passage effect" is intrinsic to baculovirus replication in cell culture, but is less critical for small laboratory-scale protein production when the number of virus passages is low (<10). It is a significant problem though for large-scale industrial production of vaccines in insect cell bioreactors (Van Lier *et al.,* 1996) and prevents the use of continous bioreactors. The major causes of loss of recombinant protein expression are (1) mutations in the *FP25K* gene, reducing the activity of the *polyhedrin* promoter (Harrison *et al.*, 1996), (2) the generation of defective interfering particles (DIs) which replicate at the expense of the full-length recombinant virus (Kool *et al.,* 1991; Pijlman *et al.*, 2001; Wickham *et al.,* 1991), (3) the intracellular accumulation of concatenated viral sequences, for example, non-*hr* (homologous repeat) origins of DNA replication) which interfere with replication of full-length genomes (Lee and Krell, 1994; Pijlman *et al.*, 2002), and (4) spontaneous deletion of the heterologous gene from the baculovirus vector. The latter aspect is especially seen in bacmid-derived vectors, which are extremely sensitive to spontaneous removal of the expression cassette, a large piece of DNA which is not under selection (Pijlman *et al.*, 2003a). To prevent the amplification of DIs, baculovirus vectors must be used at low multiplicities of infection (MOI) (de Gooijer *et al.*, 1992; Wickham *et al.*, 1991) and it is now common practice to keep the number of viral passages to a minimum and establish low-passage virus banks as seed stocks for production purposes.

In recent years, several approaches have been used to improve the stability of the baculovirus genome. The accumulation of non-*hr*–containing sequences can easily be prevented by removing this sequence from the baculovirus backbone (Pijlman *et al.*, 2002). Reducing the distance between origins of replication in the bacmid system by insertion of an extra *hr* sequence within the expression cassette also resulted in prolonged foreign gene expression in a test bioreactor (Pijlman *et al.*, 2004). In the FlashBac system, all destabilizing bacterially derived sequences are removed on recombination with the transfer vector. To prevent loss of the foreign gene cassette, a bicistronic vector was developed that contained the foreign gene and the baculovirus essential gene *GP64* on a single bicistronic transcriptional unit linked by an internal ribosome entry site (IRES). *GP64* was deleted from its original locus. In this bicistronic vector, loss of the foreign gene would automatically result in loss of expression of the essential gene, which is needed for the generation of complete virus particles as well as for DIs. GFP expression levels were kept at a high level for at least

20 passages with this vector providing dominant selection for GP64 (Pijlman *et al.*, 2006). This system awaits testing for expression of proteins of medical importance. By combining several of the methods described in this section, it is likely that genome stability will be further improved.

III. Viral Subunits Expressed in the Baculovirus System

Since its recognition as a production system for subunit vaccines (Vlak and Keus, 1990), the baculovirus–insect cell expression system has been used extensively for the expression of candidate vaccine antigens. A comprehensive overview of the viral antigens from viruses of vertebrates that have been expressed in this system is provided in Table III. Only those antigens that were tested for their ability to induce protective immune responses are included. In addition, many viral antigens have successfully been expressed in insect cells for the development of diagnostics, and to perform structural and functional studies, but these studies are excluded from this chapter. Various viral antigens ranging from capsid and envelope proteins to nonstructural proteins have been chosen for the development of subunit vaccines. These viral antigens can be divided into those that are expressed as single or oligomeric protein subunits, and those that self-assemble into VLPs. Different approaches to vaccination are described in the examples later, with special attention paid to influenza subunit vaccines.

A. Viral Envelope Proteins

Envelope proteins are synthesized as single or oligomeric subunits. The expressed envelope proteins are often functionally active and have been reported to oligomerize, an indication that they are correctly folded (Crawford *et al.*, 1999). Commonly, viral envelope glycoproteins are glycosylated in insect cells. Examples of baculovirus-produced subunit vaccine candidates (Table III) are the fusion proteins and hemagglutinins of paramyxoviruses, such as *Newcastle disease virus*, and the E proteins of *Flaviviridae,* including *West Nile virus*, dengue viruses, and CSFV. Two commercially available veterinary subunit vaccines against classical swine fever (BAYOVAC CSF E2™ and PORCILIS PESTI™) are based on the CSFV E2 glycoprotein produced in insect cells (Ahrens *et al.*, 2000; Bouma *et al.*, 1999, 2000; Depner *et al.*, 2001; van Aarle, 2003). The E2 envelope glycoprotein of CSFV was expressed as a secreted protein by removing the TMR and this resulted in a

TABLE III

VERTEBRATE IMMUNE RESPONSE STUDIES WITH VIRAL PROTEINS EXPRESSED IN THE BACULOVIRUS–INSECT CELL SYSTEM[a]

Virus family or genus	Abbreviation	Host	Antigen(s)[b]	Neutralizing antibodies[c]	T cells/ cytokines	Protection host/model	References
Asfarviridae							
African swine fever virus	ASFV	Pigs	p22, p30, p54, p72	Yes	–	No	Neilan *et al.*, 2004
			p30–p54 fusion	Yes	–	Yes	Barderas *et al.*, 2001
			HA	Yes	–	Yes	Ruiz-Gonzalvo *et al.*, 1996
Arenaviridae							
Lymphocytic choriomeningitis virus	LCMV	Humans, rodents	GP, NP	–	T cells	Yes	Bachmann *et al.*, 1994
Arteriviridae							
Porcine reproductive and respiratory syndrome virus	PRRSV	Pigs	3, 5 (7)	Yes	–	Yes	Plana Duran *et al.*, 1997
Birnaviridae							
Infectious pancreatic necrosis virus	IPNV	Fish	Structural proteins (VLP[d])	–	–	Partial	Shivappa *et al.*, 2005
Infectious bursal disease virus	IBDV	Birds	VP2 (VLP)	Yes	–	Yes	Pitcovski *et al.*, 1996; Wang *et al.*, 2000
			VP2 (VLP), VPX, PP	Yes	–	Yes	Martinez-Torrecuadrada *et al.*, 2003
			VP2 + VP3 + VP4 (chimeric)	Yes	–	Yes	Snyder *et al.*, 1994
			VP2 + VP3 + VP4	Yes	–	Yes	Vakharia *et al.*, 1994
			VP3	No	–	No	Pitcovski *et al.*, 1999
Yellowtail ascites virus	YAV	Fish	VP2, VP3, NS	Yes	–	Yes	Sato *et al.*, 2000

Bunyaviridae							
La Crosse virus	LACV	Humans	G1	Yes	–	Yes	Pekosz *et al.*, 1995
Hantaan virus	HTNV	Humans, rodents	G1, G2, NP	Yes	–	Yes	Schmaljohn *et al.*, 1990
Rift Valley fever virus	RVFV	Humans, ruminants	G1, G2	Yes	–	Yes	Schmaljohn *et al.*, 1989
Caliciviridae							
Hepatitis E virus	HEV	Humans	Capsid (VLP)	Yes	–	Yes	Li *et al.*, 2001, 2004
Norwalk virus, Genogroup I	NWV	Humans	Capsid (VLP)	Antibodies	–	–	Ball *et al.*, 1996, 1998, 1999; Guerrero *et al.*, 2001
Norwalk virus, Genogroup II	NWV	Humans	Capsid (VLP)	Antibodies	Yes	–	Nicollier-Jamot *et al.*, 2004
Circoviridae							
Chicken anaemia virus	CAV	Birds	VP1, VP2	Yes	–	Yes	Koch *et al.*, 1995
Porcine circovirus 2	PCV2	Pigs	ORF2	Yes	–	Yes	Blanchard *et al.*, 2003
Coronaviridae							
Avian infectious bronchitis virus	IBV	Chicken	S1	Yes	–	Partial	Cavanagh, 2003; Song *et al.*, 1998
Feline infectious peritonitis virus	FIPV	Cats	N	No	Yes	Yes	Hohdatsu *et al.*, 2003
SARS corona virus	SARS	Humans	Spike GP	Yes	–	Yes	Bisht *et al.*, 2005
Transmissible gastroenteritis virus	TGEV	Pigs	S + N + M	Yes	Yes	Partial	Sestak *et al.*, 1999
Deltavirus							
Hepatitis deltavirus	–	Humans/ rodents	HD Ag	No	–	–	Karayiannis *et al.*, 1993
			HD Ag p24, p27	Antibodies	–	No	Fiedler and Roggendorf, 2001

(continues)

TABLE III (*continued*)

Virus family or genus	Abbreviation	Host	Antigen(s)[b]	Neutralizing antibodies[c]	T cells/ cytokines	Protection host/model	References
Filoviridae							
Ebola virus	EBOV	Humans	GP	Yes	T cells	Partial	Mellquist-Riemenschneider *et al.*, 2003
Marburg virus	MBGV	Humans	GP	Yes	–	Yes	Hevey *et al.*, 1997
Flaviviridae							
Bovine viral diarrhea virus	BVDV	Cows	E2	Yes	–	Yes	Bolin and Ridpath, 1996
Classical swine fever virus	CSFV	Pigs	E2	Yes	–	Yes	Ahrens *et al.*, 2000; Bouma *et al.*, 1999; Hulst *et al.*, 1993
Dengue 2 virus	DEN 2	Humans	E	Yes	–	–	Kelly *et al.*, 2000
			E	Yes	–	Partial	Delenda *et al.*, 1994; Velzing *et al.*, 1999
			E	No	–	Partial	Feighny *et al.*, 1994
			NS1	Yes	–	Partial	Qu *et al.*, 1993
Dengue 4 virus	DEN 4	Humans	Cocktail	Yes	–	Yes	Zhang *et al.*, 1988
			Cocktail, E	Yes	–	Partial	Eckels *et al.*, 1994
Dengue virus 2 + 3	DEN2/3	Humans	E protein hybrid	Yes	T cells	–	Bielefeldt-Ohmann *et al.*, 1997
Japanese encephalitis virus	JEV		prME, E, NS1	Yes	–	Yes/No	Yang *et al.*, 2005
			E, NS1	Yes	–	Yes	McCown *et al.*, 1990
West Nile virus	WNV	Humans/birds	prME (VLP)	Yes	–	Yes	Qiao *et al.*, 2004
Hepatitis C virus	HCV	Humans	E1 + E2 (VLP)	Yes	T cells/ cytokines	Yes	Jeong *et al.*, 2004

St Louis encephalitis virus	SLEV	Humans	prME	Yes	–	Yes	Venugopal *et al.*, 1995
Tick-borne encaphalitis virus	TBEV	Humans	E, C	–	T cells/ cytokines	–	Gomez *et al.*, 2003
Yellow fever virus	YFV	Humans	E, E + NS1	Yes	–	Yes	Despres *et al.*, 1991
Hepadnaviridae							
Hepatitis B virus	HBV	Humans	HBsAg	Antibodies	–	–	Attanasio *et al.*, 1991
Herpesviridae							
Bovine herpesvirus 1	BHV-1	Cows	gIII	Yes	–	–	Okazaki *et al.*, 1994
			gIV	Yes	–	Yes	van Drunen Little-van den Hurk *et al.*, 1991, 1993
Canine herpesvirus	CHV	Dogs	gC	Yes	–	–	Xuan *et al.*, 1996
Equine herpesvirus 1	EHV-1	Horses	gB	Yes	–	Yes	Kukreja *et al.*, 1998
			gB, gC, gD	Yes/No	T cells	Yes/No	Packiarajah *et al.*, 1998
			gC	Yes	T cells	Yes	Stokes *et al.*, 1996a
			gC, gD	Yes	T cells	Yes	Whalley *et al.*, 1995
			gD	Yes	–	–	Foote *et al.*, 2005
			gD (DNA prime)	Yes	T cells	Yes	Ruitenberg *et al.*, 2000
			gD, gH	Yes/No	–	Yes/No	Stokes *et al.*, 1997
			gH, gL	–	–	Partial/No	Stokes *et al.*, 1996b
Feline herpes virus 1	FHV-1	Cats	gD	Yes	–	–	Maeda *et al.*, 1996
Guinea pig cytomegalovirus	GPCMV	Rodents	gB	Yes	–	Yes	Schleiss *et al.*, 2004
Herpes simpex virus 1	HSV-1	Humans	gD	Yes	T cells	Yes	Krishna *et al.*, 1989
			gE	Yes	–	–	Lin *et al.*, 2004
			gB-gI cocktail	Yes	–	Yes	Ghiasi *et al.*, 1996
			gD	Yes	–	Yes	Ghiasi *et al.*, 1991

(*continues*)

TABLE III (*continued*)

Virus family or genus	Abbreviation	Host	Antigen(s)[b]	Neutralizing antibodies[c]	T cells/ cytokines	Protection host/model	References
			gD, gG, gK	–	Cytokines	–	Ghiasi *et al.*, 1999
			gE	Yes	T cells/ cytokines	Yes	Ghiasi *et al.*, 1995
			gK	No	–	ADE[e]	Ghiasi *et al.*, 2000
			gL	No	–	No	Ghiasi *et al.*, 1994
Human cytomegalovirus	HCMV	Humans	gB	Yes	–	–	Marshall *et al.*, 2000
			IE1-pp65	–	T cells	–	Vaz-Santiago *et al.*, 2001
Phocid herpes virus 1	PhHV-1	Seals	gB	Yes	–	Yes	Harder and Osterhaus, 1997
Pseudorabies virus	PrV	Pigs	gII	Yes	–	Yes	Xuan *et al.*, 1995
			gIII	Yes	–	–	Inumaru and Yamada, 1991
Orthomyxoviridae							
Equine influenza virus	H3N8	Horses	H3	No	–	Partial	Olsen *et al.*, 1997
Human influenza A	H1N1	Humans	H1 (proteosomes)	Yes	–	Yes	Jones *et al.*, 2003
	H2N2		M2	Yes	–	Yes	Slepushkin *et al.*, 1995
	H3N2		H3 + M1 (VLP)	Yes	–	Yes	Galarza *et al.*, 2005b
	H3N2		H3	Yes	Yes	Yes	Powers *et al.*, 1995, 1997
	H3N2		H3	Yes	–	–	Treanor *et al.*, 1996
	H3N2		H3, N2	Yes	–	Yes	Brett and Johansson, 2005; Johansson, 1999

	H3N2		N2	Yes	–	Yes	Deroo *et al.*, 1996
	H6N2		N2	Yes	–	Yes	Kilbourne *et al.*, 2004
	Multiple		H1, H3	Yes	–	–	Lakey *et al.*, 1996
Avian influenza virus	H5N1	Birds	H5	No/–	–	Yes	Katz *et al.*, 2000; Swayne *et al.*, 2001
	H5N1		H5	In humans	–	–	Treanor *et al.*, 2001
	Multiple		H5, H7	Yes		Yes	Crawford *et al.*, 1999
Papovaviridae							
Bovine papillomavirus	BPV	Cows	L1	Yes	–	–	Kirnbauer *et al.*, 1992
Cottontail rabbit papillomavirus	CRPV	Rodents	L1 (VLP)	Yes	–	Yes	Breitburd *et al.*, 1995;
			L1, L1 + L2 (VLP)	Yes	–	Yes	Christensen *et al.*, 1996
Human papillomavirus 16	HPV-16	Humans	L1 (VLP)	Yes	–	–	Harro *et al.*, 2001
			L1 + L2 + E7 (VLP)	Yes	–	Yes	Greenstone *et al.*, 1998
			L1 (VLP)	–	Yes	–	Dupuy *et al.*, 1997
Paramyxoviridae							
Bovine parainfluenza virus	BPIV-3	Cattle	HN	Yes	–	Yes	Haanes *et al.*, 1997
Bovine repiratoiry syncytial virus	BRSV	Cattle	F	Yes	Yes	Yes	Sharma *et al.*, 1996
			F partial	Yes	Yes	Yes	Werle *et al.*, 1998
Human parainfluenza virus	HPIV-3	Humans	F	Low	–	Partial	Hall *et al.*, 1991
			F	Yes	–	Yes	Ray *et al.*, 1989
			HN	Yes	–	Yes	van Wyke Coelingh *et al.*, 1987
			HN (+ RSV F)	Yes	–	Yes	Du *et al.*, 1994; Homa *et al.*, 1993

(*continues*)

TABLE III *(continued)*

Virus family or genus	Abbreviation	Host	Antigen(s)[b]	Neutralizing antibodies[c]	T cells/ cytokines	Protection host/model	References
			HN-F fusion	Yes	–	Yes	Brideau *et al.*, 1993
			HN, F, HN-F	Yes	–	Yes	Lehman *et al.*, 1993
Human respiratory syncytial virus	HRSV	Humans	F (+HPIV-HN)	Yes	–	Yes	Du *et al.*, 1994; Homa *et al.*, 1993
			FG fusion	Low	–	Partial	Connors *et al.*, 1992
			FG fusion	Yes	–	Yes	Brideau *et al.*, 1989; Oien *et al.*, 1993; Wathen *et al.*, 1991
Newcastle disease virus	NDV	Birds	F	–	–	Yes	Mori *et al.*, 1994
			HN	Yes	–	Yes	Nagy *et al.*, 1991
Peste-des-petits-ruminants virus	PPRV	Ruminants	HN	Yes	Yes	–	Sinnathamby *et al.*, 2001a
Rinderpest virus	RPV	Cows	F, H	Yes	–	No	Bassiri *et al.*, 1993
			H	–	Yes	–	Sinnathamby *et al.*, 2001b
Rubella virus	RV	Humans	E2, C	–	–	No	Cusi *et al.*, 1995
Parvoviridae							
B19 virus	B19V	Humans	VP1, VP2 (VLP)	Yes	–	–	Kajigaya *et al.*, 1991
Canine parvovirus	CPV	Dogs	VP2 (VLP)	Yes	–	Yes	Lopez de Turiso *et al.*, 1992; Saliki *et al.*, 1992
Duck parvovirus	DPV	Birds	VP1, VP2 (VLP)	Yes	–	–	Le Gall-Recule *et al.*, 1996
Mink enteritis parvovius	MEV	Mink	VP2 (VLP)	Yes	–	Yes	Christensen *et al.*, 1994
Porcine parvovirus	PPV	Pigs	VP2 (VLP)	Antibodies	–	–	Martinez *et al.*, 1992

Picornaviridae							
Foot-and-mouth disease virus	FMDV	Cattle	Epitopes fused to GP64	Yes	–	Yes	Tami *et al.*, 2004
			P1–2A + part P2	–	–	Partial	Grubman *et al.*, 1993
Hepatitis A virus	HAV	Humans	polyprotein	Yes	–	–	Rosen *et al.*, 1993
Polyomaviridae							
Simian virus 40	SV40	Primates	Large T	Antibodies	–	Yes	Shearer *et al.*, 1993
				Yes	No	Yes	Bright *et al.*, 1998; Watts *et al.*, 1999
Reoviridae							
African horse sickness virus	AHSV	Horses	VP2 (VLP)	–	–	Yes	Roy and Sutton, 1998
Bluetongue virus	BTV	Sheep, cattle	VP2 (VLP)	Yes	–	Yes	Roy *et al.*, 1994
			VP2, VP5 (VLP)	Yes	–	–	Loudon *et al.*, 1991
			VP2, VP5, VP3, and VP7 (VLP)	Yes	–	Yes	French *et al.*, 1990; Pearson and Roy, 1993; Roy, 2003; van Dijk, 1993
Bovine rotavirus	BoRV	Cows	VP2 + VP4 + VP6 + VP7 (VLP)	Yes	–	Yes	Conner *et al.*, 1996a,b
Human rotavirus	HRV	Humans	VP2 + VP4 + VP6 + VP7 (VLP)	Yes	–	Yes	Conner *et al.*, 1996a
Simian rotavirus	SiRV	Primates	VP2 + VP4 + VP6 + VP7 (VLP)	Yes	–	Yes	Conner *et al.*, 1996a,b

(*continues*)

TABLE III (*continued*)

Virus family or genus	Abbreviation	Host	Antigen(s)[b]	Neutralizing antibodies[c]	T cells/ cytokines	Protection host/model	References
Retroviridae							
Feline immunodeficiency virus	FIV	Cats	gp120	Yes	–	Partial	Leutenegger *et al.*, 1998
Human immunodeficiency virus	HIV-1	Humans	p24	–	T cells	–	Fyfe *et al.*, 1993
			gp41 MEPR[f]/ PERV[g] p15E fusion	Yes	–	–	Luo *et al.*, 2006
			gp41 + V3 loop	Yes	–	–	Luo *et al.*, 1992
			gp55 (VLP)	Boost	–	–	Jaffray *et al.*, 2004
			gp55–gp120 (VLP)	Yes	–	–	Arico *et al.*, 2005
				Yes	T cells	–	Buonaguro *et al.*, 2002; Tobin *et al.*, 1997
			gp120	Antibodies	–	–	Peet *et al.*, 1997
				No	–	–	Bristow *et al.*, 1994
				–	No CTL	–	Perales *et al.*, 1995
				–	CTL	–	Doe *et al.*, 1994
			gp160	Partial	–	–	Keefer *et al.*, 1994
				No	–	–	Akerblom *et al.*, 1993
				Boost[h]	–	–	Gorse *et al.*, 1994; Graham *et al.*, 1993; Lubeck *et al.*, 1994; Montefiori *et al.*, 1992
				Boost, partial	CTL	–	Cooney *et al.*, 1993
				Antibodies	T cells	–	Lundholm *et al.*, 1994

				Memory B cells	–	–	Reuben *et al.*, 1992
				No	T cells	–	McElrath *et al.*, 1994
	HIV-1	Humans	gp160	–	T cells	–	Gorse *et al.*, 1992; Keefer *et al.*, 1991
	HIV-2	Humans	gp41 HIV-1 + HIV-2 V3 loop	Yes	–	–	Luo *et al.*, 1992
Simian immunodeficiency virus	SIV	Primates	Env on gag VLP	Yes	Yes	–	Yao *et al.*, 2000, 2002
			gp160	Boost	–	Yes	Hu *et al.*, 1992
Rhabdoviridae							
Rabies virus	RABV	Mammals	G	Yes	Yes	Yes	Prehaud *et al.*, 1989
			G	Yes	–	Yes	Fu *et al.*, 1993
			N, G	Yes	–	Yes	Drings *et al.*, 1999
Mokola virus	MOKV	Mammals	G	Antibodies	–	Yes	Tordo *et al.*, 1993

[a] Dashes in the table mean not analysed in this study.
[b] Only those antigens are included that were tested in immunization experiments.
[c] If not known whether neutralizing indicated as "antibodies."
[d] VLP, virus-like particle.
[e] ADE, antibody-dependent enhancement by nonneutralizing antibodies (resulting in chronic infections).
[f] MEPR, membrane-proximal region.
[g] PERV, porcine endogenous retrovirus.
[h] Boost, boost with baculovirus-produced recombinant protein, prime form other origin.

threefold increase in expression levels, and allowed for purification of E2 from the culture medium (Hulst *et al.*, 1993). In a similar way, the related *Bovine diarrhea virus* (BVDV) E2 protein was expressed in insect cells (Bolin and Ridpath, 1996). Recent research showed that the BVDV E2 protein needs to be glycosylated to be effectively secreted from baculovirus-infected cells (Pande *et al.*, 2005) and that the glycosylated protein was able to block BVDV infection better in an *in vitro* assay. Whether the glycosylated E2 protein also performs better as a vaccine is not known. The spike glycoprotein of the SARS coronavirus is one of the most recently expressed proteins in insect cells and protected mice against intranasal SARS infection (Bisht *et al.*, 2005).

Influenza presents a serious risk for both human and animal health. The single-stranded RNA of the influenza virus changes quickly through an accumulation of mutations and frequent recombination events, requiring annual vaccine updates (Palese, 2004). The most threatening recent example is the outbreak of avian influenza of the H5N1 serotype which has killed birds and humans in the Far East since 2003 (WHO) and which caused the first human casualties outside this area in East Turkey in January 2006. The big fear is that such an avian virus will change into a virus that can be transmitted directly from man to man, which may then lead to an influenza outbreak of pandemic dimensions (Palese, 2004). The most widely used influenza vaccines, e g., Fluzone (Sanofi Pasteur) and Fluvirin (Chiron), consist of chemically inactivated split virus or purified virus subunits. These vaccines have several disadvantages which have recently been reviewed (Cox, 2005; Cox *et al.*, 2004), including reduced efficacy in the elderly, where vaccination does reduce mortality rates but is not very effective in preventing disease. In addition, an enormous number of eggs are needed each year (one egg per dose) which will very likely lead to a shortage of vaccine in the event of a pandemic; some strains grow poorly in eggs requiring coinfections with other strains or genetic adaptations (e.g., H5N1) (Horimoto *et al.*, 2006); and these vaccines can cause strong allergic reactions in some individuals. Live, attenuated influenza vaccines have the advantage of inducing secretory and systemic immunity and are applied intranasally, preventing virus replication in the respiratory tracts (Cox *et al.*, 2004). However, all of these vaccines still need to be grown in chicken embryos, which are ironically also the target for a potentially pandemic virus like H5N1.

To overcome these drawbacks, various cell-based vaccines for influenza are under development as well as recombinant protein vaccines. Clinical trials of vaccines based on influenza virus produced in mammalian cell cultures, such as Madin Darby canine kidney (MDCK) cells, have been described (Brands *et al.*, 1999; Percheson *et al.*, 1999) and trials with influenza vaccines produced in the human retina cell line Per.C6® (Pau *et al.*, 2001) are ongoing. These products still require inactivation of the influenza virus which may reduce immunogenicity as seen for inactivated vaccines. In response to human casualties of H5 and H7 influenza viruses in Asia in the late 1990s, the immunogenicity and safety of baculovirus recombinant H5 and H7 hemagglutinin (HA) proteins was tested in chickens and resulted in 100% protection against disease symptoms (Crawford *et al.*, 1999). The immunogenicity of the baculovirus-derived H5 vaccine was subsequently evaluated in over 200 healthy human adults. The vaccine was well tolerated and provided neutralizing antibody responses equivalent to those observed in convalescent sera in ~50% of the individuals after two doses (Treanor *et al.*, 2001). A clinical trial with baculovirus-produced recombinant H3 antigens in 127 adult volunteers showed protective neutralizing antibody levels and a reduction in influenza rates in the following epidemic season compared to a placebo group (Powers *et al.*, 1995). This HA-based vaccine induced both B and T memory cells (Powers *et al.*, 1997). A clinical study of 399 individuals with an average age of 70 years was completed in 2003–2004 with an experimental vaccine (FluBlØk, Protein Sciences corporation) containing the same three HA antigen variants as present in the licensed inactivated vaccine of that flu season (Treanor *et al.*, 2006). Compared to the licensed vaccine, the recombinant vaccine produced higher antibody titers against the H3 strain, the strain responsible for the majority of influenza deaths each year (Cox, 2005). This result suggests that this vaccine can be especially useful for reduction of the annual number of influenza-related deaths in the elderly, where H3 antibody titers induced by conventional vaccines are too low to be protective. Phase III trials in healthy adults have been completed and showed a 100% protective effcicacy even against H3N2 influenza viruses (Manon Cox, personal communication) (http://www.proteinsciences.com/, Jan 2006). Preparation of a recombinant influenza virus vaccine cocktail for the coming flu season may take about 3 months to complete from the moment the new vaccine composition is announced by the World Health Organization (WHO).

The inactivated conventional vaccine and the trivalent recombinant HA-based vaccine under development are based on antibody responses against the HA surface protein and require annual modifications to the vaccine due to antigenic drift of the influenza virus. A baculovirus recombinant vaccine with both HA and neuraminidase (NA) subunits resulted in a bivalent seroconversion with antibodies against both HA and NA (Johansson, 1999). The efficacy of an H3N2 vaccine based on both HA and NA produced with a recombinant baculovirus was analyzed in a murine model and compared with a conventional killed and a live-attenuated vaccine preparation and an HA single-subunit vaccine (Brett and Johansson, 2005). The NA in the baculovirus-derived vaccine was much more immunogenic than in the conventional vaccines. The advantage of inducing an immune response to both surface proteins is illustrated by the fact that the recombinant vaccine containing both HA and NA did not only prevent infection with homotypic and closely related viruses, but also showed a strong reduction in pulmonary virus titers in infections with a more distantly related virus (H3N2 A/Panama/2007/99 versus A/Fuijan/411/2002), in contrast to a vaccine based on HA only. These results suggest that a vaccine containing intact NA tolerates more antigenic drift, thereby reducing the chance of virus escaping the immune system during the flu season.

B. Virus-like Particles

Viral capsid proteins produced in insect cells often self-assemble into VLPs. The advantage of VLPs is that they resemble the natural virus but are not infectious because they lack genetic material. VLPs are also an excellent tool for study of virus structure. VLPs can easily be purified by extraction, centrifugation, or precipitation (Brown *et al.*, 1991) and often give strong immune reactions even in the absence of adjuvants due to their particulate nature. In addition, humoral, cell-mediated, and mucosal immune responses have been reported (Roy, 1996). An example of a vaccine consisting of recombinant VLPs produced with a baculovirus vector is a patented *Canine parvovirus* vaccine (Lopez de Turiso *et al.*, 1992; Valdes *et al.*, 1999). Sometimes the expression of more than one viral coat protein is needed to make immunogenic VLPs, either due to the complexity of the capsid structure (*Bluetongue virus*: BTV, *Reoviridae*) or presence of crucial epitopes on several coat proteins. Multicomponent VLPs can be produced by using vectors with multiple promoters or by coinfections with several baculovirus vectors that each encode one or more viral proteins. One difficulty in making complex VLPs is to achieve appropriate expression levels of each protein present in the viral capsid.

The capsid protein of *Hepatitis E virus* (*Caliciviridae*) forms VLPs and these VLPs induce both systemic and mucosal immunity after oral administration in a mouse model. They also protect cynomolgus monkeys when challenged with HEV against infection and hepatitis (Li *et al.*, 2001, 2004). Infectious bursal disease (IBDV, *Birnaviridae*) of birds can also be prevented by vaccination with single component VLPs (Martinez-Torrecuadrada *et al.*, 2003; Wang *et al.*, 2000). The major capsid protein L1 of *Papovaviridae* forms VLPs and has been shown to protect cottontail rabbits against *Cottontail rabbit papillomavirus*. Combinations of the HPV-16 L1 and L2 capsid proteins and the oncogenic protein E7 protected against tumor formation in a mouse model (Greenstone *et al.*, 1998). Multivalent VLP preparations containing BTV (*Reoviridae*) VP2 subunits of various serotypes were made by coinfections of several baculovirus vectors and induced long-lasting protection in sheep (Roy *et al.*, 1994). Vaccine candidates in the form of VLPs with up to four different VPs have also been successfully produced for BTV (Pearson and Roy, 1993; van Dijk, 1993) as well as for several other *Reoviridae* (Conner *et al.*, 1996a,b). Immunization of mice with an influenza VLP containing the two matrix proteins M1 and M2, and the surface proteins HA and NA showed almost complete protection against an H3N2 virus via both intramuscular and intranasal immunization routes (Galarza *et al.*, 2005a).

The rationale for using VLPs as vaccine candidates is obvious for nonenveloped viruses, because in these viruses the capsid proteins are directly exposed to the immune system. However, they may also be useful for displaying epitopes of enveloped viruses. HIV is an enveloped virus and in this case VLPs have been produced based on gp55 (gag) to which immunogenic segments of the envelope protein gp120 were coupled (Arico *et al.*, 2005; Buonaguro *et al.*, 2002; Tobin *et al.*, 1997). An extension of these chimeric VLP-based vaccines is to use VLPs of one virus to display epitopes of heterologous proteins that do not form VLPs by themselves. Examples of such systems are *Human parvovirus B19* VLPs which carry linear epitopes in fusion with the viral VP2 protein. This system was used to display epitopes of *Murine hepatitis virus* A59 (MHV; *Coronaviridae*) and Herpes simplex virus (HSV; *Herpesviridae*) (Brown *et al.*, 1994). Such chimeric VLPs protected mice against a lethal challenge with MHV of HSV. Epitope-presenting chimeric VLPs have also been developed based on *Mouse papillomavirus* (Tegerstedt *et al.*, 2005) and *Flock house virus* VLPs (Scodeller *et al.*, 1995).

C. Inclusion of Recombinant Cytokines in the Vaccine

Mono- and oligomeric protein subunits are often less potent and need to be formulated carefully before administration to extend their half-life and to present them in a proper form to the immune system, for instance by uptake by antigen-presenting cells (APCs)

(Dertzbaugh, 1998; Schijns, 2003). Adjuvant possibilities for human application are very limited because of safety considerations and this may limit the application of monomeric subunit vaccines in humans. VLPs on the other hand have been shown to induce protection even without the addition of adjuvants (Li *et al.*, 2004; Roy, 1996). An alternative way to modulate the immune response is by the addition of recombinant cytokines as vaccine adjuvants. Cytokines can either be added separately to the vaccine or may be included in VLPs. This approach may not only modulate the magnitude but also the type of immune response (Lofthouse *et al.*, 1996). By carefully choosing which cytokine is added the immune response can be driven in a certain direction. Interferon gamma (IFN-γ) may be added to stimulate macrophages, while addition of interleukin-12 (IL-12) promotes cell-mediated adaptive immunity (Abbas and Lichtman, 2005). IL-12 can be efficiently produced with baculovirus vectors as functional dimers that shift the immunogenic balance to Th1 cells in bovine calves (Takehara *et al.*, 2002). IL-12 added to influenza VLPs enhances antibody responses but in this case VLPs alone already result in 100% protection (Galarza *et al.*, 2005a). Immune reactions to helminths involve Th2 responses. Interleukin-4 (IL-4) drives the immune response to differentiation of Th2 cells and to the production of heminth-specific IgE antibodies (Abbas and Lichtman, 2005). Addition of IL-4 may therefore be helpful for vaccines against helminths (Lofthouse *et al.*, 1996). For baculovirus-derived products the addition of cytokines has not been fully exploited, but it is commonly used for DNA vaccines. The addition of costimulators, such as B7 or CD40, to baculovirus-produced vaccines has not been reported.

D. Baculoviruses as DNA Vaccines

Baculoviral vectors with mammalian promoters driving the expression of viral genes have been used in a limited number of vaccine trials. A candidate *Pseudorabies virus* vaccine expressing its glycoprotein B from a recombinant baculovirus vector with a mammalian promoter resulted in seroconversion in immunized mice (Aoki *et al.*, 1999). Intramuscular injection with baculovirus BVs expressing the E2 glycoprotein of *Hepatitis C virus* controlled by the CMV immediate-early promoter-enhancer provided specific humoral and cellular responses (Facciabene *et al.*, 2004). Similar results were obtained with the carcinoembryonic antigen (CEA) indicating that these types of vaccines can also be effective against tumors. The addition of the *Vesicular stomatis virus* (VSV) G protein to the baculovirus envelope increased immunogenicity in this experiment, possibly by enhancement of virus fusion. BacMam™ vectors have also been used to produce mutated, attenuated influenza virus in mammalian cells by delivery of an altered *NS1* gene (Poomputsa *et al.*, 2003). Surprisingly, a baculovirus vector with the influenza hemagglutinin gene (*H1*) controlled by the chicken beta-actin promoter gave a similar level of protection as a wild-type baculovirus against a lethal influenza challenge in intranasally immunized mice (Abe *et al.*, 2003). The authors ascribe this to the induction of a strong innate immune response by the baculovirus, protecting the mice from a subsequent lethal challenge with influenza virus.

A possible drawback to the use of baculoviruses directly for vaccination and possibly also for surface display and polyhedra-incorporation vectors is the accumulation of anti-baculovirus antibodies upon repeated vaccination, resulting in rapid inactivation of subsequent vaccines of the same type. Pigs for instance have been shown to produce high levels of baculovirus-neutralizing antibodies after injection of baculovirus BVs (Tuboly *et al.*, 1993). A solution to this problem could be to design multivalent vaccines, since baculoviruses can take up a large amount of foreign DNA. Another problem with the use of baculovirus particles as vaccines is rapid degradation by the complement system which also affects gene therapy applications. Pseudotyping of BVs with the VSV glycoprotein instead of GP64 resulted in a reduction in complement inactivation (Tani *et al.*, 2003). Incorporation of human decay-accelerating factor (DAF), a complement-regulatory protein, in the BV envelope has been shown to protect baculovirus gene therapy vectors against complement-mediated inactivation (Huser *et al.*, 2001). This strategy could also be applied for vaccine purposes.

E. Combinations of Vaccine Strategies

HIV VLPs containing only gp55 (gag) have been used to boost immunization induced by a gp55 DNA vaccine (Jaffray *et al.*, 2004). In this case, a capsid protein of an enveloped virus is a reasonable subunit for vaccination because DNA vaccines are expressed intracellularly and fragments of the resulting proteins are presented by MHC complexes to induce cellular immune responses. In this way the natural situation of intracellular expression of viral genes is mimicked. T cell responses, especially the action of cytotoxic T lymphocytes (CTL) are crucial in defense against HIV. HIV combination vaccines where adenovirus, vaccinia (HIVAC-1e), or vesicular stomatitis virus vectors were used for immunization in combination with a boost with a protein subunit (gp160, gp41) have been tested in phase I trials in humans (Cooney *et al.*, 1993; Graham *et al.*, 1993; Lubeck *et al.*, 1994; Luo *et al.*, 2006; Perales *et al.*, 1995; Zheng, 1999). Viral carriers also express HIV antigens inside cells, and these antigens are displayed either by specialized APCs or other cells to the immune system. The aim of this regimen with a DNA/carrier vaccine and a protein boost regimen is therefore to induce both neutralizing antibodies and T cell responses.

F. Viral Marker Vaccines and Differential Diagnosis Technology

Nonvaccination policies exist for many animal diseases because of the risk that vaccinated animals may be protected against disease but may still be carriers of the virus. In cases of outbreaks, ring vaccination is sometimes applied, but large-scale vaccination is generally not allowed. A prerequisite for the broader use of animal vaccines is the development of marker vaccines that enable differentiation between vaccinated and infected animals. This is especially important for endemic diseases in animals where monitoring is of crucial importance to avoid spread of the virus to nonendemic regions or to other host species such as humans or wild animals. Marker vaccines also have good prospects for eradication of animal diseases in general (van Aarle, 2003). For such purposes, marker vaccines do not have to give 100% herd immunity, because reducing susceptibility and transmission can be sufficient to have a major effect in controlling animal disease (Henderson, 2005). Marker vaccines need to be accompanied by specific diagnostic tests that are commonly based on determining serum titers of a viral component that is absent in the vaccine but present in the pathogen, to which antibodies can be raised.

The baculovirus expression system has proven to be useful in producing not only the protein subunits for marker vaccines but also the recombinant polypeptides for these diagnostic tests. The commercially available CSFV vaccine based on the E2 glycoprotein is a marker vaccine because only antibodies against E2 are generated. An enzyme-linked immunosorbent assay (ELISA) test for serum antibodies against the other immunogenic surface protein E^{RNS} can be used to discriminate immunized animals from virus carriers (Langedijk *et al.*, 2001; van Aarle, 2003). Another possibility is to use only some of the epitopes of an antigen for immunization and others for the diagnostic analysis, as demonstrated for CSFV (van Rijn *et al.*, 1999). The coexistence of a subunit vaccine and a discriminative diagnostic test enabled registration of CSFV vaccines in Europe. Marker vaccines will also be very useful for immunization of poultry against avian influenza (Crawford *et al.*, 1999), where monitoring for the presence of virus is essential. Animals immunized with HA or HA/NA vaccines could be screened for antibodies against the viral matrix proteins. Similar assays have been developed for other subunit vaccines, such as one that discriminates between VSV-infected and immunized animals, where the marker vaccine is based on the glycoprotein and the assay on the nucleocapsid protein produced in insect larvae (Ahmad *et al.*, 1993).

IV. Baculovirus-Produced Vaccines Against Protozoan Parasites and Helminths

Parasites of the genera *Plasmodium, Theileria,* and *Babesia* are protozoan blood parasites causing malaria, theileriosis, and babesiosis. These parasites have a complex life cycle and are transmitted by either mosquito or tick vectors. Candidate subunit vaccines against these parasites can be roughly separated into preblood (preerythrocyte or prelymphocyte) stage vaccines, blood stage vaccines, transmission-blocking vaccines, and multistage vaccines. An overview of parasite subunits expressed in the baculovirus insect cell system for vaccine purposes is given in Table IV.

A. Plasmodium

A comprehensive record of all subunit and recombinant carrier vaccines under development for human malaria is maintained by WHO (Reed, 2005). Most of these vaccines are still in a preclinical stage but several *Plasmodium falciparum* vaccines are in phase I and phase II trials in malaria endemic countries. *Plasmodium* is transmitted in the form of sporozoites by *Anopheles* mosquitoes. A vaccine candidate based on the circumsporozoite protein (CSP) is aimed at blocking these sporozoites. Both B and T cell responses appear to be essential for protective immunity based on the CSP protein. CSP has been produced in insect cells but was minimally immunogenic when tested in 20 volunteers (Herrington *et al.*, 1992). Alternative approaches to experimental CSP vaccines, which facilitate T cell responses, are in phase II trials (Ballou *et al.*, 2004). These vaccines include CSP displayed on HBsAg VLP particles, modified vaccinia Ankara virus as recombinant carrier, or DNA vaccines.

The merozoite is the extracellular, erythrocyte-invasive form of the *Plasmodium* parasite. Merozoite surface proteins are promising

TABLE IV

VACCINE TRIALS FOR PARASITIC DISEASES BASED ON SUBUNITS EXPRESSED IN THE BACULOVIRUS–INSECT CELL SYSTEM

Pathogen	Antigen	Trial	Immunologic response	References
Protozoa				
Babesia rodhaini	P26 surface protein	Immunization of rats	40–100% protection	Igarashi *et al.*, 2000
Plasmodium berghei	Ookinete surface protein 21	Injection in mice	Antibodies, oocyst formation blocked in *Anopheles stephensi* mosquitoes	Matsuoka *et al.*, 1996
Plasmodium cynmolgi	Merozoite surface protein (MSP)-1	Challenge in primates	Protection	Perera *et al.*, 1998
Plasmodium falciparum	Circumsporozoite protein (CSP)	Human safety and immunity trial	No response to native CSP	Herrington *et al.*, 1992
Theileria parva	Sporozoite surface protein p67	Challenge in cattle	50% protection	Nene *et al.*, 1995
	Sporozoite surface protein P67 (GFp fusion, surface display)	Challenge in cattle	Upto 80% protection	Kaba *et al.*, 2004b, 2005
Theileria sergenti	P32	Challenge in cattle	Protection	Onuma *et al.*, 1997
Trypanosoma cruzi	TolT	Mice immunization/ *in vitro* inhibition by CD4 cells	50–60% reduction of parasite numbers in infected macrophages	Quanquin *et al.*, 1999
Helminths				
Fasciola hepatica	Procathepsin L3	Challenge in rats	50% protection	Dalton *et al.*, 2003
Ostertagia ostertagi	Metalloprotease 1	Challenge in cattle	No protection	De Maere *et al.*, 2005a
	Aspartyl-protease inhibitor	Challenge in cattle	No protection	De Maere *et al.*, 2005b
Schistosoma mansoni	Calpain (Sm-p80)	Challenge in mice	29–39% reduction in worms	Hota-Mitchell *et al.*, 1997

vaccine candidates due to their accessibility for antibodies and their expected role in erythrocyte invasion. The major merozoite surface protein (MSP-1) is an important prebloodstage candidate vaccine with homology to epidermal growth factor (EGF) and antibodies directed against this protein block erythrocyte invasion (Holder and Blackman, 1994). *Plasmodium cynomolgi* functions as a model system for the highly similar *Plasmodium vivax* in humans. An active, C-terminally processed form of *P. cynomolgi* MSP-1, was produced in insect cells and protected primates in a challenge experiment (Perera *et al.*, 1998). The C-terminally mature MSP-1 of *P. falciparum* was also successfully expressed in insect cells and used for ultrastructural studies (Chitarra *et al.*, 1999; Pizarro *et al.*, 2003), but has not been tested in human trials. Meanwhile, *E. coli*-expressed MSP-1 has entered phase II clinical trials (Ballou *et al.*, 2004).

Plasmodium parasites in the ookinete stage are taken up by mosquitoes and antigens specific for this stage can function as transmission-blocking subunit vaccines. The major ookinete surface antigen Pbs21 (P28) of *Plasmodium berghei* was expressed in *B. mori* larvae, preserving conformational B cell epitopes which were lost upon expression in *E. coli*. The recombinant Pbs21 antigen produced in insect cells blocked oocyte formation in *Anopheles* mosquitoes fed on immunized mice (Matsuoka *et al.*, 1996). The immunogenicity of the recombinant protein was strongly reduced when the protein was expressed as a secreted protein by removing its glycosylphosphatidylinositol (GPI) anchor signal (Martinez *et al.*, 2000). This protein provides a good example of loss of immunogenicity by expressing a membrane protein in a secreted form. Subunit vaccines based on the *Plasmodium*-induced erythrocyte membrane protein 1 (EMP-1) are aimed at blocking vertical transmission from mother to child (maternal malaria) via the placenta. This transmission involves the sequestration of *P. falciparum*-infected erythrocytes through EMP-1, which binds to chondroitin sulphate A in the placenta. EMP-1 expressed in the baculovirus system induces inhibitory antibodies which react with both homologous and heterologous EMP-1 proteins (Costa *et al.*, 2003).

B. Theileria *and* Babesia

T. parva is the causative agent of East Coast fever, a deadly cattle disease endemic in large parts of Africa. Immunization with recombinant sporozoite surface protein p67 is aimed at blocking invasion of lymphocytes by this parasite. P67 is an example of a protein that was not easy to express in a native form in insect cells as well as in many other systems. In insect cells, it was expressed at low levels and in contrast to expectations was not present on the cell surface. Similar to *E. coli*-expressed p67, it did not react with a monoclonal antibody against native p67, indicating that the folding of the protein was not correct (Nene *et al.*, 1995). Several adaptations were therefore made to the expression system. Expression of p67 coupled to the honeybee mellitin signal instead of the original signal peptide resulted in correct routing of this protein to the cell surface, but the folding was still not optimal (Kaba *et al.*, 2004a). Fusion of p67 to the C terminus of GFP drastically increased expression levels and resulted in recognition by the conformation-sensitive monoclonal antibody (Kaba *et al.*, 2002). A similar effect was also seen when parts of this protein were fused to the baculovirus GP64 glycoprotein in a surface display vector, which led to expression on the cell surface and on baculovirus BVs (Kaba *et al.*, 2003). The recombinant GFP-p67 protein and the C terminal half of p67 coupled to GP64 induced high levels of sporozoite-neutralizing serum antibodies and showed up to 80% protection against lethal *T. parva* challenge in a double-blind placebo-controlled experiment (Kaba *et al.*, 2004b, 2005). The next phase will be to evaluate the quality of these experimental vaccines under field conditions in countries in which East Coast fever is endemic.

Babesia species are a major cause of parasitemias in cattle and dogs. *Babesia divergens* is the major cause of bovine babesiosis in Europe and the increased incidence of this disease is correlated with an increase in the numbers of ticks (*Ixodus ricinus*) that transmit this parasite. *B. divergens* is also responsible for zoonotics in immunocompromised humans (Zintl *et al.*, 2003). The soluble parasite antigen (SPA) of bovine and canine *Babesia* species has been developed as a vaccine against clinical manifestations in dogs and is produced in mammalian cells infected with *Babesia* (Schetters, 2005). Several other *Babesia* antigens have been expressed in the baculovirus expression system to develop ELISA tests for diagnosis. The baculovirus-expressed *Babesia radhaini* P26 protein was shown to induce protection against the disease in rats (Igarashi *et al.*, 2000).

Theileria and *Babesia* parasites are transmitted by ioxidic ticks and in the future candidate antiparasite vaccines may be combined with vaccines directed against the tick vector (Bishop *et al.*, 2004). These vaccines may be tick antigens directly exposed to the host immune system, such as vitellin, the most abundant *B. microplus* egg protein (Tellam *et al.*, 2002) or cement proteins involved in the attachment of the tick to the skin of the host. Concealed antigens can also give good results, such as the *B. microplus* Bm86 gut antigen (Gavac™), which results in binding of antibodies taken up from immunized animals to a gut transmembrane protein (Willadsen *et al.*, 1995). Ticks are known to immunomodulate their host by secreting specific immunomodulators and transmitted parasites also profit from the reduction in immune response. Tick vaccines may therefore be aimed at reducing the chance of transmission by directly affecting the feeding process, by interacting with immunomodulators, or by reducing tick populations.

C. Trypanosoma *and* Leishmania

Chagas' disease in the Americas is caused by *Trypanosoma cruzi* and is found in humans, dogs, cats, and rodents. *T. cruzi* is a macrophage-invading protozoan, and complicating factors in the development of vaccines are immune escape and autoimmunity due to molecular mimickry (Girones *et al.*, 2005). Macrophages are also the target for *Leishmania* parasites, which use complex immune evasion strategies that affect host cell signaling (Olivier *et al.*, 2005). Immunization with the *T. cruzi* Tol A-like protein (TolT) expressed in insect cells resulted in T cell-dependent antiparasitic activity (Quanquin *et al.*, 1999). For *Leishmania* several candidate subunit vaccine antigens with protective potential have been identified, including the surface protein gp63 (Coler and Reed, 2005), but these proteins have not been expressed with baculovirus vectors in insect cells.

D. Helminths

Helminths or parasitic worms that are a serious threat to human and animal health worldwide, are divided into the Annelida (segmented worms), Platyhelminthes (flatworms including flukes), and Nematoda (roundworms). These worms have complex life cycles, which often involve more than one host. Because helminths vary widely among populations, vaccines are not competitive so far with chemical broad spectrum antihelminths (Bos and Schetters, 1990). Helminths may also modulate the immune system as exemplified by the filarial nematodes, which are present in 150–200 million humans worldwide and cause river blindness for example. These filarial nematodes are difficult to combat because they modulate T cell responses leading to chronic helminth infections (reviewed by Hoerauf *et al.*, 2005). This T cell modulation is not restricted to the response to filarial larvae but also affects allergies, the response to other pathogens and vaccines through modulation of not only antigen-specific T cells but also APCs, which affects immune responses in general.

There are a few examples of baculovirus-expressed helminth proteins but few of these have been tested in immunization studies (Table IV). However, the baculovirus expression system may be a valuable tool for these parasitic worms as illustrated by the following examples: a baculovirus-derived subunit vaccine against liver fluke (*Fasciola hepatica*) based on procathepsin L3 conferred 50% protection to rats, in contrast to the yeast-expressed protein which did not confer any protection (Reszka *et al.*, 2005). Bilharzia or schistosomiasis is caused by *Schistosoma* spp. (Platyhelminthes), blood parasites of humans in tropical areas. Several *Schistosoma* genes have been expressed in insect cells, but primarily for analysis of enzymatic functions. The large subunit of calpain (SmP80) produced in the baculovirus expression system reduced the worm burden in mice (Hota-Mitchell *et al.*, 1997). Antigens of the tapeworm *Taenia solium* have been expressed with baculovirus vectors for diagnostic purposes (Lee *et al.*, 2005; Levine *et al.*, 2004).

V. Conclusions and Prospects

There are many examples in the literature where immunization with recombinant proteins produced in the baculovirus–insect cell expression system conferred good protection against infectious disease. Subunit vaccine development begins with careful identification of the antigen, which is related to whether neutralizing adaptive immune responses are raised against the particular protein in natural infections. Once selected, the open reading frame (ORF) of the antigen is cloned while keeping flanking DNA sequences to a minimum, which can best be achieved with a proof-reading PCR enzyme. The sequence around the ATG translational start site is best modified to that of the polyhedrin or p10 gene in the wild-type baculovirus, with at least an adenosine residue at the −3 position (Chang *et al.*, 1999). Tags may be added to facilitate purification, preferably in such a way that they can subsequently be removed. Tags are not required for VLPs because they can be purified by centrifugation. Transmembrane (glyco) proteins are best produced in a secreted form by removal of TMRs to increase expression levels. However, this may occasionally affect the folding of the protein. An alternative approach is to fuse the immunogenic domains of envelope proteins to GP64. Vectors that lack chitinase and v-cathepsin genes are preferable for expression of envelope proteins. During the preparation of seed stocks, care should be taken to keep the virus passage number low and to use a multiplicity of infection of <0.1 to minimize the formation of DIs. Careful checking of the purity of recombinant bacmids or plaque purified recombinant viruses by PCR is crucial to avoid loss of recombinant virus in subsequent passages due to empty vectors that out-compete the recombinants.

Only two baculovirus-produced products are approved for veterinary practice, namely, two vaccines against classical swine fever consisting of the E2 surface glycoprotein. With these vaccines in the market (although a nonvaccination policy still exists) the confidence in this type of subunit vaccine will grow as well as the possibilities for registration, thereby increasing the likelihood that more vaccines of this kind will appear in the market. This expectation of an increasing number of products may be expanded to human applications when registration of a trivalent recombinant influenza vaccine, for which phase III clinical trials in humans in the United States have been completed recently, can be achieved. Therapeutic anticancer vaccines, such as a vaccine against prostate cancer, may be more readily accepted, in view of the severe side effects of anticancer drugs and irradiation techniques. The application of baculovirus display or BacMam™ vectors for vaccines against infectious disease may be applied in the future for animal use. Because more foreign proteins are incorporated into the vaccine than just the targeted antigen, such vaccines for human use are likely to be in the more distant future because of safety considerations.

Baculovirus expression systems compete with other cheaper production systems, which are more widely used and which scale-up more easily, such as *E. coli* and yeast (Table I), and once a good protection level is achieved there is no commercial interest to switch to the more expensive insect cell system. For those cases though, where folding or posttranslational modifications are crucial to epitope formation the baculovirus–insect cell system is a versatile expression system and many candidate vaccines have been successfully tested. Alternative methods are provided by the development of DNA vaccine technology and recombinant carrier vaccines based on vaccinia, adenovirus, or BacMam™ vectors (Table II). These methods are more prone to induce cellular immune responses than many protein subunit vaccines. The combination of a primary vaccine with an intracellular delivery system (recombinant carrier vaccines, DNA vaccines, BacMam™ vectors) followed by a boost immunization with recombinant protein subunits appears to be a promising approach, aimed at both cellular and humoral immune responses. This approach can be further strengthened through the addition of recombinant cytokines that drive the immune response in a specific direction. A major challenge now is to broaden the array of viral vaccines produced in insect cells and to develop effective vaccines against more complex organisms, such as protozoan parasites and multicellular worms, for which the baculovirus expression system also holds promise. With the increasing insight into immunology leading to new methods of vaccine production and delivery, accompanied by a wealth of genomic and proteomic data, many new generation vaccines are expected within the foreseeable future.

Acknowledgments

I thank Manon Cox, Dick Schaap, and Stephen Kaba for assistance with the sections on viral vaccines, parasite vaccines, and malaria vaccines respectively; Christina van Houte for verifying Table III, and Just Vlak for review of the manuscript and for helpful suggestions.

References

Abbas, A. K., and Lichtman, A. H. (2005). "Cellular and Molecular Immunology." Elsevier Saunders, Philadelphia, USA.

Abe, T., Takahashi, H., Hamazaki, H., Miyano-Kurosaki, N., Matsuura, Y., and Takaku, H. (2003). Baculovirus induces an innate immune response and confers protection from lethal influenza virus infection in mice. *J. Immunol.* **171:**1133–1139.

Ahmad, S., Bassiri, M., Banerjee, A. K., and Yilma, T. (1993). Immunological characterization of the VSV nucleocapsid (N) protein expressed by recombinant baculovirus in *Spodoptera exigua* larva: Use in differential diagnosis between vaccinated and infected animals. *Virology* **192:**207–216.

Ahrens, U., Kaden, V., Drexler, C., and Visser, N. (2000). Efficacy of the classical swine fever (CSF) marker vaccine Porcilis Pesti in pregnant sows. *Vet. Microbiol.* **77:**83–97.

Akerblom, L., Nara, P., Dunlop, N., Putney, S., and Morein, B. (1993). HIV experimental vaccines based on the iscom technology using envelope and GAG gene products. *Biotechnol. Ther.* **4:**145–161.

Andre, F. E. (2003). Vaccinology: Past achievements, present roadblocks and future promises. *Vaccine* **21:**593–595.

Aoki, H., Sakoda, Y., Jukuroki, K., Takada, A., Kida, H., and Fukusho, A. (1999). Induction of antibodies in mice by a recombinant baculovirus expressing pseudorabies virus glycoprotein B in mammalian cells. *Vet. Microbiol.* **68:**197–207.

Arico, E., Wang, E., Tornesello, M., Tagliamonte, M., Lewis, G. K., Marincola, F. M., Buonaguro, F. M., and Buonaguro, L. (2005). Immature monocyte derived dendritic cells gene expression profile in response to virus-like particles stimulation. *J. Transl. Med.* **3:**45.

Attanasio, R., Lanford, R. E., Dilley, D., Stunz, G. W., Notvall, L., Henderson, A. B., and Kennedy, R. C. (1991). Immunogenicity of hepatitis B surface antigen derived from the baculovirus expression vector system: A mouse potency study. *Biologicals* **19:**347–353.

Ayres, M. D., Howard, S. C., Kuzio, J., Lopez-Ferber, M., and Possee, R. D. (1994). The complete DNA sequence of *Autographa californica* nuclear polyhedrosis virus. *Virology* **202:**586–605.

Bachmann, M. F., Kundig, T. M., Freer, G., Li, Y., Kang, C. Y., Bishop, D. H., Hengartner, H., and Zinkernagel, R. M. (1994). Induction of protective cytotoxic T cells with viral proteins. *Eur. J. Immunol.* **24:**2228–2236.

Ball, J. M., Estes, M. K., Hardy, M. E., Conner, M. E., Opekun, A. R., and Graham, D. Y. (1996). Recombinant Norwalk virus-like particles as an oral vaccine. *Arch. Virol. Suppl.* **12:**243–249.

Ball, J. M., Hardy, M. E., Atmar, R. L., Conner, M. E., and Estes, M. K. (1998). Oral immunization with recombinant Norwalk virus-like particles induces a systemic and mucosal immune response in mice. *J. Virol.* **72:**1345–1353.

Ball, J. M., Graham, D. Y., Opekun, A. R., Gilger, M. A., Guerrero, R. A., and Estes, M. K. (1999). Recombinant Norwalk virus-like particles given orally to volunteers: Phase I study. *Gastroenterology* **117:**40–48.

Ballou, W. R., Arevalo-Herrera, M., Carucci, D., Richie, T. L., Corradin, G., Diggs, C., Druilhe, P., Giersing, B. K., Saul, A., Heppner, D. G., Kester, K. E., Lanar, D. E., *et al.* (2004). Update on the clinical development of candidate malaria vaccines. *Am. J. Trop. Med. Hyg.* **71:**239–247.

Barderas, M. G., Rodriguez, F., Gomez-Puertas, P., Aviles, M., Beitia, F., Alonso, C., and Escribano, J. M. (2001). Antigenic and immunogenic properties of a chimera of two immunodominant African swine fever virus proteins. *Arch. Virol.* **146:**1681–1691.

Bassiri, M., Ahmad, S., Giavedoni, L., Jones, L., Saliki, J. T., Mebus, C., and Yilma, T. (1993). Immunological responses of mice and cattle to baculovirus-expressed F and H proteins of rinderpest virus: Lack of protection in the presence of neutralizing antibody. *J. Virol.* **67:**1255–1261.

Beinart, G., Rini, B. I., Weinberg, V., and Small, E. J. (2005). Antigen-presenting cells 8015 (Provenge) in patients with androgen-dependent, biochemically relapsed prostate cancer. *Clin. Prostate Cancer* **4:**55–60.

Belyaev, A. S., and Roy, P. (1993). Development of baculovirus triple and quadruple expression vectors: Co-expression of three or four bluetongue virus proteins and the synthesis of bluetongue virus-like particles in insect cells. *Nucleic Acids Res.* **21:**1219–1223.

Bielefeldt-Ohmann, H., Beasley, D. W., Fitzpatrick, D. R., and Aaskov, J. G. (1997). Analysis of a recombinant dengue-2 virus-dengue-3 virus hybrid envelope protein expressed in a secretory baculovirus system. *J. Gen. Virol.* **78:**2723–2733.

Bishop, R., Musoke, A., Morzaria, S., Gardner, M., and Nene, V. (2004). Theileria: Intracellular protozoan parasites of wild and domestic ruminants transmitted by ixodid ticks. *Parasitology* **129**(Suppl.)**:**S271–S283.

Bisht, H., Roberts, A., Vogel, L., Subbarao, K., and Moss, B. (2005). Neutralizing antibody and protective immunity to SARS coronavirus infection of mice induced by a soluble recombinant polypeptide containing an N-terminal segment of the spike glycoprotein. *Virology* **334:**160–165.

Blanchard, P., Mahe, D., Cariolet, R., Keranflec'h, A., Baudouard, M. A., Cordioli, P., Albina, E., and Jestin, A. (2003). Protection of swine against post-weaning multisystemic wasting syndrome (PMWS) by porcine circovirus type 2 (PCV2) proteins. *Vaccine* **21:**4565–4575.

Bolin, S. R., and Ridpath, J. F. (1996). Glycoprotein E2 of bovine viral diarrhea virus expressed in insect cells provides calves limited protection from systemic infection and disease. *Arch. Virol.* **141:**1463–1477.

Bos, H. J., and Schetters, T. (1990). Is there a future for vaccines against gastrointestinal helminths? *Tijdschr. Diergeneeskd.* **115:**1102–1110.

Bouma, A., de Smit, A. J., de Kluijver, E. P., Terpstra, C., and Moormann, R. J. (1999). Efficacy and stability of a subunit vaccine based on glycoprotein E2 of classical swine fever virus. *Vet. Microbiol.* **66:**101–114.

Bouma, A., De Smit, A. J., De Jong, M. C., DeKluijver, E. P., and Moormann, R. J. (2000). Determination of the onset of the herd-immunity induced by the E2 sub-unit vaccine against classical swine fever virus. *Vaccine* **18:**1374–1381.

Brands, R., Visser, J., Medema, J., Palache, A. M., and van Scharrenburg, G. J. (1999). Influvac: A safe madin darby canine kidney (MDCK) cell culture-based influenza vaccine. *Dev. Biol. Stand.* **98:**93–100, discussion 111.

Breitburd, F., Kirnbauer, R., Hubbert, N. L., Nonnenmacher, B., Trin-Dinh-Desmarquet, C., Orth, G., Schiller, J. T., and Lowy, D. R. (1995). Immunization with viruslike particles from cottontail rabbit papillomavirus (CRPV) can protect against experimental CRPV infection. *J. Virol.* **69:**3959–3963.

Brett, I. C., and Johansson, B. E. (2005). Immunization against influenza A virus: Comparison of conventional inactivated, live-attenuated and recombinant baculovirus produced purified hemagglutinin and neuraminidase vaccines in a murine model system. *Virology* **339:**273–280.

Brideau, R. J., Walters, R. R., Stier, M. A., and Wathen, M. W. (1989). Protection of cotton rats against human respiratory syncytial virus by vaccination with a novel chimeric FG glycoprotein. *J. Gen. Virol.* **70:**2637.

Brideau, R. J., Oien, N. L., Lehman, D. J., Homa, F. L., and Wathen, M. W. (1993). Protection of cotton rats against human parainfluenza virus type 3 by vaccination with a chimeric FHN subunit glycoprotein. *J. Gen. Virol.* **74:**471–477.

Bright, R. K., Shearer, M. H., Pass, H. I., and Kennedy, R. C. (1998). Immunotherapy of SV40 induced tumours in mice: A model for vaccine development. *Dev. Biol. Stand.* **94:**341–353.

Bristow, R. G., Douglas, A. R., Skehel, J. J., and Daniels, R. S. (1994). Analysis of murine antibody responses to baculovirus-expressed human immunodeficiency virus type 1 envelope glycoproteins. *J. Gen. Virol.* **75:**2089–2095.

Brown, C. S., Van Lent, J. W., Vlak, J. M., and Spaan, W. J. (1991). Assembly of empty capsids by using baculovirus recombinants expressing human parvovirus B19 structural proteins. *J. Virol.* **65:**2702–2706.

Brown, C. S., Welling-Wester, S., Feijlbrief, M., Van Lent, J. W., and Spaan, W. J. (1994). Chimeric parvovirus B19 capsids for the presentation of foreign epitopes. *Virology* **198:**477–488.

Buonaguro, L., Racioppi, L., Tornesello, M. L., Arra, C., Visciano, M. L., Biryahwaho, B., Sempala, S. D., Giraldo, G., and Buonaguro, F. M. (2002). Induction of neutralizing antibodies and cytotoxic T lymphocytes in Balb/c mice immunized with virus-like particles presenting a gp120 molecule from a HIV-1 isolate of clade A. *Antiviral Res.* **54:**189–201.

Canales, M., Enriquez, A., Ramos, E., Cabrera, D., Dandie, H., Soto, A., Falcon, V., Rodriguez, M., and de la Fuente, J. (1997). Large-scale production in *Pichia pastoris* of the recombinant vaccine Gavac against cattle tick. *Vaccine* **15:**414–422.

Capua, I., Terregino, C., Cattoli, G., Mutinelli, F., and Rodriguez, J. F. (2003). Development of a DIVA (differentiating infected from vaccinated animals) strategy using a vaccine containing a heterologous neuraminidase for the control of avian influenza. *Avian Pathol.* **32:**47–55.

Cavanagh, D. (2003). Severe acute respiratory syndrome vaccine development: Experiences of vaccination against avian infectious bronchitis coronavirus. *Avian Pathol.* **32:**567–582.

Cha, H. J., Gotoh, T., and Bentley, W. E. (1997). Simplification of titer determination for recombinant baculovirus by green fluorescent protein marker. *Biotechniques* **23:** 782–784, 786.

Chang, M., Kuzio, J., and Blissard, G. (1999). Modulation of translational efficiency by contextual nucleotides flanking a baculovirus initiator AUG codon. *Virology* **259:** 369–383.

Chitarra, V., Holm, I., Bentley, G. A., Petres, S., and Longacre, S. (1999). The crystal structure of C-terminal merozoite surface protein 1 at 1.8 A resolution, a highly protective malaria vaccine candidate. *Mol. Cell* **3:**457–464.

Choi, J. Y., Woo, S. D., Lee, H. K., Hong, H. K., Je, Y. H., Park, J. H., Song, J. Y., An, S. H., and Kang, S. K. (2000). High-level expression of canine parvovirus VP2 using Bombyx mori nucleopolyhedrovirus vector. *Arch. Virol.* **145:**171–177.

Christensen, J., Alexandersen, S., Bloch, B., Aasted, B., and Uttenthal, A. (1994). Production of mink enteritis parvovirus empty capsids by expression in a baculovirus vector system: A recombinant vaccine for mink enteritis parvovirus in mink. *J. Gen. Virol.* **75:**149–155.

Christensen, N. D., Reed, C. A., Cladel, N. M., Han, R., and Kreider, J. W. (1996). Immunization with viruslike particles induces long-term protection of rabbits against challenge with cottontail rabbit papillomavirus. *J. Virol.* **70:**960–965.

Clark, T. G., and Cassidy-Hanley, D. (2005). Recombinant subunit vaccines: Potentials and constraints. *Dev. Biol. (Basel)* **121:**153–163.

Coler, R. N., and Reed, S. G. (2005). Second-generation vaccines against leishmaniasis. *Trends Parasitol.* **21:**244–249.

Conner, M. E., Crawford, S. E., Barone, C., O'Neal, C., Zhou, Y. J., Fernandez, F., Parwani, A., Saif, L. J., Cohen, J., and Estes, M. K. (1996a). Rotavirus subunit vaccines. *Arch. Virol.* **12**(Suppl.)**:**199–206.

Conner, M. E., Zarley, C. D., Hu, B., Parsons, S., Drabinski, D., Greiner, S., Smith, R., Jiang, B., Corsaro, B., Barniak, V., Madore, H. P., Crawford, S., *et al.* (1996b). Virus-like particles as a rotavirus subunit vaccine. *J. Infect. Dis.* **174**(Suppl. 1)**:**S88–S92.

Connors, M., Collins, P. L., Firestone, C. Y., Sotnikov, A. V., Waitze, A., Davis, A. R., Hung, P. P., Chanock, R. M., and Murphy, B. R. (1992). Cotton rats previously immunized with a chimeric RSV FG glycoprotein develop enhanced pulmonary pathology when infected with RSV, a phenomenon not encountered following immunization with vaccinia—RSV recombinants or RSV. *Vaccine* **10:**475–484.

Cooney, E. L., McElrath, M. J., Corey, L., Hu, S. L., Collier, A. C., Arditti, D., Hoffman, M., Coombs, R. W., Smith, G. E., and Greenberg, P. D. (1993). Enhanced immunity to human immunodeficiency virus (HIV) envelope elicited by a combined vaccine regimen consisting of priming with a vaccinia recombinant expressing HIV envelope and boosting with gp160 protein. *Proc. Natl. Acad. Sci. USA* **90:** 1882–1886.

Costa, F. T., Fusai, T., Parzy, D., Sterkers, Y., Torrentino, M., Douki, J. B., Traore, B., Petres, S., Scherf, A., and Gysin, J. (2003). Immunization with recombinant duffy binding-like-gamma3 induces pan-reactive and adhesion-blocking antibodies against placental chondroitin sulfate A-binding plasmodium falciparum parasites. *J. Infect. Dis.* **188:**153–164.

Cox, M. M. (2005). Cell-based protein vaccines for influenza. *Curr. Opin. Mol. Ther.* **7:** 24–29.

Cox, R. J., Brokstad, K. A., and Ogra, P. (2004). Influenza virus: Immunity and vaccination strategies. Comparison of the immune response to inactivated and live, attenuated influenza vaccines. *Scand. J. Immunol.* **59:**1–15.

Crawford, J., Wilkinson, B., Vosnesensky, A., Smith, G., Garcia, M., Stone, H., and Perdue, M. L. (1999). Baculovirus-derived hemagglutinin vaccines protect against lethal influenza infections by avian H5 and H7 subtypes. *Vaccine* **17:**2265–2274.

Cusi, M. G., Valassina, M., Bianchi, S., Wunner, W., and Valensin, P. E. (1995). Evaluation of rubella virus E2 and C proteins in protection against rubella virus in a mouse model. *Virus Res.* **37:**199–208.

Dalton, J. P., Neill, S. O., Stack, C., Collins, P., Walshe, A., Sekiya, M., Doyle, S., Mulcahy, G., Hoyle, D., Khaznadji, E., Moire, N., Brennan, G., *et al.* (2003). *Fasciola hepatica* cathepsin L-like proteases: Biology, function, and potential in the development of first generation liver fluke vaccines. *Int. J. Parasitol.* **33:**1173–1181.

de Gooijer, C. D., Koken, R. H. M., van Lier, F. L. J., Kool, M., Vlak, J. M., and Tramper, J. (1992). A structured dynamic model for the baculovirus infection process in insect cell reactor configurations. *Biotechnol. Bioeng.* **40:**537–548.

De Maere, V., Vercauteren, I., Geldhof, P., Gevaert, K., Vercruysse, J., and Claerebout, E. (2005a). Molecular analysis of astacin-like metalloproteases of *Ostertagia ostertagi*. *Parasitology* **130:**89–98.

De Maere, V., Vercauteren, I., Gevaert, K., Vercruysse, J., and Claerebout, E. (2005b). An aspartyl protease inhibitor of *Ostertagia ostertagi*: Molecular cloning, analysis of stage and tissue specific expression and vaccine trial. *Mol. Biochem. Parasitol.* **141:**81–88.

Delenda, C., Frenkiel, M. P., and Deubel, V. (1994). Protective efficacy in mice of a secreted form of recombinant dengue-2 virus envelope protein produced in baculovirus infected insect cells. *Arch. Virol.* **139:**197–207.

Depner, K. R., Bouma, A., Koenen, F., Klinkenberg, D., Lange, E., de Smit, H., and Vanderhallen, H. (2001). Classical swine fever (CSF) marker vaccine. Trial II. Challenge study in pregnant sows. *Vet. Microbiol.* **83:**107–120.

Deroo, T., Jou, W. M., and Fiers, W. (1996). Recombinant neuraminidase vaccine protects against lethal influenza. *Vaccine* **14:**561–569.

Dertzbaugh, M. T. (1998). Genetically engineered vaccines: An overview. *Plasmid* **39:** 100–113.

Despres, P., Dietrich, J., Girard, M., and Bouloy, M. (1991). Recombinant baculoviruses expressing yellow fever virus E and NS1 proteins elicit protective immunity in mice. *J. Gen. Virol.* **72:**2811–2816.

Doe, B., Steimer, K. S., and Walker, C. M. (1994). Induction of HIV-1 envelope (gp120)-specific cytotoxic T lymphocyte responses in mice by recombinant CHO cell-derived gp120 is enhanced by enzymatic removal of N-linked glycans. *Eur. J. Immunol.* **24:**2369–2376.

Drings, A., Jallet, C., Chambert, B., Tordo, N., and Perrin, P. (1999). Is there an advantage to including the nucleoprotein in a rabies glycoprotein subunit vaccine? *Vaccine* **17:**1549–1557.

Du, R. P., Jackson, G. E., Wyde, P. R., Yan, W. Y., Wang, Q., Gisonni, L., Sanhueza, S. E., Klein, M. H., and Ewasyshyn, M. E. (1994). A prototype recombinant vaccine against respiratory syncytial virus and parainfluenza virus type 3. *Biotechnology (NY)* **12:**813–818.

Dupuy, C., Buzoni-Gatel, D., Touze, A., Le Cann, P., Bout, D., and Coursaget, P. (1997). Cell mediated immunity induced in mice by HPV 16 L1 virus-like particles. *Microb. Pathog.* **22:**219–225.

Eckels, K. H., Dubois, D. R., Summers, P. L., Schlesinger, J. J., Shelly, M., Cohen, S., Zhang, Y. M., Lai, C. J., Kurane, I., Rothman, A., Hasty, S., and Howard, B. (1994). Immunization of monkeys with baculovirus-dengue type-4 recombinants containing envelope and nonstructural proteins: Evidence of priming and partial protection. *Am. J. Trop. Med. Hyg.* **50:** 472–478.

Facciabene, A., Aurisicchio, L., and La Monica, N. (2004). Baculovirus vectors elicit antigen-specific immune responses in mice. *J. Virol.* **78:**8663–8672.

Feighny, R., Burrous, J., and Putnak, R. (1994). Dengue type-2 virus envelope protein made using recombinant baculovirus protects mice against virus challenge. *Am. J. Trop. Med. Hyg.* **50:**322–328.

Fenner, F. (2000). Adventures with poxviruses of vertebrates. *FEMS Microbiol. Rev.* **24:**123–133.

Fiedler, M., and Roggendorf, M. (2001). Vaccination against hepatitis delta virus infection: Studies in the woodchuck (Marmota monax) model. *Intervirology* **44:**154–161.

Foote, C. E., Love, D. N., Gilkerson, J. R., Rota, J., Trevor-Jones, P., Ruitenberg, K. M., Wellington, J. E., and Whalley, J. M. (2005). Serum antibody responses to equine herpesvirus 1 glycoprotein D in horses, pregnant mares and young foals. *Vet. Immunol. Immunopathol.* **105:**47–57.

French, T. J., Marshall, J. J., and Roy, P. (1990). Assembly of double-shelled, viruslike particles of bluetongue virus by the simultaneous expression of four structural proteins. *J. Virol.* **64:**5695–5700.

Fu, Z. F. (1997). Rabies and rabies research: Past, present and future. *Vaccine* **15**(Suppl.)**:** S20–S24.

Fu, Z. F., Rupprecht, C. E., Dietzschold, B., Saikumar, P., Niu, H. S., Babka, I., Wunner, W. H., and Koprowski, H. (1993). Oral vaccination of racoons (Procyon lotor) with baculovirus-expressed rabies virus glycoprotein. *Vaccine* **11:**925–928.

Fyfe, L., Maingay, J. P., and Howie, S. E. (1993). Murine HIV-1 p24 specific T lymphocyte activation by different antigen presenting cells: B lymphocytes from immunized mice present core protein to T cells. *Immunol. Lett.* **35:**45–50.

Galarza, J. M., Latham, T., and Cupo, A. (2005a). Virus-like particle (VLP) vaccine conferred complete protection against a lethal influenza virus challenge. *Viral Immunol.* **18:**244–251.

Galarza, J. M., Latham, T., and Cupo, A. (2005b). Virus-like particle vaccine conferred complete protection against a lethal influenza virus challenge. *Viral Immunol.* **18:**365–372.

Gherardi, M. M., and Esteban, M. (2005). Recombinant poxviruses as mucosal vaccine vectors. *J. Gen. Virol.* **86:**2925–2936.

Ghiasi, H., Nesburn, A. B., Kaiwar, R., and Wechsler, S. L. (1991). Immunoselection of recombinant baculoviruses expressing high levels of biologically active herpes simplex virus type 1 glycoprotein D. *Arch. Virol.* **121:**163–178.

Ghiasi, H., Kaiwar, R., Slanina, S., Nesburn, A. B., and Wechsler, S. L. (1994). Expression and characterization of baculovirus expressed herpes simplex virus type 1 glycoprotein L. *Arch. Virol.* **138:**199–212.

Ghiasi, H., Wechsler, S. L., Kaiwar, R., Nesburn, A. B., and Hofman, F. M. (1995). Local expression of tumor necrosis factor alpha and interleukin-2 correlates with protection against corneal scarring after ocular challenge of vaccinated mice with herpes simplex virus type 1. *J. Virol.* **69:**334–340.

Ghiasi, H., Nesburn, A. B., and Wechsler, S. L. (1996). Vaccination with a cocktail of seven recombinantly expressed HSV-1 glycoproteins protects against ocular HSV-1 challenge more efficiently than vaccination with any individual glycoprotein. *Vaccine* **14:**107–112.

Ghiasi, H., Hofman, F. M., Cai, S., Perng, G. C., Nesburn, A. B., and Wechsler, S. L. (1999). Vaccination with different HSV-1 glycoproteins induces different patterns of ocular cytokine responses following HSV-1 challenge of vaccinated mice. *Vaccine* **17:**2576–2582.

Ghiasi, H., Perng, G. C., Nesburn, A. B., and Wechsler, S. L. (2000). Antibody-dependent enhancement of HSV-1 infection by anti-gK sera. *Virus Res.* **68:**137–144.

Girones, N., Cuervo, H., and Fresno, M. (2005). *Trypanosoma cruzi*-induced molecular mimicry and Chagas' disease. *Curr. Top. Microbiol. Immunol.* **296:**89–123.

Gomez, I., Marx, F., Saurwein-Teissl, M., Gould, E. A., and Grubeck-Loebenstein, B. (2003). Characterization of tick-borne encephalitis virus-specific human T lymphocyte responses by stimulation with structural TBEV proteins expressed in a recombinant baculovirus. *Viral Immunol.* **16:**407–414.

Gorse, G. J., Belshe, R. B., Newman, F. K., and Frey, S. E. (1992). Lymphocyte proliferative responses following immunization with human immunodeficiency virus recombinant GP160. The NIAID AIDS vaccine clinical trials network. *Vaccine* **10:**383–388.

Gorse, G. J., Frey, S. E., Patel, G., Newman, F. K., and Belshe, R. B. (1994). Vaccine-induced antibodies to native and recombinant human immunodeficiency virus type 1 envelope glycoproteins. NIAID AIDS vaccine clinical trials network. *Vaccine* **12:**912–918.

Grabherr, R., Ernst, W., Oker-Blom, C., and Jones, I. (2001). Developments in the use of baculoviruses for the surface display of complex eukaryotic proteins. *Trends Biotechnol.* **19:**231–236.

Graham, B. S., Matthews, T. J., Belshe, R. B., Clements, M. L., Dolin, R., Wright, P. F., Gorse, G. J., Schwartz, D. H., Keefer, M. C., Bolognesi, D. P., Corey, L., Stablein, D. M., *et al.* (1993). Augmentation of human immunodeficiency virus type 1 neutralizing antibody by priming with gp160 recombinant vaccinia and boosting with rgp160 in vaccinia-naive adults. The NIAID AIDS vaccine clinical trials network. *J. Infect. Dis.* **167:**533–537.

Greenstone, H. L., Nieland, J. D., de Visser, K. E., De Bruijn, M. L., Kirnbauer, R., Roden, R. B., Lowy, D. R., Kast, W. M., and Schiller, J. T. (1998). Chimeric papillomavirus virus-like particles elicit antitumor immunity against the E7 oncoprotein in an HPV16 tumor model. *Proc. Natl. Acad. Sci. USA* **95:**1800–1805.

Grubman, M. J., Lewis, S. A., and Morgan, D. O. (1993). Protection of swine against foot-and-mouth disease with viral capsid proteins expressed in heterologous systems. *Vaccine* **11:**825–829.

Guerrero, R. A., Ball, J. M., Krater, S. S., Pacheco, S. E., Clements, J. D., and Estes, M. K. (2001). Recombinant Norwalk virus-like particles administered intranasally to mice induce systemic and mucosal (fecal and vaginal) immune responses. *J. Virol.* **75:**9713–9722.

Haanes, E. J., Guimond, P., and Wardley, R. (1997). The bovine parainfluenza virus type-3 (BPIV-3) hemagglutinin/neuraminidase glycoprotein expressed in baculovirus protects calves against experimental BPIV-3 challenge. *Vaccine* **15:**730–738.

Hall, S. L., Murphy, B. R., and van Wyke Coelingh, K. L. (1991). Protection of cotton rats by immunization with the human parainfluenza virus type 3 fusion (F) glycoprotein expressed on the surface of insect cells infected with a recombinant baculovirus. *Vaccine* **9:**659–667.

Hansson, M., Nygren, P. A., and Stahl, S. (2000a). Design and production of recombinant subunit vaccines. *Biotechnol. Appl. Biochem.* **32**(Pt 2)**:**95–107.

Hansson, M., Nygren, P. A., and Stahl, S. (2000b). Design and production of recombinant subunit vaccines. *Biotechnol. Appl. Biochem.* **32:**95–107.

Harder, T. C., and Osterhaus, A. D. (1997). Molecular characterization and baculovirus expression of the glycoprotein B of a seal herpesvirus (phocid herpesvirus-1). *Virology* **227:**343–352.

Harrison, R. L., Jarvis, D. L., and Summers, M. D. (1996). The role of the AcMNPV 25K gene 'FP25' in baculovirus polh and p10 expression. *Virology* **226:**34–46.

Harro, C. D., Pang, Y. Y., Roden, R. B., Hildesheim, A., Wang, Z., Reynolds, M. J., Mast, T. C., Robinson, R., Murphy, B. R., Karron, R. A., Dillner, J., Schiller, J. T., *et al.* (2001). Safety and immunogenicity trial in adult volunteers of a human papillomavirus 16 L1 virus-like particle vaccine. *J. Natl. Cancer Inst.* **93:**284–292.

Hawtin, R. E., Zarkowska, T., Arnold, K., Thomas, C. J., Gooday, G. W., King, L. A., Kuzio, J. A., and Possee, R. D. (1997). Liquefaction of *Autographa californica* nucleopolyhedrovirus-infected insects is dependent on the integrity of virus-encoded chitinase and cathepsin genes. *Virology* **238:**243–253.

Henderson, L. M. (2005). Overview of marker vaccine and differential diagnostic test technology. *Biologicals* **33:**203–209.

Herrington, D. A., Losonsky, G. A., Smith, G., Volvovitz, F., Cochran, M., Jackson, K., Hoffman, S. L., Gordon, D. M., Levine, M. M., and Edelman, R. (1992). Safety and immunogenicity in volunteers of a recombinant *Plasmodium falciparum* circumsporozoite protein malaria vaccine produced in lepidopteran cells. *Vaccine* **10:**841–846.

Hevey, M., Negley, D., Geisbert, J., Jahrling, P., and Schmaljohn, A. (1997). Antigenicity and vaccine potential of Marburg virus glycoprotein expressed by baculovirus recombinants. *Virology* **239:**206–216.

Hoerauf, A., Satoguina, J., Saeftel, M., and Specht, S. (2005). Immunomodulation by filarial nematodes. *Parasite Immunol.* **27:**417–429.

Hohdatsu, T., Yamato, H., Ohkawa, T., Kaneko, M., Motokawa, K., Kusuhara, H., Kaneshima, T., Arai, S., and Koyama, H. (2003). Vaccine efficacy of a cell lysate with recombinant baculovirus-expressed feline infectious peritonitis (FIP) virus nucleocapsid protein against progression of FIP. *Vet. Microbiol.* **97:**31–44.

Holder, A. A., and Blackman, M. J. (1994). What is the function of MSP-I on the malaria merozoite? *Parasitol. Today* **10:**182–184.

Hom, L. G., and Volkman, L. E. (1998). Preventing proteolytic artifacts in the baculovirus expression system. *Biotechniques* **25:**18–20.

Hom, L. G., Ohkawa, T., Trudeau, D., and Volkman, L. E. (2002). *Autographa californica* M nucleopolyhedrovirus ProV-CATH is activated during infected cell death. *Virology* **296:**212–218.

Homa, F. L., Brideau, R. J., Lehman, D. J., Thomsen, D. R., Olmsted, R. A., and Wathen, M. W. (1993). Development of a novel subunit vaccine that protects cotton rats against both human respiratory syncytial virus and human parainfluenza virus type 3. *J. Gen. Virol.* **74:**1995–1999.

Horimoto, T., Takada, A., Fujii, K., Goto, H., Hatta, M., Watanabe, S., Iwatsuki-Horimoto, K., Ito, M., Tagawa-Sakai, Y., Yamada, S., Ito, H., Ito, T., *et al.* (2006). The development and characterization of H5 influenza virus vaccines derived from a 2003 human isolate. *Vaccine* **24:**3669–3676.

Hota-Mitchell, S., Siddiqui, A. A., Dekaban, G. A., Smith, J., Tognon, C., and Podesta, R. B. (1997). Protection against *Schistosoma mansoni* infection with a recombinant baculovirus-expressed subunit of calpain. *Vaccine* **15:**1631–1640.

Hu, S. L., Abrams, K., Barber, G. N., Moran, P., Zarling, J. M., Langlois, A. J., Kuller, L., Morton, W. R., and Benveniste, R. E. (1992). Protection of macaques against SIV infection by subunit vaccines of SIV envelope glycoprotein gp160. *Science* **255:**456–459.

Hulst, M. M., Westra, D. F., Wensvoort, G., and Moormann, R. J. (1993). Glycoprotein E1 of hog cholera virus expressed in insect cells protects swine from hog cholera. *J. Virol.* **67:**5435–5442.

Hunt, I. (2005). From gene to protein: A review of new and enabling technologies for multi-parallel protein expression. *Protein Expr. Purif.* **40:**1–22.

Huser, A., and Hofmann, C. (2003). Baculovirus vectors: Novel mammalian cell gene-delivery vehicles and their applications. *Am. J. Pharmacogenomics* **3:**53–63.

Huser, A., Rudolph, M., and Hofmann, C. (2001). Incorporation of decay-accelerating factor into the baculovirus envelope generates complement-resistant gene transfer vectors. *Nat. Biotechnol.* **19:**451–455.

Igarashi, I., Asaba, U., Xuan, X., Omata, Y., Saito, A., Nagasawa, H., Fujisaki, K., Suzuki, N., Iwakura, Y., and Mikami, T. (2000). Immunization with recombinant surface antigens p26 with Freund's adjuvants against *Babesia rodhaini* infection. *J. Vet. Med. Sci.* **62:**717–723.

Ikonomou, L., Schneider, Y. J., and Agathos, S. N. (2003). Insect cell culture for industrial production of recombinant proteins. *Appl. Microbiol. Biotechnol.* **62:**1–20.

Inumaru, S., and Yamada, S. (1991). Characterization of pseudorabies virus neutralization antigen glycoprotein gIII produced in insect cells by a baculovirus expression vector. *Virus Res.* **21:**123–139.

Jaffray, A., Shephard, E., van Harmelen, J., Williamson, C., Williamson, A. L., and Rybicki, E. P. (2004). Human immunodeficiency virus type 1 subtype C Gag virus-like particle boost substantially improves the immune response to a subtype C gag DNA vaccine in mice. *J. Gen. Virol.* **85:**409–413.

Jarvis, D. L. (2003). Developing baculovirus-insect cell expression systems for humanized recombinant glycoprotein production. *Virology* **310:**1–7.

Je, Y. H., Jin, B. R., Park, H. W., Roh, J. Y., Chang, J. H., Seo, S. J., Olszewski, J. A., O'Reilly, D. R., and Kang, S. K. (2003). Baculovirus expression vectors that incorporate the foreign protein into viral occlusion bodies. *Biotechniques* **34:**81–87.

Jeong, S. H., Qiao, M., Nascimbeni, M., Hu, Z., Rehermann, B., Murthy, K., and Liang, T. J. (2004). Immunization with hepatitis C virus-like particles induces humoral and cellular immune responses in nonhuman primates. *J. Virol.* **78:**6995–7003.

Johansson, B. E. (1999). Immunization with influenza A virus hemagglutinin and neuraminidase produced in recombinant baculovirus results in a balanced and broadened immune response superior to conventional vaccine. *Vaccine* **17:**2073–2080.

Jones, T., Allard, F., Cyr, S. L., Tran, S. P., Plante, M., Gauthier, J., Bellerose, N., Lowell, G. H., and Burt, D. S. (2003). A nasal proteosome influenza vaccine containing baculovirus-derived hemagglutinin induces protective mucosal and systemic immunity. *Vaccine* **21:**3706–3712.

Kaba, S. A., Nene, V., Musoke, A. J., Vlak, J. M., and van Oers, M. M. (2002). Fusion to green fluorescent protein improves expression levels of *Theileria parva* sporozoite surface antigen p67 in insect cells. *Parasitology* **125:**497–505.

Kaba, S. A., Hemmes, J. C., van Lent, J. W., Vlak, J. M., Nene, V., Musoke, A. J., and van Oers, M. M. (2003). Baculovirus surface display of *Theileria parva* p67 antigen preserves the conformation of sporozoite-neutralizing epitopes. *Protein Eng.* **16:**73–78.

Kaba, S. A., Salcedo, A. M., Wafula, P. O., Vlak, J. M., and van Oers, M. M. (2004a). Development of a chitinase and v-cathepsin negative bacmid for improved integrity of secreted recombinant proteins. *J. Virol. Methods* **122:**113–118.

Kaba, S. A., Schaap, D., Roode, E. C., Nene, V., Musoke, A. J., Vlak, J. M., and van Oers, M. M. (2004b). Improved immunogenicity of novel baculovirus-derived *Theileria parva* p67 subunit antigens. *Vet. Parasitol.* **121:**53–64.

Kaba, S. A., Musoke, A. J., Schaap, D., Schetters, T., Rowlands, J., Vermeulen, A. N., Nene, V., Vlak, J. M., and van Oers, M. M. (2005). Novel baculovirus-derived p67 subunit vaccines efficacious against East Coast fever in cattle. *Vaccine* **23:**2791–2800.

Kajigaya, S., Fujii, H., Field, A., Anderson, S., Rosenfeld, S., Anderson, L. J., Shimada, T., and Young, N. S. (1991). Self-assembled B19 parvovirus capsids, produced in a baculovirus system, are antigenically and immunogenically similar to native virions. *Proc. Natl. Acad. Sci. USA* **88:**4646–4650.

Karayiannis, P., Saldanha, J., Jackson, A. M., Luther, S., Goldin, R., Monjardino, J., and Thomas, H. C. (1993). Partial control of hepatitis delta virus superinfection by immunisation of woodchucks (Marmota monax) with hepatitis delta antigen expressed by a recombinant vaccinia or baculovirus. *J. Med. Virol.* **41:**210–214.

Kato, T., Murata, T., Usui, T., and Park, E. Y. (2005). Improvement of the production of GFP_{uv}-ß1,3-N-acetylglucosaminyltransferase 2 fusion protein using a molecular chaperone-assisted insect-cell-based expression system. *Biotechnol. Bioeng.* **89:** 424–433.

Katz, J. M., Lu, X., Frace, A. M., Morken, T., Zaki, S. R., and Tumpey, T. M. (2000). Pathogenesis of and immunity to avian influenza A H5 viruses. *Biomed. Pharmacother.* **54:**178–187.

Keefer, M. C., Bonnez, W., Roberts, N. J., Jr., Dolin, R., and Reichman, R. C. (1991). Human immunodeficiency virus (HIV-1) gp160-specific lymphocyte proliferative responses of mononuclear leukocytes from HIV-1 recombinant gp160 vaccine recipients. *J. Infect. Dis.* **163:**448–453.

Keefer, M. C., Graham, B. S., Belshe, R. B., Schwartz, D., Corey, L., Bolognesi, D. P., Stablein, D. M., Montefiori, D. C., McElrath, M. J., Clements, M. L., Gorse, G. G., Wright, P. F., *et al.* (1994). Studies of high doses of a human immunodeficiency virus type 1 recombinant glycoprotein 160 candidate vaccine in HIV type 1-seronegative humans. *AIDS Res. Hum. Retroviruses* **10:**1713–1723.

Kelly, E. P., Greene, J. J., King, A. D., and Innis, B. L. (2000). Purified dengue 2 virus envelope glycoprotein aggregates produced by baculovirus are immunogenic in mice. *Vaccine* **18:**2549–2559.

Kelso, J. M., and Yunginger, J. W. (2003). Immunization of egg-allergic individuals with egg- or chicken-derived vaccines. *Immunol. Allergy Clin. North Am.* **23:**635–648.

Kew, O. M., Wright, P. F., Agol, V. I., Delpeyroux, F., Shimizu, H., Nathanson, N., and Pallansch, M. A. (2004). Circulating vaccine-derived polioviruses: Current state of knowledge. *Bull. World Health Organ.* **82:**16–23.

Kilbourne, E. D., Pokorny, B. A., Johansson, B., Brett, I., Milev, Y., and Matthews, J. T. (2004). Protection of mice with recombinant influenza virus neuraminidase. *J. Infect. Dis.* **189:**459–461.

King, L. K., and Possee, R. D. (1992). "The Baculovirus Expression System: A Laboratory Guide." Chapman and Hall, London.

Kirnbauer, R., Booy, F., Cheng, N., Lowy, D. R., and Schiller, J. T. (1992). Papillomavirus L1 major capsid protein self-assembles into virus-like particles that are highly immunogenic. *Proc. Natl. Acad. Sci. USA* **89:**12180–12184.

Kitts, P. A., Ayres, M. D., and Possee, R. D. (1990). Linearization of baculovirus DNA enhances the recovery of recombinant virus expression vectors. *Nucleic Acids Res.* **18:**5667–5672.

Kitts, P. A., and Possee, R. D. (1993). A method for producing recombinant baculovirus expression vectors at high frequency. *Biotechniques* **14:**810–817.

Koch, G., van Roozelaar, D. J., Verschueren, C. A., van der Eb, A. J., and Noteborn, M. H. (1995). Immunogenic and protective properties of chicken anaemia virus proteins expressed by baculovirus. *Vaccine* **13:**763–770.

Kool, M., Voncken, J. W., van Lier, F. L., Tramper, J., and Vlak, J. M. (1991). Detection and analysis of *Autographa californica* nuclear polyhedrosis virus mutants with defective interfering properties. *Virology* **183:**739–746.

Kost, T. A., and Condreay, J. P. (2002). Recombinant baculoviruses as mammalian cell gene-delivery vectors. *Trends Biotechnol.* **20:**173–180.

Kost, T. A., Condreay, J. P., and Jarvis, D. L. (2005). Baculovirus as versatile vectors for protein expression in insect and mammalian cells. *Nat. Biotechnol.* **23:**567–575.

Krell, P. J. (1996). Passage effect of virus infection in insect cells. *In* "Insect Cell Cultures Fundamental and Applied Aspects," Vol. 2, pp. 125–137. Kluwer, Dordrecht, The Netherlands.

Krishna, S., Blacklaws, B. A., Overton, H. A., Bishop, D. H., and Nash, A. A. (1989). Expression of glycoprotein D of herpes simplex virus type 1 in a recombinant baculovirus: Protective responses and T cell recognition of the recombinant-infected cell extracts. *J. Gen. Virol.* **70:**1805–1814.

Kukreja, A., Walker, C., Fitzmaurice, T., Awan, A., Love, D. N., Whalley, J. M., and Field, H. J. (1998). Protective effects of equine herpesvirus-1 (EHV-1) glycoprotein B in a murine model of EHV-1-induced abortion. *Vet. Microbiol.* **62:**303–311.

Lakey, D. L., Treanor, J. J., Betts, R. F., Smith, G. E., Thompson, J., Sannella, E., Reed, G., Wilkinson, B. E., and Wright, P. F. (1996). Recombinant baculovirus influenza A hemagglutinin vaccines are well tolerated and immunogenic in healthy adults. *J. Infect. Dis.* **174:**838–841.

Langedijk, J. P. M., Middel, W. G. J., Meloen, R. H., Kramps, J. A., and de Smit, J. A. (2001). Enzyme-linked immunosorbent assay using a virus type-specific peptide based on a subdomain of envelope protein Erns for serologic diagnosis of pestivirus infections in swine. *J. Clin. Microbiol.* **39:**906–912.

Le Gall-Recule, G., Jestin, V., Chagnaud, P., Blanchard, P., and Jestin, A. (1996). Expression of muscovy duck parvovirus capsid proteins (VP2 and VP3) in a baculovirus expression system and demonstration of immunity induced by the recombinant proteins. *J. Gen. Virol.* **77:**2159–2163.

Lee, E. G., Lee, M. Y., Chung, J. Y., Je, E. Y., Bae, Y. A., Na, B. K., Kim, T. S., Eom, K. S., Cho, S. Y., and Kong, Y. (2005). Feasibility of baculovirus-expressed recombinant 10-kDa antigen in the serodiagnosis of Taenia solium neurocysticercosis. *Trans. R. Soc. Trop. Med. Hyg.* **99:**919–926.

Lee, H., and Krell, P. J. (1994). Reiterated DNA fragments in defective genomes of *Autographa californica* nuclear polyhedrosis virus are competent for AcMNPV-dependent DNA replication. *Virology* **202:**418–442.

Lehman, D. J., Roof, L. L., Brideau, R. J., Aeed, P. A., Thomsen, D. R., Elhammer, A. P., Wathen, M. W., and Homa, F. L. (1993). Comparison of soluble and secreted forms of human parainfluenza virus type 3 glycoproteins expressed from mammalian and insect cells as subunit vaccines. *J. Gen. Virol.* **74:**459–469.

Leutenegger, C. M., Hofmann-Lehmann, R., Holznagel, E., Cuisinier, A. M., Wolfensberger, C., Duquesne, V., Cronier, J., Allenspach, K., Aubert, A., Ossent, P., and Lutz, H. (1998). Partial protection by vaccination with recombinant feline immunodeficiency virus surface glycoproteins. *AIDS Res. Hum. Retroviruses* **14:**275–283.

Levine, M. Z., Calderon, J. C., Wilkins, P. P., Lane, W. S., Asara, J. M., Hancock, K., Gonzalez, A. E., Garcia, H. H., Gilman, R. H., and Tsang, V. C. (2004). Characterization, cloning, and expression of two diagnostic antigens for Taenia solium tapeworm infection. *J. Parasitol.* **90:**631–638.

Li, T., Takeda, N., and Miyamura, T. (2001). Oral administration of hepatitis E virus-like particles induces a systemic and mucosal immune response in mice. *Vaccine* **19:** 3476–3484.

Li, T. C., Suzaki, Y., Ami, Y., Dhole, T. N., Miyamura, T., and Takeda, N. (2004). Protection of cynomolgus monkeys against HEV infection by oral administration of recombinant hepatitis E virus-like particles. *Vaccine* **22:**370–377.

Lin, X., Lubinski, J. M., and Friedman, H. M. (2004). Immunization strategies to block the herpes simplex virus type 1 immunoglobulin G Fc receptor. *J. Virol.* **78:**2562–2571.

Lofthouse, S. A., Andrews, A. E., Elhay, M. J., Bowles, V. M., Meeusen, E. N., and Nash, A. D. (1996). Cytokines as adjuvants for ruminant vaccines. *Int. J. Parasitol.* **26:**835–842.

Long, G., Westenberg, M., Wang, H. L., Vlak, J. M., and Hu, Z. H. (2006). Function, oligomerization and N-linked glycosylation of the *Helicoverpa armigera* single nucleopolyhedrovius envelope fusion protein. *J. Gen. Virol.* **87:**839–846.

Lopez de Turiso, J. A., Cortes, E., Martinez, C., Ruiz de Ybanez, R., Simarro, I., Vela, C., and Casal, I. (1992). Recombinant vaccine for canine parvovirus in dogs. *J. Virol.* **66:**2748–2753.

Loudon, P. T., Hirasawa, T., Oldfield, S., Murphy, M., and Roy, P. (1991). Expression of the outer capsid protein VP5 of two bluetongue viruses, and synthesis of chimeric double-shelled virus-like particles using combinations of recombinant baculoviruses. *Virology* **182:**793–801.

Lubeck, M. D., Natuk, R. J., Chengalvala, M., Chanda, P. K., Murthy, K. K., Murthy, S., Mizutani, S., Lee, S.-G., Wade, M. S., Bhat, B. M., Bhat, R., Dheer, S. K., *et al.* (1994). Immunogenicity of recombinant adenovirus-human immunodeficiency virus vaccines in chimpanzees following intranasal administration. *AIDS Res. Hum. Retroviruses* **10:**1443–1449.

Luckow, V. A., Lee, S. C., Barry, G. F., and Olins, P. O. (1993). Efficient generation of infectious recombinant baculoviruses by site-specific transposon-mediated insertion of foreign genes into a baculovirus genome propagated in *Escherichia coli*. *J. Virol.* **67:**4566–4579.

Lundholm, P., Wahren, M., Sandstrom, E., Volvovitz, F., and Wahren, B. (1994). Autoreactivity in HIV-infected individuals does not increase during vaccination with envelope rgp160. *Immunol. Lett.* **41:**147–153.

Luo, L., Li, Y., Cannon, P. M., Kim, S., and Kang, C. Y. (1992). Chimeric gag-V3 virus-like particles of human immunodeficiency virus induce virus-neutralizing antibodies. *Proc. Natl. Acad. Sci. USA* **89:**10527–10531.

Luo, M., Yuan, F., Liu, Y., Jiang, S., Song, X., Jiang, P., Yin, X., Ding, M., and Deng, H. (2006). Induction of neutralizing antibody against human immunodeficiency virus type 1 (HIV-1) by immunization with gp41 membrane-proximal external region (MPER) fused with porcine endogenous retrovirus (PERV) p15E fragment. *Vaccine* **24:**435–442.

Ma, J. K., Barros, E., Bock, R., Christou, P., Dale, P. J., Dix, P. J., Fischer, R., Irwin, J., Mahoney, R., Pezzotti, M., Schillberg, S., Sparrow, P., *et al.* (2005). Molecular farming for new drugs and vaccines. Current perspectives on the production of pharmaceuticals in transgenic plants. *EMBO Rep.* **6:**593–599.

Maeda, K., Ono, M., Kawaguchi, Y., Niikura, M., Okazaki, K., Yokoyama, N., Tokiyoshi, Y., Tohya, Y., and Mikami, T. (1996). Expression and properties of feline herpesvirus type 1 gD (hemagglutinin) by a recombinant baculovirus. *Virus Res.* **46:**75–80.

Mao, C., Koutsky, L. A., Ault, K. A., Wheeler, C. M., Brown, D. R., Wiley, D. J., Alvarez, F. B., Bautista, O. M., Jansen, K. U., and Barr, E. (2006). Efficacy of human papillomavirus-16 vaccine to prevent cervical intraepithelial neoplasia: A randomized controlled trial. *Obstet. Gynecol.* **107:**18–27.

Marshall, G. S., Li, M., Stout, G. G., Louthan, M. V., Duliege, A. M., Burke, R. L., and Hunt, L. A. (2000). Antibodies to the major linear neutralizing domains of cytomegalovirus glycoprotein B among natural seropositives and CMV subunit vaccine recipients. *Viral Immunol.* **13:**329–341.

Martens, J. W., van Oers, M. M., van de Bilt, B. D., Oudshoorn, P., and Vlak, J. M. (1995). Development of a baculovirus vector that facilitates the generation of p10-based recombinants. *J. Virol. Methods* **52:**15–19.

Martinez, A. P., Margos, G., Barker, G., and Sinden, R. E. (2000). The roles of the glycosylphosphatidylinositol anchor on the production and immunogenicity of recombinant ookinete surface antigen Pbs21 of *Plasmodium berghei* when prepared in a baculovirus expression system. *Parasite Immunol.* **22:**493–500.

Martinez, C., Dalsgaard, K., Lopez de Turiso, J. A., Cortes, E., Vela, C., and Casal, J. I. (1992). Production of porcine parvovirus empty capsids with high immunogenic activity. *Vaccine* **10:**684–690.

Martinez-Torrecuadrada, J. L., Saubi, N., Pages-Mante, A., Caston, J. R., Espuna, E., and Casal, J. I. (2003). Structure-dependent efficacy of infectious bursal disease virus (IBDV) recombinant vaccines. *Vaccine* **21:**3342–3350.

Matsuoka, H., Kobayashi, J., Barker, G. C., Miura, K., Chinzei, Y., Miyajima, S., Ishii, A., and Sinden, R. E. (1996). Induction of anti-malarial transmission blocking immunity with a recombinant ookinete surface antigen of *Plasmodium berghei* produced in silkworm larvae using the baculovirus expression vector system. *Vaccine* **14:**120–126.

McCown, J., Cochran, M., Putnak, R., Feighny, R., Burrous, J., Henchal, E., and Hoke, C. (1990). Protection of mice against lethal Japanese encephalitis with a recombinant baculovirus vaccine. *Am. J. Trop. Med. Hyg.* **42:**491–499.

McElrath, M. J., Corey, L., Berger, D., Hoffman, M. C., Klucking, S., Dragavon, J., Peterson, E., and Greenberg, P. D. (1994). Immune responses elicited by recombinant vaccinia-human immunodeficiency virus (HIV) envelope and HIV envelope protein: Analysis of the durability of responses and effect of repeated boosting. *J. Infect. Dis.* **169:**41–47.

Mellquist-Riemenschneider, J. L., Garrison, A. R., Geisbert, J. B., Saikh, K. U., Heidebrink, K. D., Jahrling, P. B., Ulrich, R. G., and Schmaljohn, C. S. (2003). Comparison of the protective efficacy of DNA and baculovirus-derived protein vaccines for EBOLA virus in guinea pigs. *Virus Res.* **92:**187–193.

Monsma, S. A., Oomens, A. G., and Blissard, G. W. (1996). The GP64 envelope fusion protein is an essential baculovirus protein required for cell-to-cell transmission of infection. *J. Virol.* **70:**4607–4616.

Montefiori, D. C., Graham, B. S., Kliks, S., and Wright, P. F. (1992). Serum antibodies to HIV-1 in recombinant vaccinia virus recipients boosted with purified recombinant gp160. NIAID AIDS vaccine clinical trials network. *J. Clin. Immunol.* **12:**429–439.

Mori, H., Tawara, H., Nakazawa, H., Sumida, M., Matsubara, F., Aoyama, S., Iritani, Y., Hayashi, Y., and Kamogawa, K. (1994). Expression of the Newcastle disease virus (NDV) fusion glycoprotein and vaccination against NDV challenge with a recombinant baculovirus. *Avian Dis.* **38:**772–777.

Nagy, E., Krell, P. J., Dulac, G. C., and Derbyshire, J. B. (1991). Vaccination against newcastle disease with a recombinant baculovirus hemagglutinin-neuraminidase subunit vaccine. *Avian Dis.* **35:**585–590.

Neilan, J. G., Zsak, L., Lu, Z., Burrage, T. G., Kutish, G. F., and Rock, D. L. (2004). Neutralizing antibodies to African swine fever virus proteins p30, p54, and p72 are not sufficient for antibody-mediated protection. *Virology* **319:**337–342.

Nene, V., Inumaru, S., McKeever, D., Morzaria, S., Shaw, M., and Musoke, A. (1995). Characterization of an insect cell-derived Theileria parva sporozoite vaccine antigen and immunogenicity in cattle. *Infect. Immun.* **63:**503–508.

Nicollier-Jamot, B., Ogier, A., Piroth, L., Pothier, P., and Kohli, E. (2004). Recombinant virus-like particles of a norovirus (genogroup II strain) administered intranasally and orally with mucosal adjuvants LT and LT(R192G) in BALB/c mice induce specific humoral and cellular Th1/Th2-like immune responses. *Vaccine* **22:**1079–1086.

O'Reilly, D. R., Miller, L. K., and Luckow, V. A. (1992). "Baculovirus Expression Vectors, A Laboratory Manual." Oxford University Press, New York, USA.

Oien, N. L., Brideau, R. J., Thomsen, D. R., Homa, F. L., and Wathen, M. W. (1993). Vaccination with a heterologous respiratory syncytial virus chimeric FG glycoprotein demonstrates significant subgroup cross-reactivity. *Vaccine* **11:**1040–1048.

Okano, K., Vanarsdall, A. L., Mikhailov, V. S., and Rohrmann, G. F. (2006). Conserved molecular systems of the Baculoviridae. *Virology* **344:**77–87.

Okazaki, K., Honda, E., and Kono, Y. (1994). Expression of bovine herpesvirus 1 glycoprotein gIII by a recombinant baculovirus in insect cells. *J. Gen. Virol.* **75:**901–904.

Olivier, M., Gregory, D. J., and Forget, G. (2005). Subversion mechanisms by which *Leishmania* parasites can escape the host immune response: A signaling point of view. *Clin. Microbiol. Rev.* **18:**293–305.

Olsen, C. W., McGregor, M. W., Dybdahl-Sissoko, N., Schram, B. R., Nelson, K. M., Lunn, D. P., Macklin, M. D., Swain, W. F., and Hinshaw, V. S. (1997). Immunogenicity and efficacy of baculovirus-expressed and DNA-based equine influenza virus hemagglutinin vaccines in mice. *Vaccine* **15:**1149–1156.

Onuma, M., Kubota, S., Kakuda, T., Sako, Y., Asada, M., Kabeya, H., and Sugimoto, C. (1997). Control of Theileria sergenti infection by vaccination. *Trop. Anim. Health Prod.* **29:**119S–123S.

Packiarajah, P., Walker, C., Gilkerson, J., Whalley, J. M., and Love, D. N. (1998). Immune responses and protective efficacy of recombinant baculovirus-expressed glycoproteins of equine herpesvirus 1 (EHV-1) gB, gC and gD alone or in combinations in BALB/c mice. *Vet. Microbiol.* **61:**261–278.

Palese, P. (2004). Influenza: Old and new threats. *Nat. Med.* **10:**S82–S87.

Pande, A., Carr, B. V., Wong, S. Y., Dalton, K., Jones, I. M., McCauley, J. W., and Charleston, B. (2005). The glycosylation pattern of baculovirus expressed envelope protein E2 affects its ability to prevent infection with bovine viral diarrhoea virus. *Virus Res.* **114:**54–62.

Pastoret, P. P., Blancou, J., Vannier, P., and Verschueren, C. (1997). "Veterinary Vaccinology." Elsevier, Amsterdam.

Patja, A., Makinen-Kiljunen, S., Davidkin, I., Paunio, M., and Peltola, H. (2001). Allergic reactions to measles-mumps-rubella vaccination. *Pediatrics* **107:**E27.

Pau, M. G., Ophorst, C., Koldijk, M. H., Schouten, G., Mehtali, M., and Uytdehaag, F. (2001). The human cell line PER.C6 provides a new manufacturing system for the production of influenza vaccines. *Vaccine* **19:**2716–2721.

Pearson, L. D., and Roy, P. (1993). Genetically engineered multi-component virus-like particles as veterinary vaccines. *Immunol. Cell Biol.* **71:**381–389.

Peet, N. M., McKeating, J. A., Ramos, B., Klonisch, T., De Souza, J. B., Delves, P. J., and Lund, T. (1997). Comparison of nucleic acid and protein immunization for induction of antibodies specific for HIV-1 gp120. *Clin. Exp. Immunol.* **109:**226–232.

Pekosz, A., Griot, C., Stillmock, K., Nathanson, N., and Gonzalez-Scarano, F. (1995). Protection from La Crosse virus encephalitis with recombinant glycoproteins: Role of neutralizing anti-G1 antibodies. *J. Virol.* **69:**3475–3481.

Perales, M. A., Schwartz, D. H., Fabry, J. A., and Lieberman, J. (1995). A vaccinia-gp160-based vaccine but not a gp160 protein vaccine elicits anti-gp160 cytotoxic T lymphocytes in some HIV-1 seronegative vaccinees. *J. Acquir. Immune Defic. Syndr. Hum. Retrovirol.* **10:**27–35.

Percheson, P. B., Trepanier, P., Dugre, R., and Mabrouk, T. (1999). A phase I, randomized controlled clinical trial to study the reactogenicity and immunogenicity of a new split influenza vaccine derived from a non-tumorigenic cell line. *Dev. Biol. Stand.* **98:**127–132; discussion 133–124.

Perera, K. L., Handunnetti, S. M., Holm, I., Longacre, S., and Mendis, K. (1998). Baculovirus merozoite surface protein 1 C-terminal recombinant antigens are highly protective in a natural primate model for human Plasmodium vivax malaria. *Infect. Immun.* **66:**1500–1506.

Philipps, B., Rotmann, D., Wicki, M., Mayr, L. M., and Forstner, M. (2005). Time reduction and process optimization of the baculovirus expression system for more efficient recombinant protein production in insect cells. *Protein Expr. Purif.* **42:**211–218.

Pijlman, G. P., van den Born, E., Martens, D. E., and Vlak, J. M. (2001). *Autographa californica* baculoviruses with large genomic deletions are rapidly generated in infected insect cells. *Virology* **283:**132–138.

Pijlman, G. P., Dortmans, J. C., Vermeesch, A. M., Yang, K., Martens, D. E., Goldbach, R. W., and Vlak, J. M. (2002). Pivotal role of the non-hr origin of DNA replication in the genesis of defective interfering baculoviruses. *J. Virol.* **76:**5605–5611.

Pijlman, G. P., Van Schijndel, J. E., and Vlak, J. M. (2003a). Spontaneous excision of BAC vector sequences from bacmid-derived baculovirus expression vectors upon passage in insect cells. *J. Gen. Virol.* **84:**2669–2678.

Pijlman, G. P., Vermeesch, A. M., and Vlak, J. M. (2003b). Cell line-specific accumulation of the baculovirus non-hr origin of DNA replication in infected insect cells. *J. Invertebr. Pathol.* **84:**214–219.

Pijlman, G. P., de Vrij, J., van den End, F. J., Vlak, J. M., and Martens, D. E. (2004). Evaluation of baculovirus expression vectors with enhanced stability in continuous cascaded insect-cell bioreactors. *Biotechnol. Bioeng.* **87:**743–753.

Pijlman, G. P., Roode, E. C., Fan, X., Roberts, L. O., Belsham, G. J., Vlak, J. M., and van Oers, M. M. (2006). Stabilized baculovirus vector expressing a heterologous gene and GP64 from a single bicistronic transcript. *J. Biotechnol.* **123:**13–21.

Pitcovski, J., Di-Castro, D., Shaaltiel, Y., Azriel, A., Gutter, B., Yarkoni, E., Michael, A., Krispel, S., and Levi, B. Z. (1996). Insect cell-derived VP2 of infectious bursal disease virus confers protection against the disease in chickens. *Avian Dis.* **40:**753–761.

Pitcovski, J., Levi, B. Z., Maray, T., Di-Castro, D., Safadi, A., Krispel, S., Azriel, A., Gutter, B., and Michael, A. (1999). Failure of viral protein 3 of infectious bursal disease virus produced in prokaryotic and eukaryotic expression systems to protect chickens against the disease. *Avian Dis.* **43:**8–15.

Pizarro, J. C., Chitarra, V., Verger, D., Holm, I., Petres, S., Dartevelle, S., Nato, F., Longacre, S., and Bentley, G. A. (2003). Crystal structure of a Fab complex formed with PfMSP1–19, the C-terminal fragment of merozoite surface protein 1 from *Plasmodium falciparum*: A malaria vaccine candidate. *J. Mol. Biol.* **328:**1091–1103.

Plana Duran, J., Climent, I., Sarraseca, J., Urniza, A., Cortes, E., Vela, C., and Casal, J. I. (1997). Baculovirus expression of proteins of porcine reproductive and respiratory syndrome virus strain Olot/91. Involvement of ORF3 and ORF5 proteins in protection. *Virus Genes* **14:**19–29.

Poomputsa, K., Kittel, C., Egorov, A., Ernst, W., and Grabherr, R. (2003). Generation of recombinant influenza virus using baculovirus delivery vector. *J. Virol. Methods* **110:**111–114.

Possee, R. D., Thomas, C. J., and King, L. A. (1999). The use of baculovirus vectors for the production of membrane proteins in insect cells. *Biochem. Soc. Trans.* **27:**928–932.

Powers, D. C., Smith, G. E., Anderson, E. L., Kennedy, D. J., Hackett, C. S., Wilkinson, B. E., Volvovitz, F., Belshe, R. B., and Treanor, J. J. (1995). Influenza A virus vaccines containing purified recombinant H3 hemagglutinin are well tolerated and induce protective immune responses in healthy adults. *J. Infect. Dis.* **171:**1595–1599.

Powers, D. C., McElhaney, J. E., Florendo, O. A., Jr., Manning, M. C., Upshaw, C. M., Bentley, D. W., and Wilkinson, B. E. (1997). Humoral and cellular immune responses following vaccination with purified recombinant hemagglutinin from influenza A (H3N2) virus. *J. Infect. Dis.* **175:**342–351.

Prehaud, C., Takehara, K., Flamand, A., and Bishop, D. H. (1989). Immunogenic and protective properties of rabies virus glycoprotein expressed by baculovirus vectors. *Virology* **173:**390–399.

Qiao, M., Ashok, M., Bernard, K. A., Palacios, G., Zhou, Z. H., Lipkin, W. I., and Liang, T. J. (2004). Induction of sterilizing immunity against West Nile virus (WNV), by immunization with WNV-like particles produced in insect cells. *J. Infect. Dis.* **190:**2104–2108.

Qu, X., Chen, W., Maguire, T., and Austin, F. (1993). Immunoreactivity and protective effects in mice of a recombinant dengue 2 Tonga virus NS1 protein produced in a baculovirus expression system. *J. Gen. Virol.* **74:**89–97.

Quanquin, N. M., Galaviz, C., Fouts, D. L., Wrightsman, R. A., and Manning, J. E. (1999). Immunization of mice with a TolA-like surface protein of *Trypanosoma cruzi* generates CD4(+) T-cell-dependent parasiticidal activity. *Infect. Immun.* **67:** 4603–4612.

Ray, R., Galinski, M. S., and Compans, R. W. (1989). Expression of the fusion glycoprotein of human parainfluenza type 3 virus in insect cells by a recombinant baculovirus and analysis of its immunogenic property. *Virus Res.* **12:**169–180.

Reed, Z. (2005). Portfolio of candidate malaria vaccines currently in development, March 2005. Initiative on Vaccine Research, World Health Organization.

Reszka, N., Cornelissen, J. B., Harmsen, M. M., Bienkowska-Szewczyk, K., de Bree, J., Boersma, W. J., and Rijsewijk, F. A. (2005). *Fasciola hepatica* procathepsin L3 protein expressed by a baculovirus recombinant can partly protect rats against fasciolosis. *Vaccine* **23:**2987–2993.

Reuben, J. M., Liang, L., Atmar, R. L., and Greenberg, S. (1992). B-cell activation and differentiation by HIV-1 antigens among volunteers vaccinated with VaxSyn HIV-1. *J. Acquir. Immune Defic. Syndr.* **5:**719–725.

Rini, B. I. (2002). Technology evaluation: APC-8015, Dendreon. *Curr. Opin. Mol. Ther.* **4:**76–79.
Rosen, E., Stapleton, J. T., and McLinden, J. (1993). Synthesis of immunogenic hepatitis A virus particles by recombinant baculoviruses. *Vaccine* **11:**706–712.
Roy, P. (1996). Genetically engineered particulate virus-like structures and their use as vaccine delivery systems. *Intervirology* **39:**62–71.
Roy, P. (2003). Nature and duration of protective immunity to bluetongue virus infection. *Dev. Biol. (Basel)* **114:**169–183.
Roy, P., and Sutton, G. (1998). New generation of African horse sickness virus vaccines based on structural and molecular studies of the virus particles. *Arch. Virol. Suppl.* **14:**177–202.
Roy, P., Bishop, D. H., LeBlois, H., and Erasmus, B. J. (1994). Long-lasting protection of sheep against bluetongue challenge after vaccination with virus-like particles: Evidence for homologous and partial heterologous protection. *Vaccine* **12:**805–811.
Ruitenberg, K. M., Walker, C., Love, D. N., Wellington, J. E., and Whalley, J. M. (2000). A prime-boost immunization strategy with DNA and recombinant baculovirus-expressed protein enhances protective immunogenicity of glycoprotein D of equine herpesvirus 1 in naive and infection-primed mice. *Vaccine* **18:**1367–1373.
Ruiz-Gonzalvo, F., Rodriguez, F., and Escribano, J. M. (1996). Functional and immunological properties of the baculovirus-expressed hemagglutinin of African swine fever virus. *Virology* **218:**285–289.
Russell, W. C. (2000). Update on adenovirus and its vectors. *J. Gen. Virol.* **81:**2573–2604.
Saliki, J. T., Mizak, B., Flore, H. P., Gettig, R. R., Burand, J. P., Carmichael, L. E., Wood, H. A., and Parrish, C. R. (1992). Canine parvovirus empty capsids produced by expression in a baculovirus vector: Use in analysis of viral properties and immunization of dogs. *J. Gen. Virol.* **73:**369–374.
Sato, H., Nakajima, K., Maeno, Y., Kamaishi, T., Kamata, T., Mori, H., Kamei, K., Takano, R., Kudo, K., and Hara, S. (2000). Expression of YAV proteins and vaccination against viral ascites among cultured juvenile yellowtail. *Biosci. Biotechnol. Biochem.* **64:**1494–1499.
Saville, G. P., Patmanidi, A. L., Possee, R. D., and King, L. A. (2004). Deletion of the Autographa californica nucleopolyhedrovirus chitinase KDEL motif and *in vitro* and *in vivo* analysis of the modified virus. *J. Gen. Virol.* **85:**821–831.
Schetters, T. (2005). Vaccination against canine babesiosis. *Trends Parasitol.* **21:** 179–184.
Schijns, V. E. (2003). Mechanisms of vaccine adjuvant activity: Initiation and regulation of immune responses by vaccine adjuvants. *Vaccine* **21:**829–831.
Schleiss, M. R., Bourne, N., Stroup, G., Bravo, F. J., Jensen, N. J., and Bernstein, D. I. (2004). Protection against congenital cytomegalovirus infection and disease in guinea pigs, conferred by a purified recombinant glycoprotein B vaccine. *J. Infect. Dis.* **189:**1374–1381.
Schmaljohn, C. S., Parker, M. D., Ennis, W. H., Dalrymple, J. M., Collett, M. S., Suzich, J. A., and Schmaljohn, A. L. (1989). Baculovirus expression of the M genome segment of Rift Valley fever virus and examination of antigenic and immunogenic properties of the expressed proteins. *Virology* **170:**184–192.
Schmaljohn, C. S., Chu, Y. K., Schmaljohn, A. L., and Dalrymple, J. M. (1990). Antigenic subunits of Hantaan virus expressed by baculovirus and vaccinia virus recombinants. *J. Virol.* **64:**3162–3170.
Scodeller, E. A., Tisminetzky, S. G., Porro, F., Schiappacassi, M., De Rossi, A., Chiecco-Bianchi, L., and Baralle, F. E. (1995). A new epitope presenting system

displays a HIV-1 V3 loop sequence and induces neutralizing antibodies. *Vaccine* **13:**1233–1239.

Sestak, K., Meister, R. K., Hayes, J. R., Kim, L., Lewis, P. A., Myers, G., and Saif, L. J. (1999). Active immunity and T-cell populations in pigs intraperitoneally inoculated with baculovirus-expressed transmissible gastroenteritis virus structural proteins. *Vet. Immunol. Immunopathol.* **70:**203–221.

Sharma, A. K., Woldehiwet, Z., Walrevens, K., and Letteson, J. (1996). Immune responses of lambs to the fusion (F) glycoprotein of bovine respiratory syncytial virus expressed on insect cells infected with a recombinant baculovirus. *Vaccine* **14:**773–779.

Shearer, M. H., Bright, R. K., Lanford, R. E., and Kennedy, R. C. (1993). Immunization of mice with baculovirus-derived recombinant SV40 large tumour antigen induces protective tumour immunity to a lethal challenge with SV40-transformed cells. *Clin. Exp. Immunol.* **91:**266–271.

Shivappa, R. B., McAllister, P. E., Edwards, G. H., Santi, N., Evensen, O., and Vakharia, V. N. (2005). Development of a subunit vaccine for infectious pancreatic necrosis virus using a baculovirus insect/larvae system. *Dev. Biol. (Basel)* **121:** 165–174.

Sinnathamby, G., Renukaradhya, G. J., Rajasekhar, M., Nayak, R., and Shaila, M. S. (2001a). Immune responses in goats to recombinant hemagglutinin-neuraminidase glycoprotein of Peste des petits ruminants virus: Identification of a T cell determinant. *Vaccine* **19:**4816–4823.

Sinnathamby, G., Renukaradhya, G. J., Rajasekhar, M., Nayak, R., and Shaila, M. S. (2001b). Recombinant hemagglutinin protein of rinderpest virus expressed in insect cells induces cytotoxic T-cell responses in cattle. *Viral Immunol.* **14:**349–358.

Slepushkin, V. A., Katz, J. M., Black, R. A., Gamble, W. C., Rota, P. A., and Cox, N. J. (1995). Protection of mice against influenza A virus challenge by vaccination with baculovirus-expressed M2 protein. *Vaccine* **13:**1399–1402.

Smith, G. E., Summers, M. D., and Fraser, M. J. (1983). Production of human beta interferon in insect cells infected with a baculovirus expression vector. *Mol. Cell. Biol.* **3:**2156–2165.

Snyder, D. B., Vakharia, V. N., Mengel-Whereat, S. A., Edwards, G. H., Savage, P. K., Lutticken, D., and Goodwin, M. A. (1994). Active cross-protection induced by a recombinant baculovirus expressing chimeric infectious bursal disease virus structural proteins. *Avian Dis.* **38:**701–707.

Song, C. S., Lee, Y. J., Lee, C. W., Sung, H. W., Kim, J. H., Mo, I. P., Izumiya, Y., Jang, H. K., and Mikami, T. (1998). Induction of protective immunity in chickens vaccinated with infectious bronchitis virus S1 glycoprotein expressed by a recombinant baculovirus. *J. Gen. Virol.* **79:**719–723.

Stokes, A., Alber, D. G., Cameron, R. S., Marshall, R. N., Allen, G. P., and Killington, R. A. (1996a). The production of a truncated form of baculovirus expressed EHV-1 glycoprotein C and its role in protection of C3H (H-2Kk) mice against virus challenge. *Virus Res.* **44:**97–109.

Stokes, A., Alber, D. G., Greensill, J., Amellal, B., Carvalho, R., Taylor, L. A., Doel, T. R., Killington, R. A., Halliburton, I. W., and Meredith, D. M. (1996b). The expression of the proteins of equine herpesvirus 1 which share homology with herpes simplex virus 1 glycoproteins H and L. *Virus Res.* **40:**91–107.

Stokes, A., Cameron, R. S., Marshall, R. N., and Killington, R. A. (1997). High level expression of equine herpesvirus 1 glycoproteins D and H and their role in protection against virus challenge in the C3H (H-2Kk) murine model. *Virus Res.* **50:**159–173.

Streatfield, S. J., and Howard, J. A. (2003). Plant production systems for vaccines. *Expert Rev. Vaccines* **2:**763–775.

Swayne, D. E., Beck, J. R., Perdue, M. L., and Beard, C. W. (2001). Efficacy of vaccines in chickens against highly pathogenic Hong Kong H5N1 avian influenza. *Avian Dis.* **45:**355–365.

Takehara, K., Kikuma, R., Ishikawa, S., Kamikawa, M., Nagata, T., Yokomizo, Y., and Nakamura, M. (2002). Production and *in vivo* testing of a recombinant bovine IL-12 as an adjuvant for Salmonella typhimurium vaccination in calves. *Vet. Immunol. Immunopathol.* **86:**23–30.

Tami, C., Peralta, A., Barbieri, R., Berinstein, A., Carrillo, E., and Taboga, O. (2004). Immunological properties of FMDV-gP64 fusion proteins expressed on SF9 cell and baculovirus surfaces. *Vaccine* **23:**840–845.

Tani, H., Limn, C. K., Yap, C. C., Onishi, M., Nozaki, M., Nishimune, Y., Okahashi, N., Kitagawa, Y., Watanabe, R., Mochizuki, R., Moriishi, K., and Matsuura, Y. (2003). *In vitro* and *in vivo* gene delivery by recombinant baculoviruses. *J. Virol.* **77:** 9799–9808.

Tareilus, E., van Oers, M. M., and Vlak, J. (2003). Salt Enhancer Assay, European patent number 0229807.3.

Tate, C. G., Whiteley, E., and Betenbaugh, M. J. (1999). Molecular chaperones stimulate the functional expression of the cocaine-sensitive serotonin transporter. *J. Biol. Chem.* **274:**17551–17558.

Tegerstedt, K., Franzen, A. V., Andreasson, K., Joneberg, J., Heidari, S., Ramqvist, T., and Dalianis, T. (2005). Murine polyomavirus virus-like particles (VLPs) as vectors for gene and immune therapy and vaccines against viral infections and cancer. *Anticancer Res.* **25:**2601–2608.

Tellam, R. L., Kemp, D., Riding, G., Briscoe, S., Smith, D., Sharp, P., Irving, D., and Willadsen, P. (2002). Reduced oviposition of *Boophilus microplus* feeding on sheep vaccinated with vitellin. *Vet. Parasitol.* **103:**141–156.

Tessier, D. C., Thomas, D. Y., Khouri, H. E., Laliberte, F., and Vernet, T. (1991). Enhanced secretion from insect cells of a foreign protein fused to the honeybee melittin signal peptide. *Gene* **98:**177–183.

Theilmann, D. A., Blissard, G. W., Bonning, B., Jehl, J., O'Reilly, D. R., Rohrman, G. F., Thiem, S., and Vlak, J. M. (2005). "Family Baculoviridae" (C. M. Fauquet, M. A. Mayo, J. Maniloff, U. Desselberger, and L. A. Ball, eds.), 8th Ed. Elsevier/Academic Press, London, Virus Taxonomy, VIIIth Report of the ICTV.

Thomas, C. J., Brown, H. L., Hawes, C. R., Lee, B. Y., Min, M. K., King, L. A., and Possee, R. D. (1998). Localization of a baculovirus-induced chitinase in the insect cell endoplasmic reticulum. *J. Virol.* **72:**10207–10212.

Tobin, G. J., Li, G. H., Fong, S. E., Nagashima, K., and Gonda, M. A. (1997). Chimeric HIV-1 virus-like particles containing gp120 epitopes as a result of a ribosomal frameshift elicit Gag- and SU-specific murine cytotoxic T-lymphocyte activities. *Virology* **236:**307–315.

Tomiya, N., Narang, S., Lee, Y. C., and Betenbaugh, M. J. (2004). Comparing N-glycan processing in mammalian cell lines to native and engineered lepidopteran insect cell lines. *Glycoconj. J.* **21:**343–360.

Tordo, N., Bourhy, H., Sather, S., and Ollo, R. (1993). Structure and expression in baculovirus of the Mokola virus glycoprotein: An efficient recombinant vaccine. *Virology* **194:**59–69.

Treanor, J. J., Betts, R. F., Smith, G. E., Anderson, E. L., Hackett, C. S., Wilkinson, B. E., Belshe, R. B., and Powers, D. C. (1996). Evaluation of a recombinant hemagglutinin expressed in insect cells as an influenza vaccine in young and elderly adults. *J. Infect. Dis.* **173:**1467–1470.

Treanor, J. J., Wilkinson, B. E., Masseoud, F., Hu-Primmer, J., Battaglia, R., O'Brien, D., Wolff, M., Rabinovich, G., Blackwelder, W., and Katz, J. M. (2001). Safety and immunogenicity of a recombinant hemagglutinin vaccine for H5 influenza in humans. *Vaccine* **19:**1732–1737.

Treanor, J. J., Sciff, G. M., Couch, R. B., Cate, T. R., Brady, R. C., Hay, C. M., Wolff, M., She, D., and Cox, M. M. (2006). Dose-related safety and immunogenicity of a tivalent baculvoirus-expressed infleunzavirus hemagglutinin vaccine in elderly adults. *J. Inf. Dis.* **193:**1223–1228.

Tuboly, T., Nagy, E., and Derbyshire, J. B. (1993). Potential viral vectors for the stimulation of mucosal antibody responses against enteric viral antigens in pigs. *Res. Vet. Sci.* **54:**345–350.

Vakharia, V. N., Snyder, D. B., Lutticken, D., Mengel-Whereat, S. A., Savage, P. K., Edwards, G. H., and Goodwin, M. A. (1994). Active and passive protection against variant and classic infectious bursal disease virus strains induced by baculovirus-expressed structural proteins. *Vaccine* **12:**452–456.

Valdes, E., Vela, C., and Alvarez, J. I. (1999). Empty canine parvovirus capsids having CPV recombinant VP2 and vaccines having such capsids. United States patent number 5,882,652.

Valenzuela, P., Medina, A., Rutter, W. J., Ammerer, G., and Hall, B. D. (1982). Synthesis and assembly of hepatitis B virus surface antigen particles in yeast. *Nature* **298:**347–350.

van Aarle, P. (2003). Suitability of an E2 subunit vaccine of classical swine fever in combination with the E(rns)-marker-test for eradication through vaccination. *Dev. Biol. (Basel)* **114:**193–200.

van Dijk, A. A. (1993). Development of recombinant vaccines against bluetongue. *Biotechnol. Adv.* **11:**1–12.

van Drunen Littel-van den Hurk, S., Parker, M. D., Fitzpatrick, D. R., Zamb, T. J., van den Hurk, J. V., Campos, M., Harland, R., and Babiuk, L. A. (1991). Expression of bovine herpesvirus 1 glycoprotein gIV by recombinant baculovirus and analysis of its immunogenic properties. *J. Virol.* **65:**263–271.

van Drunen Littel-van den Hurk, S., Parker, M. D., Massie, B., van den Hurk, J. V., Harland, R., Babiuk, L. A., and Zamb, T. J. (1993). Protection of cattle from BHV-1 infection by immunization with recombinant glycoprotein gIV. *Vaccine* **11:**25–35.

Van Lier, F. L. J., van den Hombergh, J. P. T. W., de Gooijer, C. D., den Boer, M. M., Vlak, J. M., and Tramper, J. (1996). Long-term semi-continuous production of recombinant baculovirus protein in a repeated (fed-) batch two-stage reactor system. *Enzyme Microb. Technol.* **18:**460–466.

Van Oers, M. M., and Vlak, J. M. (1997). The baculovirus 10-kDa protein. *J. Invertebr. Pathol.* **70:**1–17.

van Oers, M. M., Thomas, A. A., Moormann, R. J., and Vlak, J. M. (2001). Secretory pathway limits the enhanced expression of classical swine fever virus E2 glycoprotein in insect cells. *J. Biotechnol.* **86:**31–38.

van Oirschot, J. T. (1999). Diva vaccines that reduce virus transmission. *J. Biotechnol.* **73:**195–205.

van Rijn, P. A., van Gennip, H. G., and Moormann, R. J. (1999). An experimental marker vaccine and accompanying serological diagnostic test both based on envelope glycoprotein E2 of classical swine fever virus (CSFV). *Vaccine* **17:**433–440.

van Wyke Coelingh, K. L., Murphy, B. R., Collins, P. L., Lebacq-Verheyden, A. M., and Battey, J. F. (1987). Expression of biologically active and antigenically authentic parainfluenza type 3 virus hemagglutinin-neuraminidase glycoprotein by a recombinant baculovirus. *Virology* **160:**465–472.

Vaz-Santiago, J., Lule, J., Rohrlich, P., Jacquier, C., Gibert, N., Le Roy, E., Betbeder, D., Davignon, J. L., and Davrinche, C. (2001). *Ex vivo* stimulation and expansion of both CD4(+) and CD8(+) T cells from peripheral blood mononuclear cells of human cytomegalovirus-seropositive blood donors by using a soluble recombinant chimeric protein, IE1-pp65. *J. Virol.* **75:**7840–7847.

Velzing, J., Groen, J., Drouet, M. T., van Amerongen, G., Copra, C., Osterhaus, A. D., and Deubel, V. (1999). Induction of protective immunity against Dengue virus type 2: Comparison of candidate live attenuated and recombinant vaccines. *Vaccine* **17:**1312–1320.

Venugopal, K., Jiang, W. R., and Gould, E. A. (1995). Immunity to St. Louis encephalitis virus by sequential immunization with recombinant vaccinia and baculovirus derived PrM/E proteins. *Vaccine* **13:**1000–1005.

Vlak, J. M., and Keus, R. J. (1990). Baculovirus expression vector system for production of viral vaccines. *Adv. Biotechnol. Processes* **14:**91–128.

Vlak, J. M., Schouten, A., Usmany, M., Belsham, G. J., Klinge-Roode, E. C., Maule, A. J., Van Lent, J. W., and Zuidema, D. (1990). Expression of cauliflower mosaic virus gene I using a baculovirus vector based upon the p10 gene and a novel selection method. *Virology* **179:**312–320.

Vlak, J. M., de Gooijer, C. D., Tramper, J., and Miltenburger, H. G. (eds.) (1996). Current applications of cell culture engineering. *In* "Insect Cell Cultures, Fundamental and Applied Aspects," Vol. 2. Kluwer, Dordrecht, The Netherlands.

Wang, H., Deng, F., Pijlman, G. P., Chen, X., Sun, X., Vlak, J. M., and Hu, Z. (2003). Cloning of biologically active genomes from a Helicoverpa armigera single-nucleocapsid nucleopolyhedrovirus isolate by using a bacterial artificial chromosome. *Virus Res.* **97:**57–63.

Wang, M. Y., Kuo, Y. Y., Lee, M. S., Doong, S. R., Ho, J. Y., and Lee, L. H. (2000). Self-assembly of the infectious bursal disease virus capsid protein, rVP2, expressed in insect cells and purification of immunogenic chimeric rVP2H particles by immobilized metal-ion affinity chromatography. *Biotechnol. Bioeng.* **67:**104–111.

Wathen, M. W., Kakuk, T. J., Brideau, R. J., Hausknecht, E. C., Cole, S. L., and Zaya, R. M. (1991). Vaccination of cotton rats with a chimeric FG glycoprotein of human respiratory syncytial virus induces minimal pulmonary pathology on challenge. *J. Infect. Dis.* **163:**477–482.

Watts, A. M., Shearer, M. H., Pass, H. I., Bright, R. K., and Kennedy, R. C. (1999). Comparison of simian virus 40 large T antigen recombinant protein and DNA immunization in the induction of protective immunity from experimental pulmonary metastasis. *Cancer Immunol. Immunother.* **47:**343–351.

Werle, B., Bourgeois, C., Alexandre, A., Massonneau, V., and Pothier, P. (1998). Immune response to baculovirus expressed protein fragment amino acids 190–289 of respiratory syncytial virus (RSV) fusion protein. *Vaccine* **16:**1127–1130.

Weyer, U., and Possee, R. D. (1991). A baculovirus dual expression vector derived from the *Autographa californica* nuclear polyhedrosis virus polyhedrin and p10 promoters: Co-expression of two influenza virus genes in insect cells. *J. Gen. Virol.* **72:**2967–2974.

Whalley, J. M., Love, D. N., Tewari, D., and Field, H. J. (1995). Characteristics of equine herpesvirus 1 glycoproteins expressed in insect cells. *Vet. Microbiol.* **46:**193–201.

Wickham, T. J., Davis, T., Granados, R. R., Hammer, D. A., Shuler, M. L., and Wood, H. A. (1991). Baculovirus defective interfering particles are responsible for variations in recombinant protein production as a function of multiplicity of infection. *Biotechnol. Lett.* **13:**483–488.

Willadsen, P., Bird, P., Cobon, G. S., and Hungerford, J. (1995). Commercialisation of a recombinant vaccine against Boophilus microplus. *Parasitology* **110**(Suppl.):S43–S50.

Wilson, R., Je, Y. H., Bugeon, L., Straschil, U., O'Reilly, D. R., and Olszweski, J. A. (2005). Display of foreign proteins using recombinant baculovirus occlusion bodies: A novel vaccination tool. Abstract 174, 38th Annual Meeting of the Society of Invertebrate Virology, Anchorage, Alaska, 2005.

Xuan, X., Nakamura, T., Ihara, T., Sato, I., Tuchiya, K., Nosetto, E., Ishihama, A., and Ueda, S. (1995). Characterization of pseudorabies virus glycoprotein gII expressed by recombinant baculovirus. *Virus Res.* **36:**151–161.

Xuan, X., Maeda, K., Mikami, T., and Otsuka, H. (1996). Characterization of canine herpesvirus glycoprotein C expressed by a recombinant baculovirus in insect cells. *Virus Res.* **46:**57–64.

Yang, D. K., Kweon, C. H., Kim, B. H., Lim, S. I., Kwon, J. H., Kim, S. H., Song, J. Y., and Han, H. R. (2005). Immunogenicity of baculovirus expressed recombinant proteins of Japanese encephalitis virus in mice. *J. Vet. Sci.* **6:**125–133.

Yao, Q., Kuhlmann, F. M., Eller, R., Compans, R. W., and Chen, C. (2000). Production and characterization of simian–human immunodeficiency virus-like particles. *AIDS Res. Hum. Retroviruses* **16:**227–236.

Yao, Q., Vuong, V., Li, M., and Compans, R. W. (2002). Intranasal immunization with SIV virus-like particles (VLPs) elicits systemic and mucosal immunity. *Vaccine* **20:**2537–2545.

Yoshida, S., Kondoh, D., Arai, E., Matsuoka, H., Seki, C., Tanaka, T., Okada, M., and Ishii, A. (2003). Baculovirus virions displaying *Plasmodium berghei* circumsporozoite protein protect mice against malaria sporozoite infection. *Virology* **316:**161–170.

Young, L. S., Searle, P. F., Onion, D., and Mautner, V. (2006). Viral gene therapy strategies: From basic science to clinical applications. *J. Pathol.* **208:**299–318.

Zhang, L., Wu, G., Tate, C. G., Lookene, A., and Olivecrona, G. (2003). Calreticulin promotes folding/dimerization of human lipoprotein lipase expressed in insect cells (Sf21). *J. Biol. Chem.* **278:**29344–29351.

Zhang, Y. M., Hayes, E. P., McCarty, T. C., Dubois, D. R., Summers, P. L., Eckels, K. H., Chanock, R. M., and Lai, C. J. (1988). Immunization of mice with dengue structural proteins and nonstructural protein NS1 expressed by baculovirus recombinant induces resistance to dengue virus encephalitis. *J. Virol.* **62:**3027–3031.

Zheng, R. (1999). Technology evaluation: HIVAC-1e. *Curr. Opin. Mol. Ther.* **1:**121–125.

Zintl, A., Mulcahy, G., Skerrett, H. E., Taylor, S. M., and Gray, J. S. (2003). *Babesia divergens*, a bovine blood parasite of veterinary and zoonotic importance. *Clin. Microbiol. Rev.* **16:**622–636.

BACULOVIRUSES AND MAMMALIAN CELL-BASED ASSAYS FOR DRUG SCREENING

J. Patrick Condreay,* Robert S. Ames,† Namir J. Hassan,‡ Thomas A. Kost,* Raymond V. Merrihew,§ Danuta E. Mossakowska,‡ David J. Pountney,‡ and Michael A. Romanos¶

*Department of Gene Expression and Protein Biochemistry
GlaxoSmithKline Discovery Research, Research Triangle Park, North Carolina 27709
†Department of Gene Expression and Protein Biochemistry
GlaxoSmithKline Discovery Research, King of Prussia, Pennsylvania 19406
‡Department of Gene Expression and Protein Biochemistry
GlaxoSmithKline Discovery Research, Harlow, Essex CM19 5AW, United Kingdom
§Department of Screening and Compound Profiling
GlaxoSmithKline Discovery Research, Research Triangle Park, North Carolina 27709
¶Department of Gene Expression and Protein Biochemistry
GlaxoSmithKline Discovery Research, Stevenage, Herts SG1 2NY, United Kingdom

I. Introduction to BacMam

Approximately 10 years ago, a new use for recombinant baculoviruses was revealed with the appearance of two publications demonstrating the delivery of recombinant gene expression cassettes containing reporter genes under control of mammalian cell-active promoters to mammalian cells primarily of liver origin. Subsequent work has shown

0065-3527/06 $35.00
DOI: 10.1016/S0065-3527(06)68007-X

that a number of cell types are susceptible to transduction. This system, which we refer to as BacMam, has found utility in a number of laboratories. In this chapter we will focus on the application of BacMam for configuring cell-based assays for automated screening of chemical libraries.

A. History

Early investigations of mammalian cells exposed to baculoviruses concluded that viral DNA could be detected in the cells, however, no evidence of viral replication or gene expression was found (Tjia *et al.*, 1983; Volkman and Goldsmith, 1983). Addition of a reporter gene under transcriptional control of a mammalian cell-active promoter (RSV LTR) to the viral DNA led to an initial demonstration of low levels of gene expression in a mouse cell line (Carbonell *et al.*, 1985). However, it was later reported that the low levels of reporter protein activity seen in mouse L929 and human A549 cells were primarily due to enzyme activity that was carried in the virion particles, and not to *de novo* synthesis in the mammalian cells (Carbonell and Miller, 1987).

In the mid-1990s, two laboratories independently demonstrated the ability of baculovirus vectors containing expression cassettes controlled by either the *Rous sarcoma virus* (RSV) promoter or the immediate early promoter of cytomegalovirus (CMV) to mediate gene delivery and expression in mammalian cells (Boyce and Bucher, 1996; Hofmann *et al.*, 1995). These initial reports showed efficient gene expression in cells derived from liver tissue, including primary cells, however, very little expression was noted in other cell types. Subsequently, other laboratories demonstrated utility of this system for gene expression in a large array of nonhepatic cell lines (Condreay *et al.*, 1999; Shoji *et al.*, 1997). In general, it appears that many mammalian cell types are able to take up baculovirus and express genes under the control of mammalian cell-active promoters. One broad category of cells that are not efficiently transduced is those of hematopoietic lineages (Condreay *et al.*, 1999). Recent reviews contain summaries of the current knowledge of cell types that are susceptible to BacMam transduction (Kost and Condreay, 2002; Kost *et al.*, 2005).

For most applications, the only modification that is made to the virus is the addition of a mammalian gene cassette. No changes are necessary to the outside of the virion. The mechanisms behind viral entry and uncoating are still largely uncharacterized. The viral membrane glycoprotein, gp64, appears to be necessary for the virus to be taken up by cells (Hofmann *et al.*, 1998; Tani *et al.*, 2001). There does not appear to be a consensus as to the nature of the cell surface molecule that serves as a receptor for the virus (Duisit *et al.*, 1999; Hofmann *et al.*, 1995; Tani *et al.*, 2001) and, in fact, this receptor may differ between cell types. Once the virus enters a cell, the viral nucleocapsid can be detected in the nucleus of cells (van Loo *et al.*, 2001). There is evidence that cell types that are not susceptible to BacMam transduction are deficient in delivering the viral nucleocapsid to the nucleus (Barsoum *et al.*, 1997; Kukkonen *et al.*, 2003).

B. Features and Advantages

Development of cell-based assays for compound screening, especially in automated, high-throughput environments, is generally thought to require the development of a cell line that stably expresses a molecular target and yields a reliable readout from the function of that target. While stable cell lines may possess some advantages in terms of predictability of expression and ease of continuous culture of the line, they also have certain drawbacks. Certain gene products can be deleterious to cells when expressed constitutively, making it difficult to generate a suitable cell line. Stable cell lines also take a significant amount of time to generate and clone, and if the target contains several subunits this difficulty is multiplied. Finally, once a stable cell line is developed one no longer has flexibility in the cellular background, and one must cryostore and culture multiple cell lines.

The use of traditional transient methods, such as liposome transfection to express target proteins, is generally not considered suitable for high-throughput screening applications. The methods can be costly, have toxic effects on cells, and may require wash steps that are difficult to fit into automated protocols. We have found the BacMam system to present a number of features and advantages (Table I) that overcome the limitations of stable cell lines and make transient gene delivery for high-throughput cell-based assays a viable and desirable alternative.

Baculoviruses have been in use by many laboratories for overexpression of recombinant proteins for more than 20 years. Several systems have been generated to construct these recombinant viruses (reviewed in Condreay and Kost, 2003; Kost *et al.*, 2005), all of which are available commercially from different suppliers of biological reagents. The various systems make differing claims as to ease, speed, and cost to produce a virus, but suffice to say that recombinant baculoviruses can easily and rapidly be made with a minimum of training and using standard laboratory equipment. The accessibility of the system to virtually any laboratory involved in biomedical research is attested

TABLE I

FEATURES AND ADVANTAGES OF THE BACMAM SYSTEM FOR MAMMALIAN CELL-BASED ASSAY DEVELOPMENT

Ease of use of recombinant baculoviruses
Viruses are generated easily and rapidly
Virus production is scalable
Long-term virus stability
Transduction by simple liquid addition
Lack of overt toxicity
Consistency within and between plates
Versatility of the BacMam System
Transient expression reduces toxic effects of overexpression
Ability to deliver multiple viruses (genes)
Ability to modulate gene expression
Adaptable to a number of assay formats
Large number of susceptible cell lines

to by the thousands of references returned by a PubMed search for "recombinant baculovirus." Additionally, the baculovirus system is readily scalable (Mannix and Jarman, 2000; Meghrous *et al.*, 2005) and viruses produced in the presence of serum are stable for long periods of time provided they are protected from prolonged exposure to light (Jarvis and Garcia, 1994).

Transduction of mammalian cells by BacMam viruses is accomplished by a simple liquid addition step. Thus, the system is easily accommodated into automated processes for assay plate preparation carried out by liquid-handling robots. Methods for preparing transduced cells appear to be quite flexible. Virus can be added to attached cells followed by harvest of the cell inoculum for assay plates (Pfohl *et al.*, 2002), mixed with harvested cells, and immediately plated into assay plates (Katso *et al.*, 2005), or added to cells grown in suspension (Ramos *et al.*, 2002) with excellent results in all cases. Functional expression of gene products can be detected within 4 h of virus addition (Pfohl *et al.*, 2002) thus minimizing problems from expression of proteins whose effects are deleterious to cells (Clare, 2006). Treatment of mammalian cells with the virus does not result in any overt toxic effects to cell cultures (Clay *et al.*, 2003; compare Fig. 3C and E in Clay *et al.*, 2003). We have transduced cells with extraordinarily high amounts of virus (>1000 plaque-forming units (pfu) per cell) without observing any deleterious effects on the cultures (unpublished data).

There are a number of features of the BacMam system that facilitate optimization of assays (Fig. 1). These features will be outlined here and specific illustrations will be discussed in the target class sections. Many cell-based assays rely on expression of more than one component in the cell; a target can be composed of two or more subunits, or a reporter construct can be required for a readout of target protein activity. Multiple expression cassettes can be delivered with BacMam by transduction with multiple viruses (Ames *et al.*, 2004a; Clare, 2006; Jenkinson *et al.*, 2003; Katso *et al.*, 2005; Pfohl *et al.*, 2002). Alternatively, because of the large

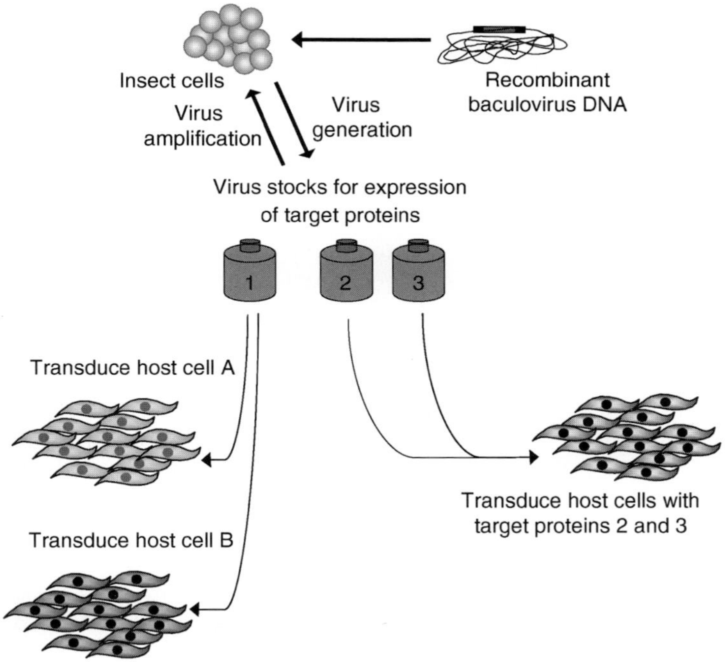

FIG 1. Versatility of the BacMam system in building cell-based assays. Recombinant viral DNA is generated by one of several commercially available systems and transfected into insect cells to generate virus stocks for different target proteins. Stocks are easily reamplified by infection of insect cells. Stocks are stored at 4 °C in the dark and are ready for use as needed. For target 1, the ability to transduce a variety of cell lines allows an investigator to easily optimize assay formats in the cell line of choice. For a multisubunit target (2/3) two (or more) viruses, each expressing one of the subunits, can simultaneously be transduced into host cells for assay of the target. The system also allows for testing of different subunits as necessary. (See Color Insert.)

capacity of baculoviruses for recombinant inserts (Cheshenko *et al.*, 2001), it is also possible to place two or more expression cassettes into a single virus (Condreay *et al.*, 1999). The amount of gene expression achieved in transduced cells can be modulated in two ways. Increases in viral multiplicity (pfu per cell) will result in increased gene expression (Ames *et al.*, 2004b; Clare, 2006; Kost and Condreay, 2002; Pfohl *et al.*, 2002), presumably by a gene dosage effect. Inhibitors of histone deacetylases, such as sodium butyrate and trichostatin A, have been shown to stimulate expression in BacMam-transduced cells (Condreay *et al.*, 1999; Spenger *et al.*, 2004). Although it is not necessarily desirable to add another pharmacological agent to a screening assay, the addition of butyrate has been enabling for certain BacMam-mediated applications (Jenkinson *et al.*, 2003; Ramos *et al.*, 2002). Delivery of genes by BacMam is remarkably consistent not only from well to well within a plate but also between plates (Jenkinson *et al.*, 2003; Katso *et al.*, 2005).

In addition to their use for automated, high-throughput screens, BacMam viruses are generally useful research tools that easily allow the investigation and optimization of alternative assays. Libraries of BacMam viruses can be generated that express different forms of protein subunits, such as altered G-proteins or isotypes of subunits (Ames *et al.*, 2004a; Pfohl *et al.*, 2002); various cofactors that modulate the activity of a protein, such as corepressors or coactivators of nuclear receptors (Boudjelal *et al.*, 2005); or proteins that will modify the cell to alter the function of a protein, such as an ion channel that will alter the membrane potential of a cell (Clare, 2006). Some of these examples will be discussed in more detail later, but having these libraries allows investigators to mix them in different combinations to better mimic responses observed *in vivo*. The ability to transduce a wide variety of cell lines and primary cell cultures (Kost and Condreay, 2002; Kost *et al.*, 2005) also facilitates configuration of assays with different characteristics. Certain osteosarcoma cell lines have been found to transduce efficiently and yield high levels of gene expression (Song and Boyce, 2001; Song *et al.*, 2003) and they have proven to be very useful for configuring screening assays (Ames *et al.*, 2004a; Clay *et al.*, 2003). Most cell lines appear to transduce well, however, optimization of parameters, such as incubation temperature, has been demonstrated to improve transduction of certain cell lines that are transduced less efficiently (Hsu *et al.*, 2004; Hu *et al.*, 2003). Fortunately a number of cell lines (Saos-2, U-2 OS, HEK293, HepG2, BHK, CV-1) transduce with only a minimum of optimization, usually only requiring a titration of virus multiplicity to obtain the desired cellular response to

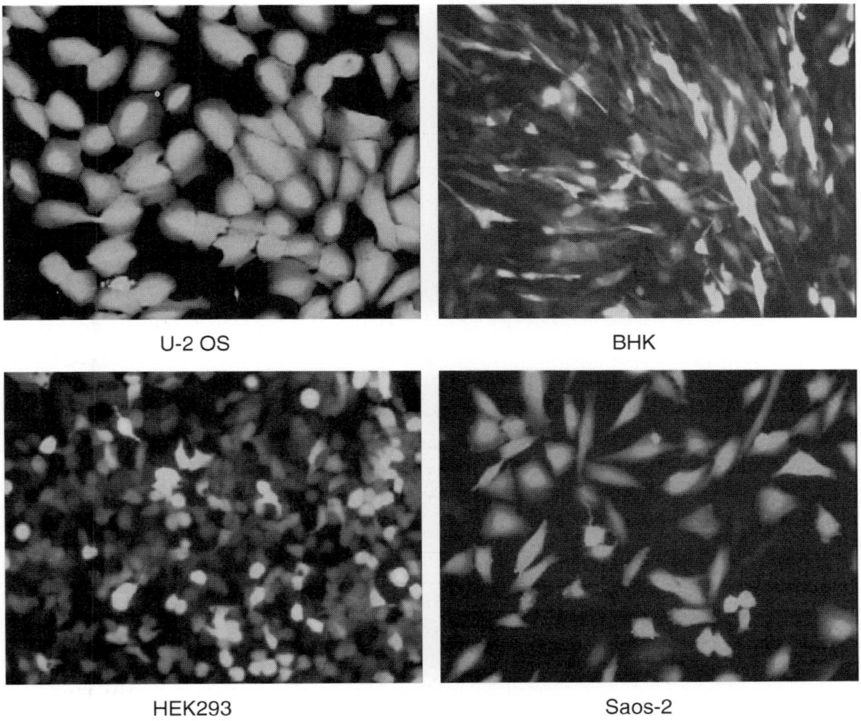

Fig 2. Mammalian cells transduced with a BacMam virus expressing green fluorescent protein (GFP). Virus as described in Condreay *et al.* (1999). Cells were transduced with 100 pfu of virus per cell and photographed 24 h after virus addition. Transduction frequency is routinely greater than 90% as measured by the number of green fluorescent cells. Reprinted with permission from Kost *et al.* (2005). (See Color Insert.)

transgene expression, thus providing screening groups with a core of cell lines to choose from for assay configuration (Fig. 2).

C. Biosafety Considerations

In general, viral systems are good choices for transient gene delivery for many of the reasons outlined earlier as advantages of the BacMam system. Baculoviruses have an excellent biosafety profile (Risk Group 1 agents), especially when compared to gene delivery vectors based on mammalian viruses (Kost *et al.*, 2006).

Baculoviruses are pathogens of arthropod species and in general have a narrow host range. They replicate in and kill their invertebrate hosts, and because of this pathogenicity their occluded form has attracted attention for use as biopesticides (Szewczyk *et al.*, 2006). This in turn raised interest in interactions between baculoviruses and nontarget (vertebrate) organisms. There is no evidence that administration of occluded baculoviruses causes disease in vertebrate organisms (Black *et al.*, 1997; Doller *et al.*, 1983). Furthermore, numerous studies with the budded form of the virus have produced no evidence of productive viral infection in vertebrate cells (Carbonell and Miller, 1987; Carbonell *et al.*, 1985; Hartig *et al.*, 1992; Tjia *et al.*, 1983; Volkman and Goldsmith, 1983) and viral promoters have little or no function in mammalian cells (Fujita *et al.*, 2006; Tjia *et al.*, 1983). Therefore, baculoviruses do not present the risk of reconstitution of replication competent viruses that are a concern with other viral vectors. Additionally, the budded form is noninfectious for the natural insect host (Jarvis, 1997) and thus represents a reduced environmental risk from an accidental release.

The ability of baculoviruses to deliver genes to mammalian cells while being inherently nonreplicative in these cells has also raised interest in using the virus for *in vivo* applications such as gene therapy. However, baculoviruses are rapidly inactivated by serum complement and thus are not suitable for systemic delivery (Hofmann *et al.*, 1998). Methods have been devised to overcome this characteristic (reviewed in Kost *et al.*, 2005), but these methods rely on some sort of modification to the envelope glycoprotein of the virus (see also chapter by Hu, this volume, pp. 287–320). The application of baculoviruses for configuring cell-based assays for high-throughput screening does not require modification of the virus in any way except for inclusion of the mammalian cell-active expression cassette and thus presents a safe viral-based gene transfer tool.

II. Nuclear Receptors

Nuclear receptors are ligand-modulated transcription factors and represent one of the major target classes for drug screening. As a family, nuclear receptors are attractive to the pharmaceutical industry due to their broad disease associations and their ability to interact with small lipophilic molecules capable of modulating transcriptional activity (McDonnell *et al.*, 1993). Examples of marketed drugs that target nuclear receptors include the peroxisome proliferator-activated

It also provides the capability of titrating protein expression, which is important in the design of assays that will discriminate between agonists, inverse agonists, and modulators.

III. Transporters

Transporters are a diverse group of large membrane proteins found in all living cells that mediate a large number of functions. Examples include allowing entry of essential nutrients into the cell, regulation of metabolite concentrations, removal of toxins and drugs from within the cell, and maintenance of critical ion gradients. Transport is accomplished by a variety of ways including facilitated diffusion, primary active transport, and secondary active transport (Griffith and Sansom, 1998).

Many transporters are the targets of therapeutic intervention. In addition, transporters are involved in absorption, distribution, metabolism, and elimination of drugs (ADME), and thus kinetic and pharmacological investigation is essential to the understanding of drug interactions. Unfortunately, these proteins tend to be expressed at low levels in native cells such that characterisation is dependent on heterologous expression of the transporter. This in turn can often be troublesome with low expression, expression of inactive protein, slow cell growth, and cell toxicity being common problems. Examples in which such problems have been encountered and overcome include the serotonin transporter (SERT) (Tate *et al.,* 2003), and the glucose transporters GLUT1 and GLUT4 (Wieczorke *et al.*, 2003).

The complexities of good functional expression of transporters and the variety of expression systems used have been summarized elsewhere (Pritchard and Miller, 2005). Briefly, the ideal heterologous expression system needs to be generated rapidly, be readily characterized in a variety of cellular backgrounds, give reproducible expression, be able to regulate the level of expression, be scalable to support high-throughput screening, and to enable multiple transporters to be analyzed within one cell. Many of these requirements can be achieved using BacMam transduction of mammalian cells.

A. *Neurotransmitter Transporters*

Aminobutyric acid (GABA) is an important inhibitory neurotransmitter and its metabolism is implicated in several neurological and psychiatric disorders. There are four principal GABA transporters

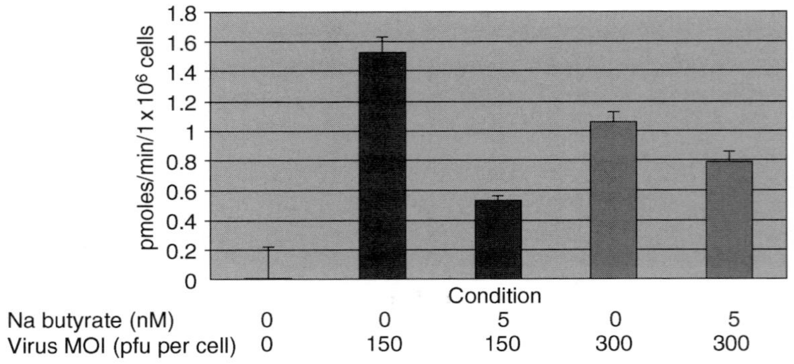

FIG 5. BacMam-mediated expression of GAT-1 in HEK293 cells. Cells were incubated with GAT-1 BacMam at the indicated MOI with and without 5 mM Na butyrate. Specific uptake of [H^3]-GABA was calculated 48 h following transduction and is expressed as pmoles/min/1 $\times$ 10^6 cells.

termed GAT-1, GAT-2, GAT-3, and BGT-1, which all belong to the SLC6 family of Na^+/Cl^--dependent neurotransmitter transporters (Gonzalez and Robinson, 2004). In our experience, the generation of GAT-1 stable cell lines in HEK293 cells was problematic; cells were extremely slow growing and exhibited very low specific uptake rates of the ligand [H^3]-GABA. However, BacMam technology enabled functional expression of GAT-1 in HEK293 cells (Fig. 5) as measured in a whole cell radioactive uptake assay. Furthermore, the level of expression could be modulated by the multiplicity of infection (MOI). High MOIs or addition of sodium butyrate (the histone deacetylase inhibitor) lead to reduced [H^3]-GABA uptake as a result of cell death possibly due to toxicity as expression levels may have exceeded the maximum levels tolerated by the cell. The ability to accurately titrate and control the level of expression enabled selection of optimal conditions to suit the assay platform.

The flexibility of BacMam enables the use of one reagent to transduce a number of different cell backgrounds including HEK293, COS-7, SH-SY5Y, U-2 OS, and LLC-PK1 cells. In parallel to investigation of the HEK293 cell line, LLC-PK1 cells were transduced with BacMam and functional GAT-1 expression evaluated at 24 and 48 h (Fig. 6). At the optimal expression conditions, the calculated pIC50 value of 5.3 μM in LLC-PK1 cells for nipecotic acid, an inhibitor of GAT-1, was within the expected range (Soudijn and van Wijngaarden, 2000).

The rapid generation of BacMams (2–3 weeks per virus in a volume sufficient to do functional characterization) and the reproducibility of

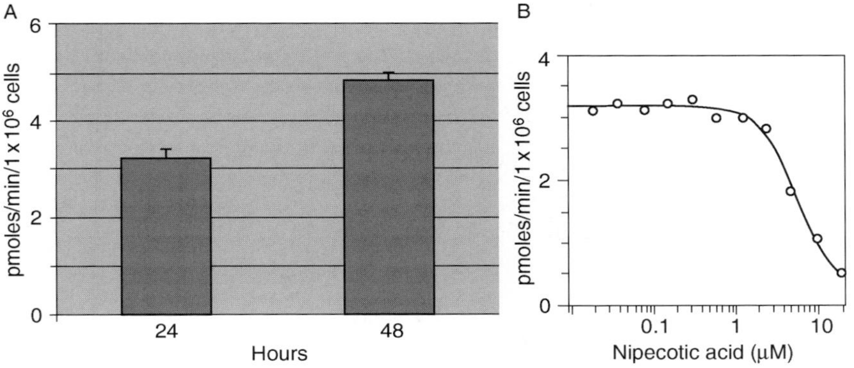

FIG 6. BacMam-mediated expression of GAT-1 in LLC-PK1 cells. (A) LLC-PK1 cells were incubated with GAT-1 BacMam at an MOI of 150. Uptake of [H^3]-GABA was measured at 24 and 48 h following transduction. (B) IC50 inhibition curves were calculated using the specific inhibitor nipecotic acid.

expression have enabled us to direct functional expression of related members of the SLC6 family, including GAT-2, GAT-3, BGT-1, SERT, DAT, and NET. We are currently investigating the possibility of using BacMam to express several members of the SLC6 family in the same cell line to enable simultaneous neurotransmitter reuptake measurements.

B. *Transporters Involved in ADME*

One of the major challenges of the drug discovery process is the advancement of safer and more efficacious drugs from the discovery phase through to clinical development. Transporters play a crucial role in this and have been shown to be involved in ADME, pharmacokinetics, and drug–drug interactions (Ayrton and Morgan, 2001; Chan *et al.,* 2004; Mizuno *et al.,* 2003). Family members of both the ATP-binding cassette (ABC) and solute carrier (SLC) are involved in ADME and include P-gp, organic anion transporter (OAT), and organic anion-transporting polypeptide (OATP) subfamilies. Unfortunately, as with other transporters, heterologous expression remains challenging in many instances and the limitations of the various expression systems have been documented elsewhere (Pritchard and Miller, 2005). To complicate matters further, some of these transporters are localized to a certain part of the plasma membrane in polarized cells and to fully characterize their function, heterologous expression will need to be developed in relevant polarized cell lines. We have shown that BacMam can provide an alternative system for expression of ABC and SLC transporters with attractive characteristics (Hassan *et al.*, 2006).

The ABC transporters are classified as primary active transporters and drive the movement of substrates by ATP hydrolysis. They play an essential role in the efflux of clinically important drugs including anticancer, antiarrythmic, and antibiotics (Chan *et al.,* 2004). The use of BacMam recombinant baculovirus for expression of ABC transporters was illustrated using breast cancer resistance protein (BCRP) as a model example (Hassan *et al.*, 2006). We demonstrate that with the BacMam system we could exploit titratable expression, transduction of a range of mammalian cell lines, expression of multiple transporters in a single cell line, and configuration of functional expression to a high-throughput platform. Some of these characteristics are illustrated in Fig. 7. Furthermore, we have observed expression of BCRP in the polarized LLC-PK1 (Hassan *et al.*, 2006) cell line, which demonstrates that BacMam can be used to transduce polarized cells. BCRP itself is expressed on the apical membrane of various epithelial cells (Xia *et al.,* 2005). In a separate set of experiments, BacMam-mediated gene delivery also proved successful for an ABC family member, BSEP (unpublished data).

The SLC superfamily of membrane proteins encompasses the secondary active transporters important in drug pharmacokinetics. These include the OAT and OATP subfamilies. The application of BacMam-mediated gene delivery to the expression of OATP and OAT family

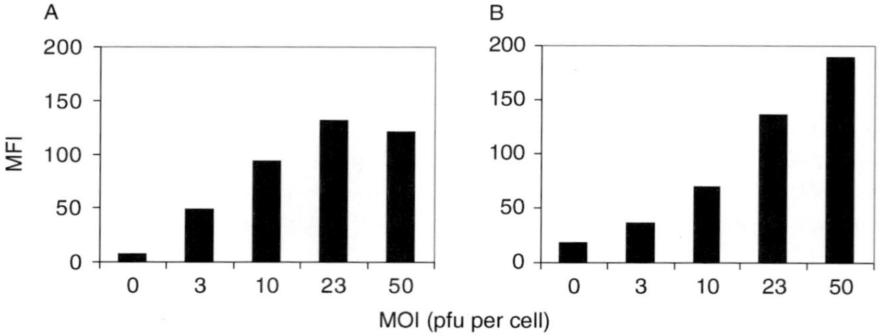

FIG 7. Modulation of BCRP cell-surface expression in BacMam-BCRP transduced mammalian cells. HEK293 (A) and U-2 OS (B) cells were transduced with BacMam-BCRP at the indicated MOI. BCRP expression was detected by flow cytometry and the mean fluorescence intensity (MFI) values of the histograms were plotted.

VI. Viral Targets

A. Hepatitis B virus *Assays*

Baculovirus-mediated gene delivery has proven to be a valuable tool for viral assay development based on the ability of recombinant baculoviruses containing mammalian viral genomes to effectively launch viral infections in transduced mammalian cells. This approach is exemplified by the efficient delivery of the *Hepatitis B virus* (HBV) genome into hepatocytes. The study of HBV replication and the effects of antiviral agents have been hampered by the lack of efficient cell culture systems. HBV cannot be propagated by infection of permanent cell lines and infection of primary hepatocyte cultures has proven difficult. Thus, most *in vitro* studies have been conducted using either transient transfection of HBV DNA sequences or stable cell lines containing integrated copies of the HBV genome such as the HepG2.2.15 line (Sells *et al.*, 1987).

The application of baculovirus-mediated gene delivery as an alternative approach to studying HBV replication was first reported by Delaney and Isom (1998). A baculovirus was generated that contained a 1.3 genome length HBV construct. In this baculovirus, HBV gene expression was driven exclusively from endogenous HBV promoters. The virus was used successfully to transduce the HepG2 hepatocyte cell line. HBV transcripts, intracellular and secreted HBV antigens, and the presence of high levels of intracellular replicative intermediates were produced, including covalently closed circular (CCC) DNA. In addition, density gradient analysis of extracellular HBV DNA indicated that the HBV DNA is primarily contained in enveloped virions. This system for HBV delivery provides a number of significant advantages as compared to transient transfection approaches and stable cell line development (Isom *et al.*, 2004). It has been used to study the effect of antiviral drugs on wild-type HBV strains (Abdelhamed *et al.*, 2002, 2003; Delaney *et al.*, 1999) and drug resistant mutant viruses

using the indicator dye $DiBAC_4(3)$ showing membrane hyperpolarization when the channel is opened with diazoxide (DZ) and subsequent repolarization when inhibited with glyburide (GLY). (B) Saturable binding of the inhibitor glyburide to membranes prepared from transduced cells. (C) Whole-cell patch clamp measurements of membrane currents show potassium selective, inward-rectifying currents in response to increasing doses of DZ. Data reproduced from Pfohl *et al.* (2002) by permission of Taylor & Francis Group, LLC., http://www.taylorandfrancis.com.

(Angus *et al.*, 2003; Chen *et al.*, 2003; Delaney *et al.*, 2001; Gaillard *et al.*, 2002). The baculovirus HBV delivery approach has also been used to study the role of the HBV X protein in human hepatocytes (Zhang *et al.*, 2004). By virtue of its ease of use, ability to rapidly generate mutant HBVs, and efficient transduction of hepatocytes, the baculovirus-mediated HBV delivery system has been shown to provide significant advantages over existing HBV cell-based assays.

B. Hepatitis C virus *Assays*

The expression of the *Hepatitis C virus* (HCV) genome in mammalian cells using recombinant baculovirus transduction was first reported by Fipaldini *et al.* (1999). In this system, the production of properly processed HCV proteins was observed, however, HCV replication was not detectable. The authors suggested that the HCV cDNA used to construct the baculovirus was probably not infectious. HCV cDNA constructs have also been delivered to human hepatocytes by baculoviruses employing a tetracycline-regulatable system (McCormick *et al.*, 2002, 2004). Although the production of viral proteins and replication competent HCV transcripts could be detected in transduced cells, no evidence was presented regarding the production of infectious HCV. A study described the production of HCV-like particles by hepatocytes transduced with a recombinant baculovirus carrying the HCV core to NS2 region regulated by a heterologous promoter (Matsuo *et al.*, 2006). Although none of these studies have demonstrated the production of infectious HCV following baculovirus transduction, the approach is in its infancy and may provide a useful complement to recently described cell culture systems employing the unique HCV genotype 2a replicon (JFH1) (Zeisel and Baumert, 2006).

C. Other Viral Applications

In addition to the expanding role of baculovirus gene delivery for the development of HBV and HCV cell-based models, other useful baculovirus applications have been demonstrated. Baculoviruses expressing herpes simplex virus 1 virus proteins (Boutell *et al.*, 2005; Poon and Roizman, 2005; Zhou and Roizman, 2002) and cytomegalovirus proteins (Dwarakanath *et al.*, 2001; Kronschnabi and Stamminger, 2003) have been used in *trans*-complementation studies to investigate the function of viral proteins. The low level of cytotoxicity, ease of use, and high efficiency of baculovirus transduction in many cell types makes the virus a very useful vector for such studies.

A novel development has been the application of baculovirus-infected insect cells for the production of recombinant adeno-associated virus (AAV) vectors. This approach, first described by Urabe *et al.* (2002), provides an attractive alternative to widely used transfection techniques for the efficient production of large quantities of recombinant AAV (Kohlbrenner *et al.*, 2005; Meghrous *et al.*, 2005; Urabe *et al.*, 2006).

Baculoviruses have also been engineered for the efficient delivery of short hairpin RNA (shRNA), adding to the available vector arsenal for RNA interference studies (Nicholson *et al.*, 2005; Ong *et al.*, 2005). Lu *et al.* (2006) have used this approach to effectively inhibit the replication of arterivirus porcine reproductive and respiratory syndrome virus in cultured cells.

VII. Conclusions

Examples have been provided from a wide range of target molecules demonstrating that the ease and versatility of BacMam-mediated gene delivery make it an excellent alternative to stable cell lines for the development of cell-based assays for drug screening. Simple liquid addition of the virus is amenable to the automated liquid-handling platforms used in these screens. The ability to "dial in" the level of gene expression and/or deliver multiple subunits of a target protein are hallmarks of the system.

The wide range of cells that serve as efficient hosts for BacMam transduction is an additional strength of the system. However, for ion channel applications, one of the preferred host cell lines is the CHO cell line, which yields variable transduction efficiencies and levels of gene expression. This variability is not such a difficulty for assay readouts that rely on responses from a population of cells, but for single-cell assays (e.g., electrophysiological analysis) having a significant number of cells that yield poor or no responses can be problematic. Inhibitors of histone deacetylase have been shown to enhance levels of gene expression in BacMam-transduced CHO cells (Condreay *et al.*, 1999), as have transcriptional transactivators that act on the CMV promoter (Ramos *et al.*, 2002). The report of an avian adenovirus gene product that acts as a histone deacetylase inhibitor and enhances gene expression in CHO cells (Hacker *et al.*, 2005) suggests that modifications may be made to CHO cells to obtain new derivatives that will act as efficient hosts for BacMam transduction.

BacMam affords the ability to tailor an assay to better mimic a response seen in native tissues. Examples include the work described earlier of Behm *et al.* (2006) with the urotensin II receptor and the use of the TREK ion channel to alter the resting potential of the cell (Clare, 2006). An improvement on this approach of using established cell lines would be to use primary cells or to establish lines that retain particular desirable characteristics of a specific tissue (Horrocks *et al.*, 2003). BacMam has been shown to transduce primary cells quite efficiently (Kost and Condreay, 2002; Kost *et al.*, 2005), yet these cells are not always readily or abundantly available. By inclusion of an expression cassette containing a dominant selectable marker in a BacMam virus and subsequent transduction of cells, it has been shown that one can select for cells that retain expression of a linked expression cassette (Condreay *et al.*, 1999). These new cell lines have integrated a portion of the input viral DNA into their genomes and are stable for expression for multiple generations (Merrihew *et al.*, 2001). Thus, transduction of primary cells with BacMam viruses that direct the expression of a selectable marker and immortalization genes, such as large T antigen (Horrocks *et al.*, 2003), may provide an efficient and gentle method to establish cell lines from primary tissues.

Implementing BacMam-based assays on a large scale requires attention to the logistical considerations associated with handling of large numbers of biological agents. These considerations include: optimized virus production methods, reproducible virus titration, quality control assays, long-term storage/inventory systems, and distribution. Virus stocks prepared and stored in the presence of fetal bovine serum at 4 °C in the dark are quite stable (Jarvis and Garcia, 1994). Little or no reduction in functional activity has been observed in virus stocks stored for periods of a year or longer (unpublished results). Virus stocks produced in serum-free medium are less stable and require the addition of stabilizing agents to maintain titer (Jorio *et al.*, 2006). It is also important to consider the potential application of automated liquid-handling instruments early in the assay development process to facilitate the transition of laboratory-scale assays into the high-throughput screening environment.

Acknowledgments

The authors wish to thank all of their colleagues at GlaxoSmithKline who have contributed to the development and implementation of the BacMam system. This work was made possible by the continuing support of John Reardon.

References

Abdelhamed, A. M., Kelley, C. M., Miller, T. G., Furman, P. A., and Isom, H. C. (2002). Rebound of hepatitis B virus replication in HepG2 cells after cessation of antiviral treatment. *J. Virol.* **76:**8148–8160.

Abdelhamed, A. M., Kelley, C. M., Miller, T. G., Furman, P. A., Cable, E. E., and Isom, H. C. (2003). Comparison of anti-hepatitis B virus activities of lamivudine and clevudine by a quantitative assay. *Antimicrob. Agents Chemother.* **47:**324–336.

Ames, R. A., Nuthulaganthi, P., Fornwald, J. A., Shabon, U., van-der-Keyl, H. K., and Elshourbagy, N. A. (2004a). Heterologous expression of G protein-coupled receptors in U-2 OS osteosarcoma cells. *Receptors Channels* **10:**117–124.

Ames, R. S., Fornwald, J. A., Nuthulaganthi, P., Trill, J. J., Foley, J. J., Buckley, P. T., Kost, T. A., Wu, Z., and Romanos, M. A. (2004b). BacMam recombinant baculoviruses in G protein-coupled receptor drug discovery. *Receptors Channels* **10:**99–107.

Angus, P., Vaughan, R., Xiong, S., Delaney, W., Gibbs, C., Brosgart, C., Colledge, D., Edwards, R., Ayres, A., Barthlomeusz, A., and Locarnini, S. (2003). Resistance to adefovir dipivoxil therapy associated with the selection of a novel mutation in the HBV polymerase. *Gastroenterology* **125:**292–297.

Ashcroft, F. M. (2000). "Ion Channels and Disease: Channelopathies." Academic Press, San Diego.

Ayrton, A., and Morgan, P. (2001). Role of transport proteins in drug absorption distribution and excretion. *Xenobiotica* **31:**469–497.

Barsoum, J., Brown, R., McKee, M., and Boyce, F. M. (1997). Efficient transduction of mammalian cells by a recombinant baculovirus having the vesicular stomatitis virus G glycoprotein. *Hum. Gene Ther.* **8:**2011–2018.

Bass, A. S., Tomaselli, G., Bullingham, R., III, and Kinter, L. B. (2005). Drug effects on ventricular repolarization: A critical evaluation of the strengths and weaknesses of current methodologies and regulatory practices. *J. Pharmacol. Toxicol. Methods* **52:**12–21.

Beato, M., Candau, R., Chavez, S., Mows, C., and Truss, M. (1996). Interaction of steroid hormone receptors with transcription factors involves chromatin remodelling. *J. Steroid Biochem.* **56:**47–59.

Behm, D. J., Stankus, G., Doe, C. P. A., Willette, R. N., Sarau, H. M., Foley, J. J., Schmidt, D. B., Nuthulaganti, P., Fornwald, J. A., Ames, R. S., Lambert, D. G., Calo, G. *et al.* (2006). The peptidic urotensin-II (UT) ligand GSK248451 possesses less intrinsic activity than the low efficacy partial agonists SB-710411 and urantide in native mammalian tissues and recombinant cell systems. *Br. J. Pharmacol.* **148:**173–190.

Black, B. C., Brennan, L. A., Dierks, P. M., and Gard, I. E. (1997). Commercialization of baculoviral insecticides. *In* "The Baculoviruses" (L. K. Miller, ed.), pp. 341–387. Plenum Press, New York.

Boudjelal, M., Mason, S. J., Katso, R. M., Fleming, J. M., Parham, J. H., Condreay, J. P., Merrihew, R. V., and Cairns, W. (2005). The application of BacMam technology in nuclear receptor drug discovery. *Biotechnol. Ann. Rev.* **11:**101–125.

Boutell, C., Canning, M., Orr, A., and Everett, R. D. (2005). Reciprocal activities between herpes simplex virus type 1 regulatory protein ICP0, a ubiquitin E3 ligase, and ubiquitin-specific protease USP7. *J. Virol.* **79:**12342–12354.

Boyce, F. M., and Bucher, N. L. R. (1996). Baculovirus-mediated gene transfer into mammalian cells. *Proc. Natl. Acad. Sci. USA* **93:**2348–2352.

Carbonell, L. F., and Miller, L. K. (1987). Baculovirus interaction with non-target organisms: A virus-borne reporter gene is not expressed in two mammalian cell lines. *Appl. Environ. Microbiol.* **53:**1412–1417.

Carbonell, L. F., Klowden, M. J., and Miller, L. K. (1985). Baculovirus-mediated expression of bacterial genes in dipteran and mammalian cells. *J. Virol.* **56:**153–160.

Cersosimo, R. J. (2003). Tamoxifen for prevention of breast cancer. *Ann. Pharmacother.* **37:**268–273.

Chan, L. M. S., Lowes, S., and Hirst, B. H. (2004). The ABCs of drug transport in intestine and liver: Efflux proteins limiting drug absorption and bioavailability. *Eur. J. Pharm. Sci.* **21:**25–51.

Chen, R. Y., Edwards, R., Shaw, T., Colledge, D., Delaney, W. E., IV, Isom, H., Bowden, S., Desmond, P., and Locarnini, S. A. (2003). Effect of the G1896A precore mutation on drug sensitivity and replication yield of lamivudine-resistant HBV *in vitro*. *Hepatology* **37:**27–35.

Cheshenko, N., Krougliak, N., Eisensmith, R. C., and Krougliak, V. A. (2001). A novel system for the production of fully deleted adenovirus vectors that does not require helper adenovirus. *Gene Ther.* **8:**846–854.

Clare, J. J. (2006). Functional expression of ion channels in mammalian systems. *In* “Expression and Analysis of Recombinant Ion Channels” (J. J. Clare and D. J. Trezise, eds.), pp. 79–109. Wiley-VCH, Weinheim.

Clay, W. C., Condreay, J. P., Moore, L. B., Weaver, S. L., Watson, M. A., Kost, T. A., and Lorenz, J. J. (2003). Recombinant baculoviruses used to study estrogen receptor function in human osteosarcoma cells. *Assay Drug Dev. Technol.* **1:**801–810.

Condreay, J. P., and Kost, T. A. (2003). Virus-based vectors for gene expression in mammalian cells: Baculovirus. *In* “Gene Transfer and Expression in Mammalian Cells” (S. C. Makrides, ed.), pp. 137–149. Elsevier Science, Amsterdam.

Condreay, J. P., Witherspoon, S. M., Clay, W. C., and Kost, T. A. (1999). Transient and stable gene expression in mammalian cells transduced with a recombinant baculovirus vector. *Proc. Natl. Acad. Sci. USA* **96:**127–132.

Delaney, W. E., IV, and Isom, H. C. (1998). Hepatitis B virus replication in human HepG2 cells mediated by hepatitis B virus recombinant baculovirus. *Hepatology* **28:** 1134–1146.

Delaney, W. E., IV, Miller, T. G., and Isom, H. C. (1999). Use of the hepatitis B virus recombinant baculovirus-HepG2 system to study the effects of (−)-beta-2′,3′-dideoxy-3′-thiacytidine on replication of hepatitis B virus and accumulation of closed circular DNA. *Antimicrob. Agents Chemother.* **43:**2017–2026.

Delaney, W. E., IV, Edwards, R., Colledge, D., Shaw, T., Torresi, J., Miller, T. G., Isom, H. C., Bock, C. T., Manns, M. P., Trautwein, C., and Locarnini, S. (2001). Cross-resistance testing of antihepadnaviral compounds using novel recombinant baculoviruses which encode drug-resistant strains of hepatitis B virus. *Antimicrob. Agents Chemother.* **45:**1705–1713.

Doller, G., Groner, A., and Straub, O. C. (1983). Safety evaluation of nuclear polyhedrosis virus replication in pigs. *Appl. Environ. Microbiol.* **45:**1229–1233.

Drews, J. (2000). Drug discovery: A historical perspective. *Science* **287:**1960–1964.

Duisit, G., Saleun, S., Douthe, S., Barsoum, J., Chadeuf, G., and Moullier, P. (1999). Baculovirus vector requires electrostatic interactions including heparan sulfate for efficient gene transfer in mammalian cells. *J. Gen. Med.* **1:**93–102.

Dwarakanath, R. S., Clark, C. L., McElroy, A. K., and Spector, D. H. (2001). The use of recombinant baculoviruses for sustained expression of human cytomegalovirus immediate early proteins in fibroblasts. *Virology* **284:**297–307.

Fipaldini, C., Bellei, B., and LaMonica, N. (1999). Expression of hepatitis C virus cDNA in a human hepatoma cell line mediated by a hybrid baculovirus-HCV vector. *Virology* **255:**302–311.

Fujita, R., Matsuyama, T., Yamagishi, J., Sahara, K., Asano, S., and Bando, H. (2006). Expression of *Autographa californica* multiple nucleopolyhedrovirus genes in mammalian cells and upregulation of the host β-actin gene. *J. Virol.* **80:**2390–2395.

Gaillard, R. K., Barnard, J., Lopez, V., Hodges, P., Bourne, E., Johnson, L., Allen, M. I., Condreay, P., Miller, W. H., and Condreay, L. D. (2002). Kinetic analysis of wild-type and YMDD mutant hepatitis B virus polymerases and effects of deoxyribonucleotide concentrations on polymerase activity. *Antimicrob. Agents Chemother.* **46:** 1005–1013.

Galperin, M. Y. (2004). The molecular biology database collection: 2004 update. *Nucl. Acids Res.* **32**(database Issue)**:**D3–D22.

Gonzalez, M. I., and Robinson, M. B. (2004). Neurotransmitter transporters: Why dance with so many partners? *Curr. Opin. Pharmacol.* **4:**30–35.

Griffith, J. K., and Sansom, C. E. (1998). "The Transporter Facts Book." Academic Press, San Diego.

Hacker, D. L., Derow, E., and Wurm, F. M. (2005). The CELO adenovirus Gam1 protein enhances transient and stable recombinant protein expression in Chinese hamster ovary cells. *J. Biotechnol.* **117:**21–29.

Hartig, P. C., Cardon, M. C., and Kawanishi, C. Y. (1992). Effect of baculovirus on selected vertebrate cells. *Dev. Biol. Stand.* **76:**313–317.

Hassan, N. J., Pountney, D. J., Ellis, C., and Mossakowska, D. (2006). BacMam recombinant baculovirus in transporter expression: A study of BCRP and OATP1B1. *Protein Expr. Purif.* **47:**591–598.

Hofmann, C., Sandig, V., Jennings, G., Rudolph, M., Schlag, P., and Strauss, M. (1995). Efficient gene transfer into human hepatocytes by baculovirus vectors. *Proc. Natl. Acad. Sci. USA* **92:**10099–10103.

Hofmann, C., Lehnert, W., and Strauss, M. (1998). The baculovirus vector system for gene delivery into hepatocytes. *Gene Ther. Mol. Biol.* **1:**231–239.

Hopkins, A. L., and Groom, C. R. (2002). The druggable genome. *Nat. Rev. Drug Discov.* **1:**727–730.

Horrocks, C., Halse, R., Suzuki, R., and Shepherd, P. R. (2003). Human cell systems for drug discovery. *Curr. Opin. Drug Discov. Devel.* **6:**570–575.

Hsu, C. S., Ho, Y. C., Wang, K. C., and Hu, Y. C. (2004). Investigation of optimal transduction conditions for baculovirus-mediated gene delivery into mammalian cells. *Biotechnol. Bioeng.* **88:**42–51.

Hu, Y. C., Tsai, T. C., Chang, Y. J., and Huang, J. H. (2003). Enhancement and prolongation of baculovirus-mediated expression in mammalian cells: Focuses on strategic infection and feeding. *Biotechnol. Prog.* **19:**373–379.

Isom, H. C., Abdelhamed, A. M., Bilello, J. P., and Miller, T. G. (2004). Baculovirus-mediated gene transfer for the study of hepatitis B virus. *Methods Mol. Biol.* **96:**219–237.

Jarvis, D. L. (1997). Baculovirus expression vectors. *In* "The Baculoviruses" (L. K. Miller, ed.), pp. 389–431. Plenum Press, New York.

Jarvis, D. L., and Garcia, A. (1994). Long-term stability of baculoviruses stored under various conditions. *Biotechniques* **16:**508–513.

Jenkinson, S., McCoy, D. C., Kerner, S. A., Ferris, R. G., Lawrence, W. K., Clay, W. C., Condreay, J. P., and Smith, C. D. (2003). Development of a novel high-throughput surrogate assay to measure HIV envelope/CCR5/CD4-mediated viral/cell fusion using BacMam baculovirus technology. *J. Biomol. Screen.* **8:**463–470.

Jorio, H., Tran, R., and Kamen, A. (2006). Stability of serum-free and purified baculovirus stocks under various storage conditions. *Biotechnol. Prog.* **22:**319–325.

Katso, R. M., Parham, J. H., Caivano, M., Clay, W. C., Condreay, J. P., Gray, D. W., Lindley, K. M., Mason, S. J., Rieger, J., Wakes, N. C., Cairns, W. J., and Merrihew, R. V. (2005). Evaluation of cell-based assays for steroid nuclear receptors delivered by recombinant baculoviruses. *J. Biomol. Screen.* **10:**715–724.

Kohlbrenner, E., Aslanidi, G., Nash, K., Shklyaev, S., Campbell-Thompson, M., Byrne, B. J., Snyder, R. O., Muzyczka, N., Warrington, K. H., Jr., and Zolotukhin, S. (2005). Successful production of pseudotyped rAAV vectors using a modified baculovirus expression system. *Mol. Ther.* **12:**1217–1225.

Kost, T. A., and Condreay, J. P. (2002). Recombinant baculoviruses as mammalian cell gene-delivery vectors. *Trends Biotechnol.* **20:**173–180.

Kost, T. A., Condreay, J. P., and Jarvis, D. L. (2005). Baculovirus as versatile vectors for protein expression in insect and mammalian cells. *Nat. Biotechnol.* **23:**567–575.

Kost, T. A., Condreay, J. P., and Mickleson, C. A. (2006). Biosafety and viral gene transfer vectors. *In* "Biological Safety: Principles and Practices" (D. O. Fleming and D. L. Hunt, eds.), 4th Ed., pp. 509–529. ASM Press, Washington, DC.

Kraus, W. L., and Wong, J. (2002). Nuclear receptor-dependent transcription with chromatin. Is it all about enzymes? *Eur. J. Biochem.* **269:**2275–2283.

Kronschnabi, M., and Stamminger, T. (2003). Synergistic induction of intercellular adhesion molecule-1 by the human cytomegalovirus transactivators IE2p86 and pp71 is mediated via an Sp-1 binding site. *J. Gen. Virol.* **84:**61–73.

Kukkonen, S. P., Airenne, K. J., Marjomäki, V., Laitinen, O. H., Lehtolainen, P., Kankaanpää, P., Mähönen, A. J., Räty, J. K., Nordlund, H. R., Oker-Blom, C., Kulomaa, M. S., and Ylä-Herttuala, S. (2003). Baculovirus capsid display: A novel tool for transduction imaging. *Mol. Ther.* **8:**853–862.

Lehmann, J. M., Moore, L. B., Smith-Oliver, T. A., Wilkison, W. O., Willson, T. M., and Kliewer, S. A. (1995). An antidiabetic thiazolidinedione is a high affinity ligand for peroxisome proliferator-activated receptor gamma (PPAR gamma). *J. Biol. Chem.* **270:**12953–12956.

Lu, L., Ho, Y., and Kwang, J. (2006). Surpression of porcine arterivirus replication by baculovirus-delivered shRNA targeting nucleoprotein. *Biochem. Biophys. Res. Commun.* **340:**1178–1183.

Magga, J., Bart, G., Oker-Blom, C., Kukkonen, J. P., Akerman, K. E., and Nasman, J. (2006). Agonist potency differentiates G protein activation and Ca(2+) signalling by the orexin receptor type 1. *Biochem. Pharmacol.* **71**(6):827–836.

Mannix, C., and Jarman, R. F. (2000). A guide to successful scale-up of the baculovirus expression system. *In* "Cell Engineering, Vol. 2: Transient Expression" (M. Al-Rubeai, ed.), pp. 43–55. Kluwer Academic Publishers, Dordrecht.

Matsuo, E., Tani, H., Lim, C. K., Komoda, Y., Okamoto, T., Miyamoto, H., Moriishi, K., Yagi, S., Patel, A. H., Miyamura, T., and Matsuura, Y. (2006). Characterization of HCV-like particles produced in a human hepatoma cell line by a recombinant baculovirus. *Biochem. Biophys. Res. Commun.* **340:**200–208.

McCormick, C. J., Rowlands, D. J., and Harris, M. (2002). Efficient delivery and regulable expression of hepatitis C virus full-length and minigenome constructs in hepatocyte-derived cell lines using baculovirus vectors. *J. Gen. Virol.* **83:**383–394.

McCormick, C. J., Challinor, L., MacDonald, A., Rowlands, D. J., and Harris, M. (2004). Introduction of replication-competent hepatitis C virus transcripts using a tetracycline-regulable baculovirus delivery system. *J. Gen. Virol.* **85:**429–439.

McDonnell, D. P., Vegeto, E., and Gleeson, M. A. (1993). Nuclear hormone receptors as targets for new drug discovery. *Biotechnology* **11:**1256–1261.

Meghrous, J., Aucoin, M. G., Jacob, D., Chahal, P. S., Arcand, N., and Kamen, A. A. (2005). Production of recombinant adeno-associated viral vectors using a baculovirus/ insect cell suspension culture system: From shake flasks to a 20 L bioreactor. *Biotechnol. Prog.* **21:**154–160.

Meier, C. A. (1997). Regulation of gene expression by nuclear hormone receptors. *J. Recept. Signal Transduct. Res.* **17:**319–335.

Merrihew, R. V., Clay, W. C., Condreay, J. P., Witherspoon, S. M., Dallas, W. S., and Kost, T. A. (2001). Chromosomal integration of transduced recombinant baculovirus DNA in mammalian cells. *J. Virol.* **75:**903–909.

Merrihew, R. V., Kost, T. A., and Condreay, J. P. (2004). Baculovirus-mediated gene delivery into mammalian cells. *Method. Mol. Biol.* **246:**355–365.

Mizuno, N., Niwa, T., Yotsumoto, Y., and Sugiyama, Y. (2003). Impact of drug transporter studies on drug discovery and development. *Pharmacol. Rev.* **55:**425–461.

Nicholson, L. J., Philippe, M., Paine, A. J., Mann, D. A., and Dolphin, C. T. (2005). RNA interference mediated in human primary cells via recombinant baculoviral vectors. *Mol. Ther.* **11:**638–644.

Ong, S. T., Li., F., Du, J., Tan, Y. W., and Wang, S. (2005). Hybrid cytomegalovirus enhancer-h1 promoter-based plasmid baculovirus vectors mediate effective RNA interference. *Hum. Gene Ther.* **16:**1404–1412.

Pfohl, J. L., Worley, J. F., III, Condreay, J. P., An, G., Apolito, C. J., Kost, T. A., and Truax, J. F. (2002). Titration of K_{ATP} channel expression in mammalian cells utilizing recombinant baculovirus transduction. *Receptors Channels* **8:**99–111.

Poon, A. P. W., and Roizman, B. (2005). Herpes simplex virus 1 ICP22 regulates the accumulation of a shorter mRNA and of a truncated U_s3 protein kinase that exhibits altered functions. *J. Virol.* **79:**8470–8479.

Pritchard, J. B., and Miller, D. S. (2005). Expression systems for cloned xenobiotic transporters. *Toxicol. App. Pharmacol.* **204:**256–262.

Ramos, L., Kopec, L. A., Sweitzer, S. M., Fornwald, J. A., Zhao, H., McAllister, P., McNulty, D. E., Trill, J. J., and Kane, J. F. (2002). Rapid expression of recombinant proteins in modified CHO cells using the baculovirus system. *Cytotechnology* **38:**37–41.

Schroeder, K. S., and Neagle, B. D. (1996). FLIPR: A new instrument for accurate, high throughput optical screening. *J. Biomol. Screen.* **1:**75–80.

Sells, M. A., Chen, M. L., and Acs, G. (1987). Production of hepatitis B virus particles in HepG2 cells transfected with cloned hepatitis B virus DNA. *Proc. Natl. Acad. Sci. USA* **84:**1005–1009.

Shoji, I., Aizaki, H., Tani, H., Ishii, K., Chiba, T., Saito, I., Miyamura, T., and Matsuura, Y. (1997). Efficient gene transfer into various mammalian cells, including non-hepatic cells, by baculovirus vectors. *J. Gen. Virol.* **78:**2657–2664.

Song, S. U., and Boyce, F. M. (2001). Combination treatment for osteosarcoma with baculoviral vector mediated gene therapy (p53) and chemotherapy (adriamycin). *Exp. Mol. Med.* **33:**46–53.

Song, S. U., Shin, S. H., Kim, S. K., Choi, G. S., Kim, W. C., Lee, M. H., Kim, S. J., Kim, I. H., Choi, M. S., Hong, W. J., and Lee, K. H. (2003). Effective transduction of osteogenic sarcoma cells by a baculovirus vector. *J. Gen. Virol.* **84:**697–703.

Soudijn, W., and van Wijngaarden, I. (2000). The GABA transporter and its inhibitors. *Curr. Med. Chem.* **7:**1063–1079.

Spenger, A., Ernst, W., Condreay, J. P., Kost, T. A., and Grabherr, R. (2004). Influence of promoter choice and trichostatin A treatment on expression of baculovirus delivered genes in mammalian cells. *Prot. Expr. Purif.* **38:**17–23.

Szewczyk, B., Hoyos-Carvajal, L., Paluszek, M., Skrzecz, I., and Lobo de Souza, M. (2006). Baculoviruses—re-emerging biopesticides. *Biotechnol. Adv.* **24:**143–160.

Tani, H., Nishijima, M., Ushijima, H., Miyamura, T., and Matsuura, Y. (2001). Characterization of cell-surface determinants important for baculovirus infection. *Virology* **279:**343–353.

Tate, C. G., Haase, J., Baker, C., Boorsma, M., Magnani, F., Vallis, Y., and Williams, D. C. (2003). Comparison of seven different heterologous protein expression systems for the production of serotonin transporter. *Biochim. Biophysica Acta* **1610:**141–153.

Tjia, S. T., zu Altenschildesche, G. M., and Doerfler, W. (1983). *Autographa californica* polyhedrosis virus (AcNPV) DNA does not persist in mass cultures of mammalian cells. *Virology* **125:**107–117.

Urabe, M., Ding, C., and Kotin, R. M. (2002). Insect cells as a factory to produce adeno-associated virus type 2 vectors. *Hum. Gene Ther.* **13:**1935–1943.

Urabe, M., Nakakura, T., Xin, K. Q., Obara, Y., Mizukami, H., Kume, A., Kotin, R. M., and Ozawa, K. (2006). Scalable generation of high-titer recombinant adeno-associated virus type 5 in insect cells. *J. Virol.* **80:**1874–1885.

van Loo, N., Fortunati, E., Ehlert, E., Rabelink, M., Grosveld, F., and Scholte, B. J. (2001). Baculovirus infection of nondividing mammalian cells: Mechanisms of entry and nuclear transport of capsids. *J. Virol.* **75:**961–970.

Ververeli, K., and Chipps, B. (2004). Oral corticosteroid-sparing effects of inhaled corticosteroids in the treatment of persistent and acute asthma. *Ann. Allerg. Asthma Im.* **92:**512–522.

Volkman, L. E., and Goldsmith, P. A. (1983). *In vitro* survey of Autographa californica nuclear polyhedrosis virus interaction with nontarget vertebrate host cells. *Appl. Environ. Microbiol.* **45:**1085–1093.

Wellington, K. (2005). Rosiglitazone/metformin. *Drugs* **65:**1581–1592.

Whitfield, G. K., Jurutka, P. W., Haussler, C. A., and Haussler, M. R. (1999). Steroid hormone receptors: Evolution, ligands, and molecular basis of biologic function. *J. Cell. Biochem.* **32–33**(Suppl.):110–122.

Wieczorke, R., Dlugai, S., Krampe, S., and Boles, E. (2003). Characterisation of mammalian GLUT glucose transporters in a heterologous yeast expression system. *Cell Physiol. Biochem.* **13:**123–134.

Xia, C. Q., Yang, J. J., and Gan, L.-S. (2005). Breast cancer resistance protein in pharmacokinetics and drug-drug interactions. *Expert Opin. Drug Metab. Toxicol.* **1:**595–611.

Zeisel, M. B., and Baumert, T. F. (2006). Production of infectious hepatitis C virus in tissue culture: A breakthrough for basic and applied research. *J. Hepatology* **44:** 436–439.

Zhang, J. H., Chung, T. D., and Oldenburg, K. R. (1999). A simple statistical parameter for use in evaluation and validation of high throughput screening assays. *J. Biomol. Screen.* **4:**67–73.

Zhang, Z., Protzer, U., Hu, Z., Jacob, J., and Liang, T. J. (2004). Inhibition of cellular proteosome activities enhances hepadnavirus replication in an HBX-independent manner. *J. Virol.* **78:**4566–4572.

Zheng, W., Spencer, R. H., and Kiss, L. (2004). High throughput assay technologies for ion channel drug discovery. *Assay Drug Dev. Technol.* **2:**543–552.

Zhou, G., and Roizman, B. (2002). Truncated forms of glycoprotein D of herpes simplex virus 1 capable of blocking apoptosis and of low-efficiency entry into cells form a heterodimer dependent on the presence of a cysteine located in the shared transmembrane domains. *J. Virol.* **76:**11469–11475.

BACULOVIRUS VECTORS FOR GENE THERAPY

Yu-Chen Hu

Department of Chemical Engineering, National Tsing Hua University
Hsinchu, Taiwan 300

Abstract

Since the discovery that baculoviruses can efficiently transduce mammalian cells, baculoviruses have been extensively studied as potential vectors for both *in vitro* and *in vivo* gene therapy. This chapter reviews the history of this research area, cells permissive to baculovirus transduction, factors influencing transduction and transgene expression, efforts to improve transduction, mechanisms of virus entry and intracellular trafficking, applications for *in vivo* and *ex vivo* gene therapy, as well as advantages, limitations, and safety issues concerning use of baculoviruses as gene therapy vectors. Recent progress and efforts directed toward overcoming existing bottlenecks are emphasized.

I. Baculovirus Transduction of Mammalian Cells

A. Historical Overview

Autographa californica multiple nucleopolyhedrovirus (AcMNPV) is one of the most well-studied baculoviruses. In 1983, baculoviruses

0065-3527/06 $35.00
DOI: 10.1016/S0065-3527(06)68008-1

were first explored as vectors for the expression of human interferon β (IFN-β) in insect cells (Smith *et al.*, 1983). Since then, the potential of the baculovirus–insect cell expression system has been fully exploited for the production of numerous recombinant proteins (reviewed in Beljelarskaya, 2002; Luckow and Summers, 1988; Patterson *et al.*, 1995) and baculovirus research advanced insect cell culture as a force in the field of biotechnology. One of the reasons for the increasing popularity of the baculovirus–insect cell expression system is safety, because baculoviruses are regarded as nonpathogenic to humans, and the baculovirus host range is restricted to insects and invertebrates.

In 1983, however, Tjia *et al.* (1983) first found that baculoviruses can be internalized by mammalian cells and at least some of the viral DNA reached the nucleus. The nuclear DNA, however, did not persist and there was no evidence that baculovirus DNA was transcribed in mammalian cells (Tjia *et al.*, 1983). Later, Volkman and Goldsmith (1983) demonstrated that baculoviruses can be internalized by nontarget vertebrate cells such as human lung carcinoma cell line A427. Carbonell *et al.* (1985) further confirmed that baculoviruses entered mammalian cells and mediated very low-level expression of *Escherichia coli* chloramphenicol acetyltransferase (CAT) under the control of polyhedrin and *Rous sarcoma virus* (RSV) promoters. However, the significance of these findings was not widely noted until a decade later.

In the mid-1990s, two pioneer groups reported that recombinant baculoviruses harboring a cytomegalovirus (CMV) promoter-luciferase gene cassette (Hofmann *et al.*, 1995) or an RSV long terminal repeat (LTR) promoter-β-galactosidase (β-gal) gene cassette (Boyce and Bucher, 1996) efficiently expressed the reporter genes in mammalian cells. The data suggested a strong preference of baculovirus vectors to transduce hepatocytes of different origins (e.g., human and rabbit), because efficient transduction and high-level expression were only observed in primary hepatocytes and hepatoma cells. Significantly, lower reporter gene expression was observed in several other cell lines (e.g., COS-1 and 293) and little to no expression was observed in other cell types including A549, CHO, NIH-3T3, and CV-1 cells. The authors suggested that the block to expression in less susceptible cells might be subsequent to viral entry, rather than the ability to be internalized by the target cells, because high- and low-expressing cell lines internalized similar amounts of virus (Boyce and Bucher, 1996). One factor accounting for the low apparent transduction efficiency in certain cell types is promoter strength; Shoji *et al.* (1997) showed that cells that were not transduced by a baculovirus-expressing β-gal under the control of the CMV promoter could be efficiently transduced by a baculovirus expressing the same reporter protein under the transcriptional control of the stronger CAG promoter. Furthermore, Shoji *et al.* (1997) compared the gene expression by baculovirus and adenovirus vectors using the same expression unit and observed the same level of expression in HepG2, HeLa, and COS-7 cells by both vectors. They even demonstrated efficient expression and proper processing of hepatitis C virus (HCV) protein mediated by a baculovirus vector. These pioneering studies paved the way for use of baculoviral vectors as tools for gene delivery into mammalian cells.

B. Cells Permissive to Baculovirus Transduction

The list of cells permissive to baculovirus transduction has rapidly expanded. These cells include cell lines originating from cells of human (e.g., HeLa, Huh-7, HepG2, keratinocytes, bone marrow fibroblasts), rodent (e.g., CHO, BHK), porcine (e.g., CPK, PK15), bovine (e.g., BT), and even fish (e.g., EPC, CHH-1) origin (Table I). In addition, baculoviruses are capable of transducing nondividing cells, such as PK1 cells arrested in S phase (van Loo *et al.*, 2001). Transduction of primary cells, such as human neural cells (Sarkis *et al.*, 2000), pancreatic islet cells (Ma *et al.*, 2000), and rat articular chondrocytes (Ho *et al.*, 2004), has also been observed. In addition, Wagle and Jesuthasan (2003) showed that baculoviruses successfully transduced the embryos of zebrafish. EphrinB2a is normally expressed in the posterior region of developing somites and baculovirus-mediated misexpression by injection of the baculovirus expressing ephrinB2a into specific tissues, caused abnormal somite boundary formation (Wagle and Jesuthasan, 2003). Moreover, we demonstrated that baculoviruses are capable of transducing mesenchymal stem cells (MSC) derived from human umbilical cord blood and bone marrow (Ho *et al.*, 2005), as well as MSC-derived adipogenic, osteogenic, and chondrogenic progenitor cells (Ho *et al.*, 2006). Despite the rapidly growing list of cells permissive to baculovirus transduction, however, baculovirus transduction of cell lines of hematopoietic origin, such as U937, K562, Raw264.7 (Condreay *et al.*, 1999), LCL-cm, and Raji (Cheng *et al.*, 2004), is inefficient. As the spectrum of cell types permissive to baculovirus transduction expands, the potential applications of baculovirus vectors are receiving increasing attention. Table II lists some of the applications of baculoviruses that have been explored (see also chapter by Condreay *et al.*, pp. 255–286; van Oers, pp. 193–253; and Mäkelä and Oker-Blom, pp. 91–112, this volume).

TABLE I
SOME CELL TYPES THAT ARE PERMISSIVE TO BACULOVIRUS TRANSDUCTION

Cell type	References
Human cells	
HeLa	Boyce and Bucher, 1996; Condreay *et al.*, 1999; Hofmann *et al.*, 1995
Huh-7	Boyce and Bucher, 1996; Condreay *et al.*, 1999; Hofmann *et al.*, 1995
HepG2	Boyce and Bucher, 1996; Hofmann *et al.*, 1995
HEK293	Sollerbrant *et al.*, 2001
WI38	Condreay *et al.*, 1999
MRC5	Palombo *et al.*, 1998; Yap *et al.*, 1998
MG63	Condreay *et al.*, 1999
ECV-304	Airenne *et al.*, 2000
HUVEC	Kronschnabl *et al.*, 2002
PC3 (prostate cancer)	Stanbridge *et al.*, 2003
KATO-III (gastric cancer)	Shoji *et al.*, 1997
Osteosarcoma SAOS-2	Condreay *et al.*, 1999; Song *et al.*, 2003
Pancreatic β cells	Ma *et al.*, 2000
Keratinocytes	Condreay *et al.*, 1999
Bone marrow fibroblast	Condreay *et al.*, 1999
Primary foreskin fibroblast	Dwarakanath *et al.*, 2001
Primary neural cells	Sarkis *et al.*, 2000
Primary hepatocytes	Boyce and Bucher, 1996; Hofmann *et al.*, 1995
Mesenchymal stem cells	Ho *et al.*, 2005
Nonhuman primate cells	
COS-7	Condreay *et al.*, 1999
Vero	Poomputsa *et al.*, 2003
CV-1	Tani *et al.*, 2001
Porcine cells	
CPK	Aoki *et al.*, 1999
FS-L3	Shoji *et al.*, 1997
PK15	Aoki *et al.*, 1999
Bovine cells	
MDBK	Aoki *et al.*, 1999
BT	Aoki *et al.*, 1999

(*continues*)

TABLE I (*continued*)

Cell type	References
Rodent cells	
L929	Airenne *et al.*, 2000; Cheng *et al.*, 2004
PC12	Shoji *et al.*, 1997
CHO	Condreay *et al.*, 1999; Hu *et al.*, 2003a
BHK	Condreay *et al.*, 1999; Hu *et al.*, 2003a
Rat hepatic stellate cells	Gao *et al.*, 2002
Mouse pancreatic β cells	Ma *et al.*, 2000
Primary rat hepatocytes	Boyce and Bucher, 1996
Primary mouse osteoblasts and osteoclast	Tani *et al.*, 2003
Rat articular chondrocyte	Ho *et al.*, 2004
Fish cells	
EPC	Leisy *et al.*, 2003
CHH-1	Leisy *et al.*, 2003
Embryo	Wagle and Jesuthasan, 2003
Rabbit cells	
Rabbit aortic smooth muscle	Raty *et al.*, 2004
RK 13 (kidney)	Nakamichi *et al.*, 2002

TABLE II

APPLICATIONS OF THE BACULOVIRUS/MAMMALIAN CELL SYSTEM

Application	References
In vitro and *in vivo* gene therapy	Airenne *et al.*, 2000; Boyce and Bucher, 1996; Hofmann *et al.*, 1995; Pieroni *et al.*, 2001; Sarkis *et al.*, 2000; Tani *et al.*, 2003
Cell-based assays	Ames *et al.*, 2004; Jenkinson *et al.*, 2003; Katso *et al.*, 2005
Studies of gene function	Clay *et al.*, 2003; Pfohl *et al.*, 2002
Studies of virology	Delaney and Isom, 1998; Dwarakanath *et al.*, 2001; Lopez *et al.*, 2002; McCormick *et al.*, 2002; Zhou *et al.*, 2000
Protein production	Chen *et al.*, 2005; Ojala *et al.*, 2004; Ramos *et al.*, 2002
Virus vector production	Cheshenko *et al.*, 2001; McCormick *et al.*, 2002; Poomputsa *et al.*, 2003; Sollerbrant *et al.*, 2001
Surface display	Ernst *et al.*, 1998; Grabherr and Ernst, 2001
Vaccine candidates	Abe *et al.*, 2003; Aoki *et al.*, 1999; Facciabene *et al.*, 2004; Tami *et al.*, 2000; Yoshida *et al.*, 2003

C. *Mechanisms of Baculovirus Entry into Mammalian Cells*

1. *Importance of the Envelope Glycoprotein gp64*

The baculovirus gp64 glycoprotein is a major component of the budded virus envelope and is essential for virus entry into insect cells by receptor-mediated endocytosis (Wickham *et al.*, 1990). Following virus entry, gp64 further mediates the acid-induced endosomal escape, thus allowing for nucleocapsid transport into the cytoplasm and nucleus (Blissard and Wenz, 1992). Similarly, gp64 is essential for virus attachment and endosomal escape in mammalian cells (Hofmann *et al.*, 1998). In support of the importance of gp64 is the finding that a monoclonal antibody specific for gp64 abolishes the capability of baculovirus vectors to transduce mammalian cells (Gronowski *et al.*, 1999). A baculovirus overexpressing gp64 from an additional gp64 gene can incorporate ~1.5 times the normal amount of gp64 on the virion surface and exhibit 10- to 100-fold more reporter gene expression in a variety of mammalian cells when compared to the control baculovirus (Tani *et al.*, 2001). The importance of gp64 for virus transduction is further substantiated as a mutant virus lacking gp64 on the viral envelope failed to transduce mammalian cells (Abe *et al.*, 2005). Furthermore, gp64 of AcMNPV was shown to rescue transduction of mammalian cells by HaSNPV (*Helicoverpa armigera* single nucleopolyhedrovirus), a virus that does not transduce mammalian cells. The range of mammalian cell types transduced by HaSNPV expressing the gp64 of AcMNPV was consistent with those transduced by AcMNPV (Lang *et al.*, 2005).

2. *Surface Molecules for Virus Docking*

Although the importance of gp64 for virus entry has been documented, the nature of the cell surface molecule that interacts with the virus is unclear in both insect and mammalian cells (Kukkonen *et al.*, 2003). Initially, it was suggested that baculovirus transduction was liver specific and that asialoglycoprotein could be involved in virus binding (Boyce and Bucher, 1996; Hofmann *et al.*, 1995). However, van Loo *et al.* (2001) showed that Pk1 cells, which do not express asialoglycoprotein receptors, can be successfully transduced, and hence asialoglycoprotein is not a key determinant. It was also shown that electrostatic interactions may be necessary for baculovirus binding to the mammalian cell surface because preincubation of 293 cells with polybrene, a cationic compound that neutralizes negatively charged epitopes on the cell membrane, resulted in a rapid decrease in virus binding (Duisit *et al.*, 1999). The same group also suggested that heparan sulfate may act as an important docking motif for baculovirus binding because removal of heparan sulfate from the cell surface by heparanase I or III prior to transduction reduced transgene (*LacZ*) expression by ≈50% (Duisit *et al.*, 1999). Aside from heparan sulfate, phospholipids on the cell surface were suggested to serve as an important docking point for gp64, thus facilitating viral entry into mammalian cells (Tani *et al.*, 2001).

On the other hand, by transient depletion of calcium using EGTA pretreatment, Bilello *et al.* (2003) demonstrated that paracellular junction complexes are important barriers for baculoviral entry into primary hepatocytes. In contrast, we found that EGTA treatment of Huh-7 cells and chondrocytes does not significantly enhance transduction efficiencies although disruption of cell junctions was apparent (unpublished data). Despite the discrepancies in identification of the surface receptors, multiple lines of evidence suggest that baculoviruses are internalized by endocytosis (Condreay *et al.*, 1999; van Loo *et al.*, 2001). By electron and confocal microscopy, Matilainen *et al.* confirmed that baculoviruses enter HepG2 cells via clathrin-mediated endocytosis. However, baculovirus attachment to clathrin-coated pits seemed to be a relatively rare phenomenon, and therefore other internalization mechanisms (possibly via macropinocytosis) may also exist (Matilainen *et al.*, 2005). Virus attachment does not appear to be limiting, because baculoviruses can efficiently bind to NIH-3T3, a cell line less susceptible to baculovirus transduction, even at 4 °C (Stanbridge *et al.*, 2003).

3. *Endosomal Escape and Intracellular Trafficking*

Having entered the mammalian cell, budded virus is transported to the endosome, followed by acid-induced endosomal escape of the nucleocapsid mediated by gp64. Endosomal escape was first uncovered by treating baculovirus-transduced mammalian cells with a lysosomotropic agent (e.g., chloroquine), which inhibits endosomal maturation and subsequent baculovirus-mediated gene expression (Boyce and Bucher, 1996; Hofmann *et al.*, 1995). The importance of endosomal escape was further confirmed by treating HepG2 cells with monensin, which blocked early endosome acidification and trapped the nucleocapsids in the endosome (Kukkonen *et al.*, 2003). Therefore, it is generally assumed that escape from the endosomes blocks baculovirus transduction of some mammalian cells (Barsoum *et al.*, 1997; Boyce and Bucher, 1996). However, Kukkonen *et al.* (2003) suggested that the block may lie not in escape from the endosome, but rather in cytoplasmic trafficking or nuclear import of the nucleocapsids. In cells nonpermissive to baculovirus-mediated transduction (e.g., EAHY, MG63, and NHO cells), virus is internalized and routed to the endosome 30 min posttransduction, and escapes from the endosome by 4 h posttransduction, but the nucleocapsid does not enter the nucleus efficiently. Accordingly, no detectable transgene expression is observed even with a very high virus load. In contrast, baculoviruses are capable of entering HepG2 cells (which are highly permissive to baculovirus transduction), escaping from the endosome and entering the nucleus 4 h after transduction. Consistent with this notion is that direct injection of nucleocapsids into the cytoplasm does not affect the translocation of nucleocapsids into the nucleus, demonstrating that endosomal escape is not necessarily a critical step (Salminen *et al.*, 2005). Nucleocapsids are transported into different subcellular compartments in different cells (Abe *et al.*, 2005). In 293T cells, the nucleocapsids reached the nucleus where the transgene was efficiently transcribed following uncoating. However, in the nonpermissive macrophage RAW264.7 cells, the nucleocapsids appeared to be trapped by the phagocytic pathway, and degraded viral DNA was then transported into toll-like receptor 9 (TLR9)-expressing intracellular compartments (Abe *et al.*, 2005).

In the cytoplasm, the nucleocapsids seem to induce the formation of actin filaments which probably facilitate the transport of nucleocapsids into the nucleus. Cytochalasin D, an agent causing reversible depolymerization of actin filaments, strongly inhibits reporter gene expression, but does not prevent the uptake of enveloped virions inside cytoplasmic vesicles, or prevent their escape into the cytoplasm (van Loo *et al.*, 2001). More recently, it was shown that disintegration of microtubules by microtubule-depolymerizing agents (e.g., nocodazole and vinblastine) significantly enhanced the nuclear transport of virus and subsequent transgene expression in HepG2 cells (Salminen *et al.*, 2005), suggesting that intact microtubules constituted a barrier to baculovirus transport toward the nucleus. The viral genome, major capsid protein, and electron-dense capsids were also found inside the nucleus, suggesting that the nucleocapsid was transported through the nuclear pore (van Loo *et al.*, 2001). All of these studies highlight the importance of intracellular trafficking for transgene expression. The proposed route of baculovirus entry and intracellular trafficking is illustrated in Fig. 1.

D. Transduction Efficiency and Level of Transgene Expression

1. Dependence on Cell Types

Baculovirus transduction efficiencies vary considerably according to the cell type and can range from 95% for BHK cells (Wang *et al.*, 2005) to lower than 10% for NIH-3T3 cells (Cheng *et al.*, 2004). Baculovirus

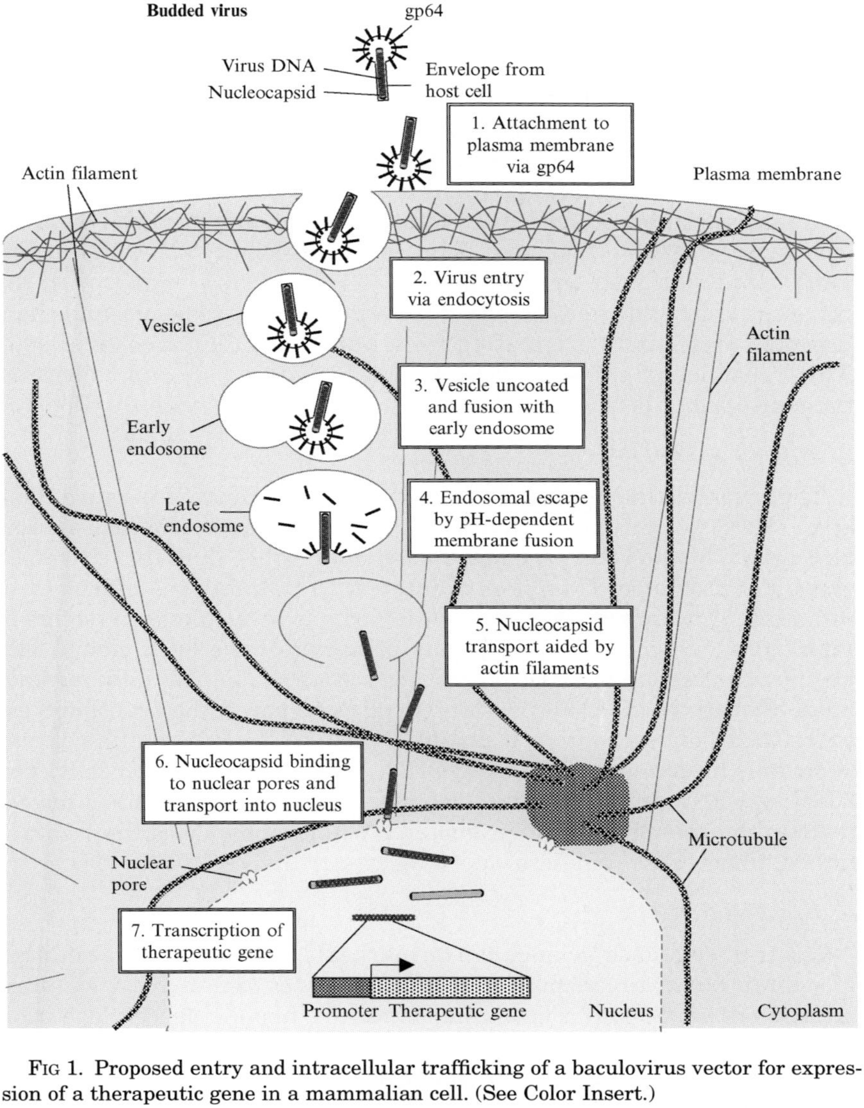

FIG 1. Proposed entry and intracellular trafficking of a baculovirus vector for expression of a therapeutic gene in a mammalian cell. (See Color Insert.)

transduction of hepatocytes (e.g., HepG2, Huh-7) is particularly efficient, with efficiencies of up to 80% (Wang *et al.*, 2005). Because plasmid transfection of hepatocytes, a common approach for gene delivery into liver cells, is notoriously difficult (with a typical delivery efficiency of 5–30%), the highly efficient baculovirus-mediated gene delivery has been exploited to study hepatitis B virus replication in HepG2 cells (Delaney and Isom, 1998) and to produce hepatitis delta virus-like particles (HDV VLP) in hepatocytes (Wang *et al.*, 2005).

The transduction efficiency may also be dependent on cellular differentiation state because transduction efficiency is only ≈30% for undifferentiated human neural progenitor cells, but can be up to ≈55% for differentiated neural cells at a multiplicity of infection (MOI) of 25 (Sarkis *et al.*, 2000). Likewise, transduction efficiency (21–90%), transgene expression level and duration (7–41 days) vary widely with the differentiation state and lineage of the adipogenic, osteogenic, and chondrogenic progenitors originating from human MSCs (Ho *et al.*, 2006). The transduction efficiency is very high for adipogenic and osteogenic progenitors, but is relatively low for chondrogenic progenitors (Fig. 2).

2. *Effects of Promoter*

The transduction efficiency is also promoter-dependent because Shoji *et al.* (1997) demonstrated that baculovirus-mediated luciferase expression driven by the CAG promoter was tenfold higher than that driven by the CMV promoter. Thus it is of interest to examine the efficiency of different promoters of viral and cellular origins in baculovirus vectors in mammalian cells. Although various promoters have been cloned into baculovirus vectors to drive gene transcription (Table III), only recently have Spenger *et al.* (2004) systematically compared the transgene expression driven by *Simian virus 40* (SV40), CMV, RSV, and a cellular promoter (human ubiquitin C) in CHO, COS-1, and HEK293 cells. The CMV and RSV promoters were the most active in all cell lines tested, followed by the ubiquitin C promoter. SV40 promoter was the weakest among these four promoters.

3. *Effects of Drugs*

The transduction efficiency can be markedly enhanced by the addition of sodium butyrate, trichostatin A (Condreay *et al.*, 1999), or valproic acid (Hu *et al.*, 2003a). These compounds are histone deacetylase inhibitors that induce histone hyperacetylation and lead to a relaxed chromatin structure (Kramer *et al.*, 2001). The use of these drugs enhances baculovirus-mediated gene transcription, thereby highlighting the importance of the chromatin state of the baculovirus genome in the transduced cells for transgene expression. Note, however, that cytotoxicity is often associated with the use of these drugs (Hu *et al.*, 2003a) and the extent to which gene expression is upregulated is dependent on the promoter and the particular cell line (Spenger *et al.*, 2004).

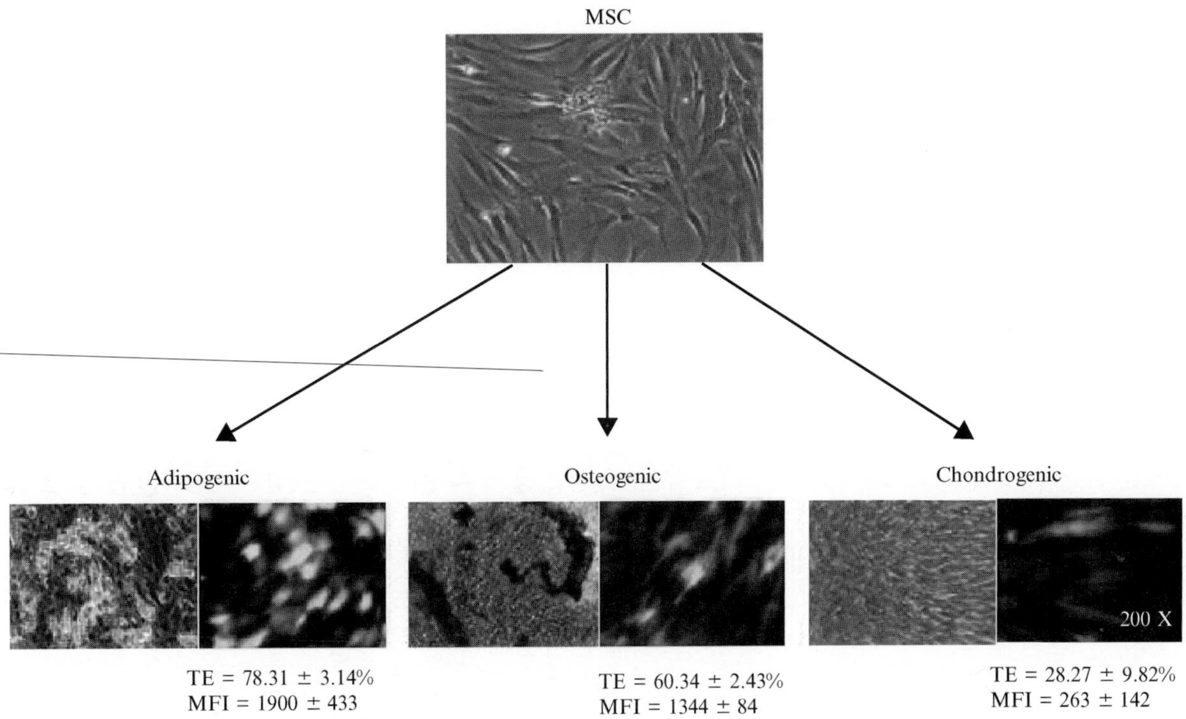

FIG 2. Baculovirus transduction of adipogenic, osteogenic, and chondrogenic progenitors. Human mesenchymal stem cells (MSC) were induced to differentiate into adipogenic, osteogenic, and chondrogenic pathways. The adipogenic, osteogenic, and chondrogenic progenitors were revealed by staining with oil red-O, von Kossa, and safranin-O staining, respectively, two weeks postinduction (left panels of each lineage pathway). The progenitors were transduced with baculoviruses expressing enhanced green fluorescent protein (EGFP) 2 weeks postinduction and exhibited different degrees of EGFP expression (right panel of each lineage pathway). The transduction efficiency (TE) and mean fluorescence intensity (MFI) were measured by flow cytometry. (See Color Insert.)

TABLE III
MAMMALIAN PROMOTERS INSERTED INTO BACULOVIRUS VECTORS

Promoter	References
Rous sarcoma virus long terminal repeat (RSV-LTR) promoter	Boyce and Bucher, 1996
Cytomegalovirus immediate early promoter CMV-IE	Hofmann *et al.*, 1995; Sollerbrant *et al.*, 2001
Simian virus 40 (SV40) promoter	Spenger *et al.*, 2004
Hybrid chicken β-actin promoter (CAG)	Shoji *et al.*, 1997; Stanbridge *et al.*, 2003
Hepatitis B virus (HBV) promoter/enhancer	Delaney and Isom, 1998
Human α-fetoprotein promoter/enhancer	Park *et al.*, 2001
Human ubiquitin C promoter	Spenger *et al.*, 2004
Hybrid neuronal promoter	Li *et al.*, 2004; Wang and Wang, 2005
Drosophila heat shock protein (*hsp70*)	Viswanathan *et al.*, 2003

4. *Effects of Baculovirus Genomic Enhancer*

Another factor influencing the transduction efficiency and expression level in certain cell lines is the activation of mammalian promoters (e.g., the CMV promoter and the heat shock promoter) by a homologous region (*hr*) in the baculovirus (AcMNPV) genome (Viswanathan *et al.*, 2003). One of the *hr* regions, *hr1*, enhances transcription from the polyhedrin and the *Drosophila* heat shock protein (*hsp70*) promoters in insect cells (e.g., Sf9) *in trans* (Venkaiah *et al.*, 2004). Yet *hr1* also functions in mammalian cells (e.g., Vero and HepG2) as an enhancer when present *in cis* and *in trans* (Viswanathan *et al.*, 2003). The upregulation of gene expression by *hr1* probably stems from binding of *hr1* with high affinity and specificity to nuclear factors in mammalian cells, thereby stimulating transcription (Viswanathan *et al.*, 2003). The insertion of an additional copy of the *hr1* region into the AcMNPV genome thus represents an attractive approach for overexpression of foreign proteins in mammalian cells (Venkaiah *et al.*, 2004). The additional *hr1* also improves the genetic stability of the bacmid-derived baculovirus and consequently prolongs expression of the heterologous protein, because spontaneous deletion of the heterologous gene(s) in the foreign bacterial artificial chromosome sequences readily occurs (Pijlman *et al.*, 2001, 2004).

5. *Transduction Protocols*

Another approach to enhancing the efficiency of baculovirus transduction of mammalian cells is to alter the transduction protocol. For transduction, typically the baculovirus is concentrated by ultracentrifugation and resuspended in phosphate-buffered saline (PBS). The cells are then incubated with the virus for 1 h at 37 °C in growth medium (e.g., DMEM) (Boyce and Bucher, 1996; Shoji *et al.*, 1997; Tani *et al.*, 2003). We developed a protocol by which incubation of unconcentrated virus (i.e., virus supernatant harvested from infected cell culture) with cells at lower temperature (e.g., 25 °C or 27 °C) for 4–8 h in PBS resulted in gene transfer into HeLa, chondrocytes (Ho *et al.*, 2004; Hsu *et al.*, 2004), and human MSC (Ho *et al.*, 2005) with efficiencies comparable or superior to those using traditional protocols. Specifically, the transduction efficiencies of human MSC derived from umbilical cord blood can be elevated from 42% (using the conventional protocol) to 73% (using the modified protocol). This protocol eliminates the need for virus ultracentrifugation, and hence not only represents a simpler approach but also reduces the chance for virus loss or inactivation during ultracentrifugation.

A key determinant for the improved transduction efficiency is the incubation medium. We found that PBS is superior to DMEM or TNM-FH (the medium for baculovirus production) in terms of transduction efficiency and transgene expression (Ho *et al.*, 2005; Hsu *et al.*, 2004). Comparison between the major components in PBS and medium revealed that $NaHCO_3$ present in DMEM or TNM-FH significantly reduced transduction efficiency (unpublished data), but the reason for this is unknown.

6. *Modifications of the Baculovirus Vector for Improved Gene Delivery*

The tropism and transduction efficiency of baculoviruses has been manipulated by modifying the envelope protein. Modification can be performed by fusing a heterologous gene in frame at the 5′ end of the gp64 gene under the control of the polyhedrin or p10 promoter. The fusion protein, after expression as an additional copy, is translocated to the plasma membrane and incorporated into the viral envelope on virus budding. Use of this approach was first demonstrated by fusion of human immunodeficiency virus-1 (HIV-1) envelope proteins to gp64 and the modified budded virus bound to the CD4 receptor on T cells (Boublik *et al.*, 1995). A similar strategy was applied to construct avidin-displaying baculoviruses, which showed a 5-fold increase in transduction efficiency in rat malignant glioma cells and a 26-fold increase in transduction efficiency in rabbit aortic smooth muscle cells compared to the wild-type baculovirus (Raty *et al.*, 2004). Baculoviruses displaying heterologous envelope proteins, such as vesicular stomatitis virus G protein (VSVG), have also been constructed. These vectors transduce human hepatoma and rat neuronal cells at efficiencies roughly 10- to 100-fold greater than baculoviruses lacking VSVG

(Barsoum *et al.*, 1997). This pseudotyped virus also transduced cell lines that are transduced at very low levels by the unmodified baculovirus, thus broadening the tropism. The enhanced transduction efficiency and wider tropism are attributed to the increased transport of baculovirus DNA into nuclei rather than to the increased binding or virus uptake (Barsoum *et al.*, 1997).

In contrast, specific targeting of baculoviruses to mammalian cells by displaying a single-chain antibody fragment specific for the carcinoembryonic antigen (CEA) or synthetic IgG-binding domains was also demonstrated (Mottershead *et al.*, 2000; Ojala *et al.*, 2001). Such viral targeting could reduce the virus dose required for *in vivo* gene therapy regimes, if baculoviruses can be engineered to bind efficiently to specific cell types.

II. Baculovirus Vectors for Gene Therapy

A. In Vivo *Gene Therapy*

Given the highly efficient gene delivery into many cell types, baculoviruses have captured increasing interest as vectors for *in vivo* gene delivery. Tissues that have been targeted include rabbit carotid artery (Airenne *et al.*, 2000), rat liver (Huser *et al.*, 2001), rat brain (Lehtolainen *et al.*, 2002; Sarkis *et al.*, 2000; Wang and Wang, 2005), mouse brain (Sarkis *et al.*, 2000), mouse skeletal muscle (Pieroni *et al.*, 2001), mouse cerebral cortex and testis (Tani *et al.*, 2003), and mouse liver (Hoare *et al.*, 2005). For baculovirus-mediated *in vivo* gene therapy, however, the complement system appears to be a significant barrier because systemic or intraportal application as well as direct injection into the liver parenchyma fail to result in detectable gene expression (Sandig *et al.*, 1996). This failure stems from inactivation of the baculovirus vector in the presence of native serum, because baculoviruses activate the classical pathway of the complement system (Hofmann *et al.*, 1998). Hoare *et al.* (2005) further showed that both classical and alternative pathways are involved in the inactivation and suggested that naturally occurring IgM antibodies with high affinity for baculoviruses may be partially responsible for the inactivation.

Various strategies have been employed to avoid complement inactivation. Hofmann and Strauss (1998) demonstrated that the survival of a baculovirus vector in human serum can be enhanced through treatment with a functional antibody-blocking complement component 5 (C5). Meanwhile, the complement inhibitor sCR1 (soluble complement receptor type 1) protects baculoviruses from serum inactivation *in vitro* and coadministration of sCR1 along with intraportal administration of the baculovirus vector leads to hepatic expression in mice (Hoare *et al.*, 2005). Inactivation of baculoviruses in human plasma and whole blood can be prevented by treatment with cobra venom factor (CVF), an inhibitor of the complement system (Hofmann and Strauss, 1998). By injecting CVF into mice one day prior to baculovirus administration, Sarkis *et al.* (2000) demonstrated that the baculovirus was not inactivated by the complement system and could transduce neural cells (mainly astrocytes) *in vivo* when directly injected into the brain of rodents. Surprisingly, they observed the same level of expression in animals without treatment of CVF, suggesting that the baculovirus was not inactivated by the complement system. These contradictory observations were probably caused by the particular immunological characteristics of the brain (Sarkis *et al.*, 2000) or differences in stereotaxic coordinates and injection pressure, or speed (Lehtolainen *et al.*, 2002).

The complement inactivation problem can also be minimized by avoiding contact of the baculovirus vectors with blood components. By using a silastic collar, transduction of adventitial cells in rabbit carotid arteries was achieved and the efficiencies were comparable to those obtained with adenoviral vectors (Airenne *et al.*, 2000). Gene expression was transient remaining high level for 1 week but disappearing by day 14, and the arterial structure and endothelium remained intact after baculovirus transduction. Baculovirus vectors have also been injected into the rodent brain where complement proteins may be absent because of the blood–brain barrier, or the complement level in the brain may be insufficient to affect gene transfer (Lehtolainen *et al.*, 2002). After *in vivo* injection into the brain, baculoviruses specifically transduced the epithelium of the choroids plexus in ventricles and the transduction efficiency was as high as 76 $\pm$ 14%. In contrast, adenovirus vectors showed preference to corpus callosum glial cells and ventricular ependymal lining. Hence, baculovirus vectors are especially useful for targeting of choroids plexus cells (Lehtolainen *et al.*, 2002).

A more cutting-edge approach to alleviate complement inactivation is the generation of complement-resistant baculoviruses by display on the viral envelope of decay-accelerating factor (DAF), a regulator that blocks complement at the central step of both the classical and alternative pathways (Huser *et al.*, 2001). Such complement-resistant baculovirus vectors allow for a substantial improvement of gene transfer into neonatal rats *in vivo* after local injection into the liver parenchyma.

Expression of the transgene (human coagulation factor IX, hFIX) was transient probably as a result of the generation of antibodies directed against the transgene product hFIX, which might lead to clearance of either expressed hFIX protein and/or positively transduced cells. Alternatively, baculoviruses can be pseudotyped by displaying VSVG on the envelope. The VSVG-modified virus enhanced gene transfer efficiencies into mouse skeletal muscle *in vivo* and the transgene expression lasted 178 days in DBA/2J mice and 35 days in BALB/c and C57BL/6 mice (Pieroni *et al.*, 2001). The VSVG-modified baculovirus also exhibited greater resistance to inactivation by animal sera and could transduce cerebral cortex and testis of mice by direct inoculation *in vivo* (Tani *et al.*, 2003).

In addition to expressing therapeutic proteins, it has been shown that transduction of Saos-2, HepG2, Huh-7, and primary human hepatic stellate cells with a baculovirus expressing shRNAs (short-hairpin RNAs) targeting lamin A/C effectively knocked down expression of the corresponding mRNA and protein (Nicholson *et al.*, 2005). More recently, baculoviruses have been used to mediate RNA interference (RNAi) using a novel hybrid promoter consisting of the CMV enhancer and polymerase III H1 promoter. The recombinant baculovirus was capable of suppressing expression of the target gene by 95% in cultured cells and by 82% *in vivo* in rat brain (Ong *et al.*, 2005). These data suggest that baculoviruses may be used as delivery vectors for triggering RNA interference for *in vivo* gene therapy.

B. Ex Vivo *Gene Therapy*

To date, most gene therapy studies using baculovirus vectors have focused on *in vivo* applications, yet relatively little is known about the potential of baculoviruses for *ex vivo* therapy. One relevant study was performed by establishing an *ex vivo* perfusion model for human liver segments (Sandig *et al.*, 1996). The recombinant baculovirus was perfused through the liver segments for 15 min and reasonable transduction rates were achieved in all perfused parts of the liver tissue. This study verified for the first time that baculovirus-mediated gene transfer is possible in liver tissue and is encouraging for future studies including *in situ* perfusion of intact livers with baculovirus vectors in animal models.

In addition, we have demonstrated highly efficient baculovirus-mediated gene transfer into articular chondrocytes (Ho *et al.*, 2004), human MSCs (Ho *et al.*, 2005), and MSC-derived progenitors (Ho *et al.*, 2006), all being candidate cell sources for the treatment of disorders in connective tissues, particularly cartilage and bone. Importantly, differentiation states of chondrocytes, MSC, and MSC-derived progenitor cells were not affected after baculovirus transduction. Further, the transduction of primary rabbit articular chondrocytes with baculoviruses expressing BMP-2 significantly improved the secretion of extracellular matrix (ECM) and promoted the expression of chondrocyte-specific genes (unpublished data), thus implicating the potential use of baculoviruses for delivery of genes encoding growth factors for cartilage and bone tissue engineering.

The ECM represents a barrier to baculovirus entry into chondrocytes because treatment of rat articular chondrocytes that had been cultured for 10 days with enzymes (hyaluronadase and heparanase) effectively removed the ECM and enhanced virus uptake and gene expression (unpublished data). Unfortunately, articular chondrocytes and osteoblasts are embedded in the ECM *in vivo*. Therefore, the dense ECM surrounding the target cells constitutes a formidable barrier for *in vivo* baculovirus-mediated gene therapy for tissue engineering. As such, *ex vivo* gene therapy may be a more appropriate choice in the context of bone and cartilage tissue engineering. For instance, differentiation of MSCs or progenitors toward a specific lineage pathway (e. g., osteogenic) may be modulated by *ex vivo* transduction with recombinant baculoviruses expressing appropriate growth factors (e.g., BMP-2). Given the efficient transduction of the partially differentiated progenitors, these cells may be transduced again with baculoviruses expressing identical (or different) factors with high efficiency, followed by seeding into scaffolds and implantation into animal models. The transduced cells may continue to express the appropriate factors *in vivo*, thereby stimulating cell differentiation and tissue (e.g., bone) regeneration in an autocrine or paracrine fashion.

III. Advantages and Limitations of Baculoviruses as Gene Therapy Vectors

A. *Advantages*

To date, vectors used for gene therapy are divided into two categories: nonviral and viral. Nonviral vectors comprise polymers (or liposomes) conjugated with polycations or other targeting molecules. However, the application of nonviral vectors is often restricted by the poor efficiency of delivery and transgene expression (Verma and Somia, 1997). In contrast, viral vectors, such as retroviral, lentiviral, adenoviral, and adeno-associated viral (AAV) vectors are in common use due to the more efficient cellular uptake and transgene expression. Despite this, each vector has intrinsic advantages and disadvantages (Table IV). For example, retroviruses can mediate integration of viral DNA into the host chromosome for permanent genetic modification; however, transcriptional silencing often occurs and results in transient expression. More critically, the random integration could lead to activation of oncogenes or inactivation of tumor suppressor genes and has resulted in unfortunate leukemia-like diseases in two X-linked SCID patients treated with retroviral vectors (Check, 2002). Lentiviral vectors, derived from *Human* or *Simian immunodeficiency virus* (HIV or SIV), are emerging vectors capable of long-term expression in dividing and nondividing cells. However, the pathogenic nature of HIV or SIV raises serious concerns about the safety of these vectors, and the production of high-titer virus stock is inefficient (Lundstrom, 2003). Adenoviruses can effectively infect dividing and nondividing cells and mediate high-level transgene expression, but the transgene expression is often transient due to the elicitation of strong humoral and cellular immunity, which has also resulted in the death of a patient (Marshall, 1999). AAV vectors can mediate sustained expression, but the packaging capacity is restricted and large-scale vector production is difficult. Furthermore, the preexisting immunity to human AAV vectors is comparable to that of adenoviral vectors (Thomas *et al.*, 2003).

In comparison with these common viral vectors, baculoviruses possess a number of advantages:

1. Lack of Toxicity and Replication

Baculovirus transduction is nontoxic to mammalian cells and does not hinder cell growth even at high MOI (Gao *et al.*, 2002; Hofmann *et al.*, 1995). Our studies again confirmed this notion because transduction with a wild-type baculovirus did not cause any observable adverse effects to chondrocytes or human MSC (Ho *et al.*, 2004, 2005). Cell proliferation, however, may be slightly retarded by transgene products, such as EGFP (Ho *et al.*, 2004), which could be toxic and might even induce apoptosis in some cells (Detrait *et al.*, 2002; Liu *et al.*, 1999). Fortunately, the cell growth rate was restored after several passages as EGFP expression attenuates (Ho *et al.*, 2005). Moreover, baculoviruses do not replicate in transduced mammalian cells (Hofmann *et al.*, 1995; Kost and Condreay, 2002; Sandig *et al.*, 1996; Shoji *et al.*, 1997).

The nonreplicative and nontoxic attributes of baculoviruses are particularly important because retroviruses, lentiviruses, and adenoviruses

TABLE IV

Comparison of Baculovirus and Other Viral Vectors

Features	Retroviral	Lentiviral	Adenoviral	AAV	Baculoviral
Ease of preparation	No	No	Yes	No	Yes
Packaging capacity	7–7.5 kb	7–7.5 kb	Up to 30 kb	3.5–4 kb	>38 kb
Route of administration	*Ex vivo*	*Ex/in vivo*	*Ex/in vivo*	*Ex/in vivo*	*Ex/in vivo*
Vector genome forms	Integrated	Integrated	Episomal	Episomal[a]	Episomal
Gene expression duration	Short	Long	Short	Long	Short
Tropism	Dividing cell	Broad	Broad	Broad	Broad
Immune response	Low	Low	High	Unknown	Unknown
Preexisting immunity	Unlikely	Unlikely	Yes	Yes	Unlikely
Safety	Integration may induce oncogenesis	Integration may induce oncogenesis	Inflammatory response, toxicity	Inflammatory response, toxicity	High[b]

[a] AAV can mediate site-specific integration into human chromosome 19, but common AAV vectors do not contain the *rep* gene and thus cannot mediate integration. However, random integration may occur.

[b] Baculoviruses are considered safe, but more studies are required to confirm the safety in *in vivo* and *ex vivo* applications.

are human pathogens, and hence emergence of replication-competent viruses (RCV) raises serious safety concerns. In contrast, baculoviruses are not pathogenic to humans, and hence the emergence of RCV is not an issue for baculovirus-meditaed gene therapy.

2. *Large Cloning Capacity*

The baculovirus (AcMNPV) genome is large (≈130 kb) and the maximum cloning capacity is at least 38 kb because the adenovirus genome has been cloned into a baculovirus vector (Cheshenko *et al.*, 2001). Such a large cloning capacity provides flexibility for multiple genes or large inserts. This flexibility is particularly advantageous in comparison with retroviral and AAV vectors whose cloning capacities are limited to 7–7.5 kb and 3.5–4 kb, respectively, and prohibit the cloning of regulatory sequences or large gene fragments (e.g., dystrophin).

3. *Ease of Production*

The production of retroviral, lentiviral, and AAV vectors requires transfection of plasmids encoding essential genes into packaging cells (Thomas *et al.*, 2003). The transfection process, however, is cumbersome, costly, and difficult to scale up. In sharp contrast, baculovirus propagation can easily be achieved by infecting insect cells in suspension culture (e.g., in spinner flasks or bioreactors) and harvesting the supernatant 3–4 days postinfection. Scale-up of the production process is straightforward because large-scale insect cell culture processes are well-established. The production phase is initiated simply by adding virus solution to cultured cells. Furthermore, the construction, propagation, and handling of baculoviruses can be performed readily in Biosafety Level 1 laboratories without the need for specialized equipment.

4. *Lack of Preexisting Immunity*

One of the problems associated with adenoviral and AAV vectors is that most people are exposed to these viruses and develop corresponding neutralizing antibodies. Circulating virus-neutralizing antibodies can preclude efficient transduction with the viral vector. In contrast, it is unlikely that people develop such preexisting immunity against baculoviruses. The use of baculovirus vectors in gene therapy, therefore, may avoid the problem of preexisting immunity.

B. Limitations

Despite the promising aspects, baculoviruses have a number of disadvantages as gene therapy vectors.

1. Transient Expression

In vitro, baculovirus-mediated expression usually lasted from 7 to 14 days for common cell lines such as CHO, HeLa, and BHK cells (Hu *et al.*, 2003a), although expression continued for 41 days in adipogenic progenitor cells (Ho *et al.*, 2006). *In vivo*, transgene expression typically declines by day 7 and disappears by day 14 (Airenne *et al.*, 2000; Lehtolainen *et al.*, 2002). The duration of *in vitro* transgene expression can be enhanced by prolonging the transduction period (e.g., upto 8 h) (Hsu *et al.*, 2004) or by supertransduction (Hu *et al.*, 2003a). Nonetheless, the extent to which expression can be prolonged is limited because expression is generally restricted to less than 1 month, which is significantly shorter than expression (in the range of months) mediated by retroviral, lentiviral, and AAV vectors.

One key difference between baculoviral and other viral vectors is that the genes carried by other vectors can persist in the host nucleus, either in an integrated or episomal form, for a longer period. However, Tjia *et al.* (1983) demonstrated that baculoviral DNA persists in the nuclei of transduced mammalian cells for only 24–48 h, as determined by Southern blot (Tjia *et al.*, 1983). We also found that the total transgene (*egfp*) copy number within baculovirus-transduced chondrocytes declined 11-fold (as determined by quantitative real-time PCR) while cell number increases 3.5-fold in 11 days, indicating that baculoviral DNA degrades over time (Ho *et al.*, 2004). The declining *egfp* copy number was concomitant with the decrease in mRNA transcription level (unpublished data) as well as fluorescence intensity.

To prolong transgene expression, Palombo *et al.* (1998) designed hybrid baculovirus–AAV vectors which contained a transgene cassette composed of the β-gal reporter gene and hygromycin resistance gene (Hyg^r) flanked by the AAV-inverted terminal repeats (ITR), which are necessary for AAV replication and integration into the host genome (Palombo *et al.*, 1998). Hybrid baculovirus–AAV vectors were derived with or without the AAV *rep* gene (whose gene products are essential for viral DNA replication and integration) cloned in different positions with respect to the baculovirus polyhedrin promoter. Transduction of 293 cells with the hybrid vector expressing the *rep* gene resulted in specific integration of ITR-flanked DNA into the AAVS1 site of chromosome 19 (Palombo *et al.*, 1998). A similar baculovirus–AAV hybrid vector incorporating an ITR-flanked luciferase gene under a neuron-specific promoter was also employed for *in vivo* studies (Wang and Wang, 2005). Even without the help of *rep* gene expression, the viral vector was able to provide transgene expression for at least 90 days when tested in rat brains. These studies demonstrate an effective methodology for engineering of baculoviral vectors for sustained transgene expression.

2. Inactivation by Serum Complement

As described earlier, contact between baculoviruses and serum complement results in rapid inactivation. Despite various attempts to minimize complement inactivation, to date the number of successful baculovirus-mediated *in vivo* gene therapy experiments in complement-competent animals is limited (Hoare *et al.*, 2005). However, the complement system is also a potent barrier to *in vivo* administration of other gene delivery systems such as liposomes (Marjan *et al.*, 1994), murine retrovirus (Takeuchi *et al.*, 1996), and various synthetic DNA complexes (Plank *et al.*, 1996).

3. Inhibition of Transduction by Intercellular Junctions

Intercellular junctions may be an additional hurdle to baculovirus-mediated gene therapy because transient disruption of these junctions by EGTA treatment prior to transduction improved gene delivery efficiency into long-term cultures of primary hepatocytes (Bilello *et al.*, 2001). Bilello *et al.* (2003) also suggested the importance of the basolateral surface for virus entry at least for some cell types. In our laboratory, however, transient disruption of cell junctions failed to effectively enhance baculovirus-mediated gene transfer into chondrocytes and HepG2 cells that were cultured to overconfluence (unpublished data), implying that other factors in addition to the paracellular junction complexes might be involved in transduction of these cells.

4. Fragility of Budded Virus

Another drawback associated with baculoviruses as gene delivery vectors is that the nucleocapsid is enveloped with lipids derived from the host cell membrane. The envelope structure is essential for virus infectivity due to the anchored gp64 (Blissard and Wenz, 1992), but it also renders virus vulnerable to mechanical force and results in relatively low virus stability, a common problem also observed for other enveloped viruses such as retrovirus (Wu *et al.*, 2000). Typically, baculoviral vectors are concentrated by ultracentrifugation after harvesting from cell culture and resuspended in PBS prior to use. However, ultracentrifugation often leads to significant loss of infectivity probably due to damage to viral envelopes. Ultracentrifugation also tends to result in virus aggregation (Barsoum, 1999). To alleviate these problems, we constructed a recombinant baculovirus with a hexahistidine (His_6) tag displayed on the viral envelope, which enables virus purification by a simple immobilized metal affinity chromatography (IMAC) (Hu *et al.*, 2003b). The IMAC methodology results in high purity (87%) and obviates the need for successive ultracentrifugation steps. However, the recovery yield in terms of infectious titer is lower than expected (<10%), probably because of damage to the viral envelope during the binding, washing, and elution steps. One possibility to alleviate virus loss during chromatographic purification steps is to display VSVG protein on the baculoviral envelope. Display of VSVG on the retrovirus envelope has been shown to enhance virus stability and the same strategy may be applied to enhancing baculovirus stability.

Besides sensitivity to mechanical force, the half-life of baculoviruses is drastically decreased from 173 h at 27 °C to 7–8 h at 37 °C (Hsu *et al.*, 2004). Such labile thermal stability, in conjunction with the tendency to be inactivated by serum complement, may further restrict the *in vivo* application of baculovirus gene delivery vectors.

IV. Safety Issues Concerning the Use of Baculoviruses for Gene Therapy

A. *DNA Integration and Viral Gene Expression*

As mentioned earlier, baculoviruses are nonpathogenic to humans and are nonreplicative in mammalian cells. Also, baculovirus DNA tends to be degraded in mammalian cells. However, Condreay *et al.* (1999) demonstrated that a recombinant baculovirus containing two expression cassettes (one harboring GFP under the control of the CMV promoter and the other harboring neomycin phosphotransferase under the control of the SV40 promoter) is capable of mediating stable expression. When transduced cells were selected with the antibiotic G418, cell lines that stably maintain the foreign expression cassettes can be obtained at high frequency and exhibit stable, high-level expression of the reporter gene for at least 25 passages. The frequency ranged from one clone in 39 transduced cells to one clone in 109 transduced cells, indicating that stable transduction is an efficient event. Stably transduced derivatives have been selected from a substantial number of cell types (e.g., CHO, Huh-7, HeLa, K562), suggesting that stable cell lines can be derived from any cell type that exhibits transient expression (Condreay *et al.*, 1999).

Such a stably expressing derivative (CHO cell) was later confirmed to stem from the integration of baculovirus DNA into the host cell genome as small, discrete single-copy fragments (Merrihew *et al.*, 2001). These fragments, ranging in size from 5 to 18 kb, had randomly distributed breakpoints outside the selected region, suggesting an illegitimate mode of integration (little or no homology between recombining DNA molecules). Such integration resulted in at least two clones that expressed GFP for up to 5 months.

Since leukemia-like conditions developed in two of the 11 SCID patients treated by retrovirus-mediated gene therapy, safety issues regarding whether and/or how vector DNA integrates into the genomic DNA are under scrutiny. Although these stably expressing cell clones are obtained under antibiotic selection, and the integration occurs in a way different from that of retroviruses (which encode a viral integrase directing nearly full-length, single-copy integration events), the possibility that baculoviruses mediate spontaneous integration into genomic DNA cannot be excluded. To date, there is no direct evidence showing that spontaneous integration of baculoviral DNA occurs in the absence of an antibiotic resistance gene and selective pressure, but extensive studies examining the state and fate of introduced viral DNA are necessary to further prove the safety of baculovirus gene therapy vectors.

Another concern regarding the use of baculoviruses is whether baculovirus endogenous genes are expressed. As long ago as 1983, Tjia *et al.* (1983) showed that baculovirus endogenous gene transcription is absent in transduced HeLa cells. Using RT-PCR, Stanbridge *et al.* (2003) assessed the expression of a number of baculovirus genes after transduction of human cells, and found no baculovirus gene transcripts in human cells. However, a study demonstrated that the baculoviral genomic early-to-late (ETL) promoter is active and able to drive reporter gene expression in mammalian cells (Liu *et al.*, 2006). Although gene expression does not equate to virus replication, the possibility that other baculoviral promoters are also active in the transduced mammalian cells cannot be excluded. Whether and how expression of baculoviral proteins at basal levels in the mammalian cells induces immune responses and how this may influence cellular gene expression and physiological state requires further investigation.

B. Immune Response and Potential as a Vaccine Vector

The "Achilles heel" of gene therapy is that immune responses used to tackle wild-type infections are activated against the vectors and/or the new transgene products (Thomas *et al.*, 2003). Although baculoviruses were found capable of entering mammalian cells as early as in 1983 (Tjia *et al.*, 1983; Volkman and Goldsmith, 1983), the host response to baculovirus uptake, either *in vitro* or *in vivo*, was not evaluated until 1999 when Gronowski *et al.* (1999) reported that administration of baculoviruses *in vitro* induced the production of IFN-α and IFN-β from human and murine cell lines. The IFN-stimulating activity of baculoviruses required live virus and was not due to the presence of viral RNA or DNA. Furthermore, administration of baculoviruses induced *in vivo* protection of mice from encephalomyocarditis virus infection (Gronowski *et al.*, 1999). A subsequent study discovered that baculovirus transduction of cultured rat hepatocytes disrupted phenobarbital (PB) gene induction, a potent transcriptional activation event characteristic of highly differentiated hepatocytes, and repressed expression of the albumin gene (Beck *et al.*, 2000). But neither cAMP nor PKA activities were affected by the virus. Baculovirus transduction also induced the expression of cytokines, such as TNF-α, IL-1α, and IL-1β, in primary rat hepatocytes, however, TNF-β, IL-2, IL-3, IL-4, IL-5, IL-6, and IFN-γ were not detected in any of the baculovirus-exposed hepatocytes (Beck *et al.*, 2000). Airenne *et al.* (2000) found that *in vivo* administration of baculovirus to rabbit carotid artery resulted in signs of inflammation. More recently, Abe *et al.* (2003) demonstrated that inoculation of baculovirus induced the secretion of inflammatory cytokines, such as TNF-α and IL-6, in a murine macrophage cell line, RAW264.7. In the same study, they also demonstrated that intranasal inoculation with a wild-type baculovirus elicited a strong innate immune response that protected mice from a lethal challenge of influenza virus (Abe *et al.*, 2003). This protective immune response was induced via the TLR9/MyD88-dependent signaling pathway (Abe *et al.*, 2005). The production of inflammatory cytokines was severely reduced in peritoneal macrophages (PECs) and splenic CD11c$^+$ dendritic cells (DCs) derived from mice deficient in MyD88 or TLR9 after stimulation with baculovirus. In contrast, a significant amount of IFN-α was still detectable in the PECs and DCs of these mice after stimulation with baculovirus, suggesting that a TLR9/MyD88-independent signaling pathway may also participate in the production of IFN-α (Abe *et al.*, 2005). The induction of cytokines required gp64, however, gp64 itself did not directly participate in the TLR-mediated immune response. Instead, the authors concluded that internalization of viral DNA via gp64-mediated membrane fusion and endosomal maturation which released the viral genome into TLR9-expressing cellular compartments were necessary for the induction of innate responses (Abe *et al.*, 2005). As mentioned earlier, membrane fusion and endosomal escape via gp64 are essential for nucleocapsid transport into the nucleus. Hence it appears that transgene expression may be coincident with induction of the TLR9/MyD88-signaling pathway.

Taken together, these findings suggest that baculoviruses may induce various immune responses *in vitro* and *in vivo* as for other viruses (e.g., adenovirus), thus raising questions as to whether this will compromise the use of baculovirus vectors for *in vivo* human gene therapy. Which cytokines are induced by baculoviruses, and how cytokines modulate cellular and humoral immunities *in vivo* are not completely understood. The question of whether baculovirus-mediated *ex vivo* gene therapy elicits the immune response is also of interest. All of these questions need to be answered with more in-depth investigations to ensure the safe application of baculoviral gene therapy vectors.

The immune response induced by baculoviruses makes it a promising candidate as a novel vaccine vehicle against infectious diseases (see chapter by van Oers, this volume, pp. 193–253). The ability of baculoviruses to induce immune responses was first exploited by Aoki *et al.* (1999), who found that a recombinant baculovirus-expressing glycoprotein gB of pseudorabies virus induced antibodies against gB protein in mice, suggesting that this recombinant baculovirus could serve as a vaccine candidate for pseudorabies. The feasibility of using baculoviruses as vaccine carriers was also demonstrated by Abe *et al.* (2003), who found that intranasal inoculation with a recombinant baculovirus expressing hemagglutinin (HA) of the influenza virus under the control of the CAG promoter elicited the innate immune response and provided mice with a high level of protection from a lethal challenge of influenza virus. The level of protection is dependent on the route of administration, and intranasal administration is considerably superior to intramuscular administration although the latter induces significantly higher anti-HA IgG levels. More recently, Facciabene *et al.* (2004) demonstrated that intramuscular injection of a baculovirus expressing carcinoembryonic antigen (CEA) induced a measurable anti-CEA–specific $CD4^+$ T cell response. The immunogenic properties of baculoviruses are not restricted to CEA because intramuscular injection of another baculovirus (Bac-E2) expressing the E2 glycoprotein of HCV induced an anti-E2 $CD8^+$ T cell response as well as the innate immune response such as natural killer (NK) cell cytolytic activity (Facciabene *et al.*, 2004). Interestingly, when Bac-E2 is pseudotyped to display VSVG on the envelope, the minimal dose required to elicit a measurable T cell response was tenfold less, indicating that the VSVG-pseudotyped Bac-E2 was a more potent vaccine carrier than the unmodified virus. This finding agrees with the previous statement that baculoviruses displaying VSVG provide for more efficient immunogen expression in transduced cells.

Baculoviruses can also provoke an immune response against an antigen when it is displayed on the viral surface. For instance, immunization with adjuvant-free baculovirus displaying rodent malaria *Plasmodium berghei* circumsporozoite protein (PbCSP) on the envelope induced high levels of antibodies and IFN-γ–secreting cells against PbCSP, and protected 60% of mice against sporozoite challenge (Yoshida *et al.*, 2003). A more recent study further showed that baculovirus displaying severe acute respiratory syndrome-coronavirus (SARS-CoV) spike protein on the envelope induced the release of IL-8 in lung cells (Chang *et al.*, 2004). These studies substantiate the potential of baculoviruses displaying immunogens as vaccine candidates.

V. Conclusions and Prospects

The broad range of mammalian cells permissive to baculovirus transduction, the nontoxic and nonreplicative nature, large packaging capacity, and ease of production make baculoviruses promising tools for gene therapy. Despite these advantages, baculoviruses are inactivated by serum complement, which restricts the application of baculovirus vectors for *in vivo* gene therapy. Additionally, the duration of transgene expression is generally short, thus baculoviruses may not be suited for long-term gene therapy unless a hybrid vector capable of integrating the expression cassette into the host genome (e.g., baculovirus–AAV) is employed. Nonetheless, baculoviruses, in conjunction with other viral vectors (e.g., adenoviral or lentiviral), may be administered sequentially to escape either preexisting or therapy-induced antiviral immunity. Additionally, baculoviruses may serve as delivery vectors for triggering RNA interference. Baculoviruses carrying tumor-suppressor or suicide genes may also be used in combination with other treatments for cancer therapy (Song and Boyce, 2001; Stanbridge *et al.*, 2003). Given the highly efficient gene transfer to chondrocytes and MSCs, baculoviruses expressing appropriate growth factors may be used for *ex vivo* genetic modification of cells prior to transplantation into animals. The growth factors, acting in either an autocrine or a paracrine fashion, potentially accelerate tissue regeneration *in vivo*. Unlike the treatment of chronic disease, it is neither necessary nor desirable for transgene expression to persist beyond the few weeks or months needed to achieve healing (Huard *et al.*, 2003; Lieberman *et al.*, 2002). As a result, long-term transgene expression is not critical in tissue engineering. Hence, the combination of baculovirus-mediated gene therapy and tissue engineering may hold great promise. Of course, to address the safety issues of employing baculoviruses in gene therapy, the DNA integration, expression of baculovirus endogenous genes, and baculovirus-induced immune responses should be investigated.

Acknowledgments

The authors gratefully acknowledge the financial support from the National Health Research Institute (grant NHRI-EX94–9412EI) and Ministry of Economic Affairs (MOEA 94A0317P2), Taiwan.

References

Abe, T., Takahashi, H., Hamazaki, H., Miyano-Kurosaki, N., Matsuura, Y., and Takaku, H. (2003). Baculovirus induces an innate immune response and confers protection from lethal influenza virus infection in mice. *J. Immunol.* **171:**1133–1139.

Abe, T., Hemmi, H., Miyamoto, H., Moriishi, K., Tamura, S., Takaku, H., Akira, S., and Matsuura, Y. (2005). Involvement of the toll-like receptor 9 signaling pathway in the induction of innate immunity by baculovirus. *J. Virol.* **79:**2847–2858.

Airenne, K. J., Hiltunen, M. O., Turunen, M. P., Turunen, A. M., Laitinen, O. H., Kulomaa, M. S., and Yla-Herttuala, S. (2000). Baculovirus-mediated periadventitial gene transfer to rabbit carotid artery. *Gene Ther.* **7:**1499–1504.

Ames, R. S., Fornwald, J. A., Nuthulaganti, P., Trill, J. J., Foley, J. J., Buckley, P. T., Kost, T. A., Wu, Z. N., and Romanos, M. A. (2004). BacMam recombinant baculoviruses in G protein-coupled receptor drug discovery. *Receptors Channels* **10:**99–107.

Aoki, H., Sakoda, Y., Jukuroki, K., Takada, A., Kida, H., and Fukusho, A. (1999). Induction of antibodies in mice by a recombinant baculovirus expressing pseudorabies virus glycoprotein B in mammalian cells. *Vet. Microbiol.* **68:**197–207.

Barsoum, J. (1999). Concentration of recombinant baculovirus by cation-exchange chromatography. *Biotechniques* **26:**834–840.

Barsoum, J., Brown, R., McKee, M., and Boyce, F. M. (1997). Efficient transduction of mammalian cells by a recombinant baculovirus having the vesicular stomatitis virus G glycoprotein. *Hum. Gene. Ther.* **8:**2011–2018.

Beck, N. B., Sidhu, J. S., and Omiecinski, C. J. (2000). Baculovirus vectors repress phenobarbital-mediated gene induction and stimulate cytokine expression in primary cultures of rat hepatocytes. *Gene. Ther.* **7:**1274–1283.

Beljelarskaya, S. N. (2002). A baculovirus expression system for insect cells. *Mol. Biol.* **36:**281–292.

Bilello, J. P., Delaney, W. E., Boyce, F. M., and Isom, H. C. (2001). Baculovirus entry into nondividing hepatocytes is enhanced by transient disruption of intercellular junctions. *Hepatology* **34:**834.

Bilello, J. P., Cable, E. E., Myers, R. L., and Isom, H. C. (2003). Role of paracellular junction complexes in baculovirus-mediated gene transfer to nondividing rat hepatocytes. *Gene Ther.* **10:**733–749.

Blissard, G., and Wenz, J. R. (1992). Baculovirus gp64 envelope glycoprotein is sufficient to mediate pH-dependent membrane fusion. *J. Virol.* **66:**6829–6835.

Boublik, Y., Di Bonito, P., and Jones, I. M. (1995). Eukaryotic virus display: Engineering the major surface glycoprotein of the *Autographa californica* nuclear polyhedrosis virus (AcNPV) for the presentation of foreign proteins on the virus surface. *Biotechnology* **13:**1079–1084.

Boyce, F. M., and Bucher, N. L. R. (1996). Baculovirus-mediated gene transfer into mammalian cells. *Proc. Natl. Acad. Sci. USA* **93:**2348–2352.

Carbonell, L. F., Klowden, M. J., and Miller, L. K. (1985). Baculovirus-mediated expression of bacterial genes in dipteran and mammalian cells. *J. Virol.* **56:**153–160.

Chang, Y. J., Liu, C. Y. Y., Chiang, B. L., Chao, Y. C., and Chen, C. C. (2004). Induction of IL-8 release in lung cells via activator protein-1 by recombinant baculovirus displaying severe acute respiratory syndrome-coronavirus spike proteins: Identification of two functional regions. *J. Immunol.* **173:**7602–7614.

Check, E. (2002). Gene therapy: A tragic setback. *Nature* **420:**116–118.

Chen, Y.-H., Wu, J.-C., Wang, K.-C., Chiang, Y.-W., Lai, C.-W., Chung, Y.-C., and Hu, Y.-C. (2005). Baculovirus-mediated production of HDV-like particles in BHK cells using a novel oscillating bioreactor. *J. Biotechnol.* **118:**135–147.

Cheng, T., Xu, C.-Y., Wang, Y.-B., Chen, M., Wu, T., Zhang, J., and Xia, N.-S. (2004). A rapid and efficient method to express target genes in mammalian cell by baculovirus. *World J. Gastroenterol.* **10:**1612–1618.

Cheshenko, N., Krougliak, N., Eisensmith, R. C., and Krougliak, V. A. (2001). A novel system for the production of fully deleted adenovirus vectors that does not require helper adenovirus. *Gene Ther.* **8:**846–854.

Clay, W. C., Condreay, J. P., Moore, L. B., Weaver, S. L., Watson, M. A., Kost, T. A., and Lorenz, J. J. (2003). Recombinant baculoviruses used to study estrogen receptor function in human osteosarcoma cells. *Assay Drug Dev. Technol.* **1:**801–810.

Condreay, J. P., Witherspoon, S. M., Clay, W. C., and Kost, T. A. (1999). Transient and stable gene expression in mammalian cells transduced with a recombinant baculovirus vector. *Proc. Natl. Acad. Sci. USA* **96:**127–132.

Delaney, W. E., and Isom, H. C. (1998). Hepatitis B virus replication in human HepG2 cells mediated by hepatitis B virus recombinant baculovirus. *Hepatology* **28:** 1134–1146.

Detrait, E. R., Bowers, W. J., Halterman, M. W., Giuliano, R. E., Bennice, L., Federoff, H. J., and Richfield, E. K. (2002). Reporter gene transfer induces apoptosis in primary cortical neurons. *Mol. Ther.* **5:**723–730.

Duisit, G., Saleun, S., Douthe, S., Barsoum, J., Chadeuf, G., and Moullier, P. (1999). Baculovirus vector requires electrostatic interactions including heparan sulfate for efficient gene transfer in mammalian cells. *J. Gene Med.* **1:**93–102.

Dwarakanath, R. S., Clark, C. L., McElroy, A. K., and Spector, D. H. (2001). The use of recombinant baculoviruses for sustained expression of human cytomegalovirus immediate early proteins in fibroblasts. *Virology* **284:**297–307.

Ernst, W., Grabherr, R., Wegner, D., Borth, N., Grassauer, A., and Katinger, H. (1998). Baculovirus surface display: Construction and screening of a eukaryotic epitope library. *Nucleic Acids Res.* **26:**1718–1723.

Facciabene, A., Aurisicchio, L., and La Monica, N. (2004). Baculovirus vectors elicit antigen-specific immune responses in mice. *J. Virol.* **78:**8663–8672.

Gao, R., McCormick, C. J., Arthur, M. J. P., Ruddle, R., Oakley, F., Smart, D. E., Murphy, F. R., Harris, M. P. G., and Mann, D. A. (2002). High efficiency gene transfer into cultured primary rat and human hepatic stellate cells using baculovirus vectors. *Liver* **22:**15–22.

Grabherr, R., and Ernst, W. (2001). The Baculovirus expression system as a tool for generating diversity by viral surface display. *Comb. Chem. High Throughput Screen.* **4:**185–192.

Gronowski, A. M., Hilbert, D. M., Sheehan, K. C. F., Garotta, G., and Schreiber, R. D. (1999). Baculovirus stimulates antiviral effects in mammalian cells. *J. Virol.* **73:** 9944–9951.

Ho, Y.-C., Chen, H.-C., Wang, K.-C., and Hu, Y.-C. (2004). Highly efficient baculovirus-mediated gene transfer into rat chondrocytes. *Biotechnol. Bioeng.* **88:**643–651.

Ho, Y.-C., Chung, Y.-C., Hwang, S.-M., Wang, K.-C., and Hu, Y.-C. (2005). Transgene expression and differentiation of baculovirus-transduced human mesenchymal stem cells. *J. Gene Med.* **7:**860–868.

Ho, Y.-C., Lee, H.-P., Hwang, S.-M., Lo, W.-H., Chen, H.-C., Chung, C.-K., and Hu, Y.-C. (2006). Baculovirus transduction of human mesenchymal stem cell-derived progenitor cells: Variation of transgene expression with cellular differentiation states. *Gene Ther.* (in press).

Hoare, J., Waddington, S., Thomas, H. C., Coutelle, C., and McGarvey, M. J. (2005). Complement inhibition rescued mice allowing observation of transgene expression following intraportal delivery of baculovirus in mice. *J. Gene Med.* **7:**325–333.

Hofmann, C., and Strauss, M. (1998). Baculovirus-mediated gene transfer in the presence of human serum or blood facilitated by inhibition of the complement system. *Gene Ther.* **5:**531–536.

Hofmann, C., Sandig, V., Jennings, G., Rudolph, M., Schlag, P., and Strauss, M. (1995). Efficient gene-transfer into human hepatocytes by baculovirus vectors. *Proc. Natl. Acad. Sci. USA* **92:**10099–10103.

Hofmann, C., Lehnet, W., and Strauss, M. (1998). The baculovirus system for gene delivery into hepatocytes. *Gene Ther. Mol. Biol.* **1:**231–239.

Hsu, C.-S., Ho, Y.-C., Wang, K.-C., and Hu, Y.-C. (2004). Investigation of optimal transduction conditions for baculovirus-mediated gene delivery into mammalian cells. *Biotechnol. Bioeng.* **88:**42–51.

Hu, Y.-C., Tsai, C.-T., Chang, Y.-J., and Huang, J.-H. (2003a). Enhancement and prolongation of baculovirus-mediated expression in mammalian cells: Focuses on strategic infection and feeding. *Biotechnol. Prog.* **19:**373–379.

Hu, Y.-C., Tsai, C.-T., Chung, Y.-C., Lu, J.-T., and Hsu, J. T.-A. (2003b). Generation of chimeric baculovirus with histidine-tags displayed on the envelope and its purification using immobilized metal affinity chromatography. *Enzyme Microb. Technol.* **33:** 445–452.

Huard, J., Li, Y., Peng, H. R., and Fu, F. H. (2003). Gene therapy and tissue engineering for sports medicine. *J. Gene Med.* **5:**93–108.

Huser, A., Rudolph, M., and Hofmann, C. (2001). Incorporation of decay-accelerating factor into the baculovirus envelope generates complement-resistant gene transfer vectors. *Nat. Biotechnol.* **19:**451–455.

Jenkinson, S., McCoy, D., Kerner, S., Fox, T., Ferris, R., Lawrence, W., Condreay, P., Clay, W., and Smith, C. (2003). Development of a novel surrogate assay to measure HIV envelope/CCR5/CD4-mediated viral/cell fusion using BacMam baculovirus technology. *FASEB J.* **17:**A665–A665.

Katso, R. M., Parham, J. H., Caivano, M., Clay, W. C., Condreay, J. P., Gray, D. W., Lindley, K. M., Mason, S. J., Rieger, J., Wakes, N. C., Cairns, W. J., and Merrhiew, R. V.

(2005). Evaluation of cell-based assays for steroid nuclear receptors delivered by recombinant baculoviruses. *J. Biomol. Screen* **10:**715–724.
Kost, T. A., and Condreay, J. P. (2002). Recombinant baculoviruses as mammalian cell gene delivery vectors. *Trends Biotechnol.* **20:**173–180.
Kramer, O. H., Gottlicher, M., and Heinzel, T. (2001). Histone deacetylase as a therapeutic target. *Trends Endocrinol. Metab.* **12:**294–300.
Kronschnabl, M., Marschall, M., and Stamminger, T. (2002). Efficient and tightly regulated expression systems for the human cytomegalovirus major transactivator protein IE2p86 in permissive cells. *Virus Res.* **83:**89–102.
Kukkonen, S. P., Airenne, K. J., Marjomaki, V., Laitinen, O. H., Lehtolainen, P., Kankaanpaa, P., Mahonen, A. J., Raty, J. K., Nordlund, H. R., Oker-Blom, C., Kulomaa, M. S., and Yla-Herttuala, S. (2003). Baculovirus capsid display: A novel tool for transduction imaging. *Mol. Ther.* **8:**853–862.
Lang, C. Y., Song, J. H., and Chen, X. W. (2005). The GP64 protein of *Autographa californica* multiple nucleopolyhedrovirus rescues *Helicoverpa armigera* nucleopolyhedrovirus transduction in mammalian cells. *J. Gen. Virol.* **86:**1629–1635.
Lehtolainen, P., Tyynela, K., Kannasto, J., Airenne, K. J., and Yla-Herttuala, S. (2002). Baculoviruses exhibit restricted cell type specificity in rat brain: A comparison of baculovirus- and adenovirus-mediated intracerebral gene transfer *in vivo*. *Gene Ther.* **9:**1693–1699.
Leisy, D. J., Lewis, T. D., Leong, J. A. C., and Rohrmann, G. F. (2003). Transduction of cultured fish cells with recombinant baculoviruses. *J. Gen. Virol.* **84:**1173–1178.
Li, Y., Wang, X., Guo, H., and Wang, S. (2004). Axonal transport of recombinant baculovirus vectors. *Mol. Ther.* **10:**1121–1129.
Lieberman, J. R., Ghivizzani, S. C., and Evans, C. H. (2002). Gene transfer approaches to the healing of bone and cartilage. *Mol. Ther.* **6:**141–147.
Liu, H. S., Jan, M. S., Chou, C. K., Chen, P. H., and Ke, N. J. (1999). Is green fluorescent protein toxic to the living cells? *Biochem. Biophys. Res. Commun.* **260:**712–717.
Liu, Y. K., Chu, C. C., and Wu, T. Y. (2006). Baculovirus ETL promoter acts as a shuttle promoter between insect cells and mammalian cells. *Acta. Pharmacol. Sin.* **27:** 321–327.
Lopez, P., Jacob, R. J., and Roizman, B. (2002). Overexpression of promyelocytic leukemia protein precludes the dispersal of ND10 structures and has no effect on accumulation of infectious herpes simplex virus 1 or its proteins. *J. Virol.* **76:**9355–9367.
Luckow, V. A., and Summers, M. D. (1988). Trends in the development of baculovirus expression vectors. *Bio/Technology* **6:**47–55.
Lundstrom, K. (2003). Latest development in viral vectors for gene therapy. *Trends Biotechnol.* **21:**117–122.
Ma, L., Tamarina, N., Wang, Y., Kuznetsov, A., Patel, N., Kending, C., Hering, B. J., and Philipson, L. H. (2000). Baculovirus-mediated gene transfer into pancreatic islet cells. *Diabetes* **49:**1986–1991.
Marjan, J., Xie, Z. C., and Devine, D. V. (1994). Liposome-induced activation of the classical complement pathway does not require immunoglobulin. *BBA-Biomembranes* **1192:**35–44.
Marshall, E. (1999). Gene therapy death prompts review of adenovirus vector. *Science* **286:**2244–2245.
Matilainen, H., Rinne, J., Gilbert, L., Marjomaki, V., Reunanen, H., and Oker-Blom, C. (2005). Baculovirus entry into human hepatoma cells. *J. Virol.* **79:**15452–15459.
McCormick, C. J., Rowlands, D. J., and Harris, M. (2002). Efficient delivery and regulable expression of hepatitis C virus full-length and minigenome constructs in hepatocyte-derived cell lines using baculovirus vectors. *J. Gen. Virol.* **83:**383–394.

Merrihew, R. V., Clay, W. C., Condreay, J. P., Witherspoon, S. M., Dallas, W. S., and Kost, T. A. (2001). Chromosomal integration of transduced recombinant baculovirus DNA in mammalian cells. *J. Virol.* **75:**903–909.

Mottershead, D. G., Alfthan, K., Ojala, K., Takkinen, K., and Oker-Blom, C. (2000). Baculoviral display of functional scFv and synthetic IgG-binding domains. *Biochem. Biophys. Res. Commun.* **275:**84–90.

Nakamichi, K., Matsumoto, Y., Tohya, Y., and Otsuka, H. (2002). Induction of apoptosis in rabbit kidney cell under high-level expression of bovine herpesvirus 1 U(s)ORF8 product. *Intervirology* **45:**85–93.

Nicholson, L. J., Philippe, M., Paine, A. J., Mann, D. A., and Dolphin, C. T. (2005). RNA interference mediated in human primary cells via recombinant baculoviral vectors. *Mol. Ther.* **11:**638–644.

Ojala, K., Mottershead, D. G., Suokko, A., and Oker-Blom, C. (2001). Specific binding of baculoviruses displaying gp64 fusion proteins to mammalian cells. *Biochem. Biophys. Res. Commun.* **284:**777–784.

Ojala, K., Tikka, P. J., Kautto, L., Kapyla, P., Marjomaki, V., and Oker-Blom, C. (2004). Expression and trafficking of fluorescent viral membrane proteins in baculovirus-transduced BHK cells. *J. Biotechnol.* **114:**165–175.

Ong, S. T., Li, F., Du, J., Tan, Y. W., and Wang, S. (2005). Hybrid cytomegalovirus enhancer H-1 promoter-based plasmid and baculovirus vectors mediate effective RNA interference. *Hum. Gene. Ther.* **16:**1404–1412.

Palombo, F., Monciotti, A., Recchia, A., Cortese, R., Ciliberto, G., and La Monica, N. (1998). Site-specific integration in mammalian cells mediated by a new hybrid baculovirus-adeno-associated virus vector. *J. Virol.* **72:**5025–5034.

Park, S. W., Lee, H. K., Kim, T. G., Yoon, S. K., and Paik, S. Y. (2001). Hepatocyte-specific gene expression by baculovirus pseudotyped with vesicular stomatitis virus envelope glycoprotein. *Biochem. Biophys. Res. Commun.* **289:**444–450.

Patterson, R. M., Selkirk, J. K., and Merrick, B. A. (1995). Baculovirus and insect cell gene expression-review of baculovirus biotechnology. *Environ. Health Perspect.* **103:** 756–759.

Pfohl, J. L., Worley, J. F., Condreay, J. P., An, G., Apolito, C. J., Kost, T.A, and Truax, J. F. (2002). Titration of K-ATP channel expression in mammalian cells utilizing recombinant baculovirus transduction. *Receptors Channels* **8:**99–111.

Pieroni, L., Maione, D., and La Monica, N. (2001). *In vivo* gene transfer in mouse skeletal muscle mediated by baculovirus vectors. *Hum. Gene. Ther.* **12:**871–881.

Pijlman, G. P., van den Born, E., Martens, D. E., and Vlak, J. M. (2001). *Autographa californica* baculoviruses with large genomic deletions are rapidly generated in infected insect cells. *Virology* **283:**132–138.

Pijlman, G. P., de Vrij, J., van den End, F. J., Vlak, J. M., and Martens, D. E. (2004). Evaluation of baculovirus expression vectors with enhanced stability in continuous cascaded insect-cell bioreactors. *Biotechnol. Bioeng.* **87:**743–753.

Plank, C., Mechtler, K., Szoka, F. C., and Wagner, E. (1996). Activation of the complement system by synthetic DNA complexes: A potential barrier for intravenous gene delivery. *Hum. Gene. Ther.* **7:**1437–1446.

Poomputsa, K., Kittel, C., Egorov, A., Ernst, W., and Grabherr, R. (2003). Generation of recombinant influenza virus using baculovirus delivery vector. *J. Virol. Methods* **110:** 111–114.

Ramos, L., Kopec, L. A., Sweitzer, S. M., Fornwald, J. A., Zhao, H. Z., McAllister, P., McNulty, D. E., Trill, J. J., and Kane, J. F. (2002). Rapid expression of recombinant proteins in modified CHO cells using the baculovirus system. *Cytotechnology* **38:**37–41.

Raty, J. K., Airenne, K. J., Marttila, A. T., Marjomaki, V., Hytonen, V. P., Lehtolainen, P., Laitinen, O. H., Mahonen, A. J., Kulomaa, M. S., and Yla-Herttuala, S. (2004). Enhanced gene delivery by avidin-displaying baculovirus. *Mol. Ther.* **9:**282–291.

Salminen, M., Airenne, K. J., Rinnankoski, R., Reimari, J., Valilehto, O., Rinne, J., Suikkanen, S., Kukkonen, S., Yla-Herttuala, S., Kulomaa, M. S., and Vihinen-Ranta, M. (2005). Improvement in nuclear entry and transgene expression of baculoviruses by disintegration of microtubules in human hepatocytes. *J. Virol.* **79:** 2720–2728.

Sandig, V., Hofmann, C., Steinert, S., Jennings, G., Schlag, P., and Strauss, M. (1996). Gene transfer into hepatocytes and human liver tissue by baculovirus vectors. *Hum. Gene. Ther.* **7:**1937–1945.

Sarkis, C., Serguera, C., Petres, S., Buchet, D., Ridet, J. L., Edelman, L., and Mallet, J. (2000). Efficient transduction of neural cells *in vitro* and *in vivo* by a baculovirus-derived vector. *Proc. Natl. Acad. Sci. USA* **97:**14638–14643.

Shoji, I., Aizaki, H., Tani, H., Ishii, K., Chiba, T., Saito, I., Miyamura, T., and Matsuura, Y. (1997). Efficient gene transfer into various mammalian cells, including non-hepatic cells, by baculovirus vectors. *J. Gen. Virol.* **78:**2657–2664.

Smith, G. E., Summers, M. D., and Fraser, M. J. (1983). Production of human beta interferon in insect cells infected with a baculovirus expression vector. *Mol. Cell. Biol.* **3:**2156–2165.

Sollerbrant, K., Elmen, J., Wahlestedt, C., Acker, J., Leblois-Prehaud, H., Latta-Mahieu, M., Yeh, P., and Perricaudet, M. (2001). A novel method using baculovirus-mediated gene transfer for production of recombinant adeno-associated virus vectors. *J. Gen. Virol.* **82:**2051–2060.

Song, S. U., and Boyce, F. M. (2001). Combination treatment for osteosarcoma with baculoviral vector mediated gene therapy (p53) and chemotherapy (adriamycin). *Exp. Mol. Med.* **33:**46–53.

Song, S. U., Shin, S. H., Kim, S. K., Choi, G. S., Kim, W. C., Lee, M. H., Kim, S. J., Kim, I. H., Choi, M. S., Hong, Y. J., and Lee, K. H. (2003). Effective transduction of osteogenic sarcoma cells by a baculovirus vector. *J. Gen. Virol.* **84:**697–703.

Spenger, A., Ernst, W., Condreay, J. P., Kost, T. A., and Grabherr, R. (2004). Influence of promoter choice and trichostatin A treatment on expression of baculovirus delivered genes in mammalian cells. *Protein. Expr. Purif.* **38:**17–23.

Stanbridge, L. J., Dussupt, V., and Maitland, N. J. (2003). Baculoviruses as vectors for gene therapy against human prostate cancer. *J. Biomed. Biotechnol.* **2003:**79–91.

Takeuchi, Y., Porter, C. D., Strahan, K. M., Preece, A. F., Gustafsson, K., Cosset, F. L., Weiss, R. A., and Collins, M. K. L. (1996). Sensitization of cells and retroviruses to human serum by (alpha 1–3) galactosyltransferase. *Nature* **379:**85–88.

Tami, C., Farber, M., Palma, E. L., and Taboga, O. (2000). Presentation of antigenic sites from foot-and-mouth disease virus on the surface of baculovirus and in the membrane of infected cells. *Arch. Virol.* **145:**1815–1828.

Tani, H., Nishijima, M., Ushijima, H., Miyamura, T., and Matsuura, Y. (2001). Characterization of cell-surface determinants important for baculovirus infection. *Virology* **279:**343–353.

Tani, H., Limn, C. K., Yap, C. C., Onishi, M., Nozaki, M., Nishimune, Y., Okahashi, N., Kitagawa, Y., Watanabe, Y., Mochizuki, R., Moriishi, K., and Matsuura, Y. (2003). *In vitro* and *in vivo* gene delivery by recombinant baculoviruses. *J. Virol.* **77:** 9799–9808.

Thomas, C. E., Ehrhardt, A., and Kay, M. A. (2003). Progress and problems with the use of viral vectors for gene therapy. *Nature Rev. Genet.* **4:**346–358.

Tjia, S. T., Altenschildesche, G. M. Z., and Doerfler, W. (1983). *Autographa californica* nuclear polyhedrosis virus (AcNPV) DNA does not persist in mass cultures of mammalian cells. *Virology* **125:**107–117.

van Loo, N. D., Fortunati, E., Ehlert, E., Rabelink, M., Grosveld, F., and Scholte, B. J. (2001). Baculovirus infection of nondividing mammalian cells: Mechanisms of entry and nuclear transport of capsids. *J. Virol.* **75:**961–970.

Venkaiah, B., Viswanathan, P., Habib, S., and Hasnain, S. E. (2004). An additional copy of the homologous region (hr1) sequence in the *Autographa californica* multinucleocapsid polyhedrosis virus genome promotes hyperexpression of foreign genes. *Biochemistry* **43:**8143–8151.

Verma, I. M., and Somia, N. (1997). Gene therapy-promises, problems and prospects. *Nature* **389:**239–242.

Viswanathan, P., Venkaiah, B., Kumar, M. S., Rasheedi, S., Vrati, S., Bashyam, M. D., and Hasnain, S. E. (2003). The homologous region sequence (hr1) of *Autographa californica* multinucleocapsid polyhedrosis virus can enhance transcription from non-baculoviral promoters in mammalian cells. *J. Biol. Chem.* **278:**52564–52571.

Volkman, L. E., and Goldsmith, P. A. (1983). *In vitro* study of *Autographa californica* nuclear polyhedrosis virus interaction with nontarget vertebrate host cells. *Appl. Environ. Microbiol.* **45:**1085–1093.

Wagle, M., and Jesuthasan, S. (2003). Baculovirus-mediated gene expression in zebrafish. *Marine Biotechnol.* **5:**58–63.

Wang, C.-Y., and Wang, S. (2005). Adeno-associated virus inverted terminal repeats improve neuronal transgene expression mediated by baculoviral vectors in rat brain. *Hum. Gene. Ther.* **16:**1219–1226.

Wang, K.-C., Wu, J.-C., Chung, Y.-C., Ho, Y.-C., Chang, M. D., and Hu, Y.-C. (2005). Baculovirus as a highly efficient gene delivery vector for the expression of hepatitis delta virus antigens in mammalian cells. *Biotechnol. Bioeng.* **89:**464–473.

Wickham, T. J., Granados, R. R., Wood, H. A., Hammer, D. A., and Shuler, M. L. (1990). General analysis of receptor-mediated viral attachment to cell surfaces. *Biophys. J.* **58:**1501–1516.

Wu, T. Y., Liono, L., Chen, S. L., Chen, C. Y., and Chao, Y. C. (2000). Expression of highly controllable genes in insect cells using a modified tetracycline-regulated gene expression system. *J. Biotechnol.* **80:**75–83.

Yap, C. C., Ishii, K., Aizaki, H., Tani, H., Aoki, Y., Ueda, Y., Matsuura, Y., and Miyamura, T. (1998). Expression of target genes by coinfection with replication-deficient viral vectors. *J. Gen. Virol.* **79:**1879–1888.

Yoshida, S., Kondoh, D., Arai, E., Matsuoka, H., Seki, C., Tanaka, T., Okada, M., and Ishii, A. (2003). Baculovirus virions displaying *Plasmodium berghei* circumsporozoite protein protect mice against malaria sporozoite infection. *Virology* **316:**161–170.

Zhou, G. Y., Galvan, V., Campadelli-Fiume, G., and Roizman, B. (2000). Glycoprotein D or J delivered in trans blocks apoptosis in SK-N-SH cells induced by a herpes simplex virus 1 mutant lacking intact genes expressing both glycoproteins. *J. Virol.* **74:**11782–11791.

SECTION III

INSECT PEST MANAGEMENT

ADVANCES IN VIRUS RESEARCH, VOL 68

GENETICALLY MODIFIED BACULOVIRUSES: A HISTORICAL OVERVIEW AND FUTURE OUTLOOK

A. Bora Inceoglu, S. George Kamita, and Bruce D. Hammock

Department of Entomology and Cancer Research Center
University of California, Davis, California 95616

Abstract

The concept of using genetic engineering to improve the natural insecticidal activity of baculoviruses emerged during the 1980s. Both academic and industrial laboratories have since invested a great deal of effort to generate genetically modified (GM) or recombinant baculoviruses with dramatically improved speeds of kill. Optimal production methodologies and formulations have also been developed, and the safety and ecology of the recombinant baculoviruses have been thoroughly investigated. Unfortunately, the initial excitement that was generated by these technologies was tempered when industry made a critical decision to not complete the registration process of GM baculoviruses for pest insect control. In this chapter, we summarize the developments in the field from a historical perspective and provide our opinions as to the current status and future potential of the technology. We will argue that GM baculoviruses are valuable and viable tools for pest insect control both alone and in combination with wild-type viruses. We believe that these highly effective biopesticides still

0065-3527/06 $35.00
DOI: 10.1016/S0065-3527(06)68009-3

have a bright future in modern agriculture as public awareness and acceptance of GM organisms, including GM baculoviruses, increases.

I. Introduction

During the last century, the development of synthetic chemical insecticides and other advances have transformed agriculture from small, family-run operations to large, global-scale operations. With this dramatic increase in scale, damage by pest insects has surged, and this in turn has sometimes led to problems associated with the overuse (and in some cases unnecessary use) of pesticides. The primary problems associated with synthetic chemical pesticides include cost, detrimental effects on nontarget organisms, and the development of resistance. In terms of the natural control of insect populations, written records of pathogens that decimate insect populations have existed for centuries (Tanada and Kaya, 1993). More specifically, the application of insect pathogenic microorganisms has been an environmentally benign method of pest insect control. Baculoviruses, for example, have been used against insect pests of forests since the 1930s (Bird and Burk, 1961).

During the 1980s, Keeley and Hayes (1987), Maeda (1989), Menn and Borkovec (1989), Miller *et al.* (1983), and others developed the concept of genetically modifying the baculovirus to improve its endogenous insecticidal activity. Enthusiasm in both academia and industry quickly moved the concept from an "idea" to a fully developed product, a one of a kind bioinsecticide. Our laboratory has taken part in this process starting from the conceptual beginnings to the time when industry, at least in the United States, made the decision to abandon their efforts to register and implement this class of green insecticide. In this chapter, we summarize from a historical point of view recombinant baculovirus technology as it pertains to improving the endogenous insecticidal activity of the baculovirus. We also provide an analysis of what we believe are the most significant developments in the field and discuss how these developments might be implemented under the current status of the technology.

Several reviews have covered the use, development, and ecology of natural and genetically modified (GM) baculoviruses as biopesticides (Black *et al.*, 1997; Bonning and Hammock, 1996; Bonning *et al.*, 2002; Copping and Menn, 2000; Cory and Myers, 2003; Hammock *et al.*, 1993; Harrison and Bonning, 2000a; Inceoglu *et al.*, 2001; Kamita *et al.*, 2005a; McCutchen and Hammock, 1994; Miller, 1995; Wood, 1996).

II. Biology of Baculoviruses

Insect pathogenic viruses are classified into 12 viral families of which *Baculoviridae* is the most intensely studied (Blissard *et al.*, 2000; Tanada and Kaya, 1993). Baculoviruses are the most ubiquitous of the more than 20 known groups of insect pathogenic viruses. The baculovirus nucleocapsid is rod shaped and enveloped, and contains a single large, covalently closed, double-stranded DNA genome. Baculoviruses are classified into two genera, nucleopolyhedrovirus (NPV) and granulovirus (GV). The NPVs are further segregated into groups I and II based on the phylogenetic relationships of 20 distinguishing genes (Herniou *et al.*, 2001). The baculovirus produces two types of progeny, the budded virus (BV) and the occluded virus (OV) (Granados and Federici, 1986; Miller, 1997). The OVs of the NPV and GV are termed polyhedron (plural = polyhedra) and granule, respectively. Each granule occludes a single virion, whereas the polyhedron occludes multiple virions. Additionally, the NPV virion can contain a single (S morphotype) or multiple (M morphotype) nucleocapids. BVs are produced during an early stage of infection as the nucleocapsid buds through the plasma membrane. BVs are responsible for the systemic or cell-to-cell spread of the virus within an infected insect. OVs are produced during a late stage of infection when the progeny nucleocapsids are directed to the nucleus (or maintained in the cytoplasm in the case of GVs), obtain an envelope, and are subsequently occluded. OVs are responsible for the horizontal or larva-to-larva transmission of the virus. Although relatively stable against environmental factors, the OVs are sensitive to the alkaline insect gut fluid that contains enzymes that break down the crystalline protein matrix, resulting in release of the occlusion-derived virions (ODVs). The extraordinary characteristic of producing two types of progeny (BVs and OVs) makes the baculovirus adept at swiftly infecting and taking over insect cells and then remaining dormant in the environment for extended periods of time following release from the dead host. Fortunately, not many mammalian viruses are as successful and prevailing as baculoviruses in infecting insects.

Baculoviruses are naturally found on leaves and in the soil. For example, a typical portion of cole slaw composed of 100 cm^2 of cabbage from an epizootic plot may contain around 1.12×10^8 OVs (Heimpel *et al.*, 1973). The NPV replication cycle begins when a susceptible host ingests a polyhedron or polyhedra resulting in the release of hundreds of ODVs in the gut. The released ODVs then pass through the peritrophic matrix and enter midgut cells. Following direct fusion to the midgut cell, the nucleocapsids are released, uncoat, and either initiate viral replication in the cell or pass directly through the cell (Keddie *et al.*, 1989) and infect other cells such as tracheal epithelium cells, or enter the hemocoel (Bonning, 2005). About 5–7 days following the ingestion of NPVs (generally 7 to greater than 14 days in the case of GV infection), the infected host continues to feed and finally succumbs to the virus and dies. Just prior to death, the infected caterpillar exhibits enhanced locomotory activity that is activated by light (Kamita *et al.*, 2005b), a behavior that putatively enhances the dispersal of the virus (Cory and Myers, 2004; Goulson, 1997). Prior to death, the infected caterpillar also appears swollen due to the immense quantities of progeny baculoviruses that are produced. Genetic modification of baculoviruses has been performed almost exclusively using NPVs. This is because NPVs show faster speeds of kill in comparison to GVs, and continuous cell lines that support high-level production of progeny are available only for the NPVs. Detailed information on the biology of baculoviruses has been reviewed elsewhere (Bonning, 2005).

III. Baculoviruses as Insecticides

The single most important task in agricultural pest control is the ability to sustain pest insect population levels below the economic injury threshold in a cost-effective manner so that the costs of pest control operations justify the income generated. With this criterion, few biocontrol agents have so far been deemed successful. Baculoviruses have several inherent advantages as biological pesticides. They are naturally occurring pathogens that are highly specific to insects and closely related arthropods. They are safe in terms of pathogenicity against vertebrates (e.g., mammals, birds, fish, amphibians, reptiles). Moreover, they are benign in terms of pathogenicity against beneficial organisms that naturally suppress pest insect populations. Baculoviruses clearly play an important role in the natural control of insect populations. Despite a great diversity of viruses infecting insects, currently registered products are exclusively from the *Baculoviridae* family. Due to their inherent insecticidal activities, natural baculoviruses (both NPVs and GVs) have been registered, and successfully used as safe and effective biopesticides for the protection of field and orchard crops and forests in the Americas, Europe, and Asia (Black *et al.*, 1997; Copping and Menn, 2000; Hunter-Fujita *et al.*, 1998; Lacey *et al.*, 2001; Moscardi, 1999; Vail *et al.*, 1999). In the early 1970s, several natural baculovirus-based pesticides were available from commercial (Elcar, Spod-X, Cyd-X, and so on) and governmental (Gypcheck, TM BioControl-1, and Neocheck-S) sources (Black *et al.*,

1997). Elcar and Spod-X were based on NPVs that are pathogenic against the heliothines and *Spodoptera* spp., respectively, whereas Cyd-X was based on a GV that is pathogenic against the devastating codling moth. Gypcheck, TM BioControl-1, and Neocheck-S were produced by the US Forest Service and based on NPVs pathogenic against *Lymantria dispar, Orgyia pseudotsugata*, and *Neodiprion sertifer*, respectively, all excellent insecticides for forest ecosystems. Natural baculovirus-based biopesticides have been especially effective for the protection of soybean and forests in South and North America, respectively. In Brazil, an NPV that is pathogenic against the velvet bean caterpillar *Anticarsia gemmatalis* (AgMNPV), the major pest of soybeans, is used for the protection of over a million hectares of this crop (Moscardi, 1999). Although successfully used for soybean and forest protection, natural baculoviruses are imperfect insecticides when judged from an agroindustrial perspective. Many crops can tolerate only minimal fruit or foliar damage. With a natural speed of kill of 5 to greater than 14 days, natural baculoviruses are no match to synthetic pyrethroids, which kill within hours of exposure.

The slow speed of kill has been addressed by modifying the baculovirus using recombinant DNA technology. Baculoviruses are also susceptible to degradation by UV light and have short field stability potentially necessitating frequent applications. In the field, several baculovirus species may also need to be coapplied to control multiple pests because of the narrow host specificity of baculoviruses. Additionally, there are potential problems with high costs associated with production and limited shelf life. All of these potential problems have been addressed, or can easily be addressed, by current technologies. The safety of baculoviruses against nontargeted organisms has been demonstrated on multiple occasions (Cory and Hails, 1997). However, there is a clear need to gain wider public acceptance of GM baculovirus biopesticides. Appropriate governmental registration is also required prior to implementation. Clearly, implementation of GM baculovirus biopesticides will require an organized team effort that can address and solve multiple problems at the administrative, laboratory, and field levels.

IV. Integration of Ideas, Recombinant Baculoviruses for Pest Control

The baculovirus offers several unique advantages as a vector for the expression of a foreign gene within insect cells and insect larvae. This potential was initially discovered during the 1980s with exciting research in the laboratories of Max Summers (Summers and Smith, 1987) and Lois Miller (1988) (see chapter by Summers, this volume, pp. 1–73). In order to express high levels of protein, these researchers took advantage of the promoter of the polyhedrin gene (*polh*) to drive expression of the foreign gene. The two groups also utilized the product of *polh*, polyhedrin (the major protein found in polyhedra), as a visual selection marker to identify recombinant baculoviruses and used cultured insect cells to isolate the recombinant baculoviruses. Both the Summers and Miller laboratories used the *Autographa californica* multiple nucleopolyhedrovirus (AcMNPV) (Vail *et al.*, 1973, 1999) as the parental baculovirus for their baculovirus expression vector systems (BEVS). AcMNPV is the baculovirus type species and basic knowledge about the biology of AcMNPV was highly instrumental in the development of BEVS. The methods for the construction and use of recombinant baculoviruses for the expression of heterologous genes have been thoroughly described and are identical to those that were later used to generate recombinant baculovirus insecticides (Merrington *et al.*, 1999; O'Reilly *et al.*, 1992; Richardson, 1995; Summers and Smith, 1987).

Carbonell *et al.* (1988) were the first to attempt to improve the insecticidal activity of a baculovirus by expressing biologically active scorpion toxin, insectotoxin-1, of *Buthus eupeus*. Three recombinant AcMNPV constructs were generated that expressed the *BeIt* gene under control of the *polh* promoter. One of the constructs (vBeIt-1) carried only the *BeIt* coding sequence, whereas the other two carried fusions of *BeIt* and a human signal peptide (vBeIt-2) or a sequence coding for the 58 N-terminal amino acid residues of AcMNPV polyhedrin (vBeIt-3). All three constructs produced high levels of *BeIt*-specific transcripts, but only the vBeIt-3 construct produced significant amounts of peptide. Unfortunately, biological activity (BeIt-specific activity) was not detected in insect bioassays using larvae of *Trichoplusia ni, Galleria mellonella*, and *Sarcophaga* with any of the constructs.

Playing critical roles in the excretion and retention of water in insects, diuretic and antidiuretic hormones regulate insect responses to changes in their environment (Coast *et al.*, 2002; Gade, 2004; Holman *et al.*, 1990). Maeda (1989) was the first to integrate the contemporary knowledge and to successfully generate a recombinant baculovirus expressing a diuretic hormone gene that disrupted the normal physiology of larvae of the silkworm *Bombyx mori*. Maeda generated a synthetic gene encoding a 41-amino acid neuropeptide hormone of the tobacco hornworm *Manduca sexta* that was designed on the basis of the codon usage of *polh* of *Bombyx mori* nucleopolyhedrovirus (BmNPV). A signal sequence for secretion from a cuticle protein (CPII) of *Drosophila melanogaster* (Meigan) (Snyder *et al.*, 1982) was also included in this gene construct. The *polh*-negative strain of the NPV from *B. mori* was used to ensure biological containment since resulting virus is unstable, of very poor oral activity, and is not known to infect wild hosts. By bioassays based on injection of BV into fifth instar larvae (as opposed to oral infection of earlier instars with polyhedra), Maeda (1989) showed that the recombinant virus (BmDH5) caused mortality about 1 day faster than the wild-type BmNPV. Although a roughly 20% improvement in speed of kill was obtained, biologically active DH was not detected in the hemolymph. This improvement in the speed of kill of BmDH5 is modest in comparison to more recent recombinant baculovirus constructs (see later), however, these studies established the groundwork for subsequent efforts. With the proof of concept by Maeda and his colleagues, the field accelerated by implementation of multiple approaches to kill insects faster using baculoviruses.

From 1989 to 1991, genes encoding juvenile hormone esterase (JHE) (Hammock *et al.*, 1990a), Bt endotoxins (Martens *et al.*, 1990; Merryweather *et al.*, 1990), and eclosion hormone (EH) (Eldridge *et al.*, 1991) were successfully expressed in recombinant baculoviruses. Of these recombinant viruses, only those expressing JHE showed an improvement in speed of kill. Our laboratory targeted JHE expression for improving the insecticidal activity of the baculovirus on the basis of two key points. First, we had established that the regulation of the titer of JH (a critical hormone for the regulation of insect development and behavior) was dependent on a JH-specific esterase and/or epoxide hydrolase (Hammock, 1985). At the time, our laboratory had more than 30 years worth of experience studying these enzymes. Second, the *jhe* gene had been cloned from the tobacco budworm *Heliothis virescens* (Hanzlik *et al.*, 1989) and was available in our laboratory. Thus, it was only natural to engineer a recombinant, polyhedrin-positive AcMNPV that would secrete JHE into the insect hemolymph in order to disrupt the normal physiology of the insect. Using an authentic, insect-derived protein to combat the insect is conceptually elegant and potentially safer. Conceptually, this reduction in JH titer should halt insect feeding.

As anticipated, our first generation recombinant baculovirus carrying the *jhe* gene expressed biologically active JHE. Larvae of *M. sexta* and *H. virescens* infected with this virus showed reduced feeding and weight gain and subsequently died slightly more quickly in comparison to control larvae infected with the wild-type AcMNPV (Eldridge *et al.*, 1992a; Hammock *et al.*, 1990a,b). Although this was a promising beginning, we later understood that JHE is rapidly cleared from the hemolymph by pericardial cell uptake (Booth *et al.*, 1992; Ichinose *et al.*, 1992a,b). It is now clear that this removal process occurs by a receptor-mediated, endocytotic, saturable mechanism that does not involve passive filtration (Bonning *et al.*, 1997a; Ichinose *et al.*, 1992a,b). Once the JHE is taken up by the pericardial cells, it is presumed to be directed to and degraded in lysosomes (Booth *et al.*, 1992). The unusually short half-life (measured in minutes) of JHE in the hemolymph is obviously a limiting factor in the insecticidal efficacy of JHE expressing recombinant baculoviruses. Several laboratories including ours are continuing to improve the *in vivo* stability of the overexpressed JHE. Although considerable effort went into expressing the JHE of the major target species, *H. virescens*, this may have been a mistake in retrospect. Some foreign proteins are quite stable when injected into caterpillars. Thus, a *trans*-specific JHE may be more active than the natural JHE of the target species.

The bacterium *Bacillus thuringiensis* (Bt) produces two major types of lepidopteran-active toxins (Aronson and Shai, 2001; Bravo *et al.*, 2005; Gill *et al.*, 1992; Schnepf *et al.*, 1998). During the early period of the development of recombinant baculoviruses for pest insect control, genes encoding the Bt protoxin were placed under a very late gene promoter and expressed by recombinant AcMNPVs (Martens *et al.*, 1990; Merryweather *et al.*, 1990). Although these recombinant AcMNPVs expressed high levels of the Bt protoxin that was subsequently cleaved in insect cells into the biologically active form, these recombinant AcMNPVs did not show improved insecticidal activity. Similar results were found in later studies that used AcMNPV (Martens *et al.*, 1995; Ribeiro and Crook, 1993, 1998) or *Hyphantria cunea* NPV (Woo *et al.*, 1998) to express biologically active Bt toxin. Considering that the site of action of the toxin is the midgut epithelial cell, these results may not be so unexpected. The use of alternative strategies in which the Bt toxin is expressed as a toxin-polyhedrin fusion that results in incorporation of the toxin into the polyhedron is discussed later. The Bt toxin also is an antifeedant. Thus, if the Bt toxin is expressed even at low levels in the polyhedron, then feeding could be reduced.

Eldridge *et al.* (1991, 1992b) hypothesized that the expression of EH at an inopportune time would induce the premature onset of eclosion and molting. Thus, a recombinant AcMNPV that expressed biologically active and secreted EH of *M. sexta* was generated. The recombinant AcMNPV, vEHEGTD, carried the *eh* gene at the ecdysteroid UDP-glucosyltransferase (*egt*) gene locus of AcMNPV. The serendipitous insertion of the *eh* gene into the *egt* locus would later prove to be uniquely advantageous as will be discussed later. Larvae of *Spodoptera frugiperda* that were injected with vEHEGTD showed median survival times (ST_{50}s) that were reduced by ~30% in comparison to control larvae injected with AcMNPV, although this improvement was most likely attributed to the deletion of the *egt* gene.

V. A New Era in Recombinant Baculoviruses, Insect-Selective Peptide Toxins

Insect-specific toxins expressed and delivered by baculoviruses defined a new era in the field of recombinant baculovirus insecticides, beginning with the first successful results obtained by the expression of a paralytic neurotoxin from the insect predatory straw itch mite *Pyemotes tritici* (Tomalski *et al.*, 1988, 1989). TxP-I induces rapid, muscle-contracting paralysis in larvae of the greater wax moth *G. mellonella* (Tomalski *et al.*, 1988, 1989). Although the mode of action of TxP-I is unknown, it is selectively toxic to lepidopteran larvae at an effective dose of about 50 ng per larvae but it is not toxic to mice at a dose of 50 mg/kg. A recombinant, occlusion-negative AcMNPV (vEV-Tox34) carrying the TxP-I–encoding gene *tox34* under a modified polyhedrin promoter P_{LSXIV} (Ooi *et al.*, 1989) was shown to paralyze or kill fifth instar larvae of *T. ni* by 2 days postinjection (Tomalski and Miller, 1991). Tomalski and Miller (1991, 1992) constructed other recombinant AcMNPVs that expressed the *tox34* gene under early (vETL-Tox34) or hybrid late/very late (vCappolh-Tox34) gene promoters. In bioassays, the ET_{50} of these recombinant AcMNPVs in neonates of *S. frugiperda* and *T. ni* was reduced by ~45% in comparison to control larvae infected with AcMNPV (Lu *et al.*, 1996; Tomalski and Miller, 1992). One surprise from these experiments was that the yield of polyhedra was reduced by ~40% in comparison to AcMNPV-infected control larvae. The authors promptly suggested that this reduction in yield may cripple the virus in terms of its ability to compete effectively with the wild-type virus in the environment (Tomalski and Miller, 1992).

By the beginning of the 1990s, the field of insect toxinology was sparsely populated. However, scientists involved with isolation of toxins from venomous animals made a major discovery, the insect-selective toxins. At the time there were no particular applications for these astonishing peptides. *Androctonus australis* insect toxin (AaIT) was the first scorpion peptide toxin that was successfully expressed by baculoviruses (Maeda *et al.*, 1991; McCutchen *et al.*, 1991; Stewart *et al.*, 1991). AaIT was originally isolated from the desert scorpion *Androctonus australis* by Eliahu Zlotkin's group (Zlotkin *et al.*, 1971). Zlotkin *et al.* (2000) have extensively reviewed this superb toxin. The AaIT peptide is a 70 amino acid, highly folded peptide with four disulfide bridges. The primary advantage of AaIT is its specificity for the insect voltage-gated sodium channel (and conversely lack of specificity against the mammal sodium channel) (Zlotkin *et al.*, 2000). The second advantage of AaIT is its potency. The potency of AaIT is at least 20-fold better than TxP-I, resulting in an effective concentration of several nanograms per insect larva. AaIT captured the scientific headlines for years as the best model peptide neurotoxin for improving the insecticidal activity of the baculovirus, although it is far more toxic to dipteran than to lepidopteran larvae.

Maeda *et al.* (1991) constructed a recombinant BmNPV carrying a synthetic *aait* gene (Darbon *et al.*, 1982) that was linked to a bombyxin signal sequence for secretion and placed under the *polh* gene promoter. Again, the *B. mori* host and virus were used for biological containment in early studies. This recombinant virus, BmAaIT, expressed biologically active AaIT that was detected in the hemolymph of BmAaIT-infected silkworm larvae. The BmAaIT-infected larvae displayed symptoms that were consistent with larvae that are injected with authentic, purified AaIT. These symptoms included body tremors and dorsal arching that are consistent with the blockage of the voltage-gated sodium channels. The BmAaIT-infected larvae ceased feeding and were paralyzed beginning at roughly 40 hours postinfection (h p.i.). By 60 h p.i., death was observed. This timing corresponds to a 40% improvement in speed of kill in comparison to control larvae infected with BmNPV.

The study by Maeda *et al.* (1991) opened the doors for the development of AaIT-expressing baculoviruses that target pest insects and two groups independently published the expression of AaIT under the very late baculoviral *p10* promoter in recombinant AcMNPVs. The two constructs, AcST-3 by Stewart *et al.* (1991) and AcAaIT by McCutchen *et al.* (1991) showed very similar efficacy with AcAaIT showing a slightly lower LD_{50} and being slightly faster in speed of kill (Fig. 1). However, such small changes in improved speed of kill could simply be due to experimental design. The defining parameter, ST_{50}, of AcAaIT-infected larvae was reduced by about 30% in comparison to control larvae infected with AcMNPV. AcAaIT is more efficient when administered by droplet feeding to the neonate larvae of *H. virescens*, with an improvement of about 45% in ST_{50} (Inceoglu and Hammock, unpublished data). By bioassay using third instar larvae of *M. sexta* (an unnatural host of AcMNPV), McCutchen *et al.* (1991) also observed that larvae infected with AcAaIT typically were paralyzed and stopped feeding many hours prior to death. The cessation of feeding, and consequent reduction in feeding damage, is important because it directly translates into increased pesticidal efficacy. The paralytic effect of AcAaIT was further

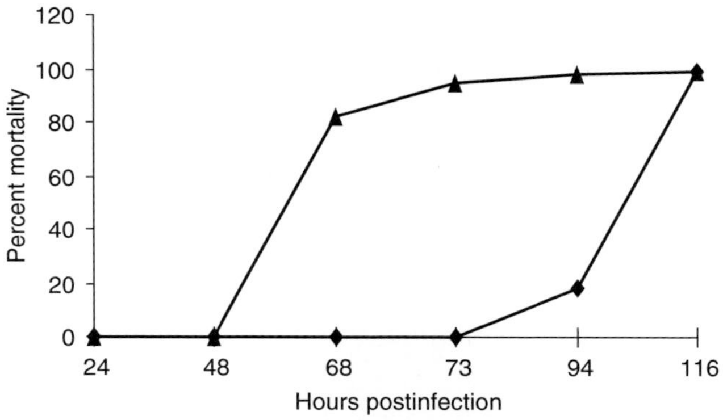

FIG 1. Time–mortality curves of wild-type AcMNPV (◆) or recombinant AaIT-expressing AcMNPV, AcAaIT (▲). The speed of kill of AcAaIT is about 40% faster than the wild-type counterpart in neonate *H. virescens*.

characterized by Hoover *et al.* (1995) using third instar *H. virescens*. They found that AcAaIT-infected larvae fall from the plant 5–11 h before death, much earlier than larvae infected with wild-type AcMNPV. Since this "knockoff" effect occurred before the induction of feeding cessation, the amount of leaf area consumed by the AcAaIT-infected larvae was 60–70% less than that consumed by AcMNPV- or mock-infected larvae. Thus, one of the key conclusions of Hoover *et al.* was that the survival time should not be the sole quantitative measure to assess the efficiency of the recombinant viruses. The increased efficiency of AaIT-expressing baculoviruses due to knockoff effects was further observed in field trials by Cory *et al.* (1994) and Sun *et al.* (2004). Another implication of the knockoff effect as pointed out by Cory *et al.* (1994) and Hoover *et al.* (1995) is that larvae falling off the plants early would lead to reduced foliage contamination because the wild-type virus-infected larvae tend to remain and die on the plant. This is yet another competitive disadvantage of the recombinant virus ensuring GM virus titer will be quickly reduced in the field.

VI. ERA OF MULTILATERAL DEVELOPMENT

By the mid 1990s, the recombinant baculovirus field had attracted still more attention and as would be expected expanded in multiple directions. Figure 2 depicts the improvements in speed of kill of

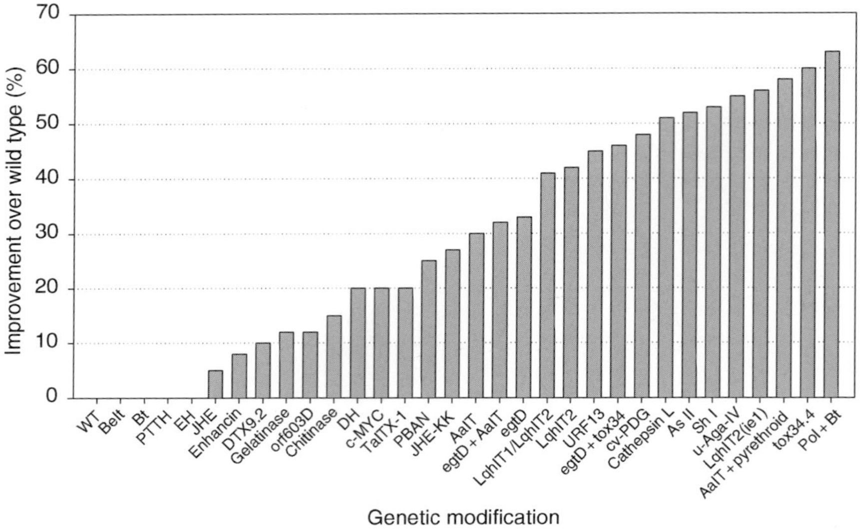

FIG 2. The speed of kill of the wild-type baculovirus can be dramatically improved by genetic modification. The genetic modification (insertion of a foreign gene or deletion of an endogenous gene) and percent improvement in speed of kill (or paralysis) relative to the wild-type virus or control virus is given. Because of differences in the parent virus, promoter, secretion signal, host strain and age, virus dose, and inoculation methods that were used, comparison between the different virus constructs is not possible. Abbreviations and reference(s): WT, wild type; BeIt, insectotoxin-1 (Carbonell *et al.*, 1988); Bt, *Bacillus thuringiensis* endotoxin (Merryweather *et al.*, 1990); PTTH, prothoracicotropic hormone (O'Reilly *et al.*, 1995); EH, eclosion hormone (Eldridge *et al.*, 1992b); JHE, juvenile hormone esterase (Hammock *et al.*, 1990a); enhancin, MacoNPV enhancin (Li *et al.*, 2003); DTX9.2, spider toxin (Hughes *et al.*, 1997); gelatinase, human gelatinase A (Harrison and Bonning, 2001); orf603D, deletion of AcMNPV *orf603* (Popham *et al.*, 1998); chitinase (Gopalakrishnan *et al.*, 1995); DH, diuretic hormone (Maeda, 1989); c-MYC, transcription factor (Lee *et al.*, 1997); TalTX-1, spider toxin (Hughes *et al.*, 1997); PBAN, pheromone biosynthesis-activating neuropeptide (Ma *et al.*, 1998); JHE-KK, stabilized JHE (Bonning *et al.*, 1997b); AaIT, scorpion *Androctonus australis* insect toxin (McCutchen *et al.*, 1991; Stewart *et al.*, 1991); egtD + AaIT, insertion of *aait* at the *egt* gene locus (Chen *et al.*, 2000); egtD, deletion of ecdysteroid UDP-glucosyltransferase (*egt*) gene (O'Reilly and Miller, 1991); LqhIT1/LqhIT2, simultaneous expression LqhIT1 and LqhIT2 (Regev *et al.*, 2003); LqhIT2, scorpion *Leiurus quinquestriatus* insect toxin 2 (Froy *et al.*, 2000); URF13, maize pore-forming protein (Korth and Levings, 1993); egtD + tox34, insertion of *tox34* under the early *DA26* promoter at the *egt* gene locus (Popham *et al.*, 1997); cv-PDG, glycosylase (Petrik *et al.*, 2003); cathepsin L (Harrison and Bonning, 2001); As II, sea anemone toxin (Prikhodko *et al.*, 1996); Sh I, sea anemone toxin (Prikhodko *et al.*, 1996); μ-Aga-IV, spider toxin (Prikhodko *et al.*, 1996); LqhIT2 (ie1), LqhIT2 under the early *ie1* promoter (Harrison and Bonning, 2000b); AaIT + pyrethroid, coapplication of low doses of AcAaIT and pyrethroid (McCutchen *et al.*, 1997); tox34.4, mite toxin (Burden *et al.*, 2000; Tomalski *et al.*, 1988); Pol + Bt, double expression of authentic polyhedrin and polyhedrin–Bt–GFP fusion proteins (Chang *et al.*, 2003) (from Kamita *et al.*, 2005a).

recombinant baculoviruses. One major direction was the identification of new insect-selective toxins with greater lepidopteran potency that could enhance speed of kill. An extension of this research direction was the development of methodologies to improve the level and timing of toxin expression, and as such crucial and significant advances in the understanding of baculovirus promoters were realized. Although numerous studies have uniformly concluded that insect-selective toxins are not dangerous, the use of genes encoding insect regulatory genes was viewed as a safer alternative to the expression of insect-selective toxins. Thus, another line of academic effort was directed toward the expression of regulatory insect hormones and enzymes with improved *in vivo* stability. This approach later expanded to include protease, chitinase, gelatinase, and other enzymes. Another simple but effective line of research was directed at deletion of endogenous baculovirus genes that kept the host insect feeding. The basis for this approach is that the baculovirus encodes genes that help the larvae to generate more mass (i.e., keep the larva feeding) such that there are more resources available for generating viral progeny. Thus, by removing these genes, the larva should prematurely stop feeding.

A. *Improved Toxins*

The rapid expansion of the recombinant baculovirus biopesticide field was driven by the early and significant success of the use of insect-selective toxins to improve the speed of kill. The process starts with mining the venom of scorpions, spiders, wasps, and other venomous animals to identify and characterize insect-selective peptides. As a natural and abundant source, arthropod-derived peptide toxins are still the first choice for this purpose. The corresponding genes of these peptide toxins are then transferred to baculovirus genomes under the control of a variety of promoters ranging from early/weak to very late/very strong expression characteristics.

The venoms of arthropods, such as scorpions, spiders, and parasitic wasps, are composed of a mixture of salts, small molecules, proteins, and peptides that are used to rapidly immobilize prey (Gordon *et al.*, 1998; Loret and Hammock, 1993; Possani *et al.*, 1999; Zlotkin, 1991; Zlotkin *et al.*, 1978, 1985). Over the years, most characterized toxins have been shown to act on major ion channels such as the Na^+, K^+, Ca^{2+}, and Cl^- channels. These channels are convenient and efficient targets for peptides because their blockage generally results in immediate paralysis. Initially, insect-selective toxins were separated into two classes based on the symptoms they produced when injected into fly larvae.

Excitatory toxins, such as AaIT from the North African scorpion *A. australis*, cause paralysis that is immediate and contractive (Zlotkin, 1991; Zlotkin *et al.*, 1971, 1985). On the other hand, depressant toxins, such as LqhIT2 from the yellow Israeli scorpion *Leiurus quinquestriatus hebraeus*, cause transient (e.g., until 5 min postinjection) contractive paralysis, followed by sustained, flaccid paralysis (Zlotkin, 1991; Zlotkin *et al.*, 1985). This classification upheld after the amino acid sequences of these peptides became available.

The popular insect-selective toxin AaIT continued to be the subject of numerous studies once it was established as a potent model peptide toxin with the pioneering work of Maeda *et al.* (1991). The *aait* gene expressed under the control of various promoters has been inserted into several baculovirus vectors, including the NPVs of the mint looper *Rachiplusia ou* (RoMNPV) (Harrison and Bonning, 2000b), cotton bollworms *Helicoverpa zea* (HzNPV) (Treacy *et al.*, 2000), and *H. armigera* (HaSNPV) (Chen *et al.*, 2000; Sun *et al.*, 2002, 2004). Insertion of *aait* into the baculovirus genome resulted in moderate to dramatic improvement in speed of kill. For example, the expression of *aait* under the late *p6.9* promoter of AcMNPV by a recombinant RoMNPV resulted in 34%, 37%, and 19% improvements in speed of kill in comparison to control larvae infected with RoMNPV when tested on neonates of *O. nubilalis, H. zea*, and *H. virescens*, respectively (Harrison and Bonning, 2000b).

The yellow Israeli scorpions *L. quinquestriatus hebraeus* and *L. quinquestriatus quinquestriatus* have also been popular sources of highly potent insecticidal toxins. The venoms of these scorpions contain both excitatory (e.g., LqqIT1, LqhIT1, and LqhIT5) and depressant (e.g., LqhIT2 and LqqIT2) insect-selective toxins (Kopeyan *et al.*, 1990; Moskowitz *et al.*, 1998; Zlotkin, 1991; Zlotkin *et al.*, 1985, 1993). Gershburg *et al.* (1998) have generated recombinant AcMNPVs expressing the excitatory LqhIT1 toxin under the very late *p10* and early *p35* gene promoters and the depressant LqhIT2 toxin under the *polh* gene promoter. These recombinant AcMNPVs show improvements in the speed of kill of up to 32% in comparison to AcMNPV. In similar experiments by Harrison and Bonning (2000b), the expression of LqhIT2 fused to a bombyxin signal sequence under the late *p6.9* or very late *p10* gene promoters resulted in ~34% decrease in median survival times compared to control larvae in neonate *H. virescens* larvae. Expression under these two promoters results in equal efficiency in terms of survival times. Similarly, when LqhIT2 gene constructs were expressed in recombinant RoMNPVs under the control of the *p6.9* or *p10* promoters of AcMNPV, these recombinants also showed similar improvements (~40%) in speed of kill in comparison to the wild-type virus when larvae of the European corn borer *O. nubilalis* Hübner and *H. zea* were used for bioassay (Harrison and Bonning, 2000b).

The expression of the insect-selective spider toxins μ-Aga-IV from *Agelenopsis aperta* (Prikhodko *et al.*, 1996), and DTX9.2 and TalTX-1 from the spiders *Diguetia canities* and *Tegenaria agrestis* (Hughes *et al.*, 1997) all result in improved speeds of kill. Likewise, two insect-selective toxins As II and Sh I from the sea anemones *Anemonia sulcata* and *Stichadactyla helianthus* resulted in 38% and 36% improvements in speed of kill in neonate *T. ni* and *S. frugiperda* larvae. Korth and Levings (1993) used their available toxin, URF13, from maize to improve the speed of kill of the baculovirus. URF13 is a mitochondrially encoded protein from maize that forms pores in the inner mitochondrial membrane (Korth *et al.*, 1991). Two recombinant occlusion-negative AcMNPVs expressing authentic or mutated URF13-encoding genes under the *polh* promoter were generated. When larvae of *T. ni* were injected with either of these viruses, all died by 60 h postinjection, however, this ~45% improvement in speed of kill apparently was not linked to the ability of the URF13 to form pores. The mechanism of this improved speed of kill appeared to involve interference of normal cellular functions (Korth and Levings, 1993).

1. Synergy Between Toxins and with Pyrethroids

The molecular target of a broad range of neurotoxins is the voltage-gated sodium channel. The sodium channels of insects and mammals are composed of at least six distinct receptor sites. The insect sodium channel has at least two additional receptors sites that are the molecular targets of insect-selective excitatory and depressant scorpion toxins (Cestele and Catterall, 2000). Several studies (Cestele and Catterall, 2000; Gordon *et al.*, 1992; Zlotkin *et al.*, 1995) have shown that depressant scorpion toxins bind to two noninteracting-binding sites (one showing high affinity and the other low affinity) on the insect sodium channel. The excitatory toxins bind only to the high-affinity receptor site (Gordon *et al.*, 1992). Herrmann *et al.* (1995) were the first to show that when excitatory and depressant toxins are simultaneously coinjected into larvae of the blowfly *Sarcophaga falculata* or *H. virescens*, the amount of toxin required to give the same paralytic response is reduced 5- to 10-fold in comparison to the amount required when only one of the toxins is injected. On the basis of this synergism, they suggested that the speed of kill of recombinant baculoviruses could be further increased by coinfecting with two or more recombinant baculoviruses each expressing toxin genes with synergistic properties or by simultaneously expressing two or more synergistic toxin genes. This hypothesis was tested by Regev *et al.* (2003) when they generated a recombinant AcMNPV (vAcLqIT1-IT2) that expressed both the excitatory LqhIT1 and depressant LqhIT2 toxins under the very late *p10* and *polh* promoters, respectively. Time-response bioassays (at an LC_{95} dose) using neonate *H. virescens* showed that the ET_{50} of vAcLqIT1-IT2 is reduced by roughly 20% in comparison to recombinant AcMNPV expressing each toxin alone or by 40% in comparison to the wild-type AcMNPV. Similar or decreased levels of synergism were found with a recombinant AcMNPV (vAcLqαIT-IT2) expressing both excitatory and depressant scorpion toxins in orally and hemocoelically infected larvae of *H. virescens, H. armigera*, and *S. littoralis*.

In addition to being the target of insect-selective scorpion toxins, the voltage-gated sodium channel is also the major target of the well known and commonly used pyrethroid class of insecticides. Although AaIT induces a neurological response similar to that evoked by the pyrethroid insecticides, the binding site on the channel of AaIT and the pyrethroids do not overlap. In fact, AaIT and other scorpion toxins that act on the voltage-gated sodium channel are synergized by pyrethroids *in vivo* allowing both recombinant baculoviruses expressing insect-selective scorpion toxins and pyrethroids to be used simultaneously (McCutchen *et al.*, 1997). Such combinations could be useful in the field. The heliothine complex is often the most resistant of the target pests to pyrethroids. If this complex can be controlled by a recombinant baculovirus, then much lower pyrethroid rates in a tank mixture can be used to control secondary pests. These lower levels of pyrethroid may be adequate to synergize the recombinant baculovirus.

2. *Improving Expression*

Following experiments that showed that the expression of the *tox34* gene from the predatory mite *P. tritici* improves insecticidal efficacy of AcMNPV, Miller's group attempted to improve TOX34 expression by placing the *tox34* gene under the late *p6.9* gene promoter. The recombinant AcMNPV (vp6.9tox34) that carried the *tox34* gene under the *p6.9* promoter expressed TxP-I at least 24 h earlier compared to expression of the *tox34* gene under the very late *p10* gene promoter. Higher toxin yield was also obtained by the earlier expression of TxP-I under the *p6.9* promoter. Lu *et al.* (1996) analyzed the occlusion-positive vp6.9tox34 virus at an LC_{95} dose and showed that the ET_{50} of this virus in neonate larvae of *S. frugiperda* and *T. ni* was reduced by nearly 60% in comparison to AcMNPV. This represents

20–30% faster paralysis in comparison to expression under the very late gene promoter. Furthermore, a variant of the TxP-I–encoding gene was recloned and expressed by Burden *et al.* (2000) under the *p10* promoter. The recombinant virus expressing this gene showed a similar reduction in the mean time to death and showed 85–95% lower yields of polyhedra per unit weight in comparison to control larvae infected with wild-type AcMNPV.

3. *Delivery of Bt Toxins*

As discussed earlier, high levels of biologically active Bt toxin are produced in insect cells by recombinant baculoviruses. However, recombinant baculoviruses expressing Bt toxin showed no improvement in virulence or decrease in the ST_{50} (Martens *et al.*, 1990; Merryweather *et al.*, 1990). This lack of efficacy most likely resulted from the site of action of Bt toxins on the surface of the midgut epithelial cell, whereas the recombinant Bt protoxin was present in the cytoplasm of cells within the insect body. Chang *et al.* (2003) have overcome the problems associated with efficiently delivering the Bt toxin genes using recombinant baculoviruses. They expressed the Bt toxin as the fusion product, polyhedrin–Cry1Ac–green fluorescent protein (GFP), in which the toxin is fused with both the polyhedrin and GFP proteins using trypsin-sensitive linkers. The recombinant AcMNPV (ColorBtrus) expressing this fusion product as well as authentic polyhedrin produced polyhedra that occlude Bt toxin and GFP, and released toxin and GFP proteins in the insect midgut. Bioassays using second or third instar larvae of the diamondback moth, *Plutella xylostella*, showed that the LD_{50} of ColorBtrus was reduced 100-fold and the ST_{50} was reduced by 60% in comparison to control larvae infected with AcMNPV (Chang *et al.*, 2003).

B. *Improved Enzymes*

1. *JHE Stability*

The speed of kill of AcMNPV was improved by about 20% by the expression of an insect-derived juvenile hormone-specific esterase (JHE) (Hammock *et al.*, 1990a). Following the initial observation of the short half-life of recombinant JHE that is injected into the hemolymph (Ichinose *et al.*, 1992b), we have improved the *in vivo* stability of JHE by mutating two lysine residues belonging to an amphipathic helix of the JHE of *H. virescens*. The mutant protein (JHE-KK) is more stable because of decreased lysosomal targeting resulting in reduced removal and/or degradation from the hemolymph (Bonning *et al.*, 1997b). Bioassays using first instar larvae of *H. virescens* or *T. ni* showed that insects infected with AcJHE-KK (AcMNPV expressing mutant JHE-KK under a strong very late viral promoter) died ~20% faster than control larvae infected with a recombinant AcMNPV expressing the authentic JHE (AcJHE) (Bonning *et al.*, 1999). However, in older instars the ST_{50} of AcJHE-KK–infected insects was only marginally reduced (Bonning *et al.*, 1999; Kunimi *et al.*, 1997).

2. Protease Expression

The baculovirus faces several barriers within the insect midgut. The various pathways by which infectious virions circumvent the midgut are illustrated by Bonning (2005). The final midgut-associated barrier to systemic infection is the basement membrane (BM) or basal lamina, a fibrous matrix composed primarily of glycoproteins, type IV collagen, and laminin that are secreted by the epithelial cells (Ryerse, 1998) with functions including structural support, filtration, and differentiation (Yurchenco and O'Rear, 1993). Harrison and Bonning (2001) have constructed recombinant AcMNPVs expressing three different proteases (rat stromelysin-1, human gelatinase A, and cathepsin L from the flesh fly *Sarcophaga peregrina*) that are known to digest BM proteins. Among these recombinant baculoviruses, the one expressing cathepsin L under the late baculovirus *p6.9* gene promoter generates a 51% faster speed of kill in comparison to AcMNPV in neonate larvae of *H. virescens*. So far, expression of BM-degrading proteases is one of the most impressive improvements in speed of kill of recombinant baculoviruses.

3. Other Enzymes

Petrik *et al.* (2003) have generated a recombinant AcMNPV, vHSA50L, that expresses an algal virus pyrimidine dimer-specific glycosylase, cv-PDG (Furuta *et al.*, 1997), that is involved in the first steps of the repair of UV-damaged DNA in an attempt to reduce UV inactivation of baculoviruses. Sunlight is known to be a major factor in the inactivation of baculoviruses in the field (Black *et al.*, 1997; Dougherty *et al.*, 1996; Ignoffo and Garcia, 1992; Ignoffo *et al.*, 1997). Although the polyhedra of vHSA50L showed no differences in UV inactivation in comparison to AcMNPV, the BV of vHSA50L were threefold more resistant. Bioassays showed that the LC_{50}s of vHSA50L and AcMNPV were significantly different in neonates of *S. frugiperda* (16-fold lower) but not *T. ni*. Consistent with the reduction in LD_{50}, LT_{50} of vHSA50L in neonates of *S. frugiperda* was reduced by ~40%.

C. Gene Deletion

Deleting an endogenous gene that results in improved speed of kill is a simple and elegant approach to improving insecticidal efficacy of the baculovirus. However, in practice, decreases in yields of the viral progeny may be a limitation of this approach. Ecdysteroids are key hormone molecules that regulate larval–pupal molting and other physiological events. Thus, the prevention of their action results in the interruption of growth or causes abnormal development and potentially death. A baculovirus-encoded enzyme called ecdysteroid UDP-glucosyltransferase (EGT) catalyzes the conjugation of sugar molecules to ecdysteroids, a process that renders the ecdysteroid inactive (O'Reilly, 1995; O'Reilly and Miller, 1989). The baculovirus gene (*egt*) that encodes EGT is found in ~90% of baculovirus genomes that have been characterized (Clarke *et al.*, 1996; Tumilasci *et al.*, 2003). Despite this wide presence, the *egt* gene is not essential for either *in vitro* or *in vivo* replication of AcMNPV (O'Reilly and Miller, 1989, 1991). Infection of larvae of *S. frugiperda* or *T. ni* with vEGTDEL, an *egt* deletion mutant of AcMNPV, gave rise to earlier mortality and reduced feeding damage (by about 40%) in comparison to AcMNPV infection (Eldridge *et al.*, 1992a; O'Reilly and Miller, 1991; Wilson *et al.*, 2000). A concurrent reduction in progeny virus yield was also found. In general, the deletion of *egt* gene homologs from the NPV genome resulted in none to moderate improvements in the speed of kill (e.g., Chen *et al.*, 2000; Pinedo *et al.*, 2003; Popham *et al.*, 1997, 1998; Slavicek *et al.*, 1999; Treacy *et al.*, 1997, 2000). This improvement in speed of kill was also dependent on the larval stage that was used for the bioassay (Bianchi *et al.*, 2000; Sun *et al.*, 2004). The *egt* minus virus was used in early field tests in Oxford, England, as an example of a virus that had a small improvement in efficacy but no novel gene added.

D. Choice of Parental Strain

There are several wild-type viruses that are infectious against a variety of pest species that can be used as the parental strain for genetic manipulations. It is possible to use a wild-type virus with strict host specificity to generate a highly host-selective insecticide, or conversely start with a baculovirus with a wider host range. This choice is also very likely to be based on the target pest. Furthermore, once a wild-type virus is selected, it is feasible to initiate or continuously implement a screening effort to isolate a natural mutant that is faster than the parent by conducting time-mortality bioassays. It should be noted though that a laboratory bioassay primarily focused on speed of kill might not reveal undesirable characteristics that may later hinder industrial manufacturing. However, early investment in such a screening effort may easily translate to advantages during the commercialization phase as exemplified by the discovery of a more virulent HzSNPV strain by DuPont scientists (Dr. L. Flexner, personal communication).

VII. Implementation of a Recombinant Baculovirus Insecticide

A. Field Trials

As improvements in the potency of baculoviruses were realized, the efficacy of these constructs was tested in the field. The earliest field trials of a GM baculovirus (occlusion-negative AcMNPVs carrying junk DNA or a *lac*Z marker gene) were performed in England during the mid to late 1980s before the faster killing recombinants became available (Black *et al.*, 1997; Levidow, 1995). In the United States, the first field trial (a 3-year study) of a GM baculovirus (a *polh* gene-deleted AcMNPV that was co-occluded with wild-type AcMNPV) had begun in 1989 (Wood *et al.*, 1994). The early field trials were designed to determine the environmental persistence of the virus. These trials revealed that the persistence of occlusion-negative constructs is exceptionally low. This, however, is less relevant from a practical point of view because occlusion-negative constructs are not likely to be applied in the field. In 1993, Cory *et al.* conducted the first field trial to test the efficacy of an occlusion-positive recombinant baculovirus that was expected to have a dramatically improved speed of kill. The AcMNPV construct that expressed AaIT, that is, AcST-3 was used for these studies (Cory *et al.*, 1994). The results of these field trials were in agreement with the laboratory experiments. Cabbage plants treated with AcST-3 and artificially inoculated with third instar larvae of *T. ni* showed 23–29% lower feeding damage in comparison to wild-type AcMNPV-treated control cabbage plants. However, the reduction in feeding damage (~50% reduction) was not as impressive as observed in the laboratory trials (Stewart *et al.*, 1991). On further investigation, Cory *et al.* (1994) determined that this resulted from a tenfold lower yield of AcST-3 progeny in comparison to AcMNPV. These studies also revealed a key biological observation, namely that the AcST-3–treated larvae were knocked off the host plant. The knockoff effect is perhaps more important than the speed of kill because larvae that are not on the plant cannot feed on the plant. Second, if the recombinant baculovirus is unable to propagate as efficiently as the wild-type virus, it will be outcompeted by the wild-type virus, resulting in an additional layer of safety. In China, multi-year-long field trials have been conducted to test the ability of an occlusion-positive *Helicoverpa armigera* NPV carrying the *aait* gene at the *egt* gene locus (HaSNPV-AaIT) to protect cotton (Sun *et al.*, 2002, 2004). These field trials showed that the yield of cotton in HaSNPV-AaIT treated plots was nearly 20% higher in comparison to wild-type HaSNPV-treated plots and similar to plots treated with chemical insecticides such as λ-cyhalothrin, endosulfan, and β-cypermethrin.

All of the field trials to date indicate that GM baculoviruses are safe and effective biological pesticides that can compete with chemical pesticides in terms of protection from pest insects and maintenance of crop yields (Fig. 3) (Kamita *et al.*, 2005a). These trials also show that

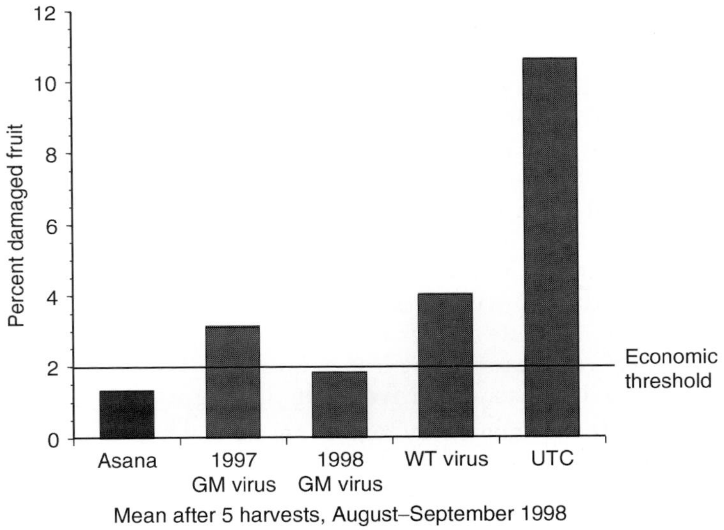

FIG 3. Tomato field trial evaluation from North Carolina (conducted by DuPont, United States). Untreated (UTC) or wild-type baculovirus-treated (WT virus) tomato plots received extensive feeding damage that was well above the economic injury threshold. The application of a synthetic pyrethroid (Asana, esfenvalerate) protected the tomato plants against insect damage. Whereas the first generation of GM baculovirus (1997 GM virus) did not profitably control against pest damage, further improvements (e.g., new parental strain, early promoter, transactivator technology, improved formulation) in the second generation (1998 GM Virus) rendered these GM baculoviruses competitive with synthetic pyrethroid treatment.

GM baculoviruses will quickly disappear from the environment and will have very little to no adverse effects against beneficial insects.

B. Production

A detailed discussion of the numerous issues regarding the commercialization of GM baculovirus insecticides including marketing, *in vivo* and *in vitro* production, formulation, storage, and public acceptance can be found elsewhere (Black *et al.*, 1997). Here, we will briefly discuss a key component of a successful biological control program: the production system. Currently, two choices are available for the production of GM baculovirus insecticides: *in vivo* production using field or insectary-reared insects and *in vitro* production through fermentation of insect cells. Production will most likely become a major factor contributing to the cost of a GM baculovirus insecticide. Therefore, the choice of *in vivo* versus *in vitro* manufacturing is undoubtedly important. *In vitro* systems are sterile, easily scalable, and afford higher predictability in product yield in addition to having the flexibility of allowing one to regulate gene expression. The end product is relatively pure and the process is inexpensive at high volumes. The maintenance of sterility, however, is a disadvantage with microbial contamination being an obvious pitfall for cell culture. The *in vitro* system requires higher initial capital and is expensive at low volumes. Appropriate cell lines that can easily be cultured in large-scale bioreactors may not be available for some key insect viruses. *In vivo* methods on the other hand afford much lower initial investment and have significantly lower operational costs. This is well exemplified by the production of baculoviruses in Brazil where the process occurs in the field. In this case sterility is less of an issue. The system, although labor intensive, involves a relatively simple operation and it is a well-proven production method for several viruses. The disadvantages of *in vivo* production include the necessity to reliably maintain and rear sufficient quantities of host insects, the occurrence of disease in the colony, limitations in scalability, impurity of the final product, and the smaller scale of the operation. An *in vivo* system may also suffer from the lack of flexibility in the regulation of gene expression. Nonetheless, as technologies advance, solutions to the limitations of both systems are likely to emerge (Fig. 4).

Data from field trials conducted by DuPont scientists clearly show that the expression of an insect-selective toxin reduces yield of virus and may have other effects on the final product (Dr. L. Flexner, personal communication). In order to circumvent these effects, DuPont scientists

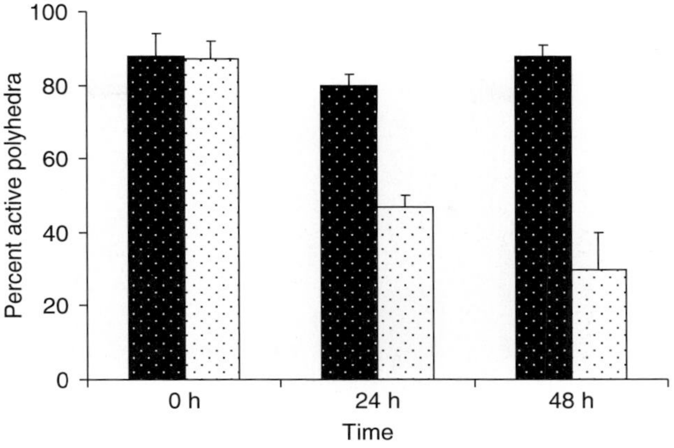

FIG 4. Improving field stability through formulation research. Polyhedra of HzSNPV formulated with titanium dioxide coaccervation and spray dried remain viable significantly longer (white bar) than the unformulated product (black bar) (figure provided by Dr. L. J. Flexner, DuPont).

have exploited the advantages of a two-phase system by incorporating an "on–off" switch based on a tetracycline transactivator gene placed into the genome of the host cells (McCutchen, US Patent No. 6322781). Briefly, the transgenic host line (insects or cultured cells) produces a protein that is able to suppress expression of the toxin gene under the control of a hybrid promoter in the recombinant baculovirus genome that regulates toxin expression. In the presence of tetracycline (i.e., during the production phase), the recombinant toxin gene is silent and infection results in normal yields of progeny virus. However, the toxin gene becomes active as soon as tetracycline is withdrawn (i.e., within the pest insect host) and the toxin gene is expressed. McCutchen disclosed that the LT_{50} of recombinant baculoviruses expressing the LqhIT2 toxin under *ie-1* promoter was increased by 45–60% under the control of transactivator repression, essentially resulting in the production of normal numbers (i.e., equivalent to the wild-type virus) of polyhedra by the recombinant virus (Fig. 5).

C. Technology Stacking

The expectation of an ideal insect control agent is that it should cost effectively and specifically kill the pest insect within several hours of application. Recombinant baculovirus technology is thoroughly

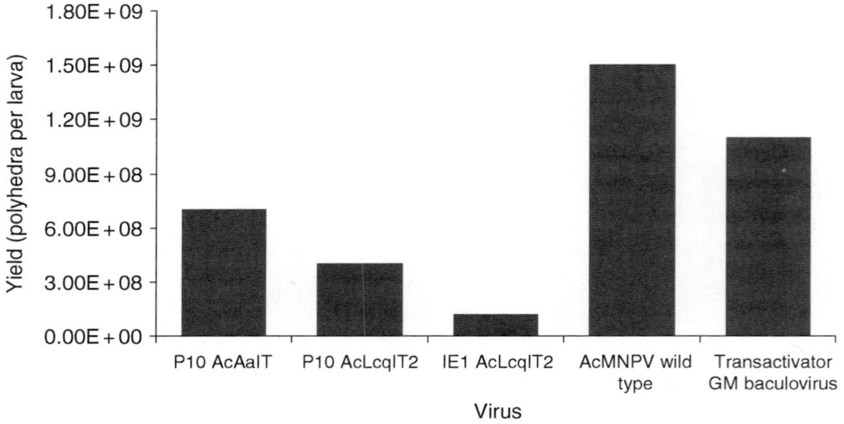

FIG 5. The yield of insect-selective toxin expressing baculovirus in the absence or presence of a tetracycline-controlled transactivator system. The speed of kill of the recombinant baculovirus is inversely correlated with the yield of polyhedra; the wild-type virus produces the highest number of polyhedra per larvae. By turning off toxin expression during the production phase, the yield of polyhedra dramatically improves (figure provided by Dr. L. J. Flexner).

validated and the speed of kill is approaching that of synthetic insecticides in the latest field trials. Improving the technology continuously should lead to faster and more efficient recombinant baculoviruses. As reviewed previously and here, there have been many advances to generate recombinant baculoviruses that kill insects considerably faster than the wild-type parent baculovirus. Therefore, the technology to generate a single baculovirus incorporating numerous exciting developments already exists, although this has not yet been attempted. In this section, we will summarize a “dream” baculovirus, one that theoretically includes several modifications so as to make it competitive against synthetic insecticides.

The dream recombinant baculovirus undoubtedly requires an efficient backbone, the baculovirus genome. This genome is required to contain at least all of the essential genes for replication but may also be supplemented with genes that increase the resistance of the virus to environmental effects such as sunlight. A smaller genome may theoretically result in a faster rate of replication in the host insect and may reduce the fitness of the recombinant, which are both desirable qualities. The deletion of the *egt* gene alone, for example, has been shown to make a faster killing recombinant baculovirus. As an alternative, one can even exclude some essential genes from the genome of the ultimate baculovirus and compensate these by providing the products of these genes in the manufacturing phase. An immediate example would be the production of a *polh*-deficient virus under the control of the tetracycline switch so that the recombinant baculovirus is packed within polyhedra when being produced but cannot form OVs in the field. This concept of supplementing genes can further be extended to other genes including the recombinant toxin genes where expression of the toxic protein can be switched off in the production phase as exemplified by DuPont technology. An obvious industrial advantage is that this *polh*-negative virus would have low environmental persistence and would need to be reapplied, making it more attractive to the manufacturers. It will be important then to assess the ability to manufacture a crippled virus profitably because these approaches commit industry to a more difficult multistage production process. Undoubtedly, many baculovirus genes are not needed for viral growth and host kill. Potentially large sections of the baculovirus genome can be removed for use of a baculovirus as a green insecticide, although this procedure would be deleterious for a biological control agent.

The potential competitive advantage of selecting a good parental virus is reviewed above in Section VI.D. Simply, a virus with a tenfold lower LD_{50} means about ten times more area can be sprayed with the same quantity of baculovirus. Once a desired genome is selected, it is then a matter of incorporating or mixing and matching the available (and of course yet undiscovered) improvements into this genome. Among the modifications that increase infectivity are enhancins and BM-degrading proteases. Although the expression of enhancins merely increases the speed of kill of baculoviruses, they significantly increase the lethality. Combining an enhancin gene with a cathepsin L from the flesh fly *Sarcophaga peregrina* that resulted in a 51% reduction in speed of kill then seems appropriate. The effect of this increased speed of kill was due in part to the ability of this recombinant to colonize the insect host more rapidly due to damaged BM.

Incorporation of a Bt toxin product into the baculovirus polyhedra has proven to be a very attractive strategy for both improving virulence and the speed of kill of the baculovirus. Even though it is unclear if this strategy can be combined with the expression of other factors, such as peptide toxins, it would be worthwhile to consider stacking this modification, at least in this theoretical section. Clearly, a recombinant virus with a very low LD_{50} is advantageous and incorporation of Bt toxins is one of the best-known ways to attain this effect. An obvious advantage of incorporating the enhancin and cathepsin genes is that their expression is not expected to impact the expression of lethal factor genes. The currently used scorpion-derived toxins largely are misfolded in insect cells. A systematic study of folding of these toxins could improve expression over tenfold. Similarly, approaches to enhance the stability of the toxin mRNA and translation is certain to improve efficacy. Therefore, in addition to modifications mentioned above, the expression of a pair of synergistically acting insect-selective scorpion toxins could be considered for our dream baculovirus. Although many other excellent insect-selective toxins have been identified, expressed in and tested as recombinant baculovirus insecticides, positive cooperativity has only been shown with a limited number of peptides. Furthermore, the cooperativity among the toxins is dependant on the insect species and thus needs to be fine-tuned. The synergy between toxins belonging to the excitatory and the depressant peptide family could be exploited for this purpose. Another important consideration is the choice of promoter for the expression of all foreign genes in recombinant baculovirus insecticides. The very late and strong *polh* or *p10* promoters have traditionally been used for most studies. However, the benefits of using weaker and earlier promoters have lately become apparent. More potent toxins or toxin combinations allow the evaluation of a wider range of promoter systems. The last group of modifications that can be incorporated into our baculovirus is auxiliary features, such as UV resistance enhancement, related to field stability and formulation. The expression of algal virus pyrimidine dimer-specific glycosylase, cv-PDG (Furuta *et al.*, 1997), involved in the repair of UV-damaged DNA improved both virulence and speed of kill. Therefore, we will recommend this modification as the last component of the GM baculovirus.

To summarize our suggestions, a genome with extensive deletions to remove all nonessential genes, expressing enhancin and cathepsin proteins combined with a pair of synergistically acting peptide toxins under the control of an early weak promoter, a Bt–polyhedrin fusion protein and expressing the algal virus pyrimidine dimer-specific glycosylase may make a better recombinant baculovirus.

VIII. Concluding Thoughts

Natural baculoviruses show poor speed of kill and limited effective range in comparison to synthetic chemical pesticides such as the pyrethroids, thus, one might think that these biological pesticides may not provide sufficient protection for crops. However, there are cases, such as forest ecosystems, in which the use of a natural baculovirus may seem less efficient in the short term but is more effective, cost efficient, and less destructive for the ecosystem over the long term. In these situations, the natural baculovirus should be the biopesticide of choice. Analyzing the pest problems in a region and implementing a customized approach that includes baculoviruses, however, is not an easy task. The best example of how this was effectively performed is in Brazilian soybean agroecosystems (Moscardi, 1999). The decision to use or to not use a biological (or chemical) pesticide should be based on the level of tolerance that a particular crop has to pest damage. In general, as the economic injury threshold of a particular crop decreases so does the likelihood that biopesticides will be used for its protection. This is a function not only of crop physiology but also of the value of the crop. As an example of physiology, the baculovirus-based pest management system is more likely to be ineffective in tropical areas because of the more rapid speed of development of the pest.

During the last 15 plus years, a large number of GM baculovirus constructs have been generated, which show variable improvements in speed of kill (Fig. 2). Undoubtedly, numerous other constructs have been generated in commercial and academic laboratories that are not found in the scientific literature. Some of the best constructs to date induce feeding cessation as early as 24 h p.i., a period of time that is sufficient to make these GM baculovirus pesticides as effective as synthetic chemical insecticides (Kamita *et al.*, 2005a). Our opinion is that the efficacy and safety of GM baculoviruses have been sufficiently proven to make GM baculovirus pesticides a viable tool in our crop protection toolbox. The benefits to society of GM baculoviruses far outnumber risks, especially in comparison to the potential risks posed by many synthetic chemical pesticides and GM crop plants that are routinely used now.

The development of GM baculoviruses for pest insect control is an element of the era of genetic engineering, an era in which the availability of potent, orally active insect-selective toxins from Bt bacteria has driven the field toward GM crops expressing Bt toxins (Aronson and Shai, 2001; Bravo *et al.*, 2005; Gill *et al.*, 1992; Schnepf *et al.*, 1998). Of course, this emphasis on one gene family may in the future prove to be problematic due to the occurrence of insect resistance. In fact, there are now alarming reports on the development of resistance against Bt-expressing plants (Gunning *et al.*, 2005). However, there is so far surprisingly little impact of this resistance considering the widespread use of the Bt gene. Whether resistance will become a major problem for Bt toxin-expressing GM plants remains to be seen.

We believe that the availability of alternative technologies including GM baculoviruses will be an essential part of successful pest management strategies. The generation of resistant insects occurs relatively quickly with synthetic chemical insecticides and has been observed with Bt toxins (Gunning *et al.*, 2005). Chemical insecticides and other agents, such as Bt toxins, are small molecules that often act on a single defined target site potentially making them more vulnerable to target site insensitivity or metabolic enzymes. In contrast, the baculovirus is a much larger agent that takes over the insect host. In both laboratory and field studies, in which insects were continuously exposed to baculoviruses, no significant and consistent level of insect resistance is found (Fuxa, 1993; Fuxa *et al.*, 1988). Although baculovirus-resistant populations may emerge in the long term, the time frame required for this should be immensely longer than that of a synthetic chemical agent.

As transfection systems improve, it becomes technically possible to put Bt toxin genes into a wider variety of crops. However, the economic and regulatory barriers to transgenic crops will limit GMOs to major crops for the foreseeable future. Thus, an advantage of the recombinant baculovirus is that it could be used on a variety of crops including high-value vegetables with much lower economic barriers. However, the greatest value of the recombinant viruses may be in developing countries. In developing countries, they can replace earlier generations of synthetic chemicals that have serious human health and environmental risks rather than compete with modern pesticides with superior green properties. There are also strategies to produce recombinant virus locally in developing countries that would avoid either importing expensive chemicals or reliance on major industrial infrastructure.

Several of the technologies that have emerged from GM baculovirus pesticide development have universal applicability in medical and general biological research. The field of toxinology, for example, has benefited greatly from the isolation and characterization of novel peptide toxins for use in improving the insecticidal efficacy of the baculovirus. The synergy between the fields of pest control and toxinology has not only caused an explosive increase in the number of novel insect-selective peptides isolated from a variety of venomous animals but also led to increased knowledge in toxin mechanism of action, selectivity, cooperativity, and target ion channel diversity in insects and mammals over the last two decades (Fig. 6). Today, most publications involving the isolation and characterization of insect-selective toxins generally mention the potential for use of these peptide toxins to further improve recombinant baculovirus insecticides.

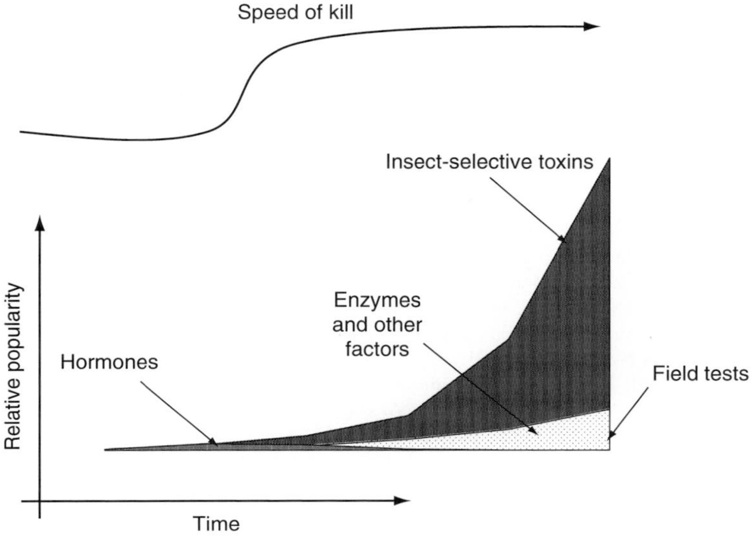

FIG 6. Trends in GM baculovirus research. Although the first generation of GM baculoviruses expressing insect hormone genes showed significant improvements in speed of kill in comparison to the wild-type parent, the expression of genes encoding enzymes that disrupt insect physiology and subsequently insect-selective toxins now dominate the field. The identification of new genes, improvements in expression technology, and advances in manufacturing and formulation have contributed dramatically to improving the speed of kill of GM baculoviruses.

The idea of expressing foreign genes by recombinant baculoviruses has undoubtedly progressed toward a superb expression system which has had and continues to have a wide impact on biological and medical research. Another outcome, also the subject of this chapter, was fine bioinsecticides. All of the studies to date indicate that GM recombinant baculoviruses can become an integral part of modern pest insect control strategies. Therefore, it is now a matter of when and who will take advantage of this technology.

ACKNOWLEDGMENTS

This work was funded in part by grants from the USDA (2003-35302-13499), NIEHS (P30 ES05707), and NIH (AI58267). This chapter is fondly dedicated to our colleagues and mentors Susumu Maeda (1950–1998) and Sean Duffey (1943–1997) who were lost while we were engaged in the development of recombinant viruses.

References

Aronson, A. I., and Shai, Y. (2001). Why *Bacillus thuringiensis* insecticidal toxins are so effective: Unique features of their mode of action. *FEMS Microbiol. Lett.* **195**(1):1–8.

Bianchi, F. J. J. A., Snoeijing, I., van der Werf, W., Mans, R. M. W., Smits, P. H., and Vlak, J. M. (2000). Biological activity of SeMNPV, AcMNPV, and three AcMNPV deletion mutants against *Spodoptera exigua* larvae (Lepidoptera: Noctuidae). *J. Invertebr. Pathol.* **75**(1):28–35.

Bird, F. T., and Burk, J. M. (1961). Artificial disseminated virus as a factor controlling the European spruce sawfly, *Diprion hercyniae* (Htg.) in the absence of introduced parasites. *Can. Entomologist* **93**:228–238.

Black, B. C., Brennan, L. A., Dierks, P. M., and Gard, I. E. (1997). Commercialization of baculoviral insecticides. *In* "The Baculoviruses" (L. K. Miller, ed.), pp. 341–387. Plenum Press, New York.

Blissard, G., Black, B., Crook, N., Keddie, B. A., Possee, R., Rohrmann, G., Theilmann, D., and Volkman, L. (2000). Family baculoviridae. *In* "Virus Taxonomy: Classification and Nomenclature of Viruses. Seventh Report of the International Committee for the Taxonomy of Viruses" (M. H. V. vanRegenmortel, C. M. Fauquet, D. H. L. Bishop, E. B. Carstens, M. K. Estes, S. M. Lemon, D. J. McGeoch, J. Maniloff, M. A. Mayo, C. R. Pringle, and R. B. Wickner, eds.), pp. 195–202. Academic Press, San Diego.

Bonning, B. (2005). Baculoviruses: Biology, biochemistry, and molecular biology. *In* "Comprehensive Molecular Insect Science" (L. I. Gilbert, K. Iatrou, and S. S. Gill, eds.), Vol. 6, pp. 233–270. 7 vols. Elsevier, Oxford.

Bonning, B. C., and Hammock, B. D. (1996). Development of recombinant baculoviruses for insect control. *Annu. Rev. Entomol.* **41**:191–210.

Bonning, B. C., Booth, T. F., and Hammock, B. D. (1997a). Mechanistic studies of the degradation of juvenile hormone esterase in *Manduca sexta*. *Arch. Insect Biochem. Physiol.* **34**:275–286.

Bonning, B. C., Ward, V. K., VanMeer, M. M. M., Booth, T. F., and Hammock, B. D. (1997b). Disruption of lysosomal targeting is associated with insecticidal potency of juvenile hormone esterase. *Proc. Natl. Acad. Sci. USA* **94**(12):6007–6012.

Bonning, B. C., Possee, R. D., and Hammock, B. D. (1999). Insecticidal efficacy of a recombinant baculovirus expressing JHE-KK, a modified juvenile hormone esterase. *J. Invertebr. Pathol.* **73**(2):234–236.

Bonning, B. C., Boughton, A. J., Jin, H., and Harrison, R. L. (2002). Genetic enhancement of baculovirus insecticides. *In* "Advances in Microbial Control of Insect Pests" (K. Upadhyay, ed.). Kluwer Academic/Plenum Publishers, New York.

Booth, T. F., Bonning, B. C., and Hammock, B. D. (1992). Localization of juvenile hormone esterase during development in normal and in recombinant baculovirus-infected larvae of the moth *Trichoplusia ni*. *Tissue Cell* **24**(2):267–282.

Bravo, A., Soberon, M., and Gill, S. S. (2005). *Bacillus thuringiensis*: Mechanisms and use. *In* "Comprehensive Molecular Insect Science" (L. I. Gilbert, K. Iatrou, and S. S. Gill, eds.), Vol. 6, pp. 175–205. 7 vols. Elsevier, Oxford.

Burden, J. P., Hails, R. S., Windass, J. D., Suner, M. M., and Cory, J. S. (2000). Infectivity, speed of kill, and productivity of a baculovirus expressing the itch mite toxin txp-1 in second and fourth instar larvae of *Trichoplusia ni*. *J. Invertebr. Pathol.* **75**(3): 226–236.

Carbonell, L. F., Hodge, M. R., Tomalski, M. D., and Miller, L. K. (1988). Synthesis of a gene coding for an insect specific scorpion neurotoxin and attempts to express it using baculovirus vectors. *Gene* **73**:409–418.

Cestele, S., and Catterall, W. A. (2000). Molecular mechanisms of neurotoxin action on voltage-gated sodium channels. *Biochimie* **82**(9–10):883–892.

Chang, J. H., Choi, J. Y., Jin, B. R., Roh, J. Y., Olszewski, J. A., Seo, S. J., O'Reilly, D. R., and Je, Y. H. (2003). An improved baculovirus insecticide producing occlusion bodies that contain *Bacillus thuringiensis* insect toxin. *J. Invertebr. Pathol.* **84**(1):30–37.

Chen, X. W., Sun, X. L., Hu, Z. H., Li, M., O'Reilly, D. R., Zuidema, D., and Vlak, J. M. (2000). Genetic engineering of *Helicoverpa armigera* single-nucleocapsid nucleopolyhedrovirus as an improved pesticide. *J. Invertebr. Pathol.* **76**(2):140–146.

Clarke, E. E., Tristem, M., Cory, J. S., and O'Reilly, D. R. (1996). Characterization of the ecdysteroid UDP-glucosyltransferase gene from *Mamestra brassicae* nucleopolyhedrovirus. *J. Gen. Virol.* **77**:2865–2871.

Coast, G. M., Orchard, I., Phillips, J. E., and Schooley, D. A. (2002). Insect diuretic and antidiuretic hormones. *Adv. Insect Physiol.* **29**:279–400.

Copping, L. G., and Menn, J. J. (2000). Biopesticides: A review of their action, applications and efficacy. *Pest Manag. Sci.* **56**:651–676.

Cory, J. S., and Hails, R. S. (1997). The ecology and biosafety of baculoviruses. *Curr. Opin. Biotech.* **8**:323–327.

Cory, J. S., and Myers, H. (2003). The ecology and evolution of insect baculovirus. *Annu. Rev. Ecol. Evol. Syst.* **34**:239–272.

Cory, J. S., and Myers, J. H. (2004). Adaptation in an insect host-plant pathogen interaction. *Ecol. Lett.* **7**(8):632–639.

Cory, J. S., Hirst, M. L., Williams, T., Halls, R. S., Goulson, D., Green, B. M., Carty, T. M., Possee, R. D., Cayley, P. J., and Bishop, D. H. L. (1994). Field trial of a genetically improved baculovirus insecticide. *Nature (London)* **370**:138–140.

Darbon, H., Zlotkin, E., Kopeyan, C., Vanrietschoten, J., and Rochat, H. (1982). Covalent structure of the insect toxin of the North African scorpion *Androctonus australis* Hector. *Int. J. Pep. Prot. Res.* **20**(4):320–330.

Dougherty, E. M., Gurthrie, K. P., and Shapiro, M. (1996). Optical brighteners provide baculvoirus activity enhancement and UV radiation protection. *Biol. Control* **7**:71–74.

Eldridge, R., Horodyski, F. M., Morton, D. B., O'Reilly, D. R., Truman, J. W., Riddiford, L. M., and Miller, L. K. (1991). Expression of an eclosion hormone gene in insect cells using baculovirus vectors. *Insect Biochem.* **21**:341–351.

Eldridge, R., O'Reilly, D. R., Hammock, B. D., and Miller, L. K. (1992a). Insecticidal properties of genetically engineered baculoviruses expressing an insect juvenile hormone esterase gene. *Appl. Environ. Microbiol.* **58**(5):1583–1591.

Eldridge, R., O'Reilly, D. R., and Miller, L. K. (1992b). Efficacy of a baculovirus pesticide expressing an eclosion hormone gene. *Biol. Control* **2**:104–110.

Froy, O., Zilberberg, N., Chejanovsky, N., Anglister, J., Loret, E., Shaanan, B., Gordon, D., and Gurevitz, M. (2000). Scorpion neurotxoins: Structure/function relationships and application in agriculture. *Pest Manag. Sci.* **56**:472–474.

Furuta, M., Schrader, J. O., Schrader, H. S., Kokjohn, T. A., Nyaga, S., McCullough, A. K., Lloyd, R. S., Burbank, D. E., Landstein, D., Lane, L., and VanEtten, J. L. (1997). Chlorella virus PBCV-1 encodes a homolog of the bacteriophage T4 UV damage repair gene *denV*. *Appl. Environ. Microbiol.* **63**(4):1551–1556.

Fuxa, J. R. (1993). Insect resistance to viruses. *In* "Parasites and Pathogens of Insects" (N. E. Beckage, S. N. Thompson, and B. A. Federici, eds.), Vol. 2, pp. 197–209. Academic Press, New York.

Fuxa, J. R., Mitchell, F. L., and Richter, A. R. (1988). Resistance of Spodoptera-frugiperda [Lep, Noctuidae] to a nuclear polyhedrosis-virus in the field and laboratory. *Entomophaga* **33**(1):55–63.

Gade, G. (2004). Regulation of intermediary metablolism and water balance of insects by neuropeptides. *Annu. Rev. Entomol.* **49:**93–113.

Gershburg, E., Stockholm, D., Froy, O., Rashi, S., Gurevitz, M., and Chejanovsky, N. (1998). Baculovirus-mediated expression of a scorpion depressant toxin improves the insecticidal efficacy achieved with excitatory toxins. *FEBS Lett.* **422**(2)**:**132–136.

Gill, S. S., Cowles, E. A., and Pietrantonio, P. V. (1992). The mode of action of *Bacillus thuringiensis* endotoxins. *Annu. Rev. Entomol.* **37:**615–636.

Gopalakrishnan, K., Muthukrishnan, S., and Kramer, K. J. (1995). Baculovirus-mediated expression of a *Manduca sexta* chitinase gene: Properties of the recombinant protein. *Insect Biochem. Mol. Biol.* **25:**255–265.

Gordon, D., Moskowitz, H., Eitan, M., Warner, C., Catterall, W. A., and Zlotkin, E. (1992). Localization of receptor sites for insect-selective toxins on sodium channels by site-directed antibodies. *Biochemistry* **31:**7622–7628.

Gordon, D., Savarin, P., Gurevitz, M., and Zinn-Justin, S. (1998). Functional anatomy of scorpion toxins affecting sodium channels. *J. Toxicol. Toxin Rev.* **17**(2)**:**131–159.

Goulson, D. (1997). *Wipfelkrankheit*: Modification of host behaviour during baculoviral infection. *Oecologia* **109:**219–228.

Granados, R. R., and Federici, B. A. (1986). "The Biology of Baculoviruses." I and II CRC Pres, Inc, Boca Raton.

Gunning, R. V., Dang, H. T., Kemp, F. C., Nicholson, I. C., and Moores, G. D. (2005). New resistance mechanism in *Helicoverpa armigera* threatens transgenic crops expressing *Bacillus thuringiensis* Cry1Ac toxin. *Appl. Environ. Microbiol.* **71:**2558–2563.

Hammock, B. D. (1985). Regulation of juvenile hormone titer: Degradation. *In* "Comprehensive Insect Physiology, Biochem., and Pharmacology" (G. A. Kerkut and L. I. Gilbert, eds.), Vol. 7, pp. 431–472. Pergamon Press, New York.

Hammock, B. D., Bonning, B. C., Possee, R. D., Hanzlik, T. N., and Maeda, S. (1990a). Expression and effects of the juvenile hormone esterase in a baculovirus vector. *Nature* **344**(6265)**:**458–461.

Hammock, B. D., Wrobleski, V., Harshman, L., Hanzlik, T., Maeda, S., Philpott, M., Bonning, B., and Possee, R. (1990b). Cloning, expression and biological activity of the juvenile hormone esterase from *Heliothis virescens*. *In* "Molecular Insect Science" (H. Hagedorn, J. Hildebrand, M. Kidwell, and J. Law, eds.), pp. 49–56. Plenum Press, New York.

Hammock, B. D., McCutchen, B. F., Beetham, J., Choudary, P. V., Fowler, E., Ichinose, R., Ward, V. K., Vickers, J. M., Bonning, B. C., Harshman, L. G., Grant, D., Uematsu, T., *et al.* (1993). Development of recombinant viral insecticides by expression of an insect-specific toxin and insect-specific enzyme in nuclear polyhedrosis viruses. *Arch. Insect Biochem. Physiol.* **22**(2–4)**:**315–344.

Hanzlik, T. N., Abdel-Aal, Y. A. I., Harshman, L. G., and Hammock, B. D. (1989). Isolation and sequencing of cDNA clones coding for juvenile hormone esterase from *Heliothis virescens*: Evidence for a catalytic mechanism of the serine carboxylesterases different from that of the serine proteases. *J. Biol. Chem.* **264**(21)**:**12419–12425.

Harrison, R. L., and Bonning, B. C. (2000a). Genetic engineering of biocontrol agents of insects. *In* "Biological and Biotechnological Control of Insect Pests" (J. E. Rechcigl and N. A. Rechcigl, eds.), pp. 243–280. Lewis Publishers, Boca Raton, Florida.

Harrison, R. L., and Bonning, B. C. (2000b). Use of scorpion neurotoxins to improve the insecticidal activity of *Rachiplusia ou multicapsid nucleopolyhedrovirus*. *Biol. Control* **17**(2)**:**191–201.

Harrison, R. L., and Bonning, B. C. (2001). Use of proteases to improve the insecticidal activity of baculoviruses. *Biol. Control* **20**(3)**:**199–209.

Heimpel, A. M., Thomas, E. D., Adams, J. R., and Smith, L. J. (1973). The presence of nuclear polyhedrosis virus of *Trichoplusia ni* on cabbage from the market shelf. *Environ. Entomol.* **2:**72–75.

Herniou, E. A., Luque, T., Chen, X., Vlak, J. M., Winstanley, D., Cory, J. S., and O'Reilly, D. (2001). Use of whole genome sequence data to infer baculovirus phylogeny. *J. Virol.* **75**(17)**:**8117–8126.

Herrmann, R., Moskowitz, H., Zlotkin, E., and Hammock, B. D. (1995). Positive cooperativity among insecticidal scorpion neurotoxins. *Toxicon* **33**(8)**:**1099–1102.

Holman, G. M., Nachman, R. J., and Wright, M. S. (1990). Insect neuropeptides. *Annu. Rev. Entomol.* **35:**201–217.

Hoover, K., Schultz, C. M., Lane, S. S., Bonning, B. C., Duffey, S. S., McCutchen, B. F., and Hammock, B. D. (1995). Reduction in damage to cotton plants by a recombinant baculovirus that knocks moribund larvae of *Heliothis virescens* off the plant. *Biol. Control* **5**(3)**:**419–426.

Hughes, P. R., Wood, H. A., Breen, J. P., Simpson, S. F., Duggan, A. J., and Dybas, J. A. (1997). Enhanced bioactivity of recombinant baculoviruses expressing insect-specific spider toxins in lepidopteran crop pests. *J. Invertebr. Pathol.* **69**(2)**:**112–118.

Hunter-Fujita, F. R., Entwistle, P. F., Evans, H. F., and Crook, N. E. (1998). "Insect Viruses and Pest Management." John Wiley & Sons, Chichester.

Ichinose, R., Kamita, S. G., Maeda, S., and Hammock, B. D. (1992a). Pharmacokinetic studies of the recombinant juvenile hormone esterase in *Manduca sexta*. *Pestic. Biochem. Physiol.* **42:**13–23.

Ichinose, R., Nakamura, A., Yamoto, T., Booth, T. F., Maeda, S., and Hammock, B. D. (1992b). Uptake of juvenile hormone esterase by pericardial cells of *Manduca sexta*. *Insect Biochem. Mol. Biol.* **22**(8)**:**893–904.

Ignoffo, C. M., and Garcia, C. (1992). Combinations of environmental factors and simulated sunlight affecting activity of inclusion bodies of the *Heliothis* (Lepidoptera, Noctuidae) nucleopolyhedrosis virus. *Environ. Entomol.* **21**(1)**:**210–213.

Ignoffo, C. M., Garcia, C., and Saathoff, S. G. (1997). Sunlight stability and rain-fastness of formulations of baculovirus *Heliothis*. *Environ. Entomol.* **26**(6)**:**1470–1474.

Inceoglu, A. B., Kamita, S. G., Hinton, A. C., Huang, Q. H., Severson, T. F., Kang, K. D., and Hammock, B. D. (2001). Recombinant baculoviruses for insect control. *Pest Manag. Sci.* **57**(10)**:**981–987.

Kamita, S. G., Kang, K.-d., Inceoglu, A. B., and Hammock, B. D. (2005a). Genetically modified baculoviruses for pest insect control. *In* "Comprehensive Molecular Insect Science" (L. I. Gilbert, K. Iatrou, and S. S. Gill, eds.), Vol. 6, pp. 271–322. 7 vols. Elsevier, Oxford.

Kamita, S. G., Nagasaka, K., Chua, J. W., Shimada, T., Mita, K., Kobayashi, M., Maeda, S., and Hammock, B. D. (2005b). A baculovirus-encoded protein tyrosine phosphatase gene induces enhanced locomotory activity in a lepidopteran host. *Proc. Natl. Acad. Sci. USA* **102**(7)**:**2584–2589.

Keddie, B. A., Aponte, G. W., and Volkman, L. E. (1989). The pathway of infection of Autographa-californica nuclear polyhedrosis-virus in an insect host. *Science* **243** (4899)**:**1728–1730.

Keeley, L. L., and Hayes, T. K. (1987). Speculations on biotechnology applications for insect neuroendocrine research. *Insect Biochem.* **17**(5)**:**639–651.

Kopeyan, C., Mansuelle, P., Sampieri, F., Brando, T., Barhraoui, E. M., Rochat, H., and Granier, C. (1990). Primary structure of scorpion anti-insect toxins isolated from the venom of *Leiurus quinquestriatus quinquestriatus*. *FEBS Lett.* **261:**424–426.

Korth, K. L., and Levings, C. S. (1993). Baculovirus expression of the maize mitochondrial protein Urf13 confers insecticidal activity in cell cultures and larvae. *Proc. Natl. Acad. Sci. USA* **90**(8)**:**3388–3392.

Korth, K. L., Kaspi, C. I., Siedow, J. N., and Levings, C. S. (1991). Urf13, a maize mitochondrial pore-forming protein, is oligomeric and has a mixed orientation in *Escherichia coli* plasma membranes. *Proc. Natl. Acad. Sci. USA* **88**(23)**:**10865–10869.

Kunimi, Y., Fuxa, J. R., and Richter, A. R. (1997). Survival times and lethal doses for wild and recombinant *Autographa californica* nuclear polyhedrosis viruses in different instars of *Pseudoplusia includens*. *Biol. Control* **9**(2)**:**129–135.

Lacey, L. A., Frutos, R., Kaya, H. K., and Vail, P. (2001). Insect pathogens as biological control agents: Do they have a future? *Biol. Control* **21**(3)**:**230–248.

Lee, S. Y., Qu, X. Y., Chen, W. B., Poloumienko, A., MacAfee, N., Morin, B., Lucarotti, C., and Krause, M. (1997). Insecticidal activity of a recombinant baculovirus containing an antisense c-myc fragment. *J. Gen. Virol.* **78:**273–281.

Levidow, L. (1995). The Oxford baculovirus controversy: Safely testing safety? *Biosci.* **45**(8)**:**545–551.

Li, Q. J., Li, L. L., Moore, K., Donly, C., Theilmann, D. A., and Erlandson, M. (2003). Characterization of *Mamestra configurata* nucleopolyhedrovirus enhancin and its functional analysis via expression in an *Autographa californica* M nucleopolyhedrovirus recombinant. *J. Gen. Virol.* **84:**123–132.

Loret, E. P., and Hammock, B. D. (1993). Structure and neurotoxicity of scorpion venom. *In* "Scorpion Biology and Research" (P. H. Brownell and G. Polis, eds.), pp. 204–233. Oxford University Press, Oxford, UK.

Lu, A., Seshagiri, S., and Miller, L. K. (1996). Signal sequence and promoter effects on the efficacy of toxin-expressing baculoviruses as biopesticides. *Biol. Control* **7**(3)**:** 320–332.

Ma, P. W. K., Davis, T. R., Wood, H. A., Knipple, D. C., and Roelofs, W. L. (1998). Baculovirus expression of an insect gene that encodes multiple neuropeptides. *Insect Biochem. Mol. Biol.* **28**(4)**:**239–249.

Maeda, S. (1989). Increased insecticidal effect by a recombinant baculovirus carrying a synthetic diuretic hormone gene. *Biochem. Biophys. Res. Comm.* **165:**1177–1183.

Maeda, S., Volrath, S. L., Hanzlik, T. N., Harper, S. A., Majima, K., Maddox, D. W., Hammock, B. D., and Fowler, E. (1991). Insecticidal effects of an insect-specific neurotoxin expressed by a recombinant baculovirus. *Virology* **184:**777–780.

Martens, J. W. M., Honee, G., Zuidema, D., Lent, J. W. M. V., Visser, B., and Vlak, J. M. (1990). Insecticidal activity of a bacterial crystal protein expressed by a recombinant baculovirus in insect cells. *Appl. Environ. Microbiol.* **56:**2764–2770.

Martens, J. W. M., Knoester, M., Weijts, F., Groffen, S. J. A., Hu, Z. H., Bosch, D., and Vlak, J. M. (1995). Characterization of baculovirus insecticides expressing tailored *Bacillus thuringiensis* Cry1a(B) crystal proteins. *J. Invertebr. Pathol.* **66**(3)**:**249–257.

McCutchen, B. F., and Hammock, B. D. (1994). Recombinant baculovirus expressing an insect-selective neurotoxin: Characterization, strategies for improvement and risk assessment. *In* "Natural and Engineered Pest Management Agents" (P. A. Hedin, J. J. Menn, and R. M. Hollingworth, eds.), pp. 348–367. American Chemical Society, Washington, DC.

McCutchen, B. F., Choundary, P. V., Crenshaw, R., Maddox, D., Kamita, S. G., Palekar, N., Volrath, S., Fowler, E., Hammock, B. D., and Maeda, S. (1991). Development of a recombinant baculovirus expressing an insect-selective neurotoxin: Potential for pest control. *Bio/Technol.* **9:**848–852.

McCutchen, B. F., Hoover, K., Preisler, H. K., Betana, M. D., Herrmann, R., Robertson, J. L., and Hammock, B. D. (1997). Interactions of recombinant and wild-type baculoviruses with classical insecticides and pyrethroid-resistant tobacco budworm (Lepidoptera: Noctuidae). *J. Econ. Entomol.* **90**(5)**:**1170–1180.

Menn, J. J., and Borkovec, A. B. (1989). Insect neuropeptides—potential new insect control agents. *J. Agric. Food Chem.* **37**(1)**:**271–278.

Merrington, C. L., King, L. A., and Posse, R. D. (1999). Baculovirus expression systems. *In* "Protein Expression: A Practical Approach" (S. J. Higgins and B. D. Hames, eds.), pp. 101–127. Oxford University Press.

Merryweather, A. T., Weyer, U., Harris, M. P. G., Hirst, M., Booth, T., and Possee, R. D. (1990). Construction of genetically engineered baculovirus insecticides containing the *Bacillus thuringiensis* subsp. *kurstaki* HD-73 delta endotoxin. *J. Gen. Virol.* **71:** 1535–1544.

Miller, L. K. (1988). Baculoviruses as gene expression vectors. *Annu. Rev. Microbiol.* **42:**177–199.

Miller, L. K. (1995). Genetically engineered insect virus pesticides: Present and future. *J. Invertebr. Pathol.* **65:**211–216.

Miller, L. K. (1997). "The Baculoviruses." The Viruses (H. Fraenkel-Conrat and R. R. Wagner, eds.) Plenum Press, New York.

Miller, L. K., Lingg, A. J., and Bulla, L. A. J. (1983). Bacterial, fungal and viral insecticides. *Science* **219:**715–721.

Moscardi, F. (1999). Assessment of the application of baculoviruses for control of Lepidoptera. *Annu. Rev. Entomol.* **44:**257–289.

Moskowitz, H., Herrmann, R., Jones, A. D., and Hammock, B. D. (1998). A depressant insect-selective toxin analog from the venom of the scorpion *Leiurus quinquestriatus hebraeus*, purification and structure/function characterization. *Eur. J. Biochem.* **254:**44–49.

Ooi, B. G., Rankin, C., and Miller, L. K. (1989). Downstream sequences augment transcription from the essential initiation site of a baculovirus polyhedrin gene. *J. Mol. Biol.* **210:**721–736.

O'Reilly, D. R. (1995). Baculovirus-encoded ecdysteroid UDP-glucosyltransferases. *Insect Biochem. Mol. Biol.* **25**(5)**:**541–550.

O'Reilly, D. R., and Miller, L. K. (1989). A baculovirus blocks insect molting by producing ecdysteroid UDP-glucosyltransferase. *Science* **245:**1110–1112.

O'Reilly, D. R., and Miller, L. K. (1991). Improvement of a baculovirus pesticide by deletion of the *egt* gene. *Bio/Technol.* **9**(11)**:**1086–1089.

O'Reilly, D. R., Miller, L. K., and Luckow, V. A. (1992). "Baculovirus Expression Vectors: A Laboratory Manual." W. H. Freeman and Co, New York.

O'Reilly, D. R., Kelly, T. J., Masler, E. P., Thyagaraja, B. S., Robson, R. M., Shaw, T. C., and Miller, L. K. (1995). Overexpression of *Bombyx mori* prothoracicotropic hormone using baculovirus vectors. *Insect Biochem. Mol. Biol.* **25:**475–485.

Petrik, D. T., Iseli, A., Montelone, B. A., Van Etten, J. L., and Clem, R. J. (2003). Improving baculovirus resistance to UV inactivation: Increased virulence resulting from expression of a DNA repair enzyme. *J. Invertebr. Pathol.* **82:**50–56.

Pinedo, F. J. R., Moscardi, F., Luque, T., Olszewski, J. A., and Ribeiro, B. M. (2003). Inactivation of the ecdysteroid UDP-glucosyltransferase (*egt*) gene of *Anticarsia gemmatalis* nucleopolyhedrovirus (AgMNPV) improves its virulence towards its insect host. *Biol. Control* **27**(3)**:**336–344.

Popham, H. J. R., Li, Y. H., and Miller, L. K. (1997). Genetic improvement of *Helicoverpa zea* nuclear polyhedrosis virus as a biopesticide. *Biol. Control* **10**(2)**:**83–91.

Popham, H. J. R., Pellock, B. J., Robson, M., Dierks, P. M., and Miller, L. K. (1998). Characterization of a variant of *Autographa californica* nuclear polyhedrosis virus with a nonfunctional ORF 603. *Biol. Control* **12:**223–230.

Possani, L. D., Becerril, B., Delepierre, M., and Tytgat, J. (1999). Scorpion toxins specific for Na+-channels. *Eur. J. Biochem.* **264**(2)**:**287–300.

Prikhodko, G. G., Robson, M., Warmke, J. W., Cohen, C. J., Smith, M. M., Wang, P. Y., Warren, V., Kaczorowski, G., VanderPloeg, L. H. T., and Miller, L. K. (1996). Properties of three baculoviruses expressing genes that encode insect-selective toxins: Mu-Aga-IV, As II, and SH I. *Biol. Control* **7**(2)**:**236–244.

Regev, A., Rivkin, H., Inceoglu, B., Gershburg, E., Hammock, B. D., Gurevitz, M., and Chejanovsky, N. (2003). Further enhancement of baculovirus insecticidal efficacy with scorpion toxins that interact cooperatively. *FEBS Lett.* **537**(1–3)**:**106–110.

Ribeiro, B. M., and Crook, N. E. (1993). Expression of full-length and truncated forms of crystal protein genes from *Bacillus thuringiensis* subsp. *kurstaki* in a baculovirus and pathogenicity of the recombinant viruses. *J. Invertebr. Pathol.* **62**(2)**:**121–130.

Ribeiro, B. M., and Crook, N. E. (1998). Construction of occluded recombinant baculoviruses containing the full-length *cry1Ab* and *cry1Ac* genes from *Bacillus thuringiensis*. *Braz. J. Med. Biol. Res.* **31**(6)**:**763–769.

Richardson, C. D. (1995). Baculovirus expression protocols. *In* "Methods in Molecular Biology." Humana Press, Totowa, New Jersey.

Ryerse, J. S. (1998). Basal laminae. *In* "Insecta" (F. W. Harrison and M. Locke, eds.), Vol. 11A, pp. 3–16. Wiley-Liss, New York.

Schnepf, E., Crickmore, N., Van Rie, J., Lereclus, D., Baum, J., Feitelson, J., Zeigler, D. R., and Dean, D. H. (1998). *Bacillus thuringiensis* and its pesticidal crystal proteins. *Microbiol. Mol. Biol. Rev.* **62**(3)**:**775.

Slavicek, J. M., Popham, H. J. R., and Riegel, C. I. (1999). Deletion of the *Lymantria dispar* multicapsid nucleopolyhedrovirus ecdysteroid UDP-glucosyl transferase gene enhances viral killing speed in the last instar of the gypsy moth. *Biol. Control* **16**(1)**:** 91–103.

Snyder, M., Hunkapiller, M., Yuen, D., Silvert, D., Fristrom, J., and Davidson, N. (1982). Cuticle protein genes of *Drosophila*—structure, organization and evolution of 4 clustered genes. *Cell* **29**(3)**:**1027–1040.

Stewart, L. M., Hirst, M., Ferber, M. L., Merryweather, A. T., Cayley, P. J., and Possee, R. D. (1991). Construction of an improved baculovirus insecticide containing an insect-specific toxin gene. *Nature* **352:**85–88.

Summers, M. D., and Smith, G. E. (1987). "A Manual of Methods for Baculovirus Vectors and Insect Cell Culture Procedures." Texas Agricultural Experiment Station Bulletin No. 1555.

Sun, X., Wang, H., Sun, X., Chen, X., Peng, C., Pan, D., Jehle, J. A., van der Werf, W., Vlak, J. M., and Hu, Z. (2004). Biological activity and field efficacy of a genetically modified *Helicoverpa armigera* single-nucleocapsid nucleopolyhedrovirus expressing an insect-selective toxin from a chimeric promoter. *Biol. Control* **29:**124–137.

Sun, X. L., Chen, X. W., Zhang, Z. X., Wang, H. L., Bianchi, J. J. A., Peng, H. Y., Vlak, J. M., and Hu, Z. H. (2002). Bollworm responses to release of genetically modified *Helicoverpa armigera* nucleopolyhedroviruses in cotton. *J. Invertebr. Pathol.* **81**(2)**:** 63–69.

Tanada, Y., and Kaya, H. K. (1993). "Insect Pathology." Academic Press, Inc., San Diego.

Tomalski, M. D., and Miller, L. K. (1991). Insect paralysis by baculovirus-mediated expression of a mite neurotoxin gene. *Nature* **352**(6330)**:**82–85.

Tomalski, M. D., and Miller, L. D. (1992). Expression of a paralytic neurotoxin gene to improve insect baculoviruses as biopesticides. *Bio/Technol.* **10:**545–549.
Tomalski, M. D., Bruce, W. A., Travis, J., and Blum, M. S. (1988). Preliminary characterization of toxins from the straw itch mite, *Pyemotes tritici*, which induce paralysis in the larvae of a moth. *Toxicon* **26**(2)**:**127–132.
Tomalski, M. D., Kutney, R., Bruce, W. A., Brown, M. R., Blum, M. S., and Travis, J. (1989). Purification and characterization of insect toxins derived from the mite, *Pyemotes tritici*. *Toxicon* **27**(10)**:**1151–1167.
Treacy, M. F., All, J. N., and Ghidiu, G. M. (1997). Effect of ecdysteroid UDP-glucosyltransferase gene deletion on efficacy of a baculovirus against *Heliothis virescens* and *Trichoplusia ni* (Lepidoptera: Noctuidae). *J. Econ. Entomol.* **90**(5)**:**1207–1214.
Treacy, M. F., Rensner, P. E., and All, J. N. (2000). Comparative insecticidal properties of two nucleopolyhedrovirus vectors encoding a similar toxin gene chimer. *J. Econ. Entomol.* **93**(4)**:**1096–1104.
Tumilasci, V. F., Leal, E., Marinho, P., Zanotto, A., Luque, T., and Wolff, J. L. C. (2003). Sequence analysis of a 5.1 kbp region of the *Spodoptera frugiperda* multicapsid nucleopolyhedrovirus genome that comprises a functional ecdysteroid UDP-glucosyltransferase (*egt*) gene. *Virus Genes* **27**(2)**:**137–144.
Vail, P. V., Jay, D. L., and Hunter, D. K. (1973). Infectivity of a nuclear polyhedrosis virus from the alfalfa looper, *Autographa californica*, after passage through alternate hosts. *J. Invertebr. Pathol.* **21:**16–20.
Vail, P. V., Hostertter, D. L., and Hoffmann, D. F. (1999). Development of the multinucleocapsid nucleopolyhedroviruses (MNPVs) infectious to loopers (Lepidoptera: Noctuidae: Plusiinae) as microbial control agents. *Integr. Pest Manag. Rev.* **4:**231–257.
Wilson, K. R., O'Reilly, D. R., Hails, R. S., and Cory, J. S. (2000). Age-related effects of the *Autographa californica* multiple nucleopolyhedrovirus *egt* gene in the cabbage looper (*Trichoplusia ni*). *Biol. Control* **19:**57–63.
Woo, S. D., Kim, W. J., Kim, H. S., Jin, B. R., Lee, Y. H., and Kang, S. K. (1998). The morphology of the polyhedra of a host range-expanded recombinant baculovirus and its parents. *Arch. Virol.* **143**(6)**:**1209–1214.
Wood, H. A. (1996). Genetically enhanced baculovirus insecticides. *In* "Molecular Biology of the Biological Control of Pests and Diseases of Plants" (M. Gunasekaran and D. J. Weber, eds.), pp. 91–104. CRC Press, Boca Raton.
Wood, H. A., Hughes, P. R., and Shelton, A. (1994). Field studies of the coocclusion strategy with a genetically altered isolate of the *Autographa californica* nuclear polyhedrosis virus. *Environ. Entomol.* **23**(2)**:**211–219.
Yurchenco, P. D., and O'Rear, J. (1993). Supramolecular organization of basement membranes. *In* "Molecular and Cellular Aspects of Basement Membranes" (D. H. Rohrback and R. Timpl, eds.), pp. 19–47. Academic Press, New York.
Zlotkin, E. (1991). Venom neurotoxins—models for selective insecticides. *Phytoparasitica* **19**(3)**:**177–182.
Zlotkin, E., Fraenkel, G., Miranda, F., and Lissitzky, S. (1971). The effect of scorpion venom on blow fly larvae; a new method for the evaluation of scorpion venom potency. *Toxicon* **8:**1–8.
Zlotkin, E., Miranda, F., and Rochat, H. (1978). Venoms of buthinae. *In* "Arthropod Venoms" (S. Bettini, ed.), pp. 317–369. Springer-Verlag, New York.
Zlotkin, E., Kadouri, D., Gordon, D., Pelhate, M., Martin, M. F., and Rochat, H. (1985). An excitatory and a depressant insect toxin from scorpion venom both affect sodium

conductance and possess a common binding site. *Arch. Biochem. Biophys.* **240**(2): 877–887.

Zlotkin, E., Gurevitz, M., Fowler, E., and Adams, M. E. (1993). Depressant insect selective neurotoxins from scorpion venom—chemistry, action, and gene cloning. *Arch. Insect Biochem. Physiol.* **22**(1–2):55–73.

Zlotkin, E., Moskowitz, H., Herrmann, R., Pelhate, M., and Gordon, D. (1995). Insect sodium channel as the target for insect-selective neurotoxins from scorpion venom. *ACS Symp. Ser.* **591:**56–85.

Zlotkin, E., Fishman, Y., and Elazar, M. (2000). AaIT: From neurotoxin to insecticide. *Biochimie* **82**(9–10):869–881.

ADVANCES IN VIRUS RESEARCH, VOL 68

DENSOVIRUSES FOR CONTROL AND GENETIC MANIPULATION OF MOSQUITOES

Jonathan Carlson,* Erica Suchman,* and Leonid Buchatsky†

*Department of Microbiology, Immunology and Pathology
Colorado State University, Fort Collins, Colorado 80523
†Department of Zoology, Taras Shevchenko' Kiev National University
Kiev, Ukraine 01033

Abstract

Mosquito densoviruses (MDV) are parvoviruses that replicate in the nuclei of mosquito cells and cause the characteristic nuclear hypertrophy (densonucleosis) that gives them their name. Several MDV that differ in pathogenicity both *in vitro* and *in vivo* have been isolated. MDV have a number of features that make them potentially attractive as biological control agents for mosquito-borne disease. They are non-enveloped and relatively stable in the environment. They are highly specific for mosquitoes and they infect and kill larvae in a dose dependent manner in the aqueous larval habitat. Infected larvae that survive to become adult mosquitoes exhibit a dose-dependent shortening of lifespan and many do not survive longer than the extrinsic incubation period for arboviruses. Thus they may have a significant impact on transmission of pathogens. Infected females can transmit the virus

0065-3527/06 $35.00
DOI: 10.1016/S0065-3527(06)68010-X

vertically by laying infected eggs in new oviposition sites. Studies on how MDV affect populations are relatively limited. Population cage studies suggest that they will persist and spread in populations and limited field studies have shown similar preimaginal mortality in wild populations to that seen in laboratory studies.

The availability of infectious clones of MDV genomes allows the development of densovirus vectors for expressing genes of interest in mosquito cells and mosquitoes. Recently short hairpin RNA expression cassettes that induce RNA interference have been inserted into densovirus genomes. These expression cassettes should be useful for both research and disease-control applications.

I. Introduction

A. *Resurgence of Mosquito-Borne Diseases*

Mosquito-borne diseases are serious problems for both human and veterinary medicine and are significant impediments to economic development for much of the world. The outbreak of *West Nile virus* on the East Coast of the United States in the summer of 1999 (Anderson *et al.*, 1999; Lanciotti *et al.*, 1999) and its subsequent inexorable spread through the United States dramatically highlights our vulnerability to such diseases in spite of an unparalleled public health system. Indeed it has been suggested that *West Nile virus* is a good model for the introduction of a biological weapon into a country to attack the human and animal populations. Equally disturbing is the reemergence of dengue (DEN) throughout the New World tropics to the southern borders of the United States. It is not only these new "emerging" diseases, such as dengue hemorrhagic fever and West Nile encephalitis, that cause concern but also the resurgence of old diseases such as malaria and yellow fever (Anderson *et al.*, 1999; Gratz, 1999; Lanciotti *et al.*, 1999).

Traditional strategies for control of mosquito-borne disease include reducing mosquito populations, altering human behavior to decrease mosquito/human contact, and vaccination. Unfortunately, these conventional methods for control of mosquito-borne diseases are rapidly becoming insufficient. The reasons for this include factors such as (1) the lack of effective vaccines, (2) increased and unplanned urbanization in the tropics, (3) the demise or neglect of public health infrastructure involving medical entomology and vector control programs due to expense and public ambivalence, and (4) the trafficking of pathogens throughout the world, especially facilitated by jet travel. Control efforts are compromised by undesirable ecological effects of control measures on nontarget species and by a throwaway societal attitude, which allows empty soda and beer cans and other containers to become breeding sites for the vectors and greatly reduces the efficacy of source reduction campaigns. The increased incidence of pesticide resistance in mosquitoes, and the lack of new pesticides further complicate control efforts. Alternative approaches are badly needed. New precisely targeted methods of control are sorely needed for mosquito vectors.

B. The Role of Densoviruses in Control of Vector-Borne Disease

Mosquito densonucleosis viruses or densoviruses (MDV) have features that make them attractive for use in integrated vector-borne disease control programs. They are relatively stable in the environment. They are exclusively targeted to mosquitoes and have the potential to spread and persist in mosquito populations by normal infection and replication mechanisms. The use of densoviruses in the fight against vector-borne disease could follow either of two potential strategies. The more conventional strategy is use of densovirus as a microbial pesticide for biological control of mosquitoes. *Aedes* densonucleosis virus (AeDNV) infection affects all lifestages of *Aedes aegypti* and significantly shortens the adult lifespan to the point that the virus has the potential to significantly modify the age structure of adult mosquito populations. This in turn should significantly reduce the vectorial capacity of the population (Suchman *et al.*, 2006). The second strategy is based on our work to develop densoviruses as transducing vectors (Afanasiev and Carlson, 2000; Carlson *et al.*, 2000). Constructs to induce interference to arboviral infection could be introduced into mosquitoes by transduction, for “immunization” or reduction of vectorial capacity for arboviruses.

II. The Biology of Mosquito Densoviruses

A. Life Cycle

Mosquito larvae are infected by MDV present in the water where female mosquitoes lay their eggs (Fig. 1). Infected larvae excrete virus into the water increasing the viral titer (Barreau *et al.*, 1996; Ledermann *et al.*, 2004) allowing for horizontal spread of the virus to other larvae. Infected larvae will either become moribund and die or become infected pupae and adults (Kittayapong *et al.*, 1999; Ledermann *et al.*, 2004). Infected adults may have a decreased lifespan and reduced fecundity as

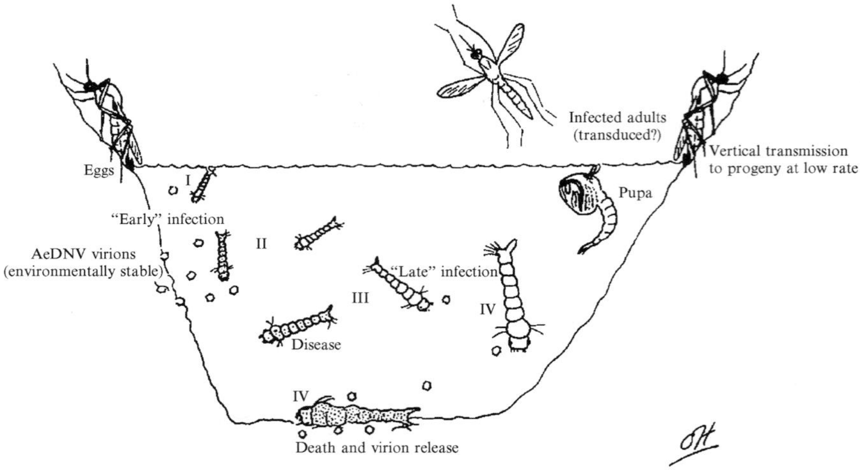

FIG 1. Mosquito densovirus life cycle. See text for details. Drawn by Steve Higgs.

measured by the number of eggs laid and egg viability (Barreau *et al.*, 1997; Suchman *et al.*, 2006). The virus is transmitted vertically, and offspring from infected mothers may go on to be infected adults (Barreau *et al.*, 1997; Buchatsky, 1989; Kittayapong *et al.*, 1999; and Suchman *et al.,* 2006). The ability to be transmitted vertically should allow for virus spread to new oviposition sites by infected females. This characteristic is potentially beneficial because containers, such as discarded soda cans and tires are often used as oviposition sites and can easily be missed in control efforts.

B. Mosquito Densovirus Isolates

Densonucleosis was first discovered by French researchers working on the caterpillar *Galleria mellonella* resulting in the isolation of *Galleria mellonella* densonucleosis virus (GmDNV; Meynadier *et al*, 1964). Densoviruses from several insect species are known, with representatives in five insect orders (Lepidoptera, Diptera, Orthoptera, Dictyoptera, and Odonata), as well as a few in the Crustaceae (Bergoin and Tijssen, 2000). The first MDV, AeDNV was discovered at Kiev National University in an *Ae. aegypti* laboratory colony that was originally collected from Southeast Asia (Lebedeva *et al.*, 1972) (Table I). MDVs have also been isolated from natural populations of *Aedes albopictus*, *Ae. aegypti,* and *Anopheles minimus* in parts of Thailand

TABLE I
MOSQUITO DENSOVIRUSES

Host	Name	Country of origin	Year	References
Ae. aegypti mosquito lab colony	AeDNV	Soviet Union	1973	Lebedeva *et al.*, 1972
Ae. pseudoscutellaris cell line Mos 61			1980	Gorziglia *et al.*, 1980
Ae. albopictus cell line C6/36	AalDNV (AaPV)	France	1993	Jousset *et al.*, 1993
Ae. albopictus cell line C6/36	APeDNV	Peru	NA[a]	Paterson *et al.*, 2005
Haemogogus equinus cell line GML-HE-12	HeDNV	United States	1995	O'Neill *et al.*, 1995
Toxorhynchites amboinensis cell line TRA-284	TaDNV	United States	1995	O'Neill *et al.*, 1995
C. theilerei cell line 1	CtDNV	United States	1995	O'Neill *et al.*, 1995
C. pipiens mosquito lab colony	CpDNV	France	2000	Jousset *et al.*, 2000
Ae. aegypti and *Ae. albopictus* in Thailand	AThDNV	Thailand	1999	Kittayapong *et al.*, 1999
An. minimus S.L. in Thailand	AThDNV	Thailand	2000	Rwegoshora *et al.*, 2000
Ae. albopictus cell line C6/36	C6/36 DNV	China	2004	Chen *et al.*, 2004

[a] Not known or not applicaple.

(Kittayapong *et al.*, 1999; Rwegoshora *et al.*, 2000); however, most of them have been isolated from mosquito cell lines (Gorziglia *et al.*, 1980; Jousset *et al.*, 1993, 2000; O'Neill *et al.*, 1995) (Table I). Although these viruses are named after the species of mosquito that was used to establish the cell culture from which the virus was isolated, the virus may not come from that mosquito. Mosquito cell lines are often used for screening of mosquito collections for arboviruses, and because most of the known MDVs cause persistent noncytopathic infections in these lines, these lines could easily be contaminated in the laboratory. Thus the species designations for these viruses must be treated with caution and may not reflect the source of virus in nature.

Most importantly, MDV host specificity is apparently restricted to mosquitoes, although not all mosquito species are infected by any given virus. For example, AeDNV has been shown to infect a number of species from the genera *Aedes, Culex,* and *Culiseta* (Table II).

However, there was no evidence of infection of *Anopheles maculipennis*. When *Anopheles gambiae* larvae were incubated with a mixture of AeDNV and green fluorescent protein (GFP)-transducing particles to allow the infection to be monitored by fluorescence (Section III.B and IV.A), only a few cells in the anal papillae became infected and

TABLE II
PATHOGENESIS OF AEDNV TO INVERTEBRATE SPECIES[a]

Animal species	Number of individuals	Developmental stage	Route of infection	Pathological effect
Insects				
Ae. aegypti	1140	Instar I–IV larvae	PO	+
Ae. albopictus	550	Instar I larvae	PO	+
Ae. togoi	450	Instar I larvae	PO	+
Ae. vexans	419	Instar I, II larvae	PO	+
Ae. geniculatus	233	Instar I larvae	PO	+
Ae. caspius dorsalis	905	Instar I, II larvae	PO	+
Ae. cantans	440	Instar II larvae	PO	+
Ae. caspius caspius	90	Instar II larvae	PO	+
C. pipiens pipiens	915	Instar I, II larvae	PO	+
C. p. molestus	641	Instar I larvae	PO	+
C. annulata	315	Instar I, II larvae	PO	+
An. maculipennis	548	Instar I, II larvae	PO	–
Chironomus sp.	142	Larvae	PO	–
M. domestica	335	Instar III, IV larvae	PO, IL	–
P. regina	210	Instar III, IV larave	PO, IL	–
A. mellifera	200	Adult	PO	–
G. mellonella	450	Instar III, IV larvae	PO, IL	–
B. mori	115	Instar III, IV larvae	PO, IL	–
A. crataegi	184	Instar III, IV larvae	PO, IL	–
M. neustria	270	Instar III, IV larvae	PO, IL	–
P. dispar	225	Instar III, IV larvae	PO, IL	–
Crustaceans				
Daphnia sp.		Adults and youth	PO	–
Cyclops sp.		Adults and youth	PO	–
Worms				
Lumbricus sp.	50	Adults	SC	–

[a] Adapted from Buchatsky *et al.*, 1997.

Abbreviations used in column "Route of infection": PO, per oral; IL, intralymphatic; SC, subcutaneous.

the virus did not disseminate to other tissues as seen with *Ae. aegypti* (Ward *et al.,* 2001a). As shown in Table II, AeDNV does not infect the larvae of flies, chironomids and lepidopterans, butterflies, bees, cockroaches, crustaceans, or worms. There was no evidence for densovirus infection of fish, birds, rats, rabbits, hamsters, or other mammals (Buchatsky, 1989; El-Far *et al.*, 2004; Fediere, 2000; Jousset *et al.*, 1993; Vasileva *et al.*, 1990).

C. Molecular Biology of Mosquito Densoviruses

The mosquito densoviruses are in the *Densovirinae* subfamily of the family *Parvoviridae*. All but one of them are in the genus *Brevidensovirus* (Afanasiev and Carlson, 2001). The lone exception is the *Culex pipiens* densonucleosis virus (CpDNV), which is in the *Densovirus* genus. As with all parvoviruses, the MDV genomes are linear single-stranded DNA molecules. In contrast to the members of the *Densovirus* genus, which have ambisense genomes with the structural and nonstructural genes coded on different strands, the members of the *Brevidensovirus* genus have both structural and nonstructural genes encoded by the same strand. In the MDV, it is primarily the negative sense strand (complementary to the mRNA) that is packaged into virions. The MDV particles are nonenveloped icosahedral capsids about 20 nm in diameter (Fig. 2).

Replication of parvoviruses takes place in the nucleus of the infected cell. The genomes of the MDV are ~4000 bases in length. Both the 5′ and 3′ ends of the MDV genome have inverted repeat sequences that are predicted to fold into T- or Y-shaped secondary structures that presumably act as origins of replication and/or as packaging signals (Fig. 2B). The 3′OH end of the genome acts as a primer, which is extended by host DNA polymerase to convert the single-stranded genome to a double-stranded replicative form. The details of MDV replication have not been elucidated but it is likely that it replicates by a rolling hairpin mechanism similar to the mammalian parvoviruses (Berns, 1996). MDV genomes are arranged with the genes for the nonstructural proteins NS1 and NS2 occupying the 5′ portion of the genome and the gene for the viral capsid protein (VP) occupying the 3′ portion of the genome. The coding sequence for the NS2 protein is completely contained within the *NS1* gene in an alternate reading frame (+1 with respect to NS1; Fig. 2B).

The mRNA for the nonstructural proteins is transcribed from a promoter P_{NS} located about 7% of the genome length from the 5′ end. The mRNA for the VP is transcribed from a promoter P_{VP} located about

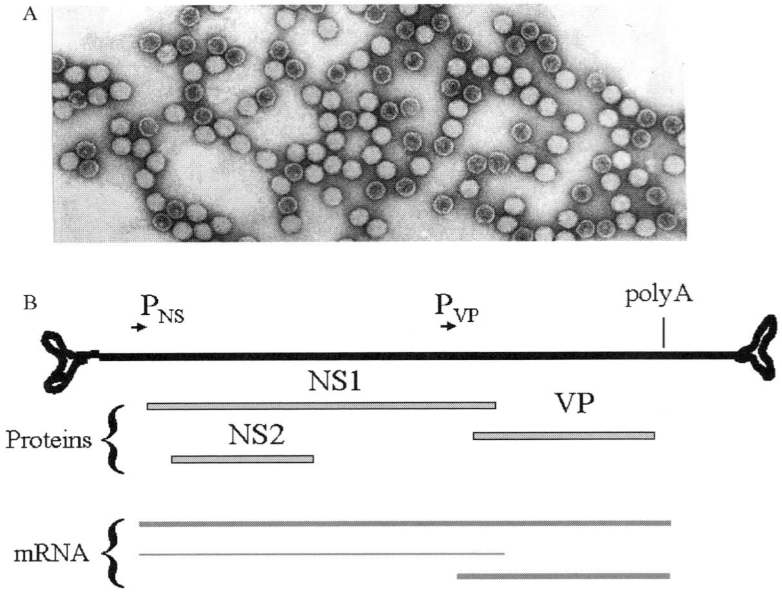

FIG 2. Mosquito densoviruses. (A) Electron micrograph of AeDNV virus particles stained with 1% uranyl acetate. (B) The genome and coding strategy of MDV with the genes for the viral proteins and the mRNAs indicated. The thickness of the lines indicates the relative abundance of the mRNA species. Promoter and polyadenylation sites for mRNA synthesis are indicated.

60% of the genome length from the 5′ end. Both mRNAs are terminated and polyadenylated at a site just downstream of the coding sequence of the capsid protein gene (Fig. 2B). There is an alternative site for polyadenylation of nonstructural protein mRNA just downstream of the termination codon for the *NS1* gene. Apparently this termination signal is rarely used, however, since a transcript of the predicted size was not seen on Northern blots of mRNA from infected cells (Ward *et al.*, 2001b). Translation of both NS1 and NS2 from the same mRNA appears to require a sequence predicted to form a stem loop secondary structure between the AUG initiation codons for the two proteins (Kimmick *et al.*, 1998). The functions of the nonstructural proteins are not well understood. NS1 is presumed to act as a nickase/helicase during replication and packaging analogous to homologs in mammalian parvoviruses. Even less is known of the function of the NS2 protein because it has little homology to the proteins of other parvoviruses. However,

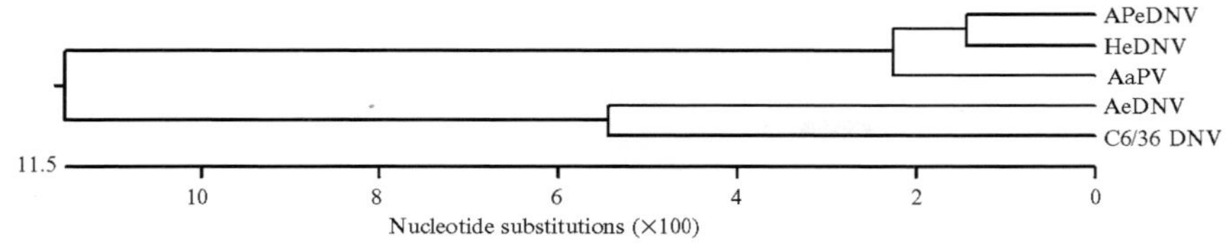

FIG 3. MDV phylogeny. The complete sequences from Genbank (accession numbers are shown in parentheses) for AeDNV (M37899), C6/36 DNV (AY095351), HeDNV (AY605055), APeDNV (AY310877), and AaPV (X74945) were aligned by Clustal W and analyzed to produce a phylogenetic tree by the MegAlign program of LaserGene (DNASTAR, Inc. Madison, Wisconsin, USA). These sequences fall into two clades with 80% identity.

NS2 is essential for the virus because mutations in *NS2* eliminate the production of viable virus (Afanasiev, B., and Azarkh, E., unpublished data). The VP protein is translated from the mRNA initiated at P_{VP}. The virus particles have two proteins VP1 and VP2 of about 40 and 38 kd, respectively. A variable proportion of VP1 is cleaved 20 amino acids from the amino terminus to produce VP2 (Flipse, M., unpublished data). The timing and significance of this cleavage is not clear.

Complete genome sequences are available for several MDV. Comparison of these sequences shows that they fall into two clades that are about 80% identical (Fig. 3). Members of the same clade have greater that 88% identity. Members of the clade containing AeDNV and C6/36 DNV have likely origins in Asia whereas members of the other clade have likely origins in Europe or the Americas.

III. Pathogenesis of Mosquito Densoviruses

A. In Vitro

A number of MDV have been isolated from different persistently infected mosquito cell lines that show no obvious signs of infection (Boublik *et al.,* 1994; Chen *et al.*, 2004; O'Neill *et al.*, 1995). For example, three different densoviruses have been isolated from persistently infected C6/36 *Ae. albopictus* cells: AaPV (renamed AalDNV) in France (Boublik *et al.,* 1994), C6/36 DNV in China (Chen *et al.*, 2004) and APeDNV in Peru (Paterson *et al.*, 2005) (Table I). The sequences of these viruses are sufficiently different that it is highly unlikely that they evolved from a single contamination event.

Although all of the DNVs are able to establish a persistent infection, they often maintain themselves by virus production in only a fraction of the cell population. We have found that the exact percentage varies between viruses and cell types. For example, in C6/36 cells infected with AeDNV, only around 1–2% of cells produce viral antigen when assayed by immunofluorescence (IFA), whereas 5–10% of cells are infected by APeDNV (Paterson *et al.*, 2005). In a report on infection of C6/36 cells with AalDNV, Burivong *et al.* (2004) found that 80–90% of the cells produced antigen during the early stages of infection, and although the cells were not lysed the percentage of cells that produced antigen gradually decreased to around 20% and remained near that level for a number of passages. The mechanism by which the characteristic fraction of antigen positive cells is maintained in persistently infected cultures is unknown. The production of defective interfering particles has been suggested as a possible mechanism (Burivong *et al.*, 2004), but this seems unlikely because persistent infections of AeDNV and APeDNV produced by transfection of cells with infectious clones of the viral genome establish characteristic persistent levels of infection within a few days of transfection. Cells infected with virus from these transfections show the same characteristic low levels of antigen-producing cells. It seems unlikely that significant quantities of defective genomes would be generated during the short course of virus production in these experiments. C6/36 cells persistently infected with AeDNV are susceptible to *Haemogogus equinus* densonucleosis virus (HeDNV) superinfection (Paterson *et al.*, 2005). This is inconsistent with AeDNV-derived defective interfering particles as the mechanism for maintaining persistence, because it is likely that both viruses use the same cellular machinery for replication and defective interfering genomes of AeDNV would be expected to interfere with HeDNV infections. The susceptibility to superinfection also argues against the establishment of an antiviral or immune state that protects the majority of cells against densovirus infection. However, in contrast, it has been shown that C6/36 cells persistently infected with AalDNV show a marked decrease in susceptibility to infection with *Dengue virus* (Burivong *et al.*, 2004), which suggests that AalDNV may induce some sort of heterologous antiviral state. It is of course possible that different MDVs are using different mechanisms to establish and maintain persistent infections.

Most MDVs establish persistent infections in C6/36 cells with no obvious observable cytopathic effect (CPE) (Burivong *et al.*, 2004; Jousset *et al.*, 1993, 2000; O'Neill *et al.*, 1995). In contrast, infection of the C6/36 cell line with HeDNV, from the *Haemagogus equinus* cell line GML-HE-12 (O'Neill *et al.*, 1995), resulted in complete destruction of the cell monolayer 4–6 days postinfection and the accumulation of substantial quantities of virus (Paterson *et al.*, 2005). Several different assays including changes in cellular morphology (membrane budding, apoptotic bodies), oligonucleosome laddering, caspase activation, and exposure of phosphatidylserine on the cellular membrane, were used to show that HeDNV, but not AeDNV is able to induce apoptosis in infected cells, which may account for the presence of CPE in HeDNV infections. The critical viral determinants for inducing or preventing apoptosis are unknown.

Work in our laboratory has shown that AeDNV, originally isolated from mosquitoes, establishes persistent infections in C6/36 cells with only about 1% of cells expressing antigen (Afanasiev *et al.*, 1994; Paterson *et al.*, 2005). Similarly, when infected with CpDNV, approximately 1% of *An. gambiae, Culex quinquefasciatus,* and *Culex tarsalis* cells were found to produce CpDNV by IFA (Jousset *et al.*, 2000). These low levels of infection of cell cultures suggest that viruses isolated from mosquitoes are not well adapted to growth in cultured cells. Viruses isolated from cell culture produce more virus in infected cell culture than viruses isolated from mosquito colonies (Paterson *et al.*, 2005). Therefore, viruses found persistently infecting cell lines may have adapted for improved replication in cell culture.

It is difficult to study AeDNV-infected cells in culture because of the low percentage of infected cells. However, using a recombinant viral-transducing genome, pANS1-GFP in which the gene for the green fluorescent protein (GFP) is fused to the 3′ end of the *NS1* gene of AeDNV (Section IV.A), it was possible to compare infected and uninfected cells by flow cytometry to detect GFP. These studies showed that cells that did not express GFP gave a typical profile for C6/36 cells with the majority of cells in the G1 phase of the cell cycle as assessed by propidium iodide fluorescence to determine DNA content. In contrast, most of the cells that had the NS1-GFP construct appeared to be arrested in the G2 phase (Paterson *et al.*, 2005). The GFP-expressing cells did not appear to be in the process of apoptosis since there was no indication of hypodiploid DNA content, and DNA isolated from GFP-expressing cells did not show the oligonucleosomal laddering indicative of apoptosis. Thus expression of the AeDNV nonstructural proteins alters the cell cycle but does not necessarily cause the CPEs associated with apoptosis. C6/36 cells have also been stably transformed with a plasmid that constitutively expresses the AeDNV *VP* gene but not the nonstructural proteins (Afanasiev and Carlson, 2000). Therefore neither the structural nor the nonstructural proteins of AeDNV alone cause apoptosis.

C6/36 cells have been adapted to a serum-free protein-free medium (Sf900-II) and spinner flasks (Suchman and Carlson, 2004). These cultures support the production of high levels of virus. Furthermore, we have been able to transfect cells in T-75 flasks, and then transfer these transfected cells to spinner flasks, and dramatically increase virus yields (Piper, J., unpublished data). This approach allows for the production of much larger quantities of the virus, at far less cost, than in the past.

B. In Vivo

1. Effects on Larvae and Pupae

In vivo laboratory studies on several MDV have revealed a spectrum of MDV-associated pathological effects. *Ae. aegypti* larvae infected by AeDNV or AalDNV lose their mobility, hang near the surface of the water or sink to the bottom, and twitch convulsively when disturbed. They become deformed and distended and lose their pigmentation and exhibit a whitish color (Barreau *et al.*, 1996; Buchatsky, 1989). Similar morphological characteristics are observed in CpDNV-infected larvae of *C. pipiens* (Jousset *et al.*, 2000). Within these larvae, multiple tissues may be infected including the fat body, imaginal discs, peritracheal cells, foregut, midgut, muscles, salivary glands, Malpighian tubules, hemocytes, neural ganglia, and gonads. The nuclei of infected cells increase two- to threefold in diameter and stain prominently with hematoxylin, Giemsa, or indirect IFA using antiserum against the virus particle (Fig. 4). Virus particles accumulate in the infected cell nuclei in paracrystalline arrays (Fig. 5). The timing of larva to pupa and pupa to adult molts may be significantly delayed. Death is often associated with the stress of molting.

Studies using AeDNV-transducing virus carrying a recombinant genome expressing a *GFP* reporter gene helped to confirm earlier histological studies and define the course of MDV replication and dissemination within *Ae. aegypti* larvae (Afanasiev *et al.*, 1999; Ward *et al.*, 2001a). The transducing virus pANS1-GFP has the *GFP* gene fused to the carboxy terminus of the NS1 protein (Section IV.A). This genome is capable of self-replication but cannot package without a source of *VP* from a helper virus such as wild-type AeDNV. On coinfection of a cell with both a transducing particle and the wild-type virus, both genomes are replicated and packaged into particles. Infected cells in living larvae can easily be seen by fluorescence microscopy making it possible to follow the course of infection in individual larvae. When larvae were

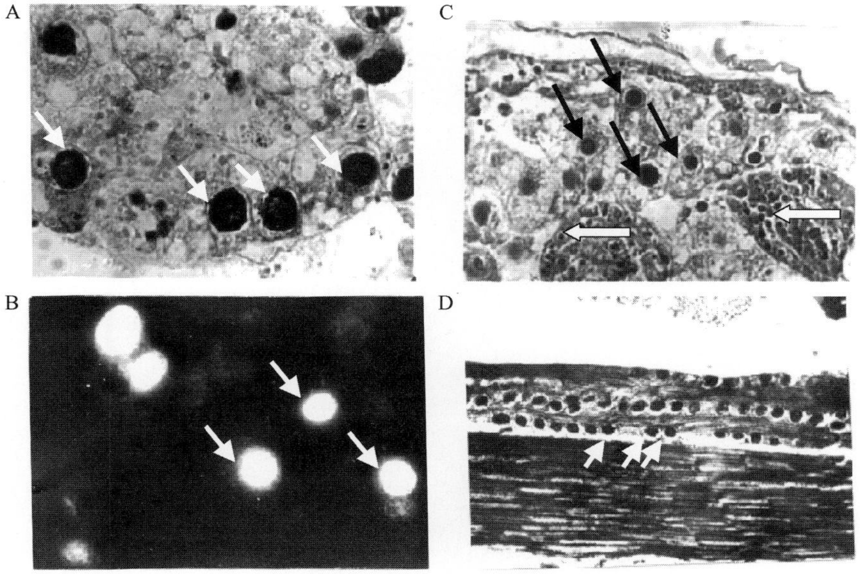

FIG 4. Densonucleosis histology. (A) Hypertrophied nuclei (arrows) in the fat body of an AeDNV-infected *Ae. aegypti* larva. Microscopic sections were stained with ferrous hematoxylin by Geidengayn. ×630. (B) Hypertrophied nuclei (arrows) in fat body of an AeDNV-infected *Ae. aegypti* larva stained by indirect immunofluorescence using rabbit antiserum to AeDNV. ×400. (C) Hypertrophied nuclei in the imaginal disks (white arrowheads) and in the fat body (black arrows) of an AeDNV-infected *Ae. aegypti* larva stained with ferrous hematoxylin. ×280. (D) Hypertrophied nuclei (arrows) in the muscle fibers of an AeDNV-infected *Ae. aegypti* pupa stained with ferrous hematoxylin. ×280.

exposed to a mixture of transducing particles and wild-type virus at 0–4 days of age, expression of the *GFP* reporter gene was observed as early as 24-h postinoculation. The anal papillae of the larvae were usually the first tissue to show fluorescence suggesting that they are the primary sites of infection (Fig. 6A). The infection subsequently spreads to the cells of the fat body, as well as many other tissues including muscle fibers and nerves (Fig. 6B). Sodium chloride concentrations in the larval-rearing water above 0.05 M inhibited virus infection, suggesting that ion concentration in the larval habitat is important for densovirus infection (Ward *et al.*, 2001a). This observation may be due to the shrinking of the anal papillae in high-salt conditions resulting in reduced ion exchange between the environment and the larval hemolymph and reduced viral entry. A larva with an infected anal papilla can

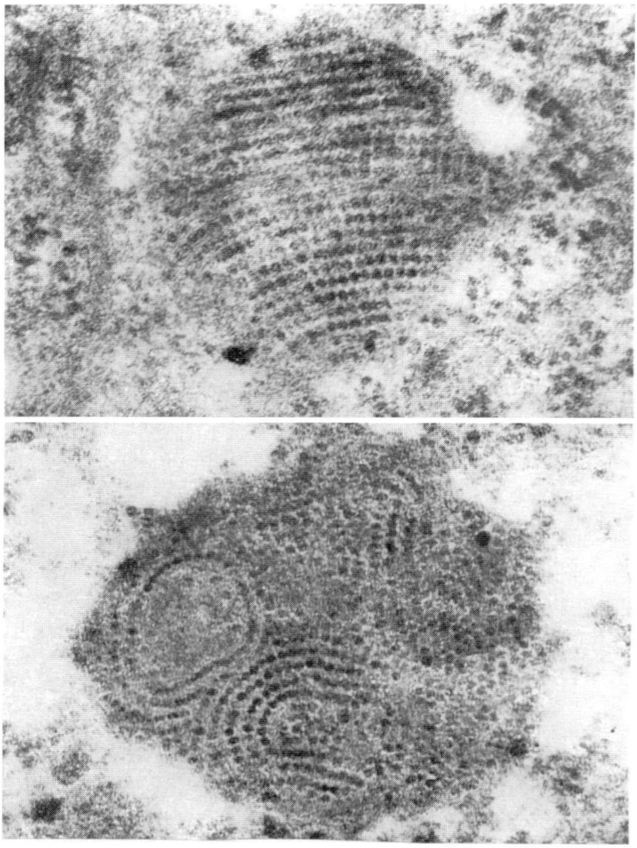

FIG 5. Intracellular arrays of MDV. Electron micrograph of paracrystalline inclusions of virions in an AeDNV-infected cell of an *Ae. aegypti* larva.

sometimes lose the infected papilla and if the infection has not yet disseminated the animal will show no further signs of infection (Ward *et al.*, 2001a). This may be associated with melanization of the infected papilla. Consistent with this, infected larvae show a 66% increase in the level of monophenol oxidase and a 15-fold increase in tyrosine aminotransferase over uninfected levels (Buchatsky *et al.*, 1997).

It is difficult to compare quantitative aspects of MDV infection determined in different laboratories for a number of reasons. Strains of mosquitoes of the same species in different laboratory colonies may vary in susceptibility to viruses, and the differences in procedures and

A

B

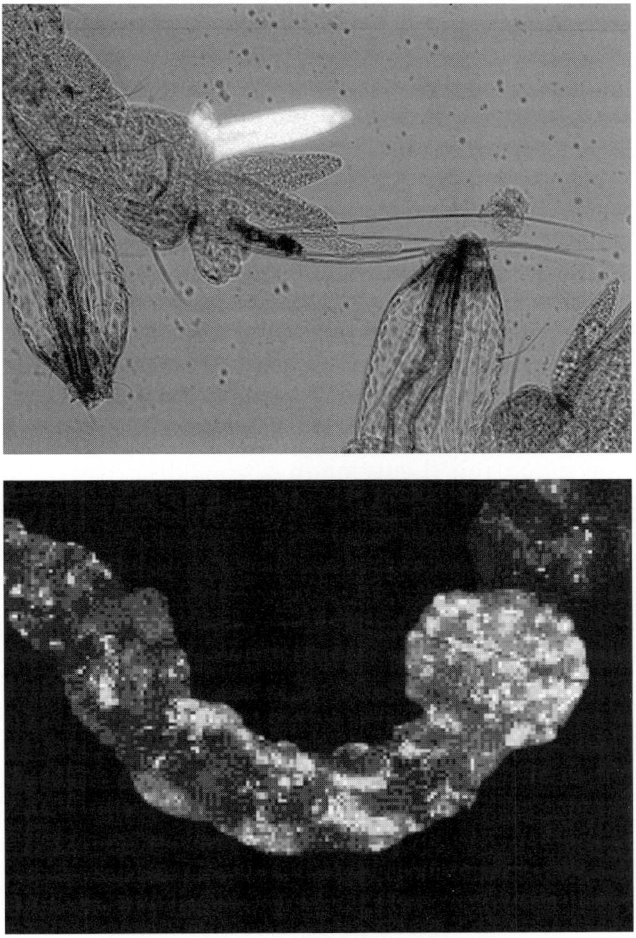

FIG 6. *Ae. aegypti* larvae infected with GFP transducing virus. (A) Infected anal papilla of an *Ae. aegypti* larva infected by the pANS1-GFP–transducing virus. (B) Larva infected with the pASN1-GFP–transducing virus showing dissemination of virus. (See Color Insert.)

facilities can affect results. The different virus strains also vary greatly in their pathogenicity *in vivo* as well as *in vitro*. The persistent nature of most densovirus infections in mosquito cell cultures makes virus quantitation by plaque assay or $TCID_{50}$ assay difficult if not impossible. The development of a quantitative PCR assay for detection of several virus strains (Ledermann *et al.*, 2004) is a step in the right direction, although the particle to infectivity ratios may vary for different viruses and the factors that influence infectivity are poorly understood. Nevertheless it is possible to make some generalizations from results reported in the literature.

Infection of first instar *Ae. aegypti* larvae with AalDNV resulted in a 97.6% mortality rate, which occurred mostly in third and fourth instar larvae (48%) and pupae (42%) (Barreau *et al.*, 1996). AeDNV-induced mortality of *Ae. aegypti* mosquitoes is dose dependant, and had a peak total mortality of 75.1% measured over 28 days at the 1×10^{11} genome equivalents per ml (geq/ml) dose as determined by quantitative real time PCR. Mortality was composed of 51.2% larval, 13.6% pupal, and 10.3% adult death (Ledermann *et al.*, 2004). Similar results were acquired for virus isolated from cell culture or by grinding up dead larvae. Infection with 1×10^{10} geq/ml AeDNV resulted in ~50% mortality in 28 days, and lower doses did not show statistically significant mortality over controls (Ledermann *et al.*, 2004). The combined larval and pupal mortality of a number of species of mosquitoes after infection with AeDNV is summarized in Table III (Buchatsky *et al*, 1997) and varies between 16% and 66%. *Aedes* Thailand densonucleosis virus (AThDNV) had a larval mortality rate of 82% in *Ae. albopictus*, 51% in *Ae. aegypti* (Kittayapong *et al.*, 1999), and 17.5% in *An. minimus* (Rwegoshora and Kittayapong, 2004). Although the CpDNV virus was first noted by mortality in a laboratory mosquito colony, the

TABLE III
THE EFFECT OF AEDNV ON DIFFERENT MOSQUITO SPECIES[a]

Mosquito species	Number of insects in the experiment	Larval and pupal mortality (%) in the experiment	Number of insects in the control	Larval and pupal mortality in the control (%)
Ae. aegypti	1140	66.4 ± 2.5	742	1.6 ± 0.5
Ae. albopictus	550	64.1 ± 2.2	398	6.0 ± 1.0
Ae. togoi	450	65.3 ± 1.2	315	4.7 ± 1.3
Ae. vexans	419	59.3 ± 2.0	335	9.2 ± 0.3
Ae. geniculatus	233	32.0 ± 3.0	123	1.6 ± 1.1
Ae. caspius	383	56.1 ± 2.5	245	18.7 ± 2.4
Ae. cantans	440	64.0 ± 8.3	320	34.3 ± 2.4
C. pipiens pipiens	915	43.2 ± 6.4	428	11.4 ± 8.3
C. p. molestus	641	16.0 ± 3.5	350	4.2 ± 1.0
C. *annulata*	315	47.6 ± 5.6	225	24.0 ± 2.8

[a] Adapted from Buchatsky *et al.*, 1997.

mortality has yet to be quantified (Jousset, *et al.*, 2000). HeDNV has shown ~10–30% mortality in *Ae. aegypti* (Ledermann *et al.*, 2004; O'Neill *et al.*, 1995) with the majority of mortality occurring during the third or fourth instar, and only 5–10% occurring during the pupal or adult stages. At the other end of the spectrum, infection of *Ae. aegypti* with APeDNV did not induce significant mortality even at the highest doses (Ledermann, *et al.*, 2004). *Toxorhynchites amboinensis* densonucleosis virus (TaDNV) also induced no mortality in *Ae. albopictus, C. quinquefasciatus,* or *An. gambiae* (O'Neill *et al.*, 1995). It is interesting to note that ability to efficiently infect cell culture or induce apoptosis or CPE in cell culture is not a good predictor for the pathogenicity of a virus in mosquitoes. HeDNV, the only known virus that produces CPE in C6/36 cells, causes only low levels of mortality in mosquitoes (Ledermann *et al.*, 2004). Viruses that have adapted to life in cell culture may lose their ability to induce mortality in mosquitoes. However, this is not always the case because AalDNV, which was isolated from a persistently infected cell culture, also induces a high rate of mortality in mosquitoes (Barreau *et al.*, 1996).

Studies by Barreau *et al.* (1996) demonstrated that several factors influence AalDNV infection. They found that the higher the density of larvae in the rearing water, the higher the virus-induced mortality. Furthermore, there was much greater mortality when infected larvae were reared at 25°C and 27°C vs 20°C, or had contact with infectious virus particles for 48 vs 36 hours. Lastly they found that larvae fed on infected cells either as 1st instar or 3rd instar larvae showed dramatic differences in mortality (78% vs 30%). Ward *et al.* (2001a) showed similar results with AeDNV when large numbers of larvae were infected in small amounts of virus containing water. However, studies with less crowding of larvae in rearing water (25 larvae/80 mls) demonstrated little difference in mortality at any stage when larvae were infected either immediately after hatching or 72 hours later (Suchman *et al.*, 2006).

2. *Effects on Adult Mosquitoes*

A high percentage of adult mosquitoes were infected following exposure to MDV as young larvae. After larval infection of *Ae. aegypti* with AalDNV, 32% of surviving adults were found to be infected by IFA of head squashes (Barreau *et al.*, 1996). Of the adults that survived AThDNV infection, 80% of *Ae. aegypti,* 33% of *Ae. albopictus,* and 33% of *An. minimus* were infected (Kittayapong *et al.*, 1999). About 96.4% of adults that survived infection with 1×10^{10} geq/ml of AeDNV were infected (Ledermann, *et al.*, 2004). Infecting larvae with 1×10^{10} geq/ml of HeDNV and AePDNV, neither of which showed significant mortality in larvae, pupae, or adults, also resulted in 70–75% infection rates in adults. Analysis of virus titers in the HeDNV- and APeDNV-infected mosquitoes by real time PCR showed similar amounts to those of AeDNV-infected mosquitos. All three viruses were excreted at similar levels into the rearing water of infected larvae, indicating that the lack of mortality was not due to a failure to initiate a successful infection. Thus, all three viruses apparently infect and replicate in mosquitoes, but the pathogenicity of the viruses differs significantly (Ledermann, *et al.*, 2004).

Virus infection also affects adult lifespan (Fig. 7). A dose-dependant decrease in adult lifespan was seen that could significantly reduce the ability of infected mosquitoes to transmit viruses, such as dengue, due to high mortality before the extrinsic incubation period was completed (Suchman *et al.*, 2006). There was a slight though significant effect of infection on adult size as assayed by wing length (Fig. 8). This result suggests a decrease in fitness that may account for the observed reduction in adult lifespan.

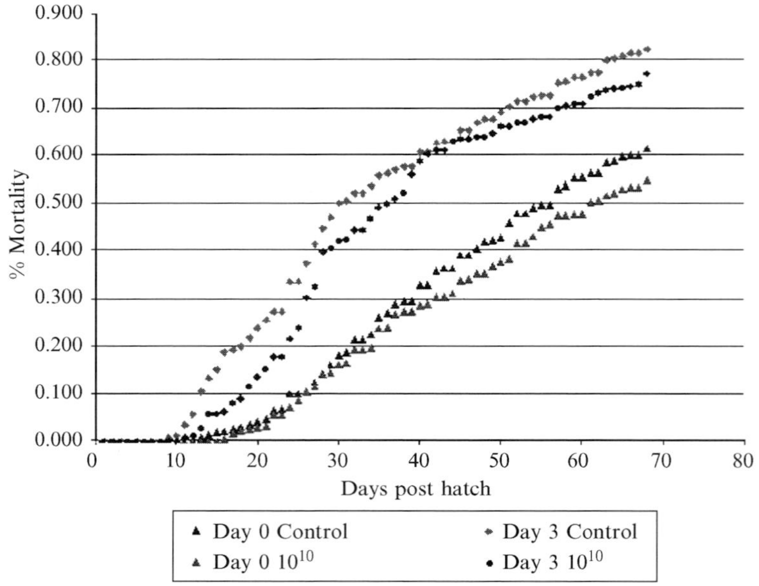

FIG 7. Effect of AeDNV infection on survival of adult *Ae. aegypti*. Two hundred larvae were infected with 10^{10} geq/ml of AeDNV either immediately after hatching or 3 days after hatching. The mortality percentage is shown for virus and control treatments. (See Color Insert.)

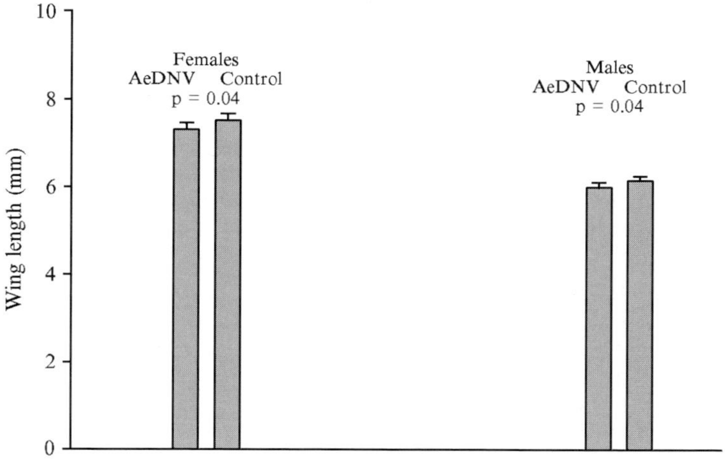

FIG 8. Effect of AeDNV infection on wing length of adult *Ae. aegypti*. The wing length of infected, and uninfected males and females was measured ($n = 240$ per treatment). The difference was found to be statistically significant in both males and females (ANOVA, $p = 0.04$).

Many of the MDV are vertically transmitted. However, each virus seems to be transmitted at different levels (perhaps reflecting in part the method of detection). Virus reproduction in the ovaries has been observed in other densovirus-infected insects (Buchatsky, 1989; Fediere, 2000), and such replication could account for vertical transmission of MDVs. The question of whether the virus is carried within the egg (transovarial transmission), or whether it adheres to the outside of the egg, or is passed with fluids deposited along with the egg has not been resolved. However, when eggs from infected females were treated with 2% formalin, 0.2% sodium hypochlorite, or 0.1% mercuric chloride, which should sterilize the surface of the eggs, the mortality of the offspring was the same as for untreated eggs. This suggests that there is at least some virus inside the egg (Buchatsky *et al.*, 1997). About 61.7% of the offspring of *Ae. aegypti* females infected with 1×10^{10} geq/ml of AeDNV immediately after hatching were infected. Nearly 60.7% of the adults that developed from these larvae were infected. Of the larvae that hatched from females infected on day 3, 42.3% were found to be infected, and 42.8% of the resulting adults were infected. However, when viral titers in infected offspring were assayed by quantitative real time PCR, only low levels of virus could be detected with an average viral titer of 1×10^5 geq/larva and a range of 2×10^4–2×10^5 geq/larva. The filial infection rate was 70% (Suchman *et al.*, 2006). Much lower rates of vertical transmission were found in females infected with the nonpathogenic APeDNV where 19% of larvae and 16% of adults were infected and the filial infection rate was 44.4% (Suchman, E., unpublished data). HeDNV, which also has low pathogenicity, was found to have a filial transmission rate of 20% (O'Neill *et al.*, 1995). AThDNV-infected *Ae. aegypti* females produced offspring with a 100% filial infection rate and 57% of the adults that emerged were infected. It should be noted that when these offspring were bred, with each generation the filial infection rate and percent of infected offspring decreased, and by the seventh generation none of the offspring were positive (Kittayapong, *et al.*, 1999). AThDNV-infected *An. minimus* also had a filial infection rate of 100% and vertical transmission rates ranged from 25–53.8% (Kittayapong, *et al.*, 1999). Similarly, when Barreau *et al.* (1997) determined the $TCID_{50}$ of AalDNV by IFA and infected females with a low (10^3–10^4 $TCID_{50}$) or high (10^6–10^7 $TCID_{50}$) dose of virus they found a 28% or 55.5% vertical transmission rate, respectively. Transmission diminished with subsequent passage and did not persist past the second generation. No other densoviruses have been characterized to this level.

Variable effects on fecundity have been noted for MDV-infected females as measured by decreased egg laying or egg viability. Data from Colorado and Kiev have shown that AeDNV infection affects reproduction. Both groups found that AeDNV-infected females laid 10–20% fewer eggs than uninfected females. (Suchman *et al.*, 2006). This is somewhat different from results reported for AalDNV and AThDNV in which there was no significant reduction in egg laying (Barreau *et al.*, 1997; Rwegoshora and Kittayapong, 2004). Both AalDNV and AeDNV infection resulted in a 10% reduction in egg hatch rate. Furthermore, it was more common for the eggs of females infected with either AeDNV or AalDNV to fail to hatch, and for survival of the larvae to be reduced (Barreau *et al.*, 1997, Suchman *et al.*, 2006). Butchatsky found that AeDNV-infected females also more often fail to lay viable eggs (Suchman *et al.*, 2006). Rwegoshora and Kittayapong found that AThDNV however had no effect on egg hatching.

C. *Effects on Mosquito Populations*

The pathogenic effects of densovirus infection on the different life stages of the mosquito reviewed earlier suggest that there could be significant consequences of MDV infection of mosquito populations.

Relatively little has been done to assess this, but we review here preliminary laboratory population cage studies, studies of MDV in natural mosquito populations in Thailand, and limited field studies of the AeDNV-based microbial pesticide Viroden conducted in the Soviet Union.

1. Population Cage Studies

In order to test the ability of densoviruses to persist and spread in a susceptible mosquito population, population cages were set up with two water containers each. One of these was used as the larval-rearing site and another as the oviposition site. Eggs were added to the larval-rearing site, and resulting adults emerged into the cage. Restrained mice were introduced into the cages periodically to provide blood meals for female mosquitoes. A funnel was inverted over the larval-rearing site container to discourage females from reentering and ovipositing in that container. Papers were placed in the oviposition containers to provide a substrate for egg laying. A fixed number of eggs were introduced into the larval containers each week to maintain the population. AeDNV was inoculated into the larval habitat containers of the experimental cages at a concentration (10^8 per ml) too low to cause noticeable pathogenesis (Ledermann *et al.*, 2004). The oviposition papers were collected weekly and eggs were counted as a measure of the population size. Water samples were collected weekly from the larval habitat and oviposition containers for assay of virus by quantitative PCR.

The amount of virus in the larval habitat containers increased over a few weeks to concentrations that are pathogenic to mosquito larvae (10^{10} per ml) and subsequently the number of eggs dropped to levels significantly lower than in the control cages. Virus appeared in the oviposition site containers as soon as the first generation of adults began laying eggs. These observations did not change for the duration of the experiment (17 weeks). When larvae that hatched from eggs collected from the virus-inoculated cages at the end of the experiment were tested for susceptibility to AeDNV, they were as susceptible to infection as control larvae. These results suggest that AeDNV, when inoculated into a larval-rearing site, can persist and increase in concentration by horizontal transmission. Virus could also spread to new containers by adults infected as a result of vertical transmission. Furthermore, development of resistance to the virus was not observed during the course of the experiment (Suchman *et al.*, unpublished data).

2. *MDV in Natural Mosquito Populations*

Very little work has been done to analyze the prevalence of densoviruses in wild populations of mosquitoes, or to determine how these viruses persist in nature. All of the published work was conducted in Thailand. Kittayapong *et al.* tested trapped adult *Ae. aegypti* and *Ae. albopictus* for AThDNV infection and found a 44.4% prevalence in *Ae. aegypti* but no infection in 79 *Ae. albopictus*. Since AThDNV exhibits a very high mortality rate (82%) in *Ae. albopictus* larvae in the laboratory, infected larvae may die. In fact, Kittayapong *et al.* (1999) hypothesized that the observed differences in AThDNV-induced mortality between *Ae. albopictus* (82%) and *Ae. aegypti* (51%) along with the widespread occurrence of the virus in natural *Ae. aegypti* populations suggests that reduced susceptibility to the virus may provide a competitive advantage to *Ae. aegypti* over *Ae. albopictus*. When environmental prevalence of AThDNV in *Ae. aegypti* adults was analyzed over a 6-month period (May–September), there was high prevalence in the spring (16–22%) peaking at 35% in July and then dropping off dramatically in August and September (6–8%). Interestingly, no females were found to be positive in August and September. This suggests that although high levels of vertical transmission are observed, both vertical and horizontal transmission of the virus play a role in maintaining the virus in natural populations. The finding that infected larvae excrete virus into the rearing water may explain how the virus is maintained when no infected females are present. Furthermore, our lab has shown that both male and female infected adults also excrete virus into water sources (Plake, E., Kleker, B., and Piper, J., unpublished data). When Rwegoshora *et al.* (2000) analyzed AThDNV infection in wild *An. minimus* populations over a 1-year period they found an overall infection rate of 18.8% in larvae and 15% in adults with significant seasonal variation. Larval infection was significantly higher when rainfall was recorded 2 months before collection. Given the potential for use of MDVs as biological control agents for disease vectors it would be beneficial to learn more about these viruses in their natural environment.

3. *Field Tests of AeDNV*

A preparation of AeDNV consisting of ground, infected *Ae. aegypti* larvae in phosphate-buffered saline and glycerol known as "Viroden" has undergone extensive testing as a microbial pesticide in Ukraine (Buchatsky *et al.*, 1987b, 1997). The effect of Viroden on larval and pupal mosquitoes was tested in field trials conducted in three different climatic zones: in Ukraine near Kiev, Russia near Tula, and Tadjikistan in the Kurgan-Tube regions (Buchatsky *et al.*, 1987a). In Ukraine, small artificial ponds (100 cm × 100 cm × 50 cm) were dug and lined with polyethylene film. Local aquatic organisms including mosquito larvae and bottom substrate were added. Mosquito densities were determined by averaging the counts from four samples from different parts of the pond. Viroden was introduced by spraying on the water surface. The numbers of mosquitoes were monitored and compared to untreated control reservoirs. When compared to control ponds, combined larval and pupal mortality ranged from 44.0% to 86.0%, thus demonstrating efficacy of Viroden against the local species *Aedes caspius, Ae. vexans, Ae. cantans, Ae. cinereus,* and *C. pipiens*. Bioassays conducted on water samples from treated reservoirs showed that active virus remained in the water for at least a year. This result means that the disease can be passed onto subsequent generations of mosquitoes, as long as new generations of larvae are present to become infected.

Similar field trials were conducted under the colder climatic conditions of the Tula region of Russia. Various natural reservoirs were treated, including swamped meadows adjacent to streams, ditches overgrown with vegetation, and old trenches. All of these reservoirs were known to serve as egg-laying sites for *Aedes communis* early in the spring and for *Culex* sp. later in the season. Twelve reservoirs with a combined surface area of 150 m^2 were treated. For each test reservoir, two–three comparable control reservoirs were selected. Treatments were started after the appearance of first instar larvae. *Ae. communis* and *Culex* sp. showed combined larval and pupal mortality in the experimental reservoirs of between 59.0% and 76.1%. Viroden did not have any noticeable toxic effect on other aquatic organisms including crustaceans, water beetles, and frogs.

Field trials were also conducted under the hot climatic conditions of the irrigated farm zone of the Kurgan-Tube region of Tadjikistan. Ten reservoirs, with a combined surface area of 45.6 m^2 and containing first and second instar larvae of *Ae. caspius* and occasionally *C. pipiens*, were used in these trials. Initial larval densities in these reservoirs reached 1400–2000 per m^2 of water surface. Combined larval and pupal mortality fluctuated between 43.0% and 73.0%, and again no toxic effect of Viroden on nontarget species was detected. In these field trials it was not possible to determine the effects of Viroden on the mortality and fecundity of adult mosquitoes, or the hatch rates of the next generation. However, because similar effects of Viroden were observed on larvae and pupae under laboratory and field conditions, the effects of Viroden on adults under natural conditions could be similar to those seen in the laboratory.

Highly virulent pathogens are similar to chemical insecticides in that their action results in extermination of a pest population. Moderately effective pathogens, such as MDV, persist in the population causing prolonged population reduction. The pathogenicity of AeDNV against the larvae of *Culicidae* and the persistence of viable virus in reservoirs suggests that densovirus-based treatments, such as Viroden, can be an effective regulator of populations of dangerous disease vectors such as mosquitoes.

IV. Densovirus-Transducing Vectors

The complete genomes of AeDNV and APeDNV have been cloned into the *Escherichia coli* plasmid vector pUC19. These clones are called pUCA and pUCP respectively. When these clones are transfected into C6/36 *Ae. albopictus* cells, the viral genes are expressed from the viral promoters and the viral genome is released from the plasmid vector (presumably by the nickase and/or helicase activities of the NS1 protein) and infection is initiated. Infectious virus particles are produced and released from the cells. The availability of these infectious clones makes it possible to construct transducing genomes, which carry genes of interest in the viral genome and when packaged into virus particles, introduce the genes into cells by infection. The major limitation of densovirus-based transducing systems is the small size of the genome and the strict size limit of the genome imposed by the icosahedral geometry of the virus particle. The brevidensovirus genomes are about 4000 nucleotides in length and genomes that exceed that size by less than ~10% can still be packaged. Genomes larger than that are not packaged (Afanasiev *et al.*, 1994).

The only *cis*-acting sequences necessary for replication and packaging are the terminal inverted repeat sequences that make up the first ~200 bases of the 5′ end ~150 bases at the 3′ end. The size of the coding region of these viruses is ~3350 bases and all of the viral gene products are necessary for viability. Therefore there is little room for the insertion of extra genes. Most of the transducing viruses described to date have been constructed by replacing viral coding sequences with the gene of interest. This results in a defective genome, which requires a helper to supply the viral proteins that are missing by complementation.

A. Replacement Vectors

Transducing genomes carrying the *E. coli* β-galactosidase gene have been constructed and described elsewhere (Afanasiev and Carlson, 2000; Afanasiev *et al.*, 1994; Carlson *et al.*, 2000). In order to be packaged essentially all of the viral-coding sequences were removed from these genomes. Such total replacement genomes require both the nonstructural and the *VP* genes from a helper virus for replication and packaging. These genomes were quite useful in demonstrating transduction *in vitro* but have not been used extensively *in vivo*.

Densovirus vectors have also been made in which only the *VP* gene has been replaced by a gene of interest. These genomes retain intact *NS1* and *NS2* genes and can excise themselves from the plasmid vector and replicate but they must be supplied with VP protein to be packaged. Care must be taken in the design of these vectors because the 5′ end of the *VP* gene overlaps the 3′ end of the *NS1* gene by 115 bases. Truncation of the NS1 protein eliminates NS1 activity and therefore any new gene must be inserted downstream of the NS1-coding sequence. The most useful of these vectors has been pANS1-GFP (originally called p7NS1-GFP; Afanasiev *et al.*, 1999) which has the *GFP* gene fused in frame to the *NS1* gene eliminating the *NS1* termination codon (Fig. 9A). This construct produces an NS1–GFP fusion protein, which seems to have all of the activities of the NS1 protein (Afanasiev *et al.*, 1999). Both pUCA and pUCP, the infectious clones of AeDNV and APeDNV, can be used as helpers for pANS1-GFP. This suggests that genomes with AeDNV ends can be packaged in APeDNV capsids and pseudotyping of MDV genomes from different clades is possible. This transducing genome when delivered to larvae as a mixture of transducing particles and helper virus allowed the progress of infections to be observed in living larvae by GFP fluorescence (Section III.B; Afanasiev *et al.*, 1999; Ward *et al.*, 2001a).

B. Nondefective Vectors

When one aligns brevidensovirus genome sequences, the most variable part of the genome is the region between the termination codon of the *VP* gene and the inverted repeat sequences that define the 3′ end of the genome. This region contains the sequences necessary for the termination of the viral mRNAs (Fig. 2B), which through deletion mapping and site-directed mutagenesis were determined to be within the first 80 bases downstream of the termination codon of the *VP* gene in AeDNV (Konet, unpublished data). Sequences immediately

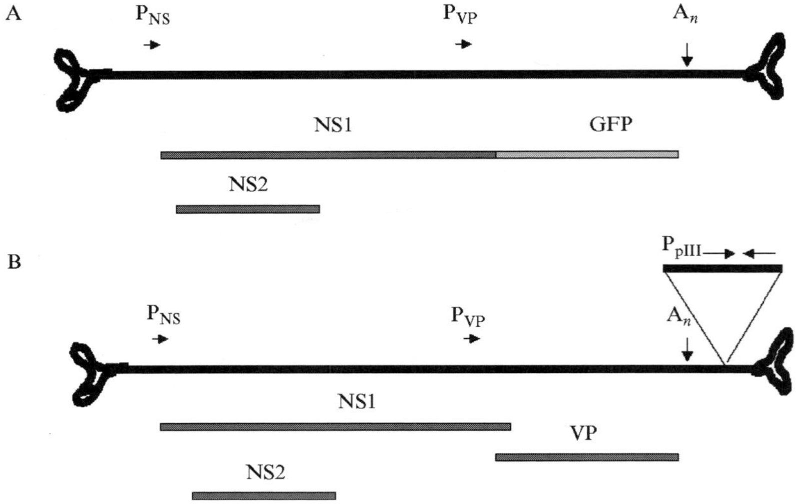

FIG 9. Transducing densovirus vectors. (A) The green fluorescent protein-transducing virus pANS1-GFP in which the *GFP* gene replaces the *VP* gene. (B) A nondefective RNAi transducing virus with an RNA Pol III–shRNA expression cassette inserted downstream of *VP*. A_n, polyA tail.

downstream of this termination region appear by deletion analysis to be dispensable for virus growth and viability and could therefore be replaced by sequences of choice. However, for packaging within particles, the size of the inserted sequence is limited to 200–300 bases. Although this size limit excludes most protein-coding genes, there is enough room for constructs encoding short hairpin RNAs (shRNA) that can be used to mediate posttranscriptional gene silencing by RNA interference (RNAi). There has been considerable success in inducing RNAi in mammalian systems with vectors that transcribe shRNAs (21–25 bp) from RNA polymerase III (Pol III) promoters (Kojima *et al.*, 2004; Lewis *et al.*, 2002).

We identified putative mosquito Pol III promoters that transcribe *U6snRNA* genes from both the *An. gambiae* and *Ae. aegypti* genomes by searching the completed genome sequences with the highly conserved *U6snRNA* sequence. The sequences upstream of the *U6snRNA* genes were amplified by PCR with primers designed from the genomic sequences, and cloned. These promoters were tested for activity by cloning an inverted repeat sequence coding for an shRNA that targets the firefly luciferase gene, downstream of the promoter. The Pol III transcripts terminate in a series of six–eight T residues placed downstream of the inverted repeat. When the shRNA expression constructs and a construct that expresses the firefly luciferase were cotransfected into Ag-55 *An. gambiae* or ATC-10 *Ae. aegypti* cells, luciferase expression was reduced by 50–90%. Thus shRNA expressed from Pol III expression constructs can mediate RNAi in mosquito cells as for mammalian cells (Konet, unpublished data).

The optimal *An. gambiae* and *Ae. aegypti* Pol III luciferase shRNA expression cassettes have been cloned into the dispensable portion of the AeDNV genome to produce nondefective expression vectors (Fig. 9B). These vectors have been transfected into C6/36 cells to produce virus, and this virus was used to infect *Ae. aegypti* larvae. Virus was recovered from surviving adults after 28 days and examined for the integrity of the expression cassette. All of the viruses retained expression cassettes that were unchanged by passage through the mosquito (Konet, D., unpublished data). This result suggests that these nondefective transducing viruses may be useful for manipulating gene expression in infected mosquitoes by shRNAs or perhaps micro-RNAs (miRNAs) expressed from Pol III expression cassettes.

V. Conclusion: Densoviruses and Vector-Borne Disease

The control of vector-borne diseases is a critical factor in improving the quality of life throughout the world and especially in the tropics. Due to the complexity of the problem this goal will only be realized by integrating and adapting multiple tools and strategies to address a wide variety of situations. Mosquito densoviruses are one of the tools that should have application in this struggle. Their specificity for mosquitoes and their stability in the environment are highly attractive. Their ability to persist and accumulate in larval habitats by horizontal transmission among larvae suggests that even a single introduction of virus at a relatively low titer can eventually build up and result in significant pathogenesis in the population. The ability of the virus to spread to new oviposition sites by vertical transmission is also attractive. A single infected female typically lays eggs at multiple sites during a gonotrophic cycle and may thereby spread the virus to sites that are difficult to find and treat by more conventional control programs. The fact that virus infection can significantly reduce the adult lifespan has the potential to greatly influence the vectorial capacity for transmission of disease. A female mosquito must acquire a pathogen in a blood meal and the pathogen must move from the midgut to the salivary glands before it can be transmitted during a subsequent blood meal. The time required for this process, the extrinsic incubation period, is typically 10–15 days depending on the pathogen and other factors. If densovirus infection reduces the adult lifespan to less than the extrinsic incubation period, pathogen transmission should be greatly reduced. Thus, even though the mosquito population may not be eliminated by densovirus infection, disease transmission may be slowed.

The possibility of using genetically engineered MDV for enhanced efficacy as a biological control agent is also attractive and could allow for alternative approaches. The prospect of genetic manipulation of mosquitoes with nondefective transducing viruses expressing shRNA or miRNA seems feasible and could potentially be developed into novel applications for combating mosquito-borne diseases. These would seem to be more politically palatable than the release of genetically modified mosquitoes.

Although MDV has a number of attractive attributes as a microbial control agent, there are a number of questions that remain to be answered before the maximum value of MDV can be realized. Are densoviruses common in natural populations and how are they maintained in nature? Do MDV have a role in determining the structure and the dynamics of mosquito populations? Thus far the only reports of densoviruses in wild mosquito populations are from Thailand (Kittayapong *et al.*, 1999), but no comprehensive survey of wild populations has been conducted elsewhere. The multiple isolates of MDV from cell culture and laboratory mosquito colonies would suggest that MDV may be relatively common in the field. There is a great need for more work to be done in the field. What are the mechanisms and molecular determinants of, for example, viral host range, tissue tropism, pathogenesis, and vertical transmission? Much remains to be done on the basic virology and pathogenesis of these viruses in the mosquito. Do mosquitoes develop resistance to MDVs and why do different strains of virus differ so dramatically in pathogenicity? Is there an innate antiviral immune response against DNA viruses in mosquitoes? The emerging picture of immunity in insects is one of unexpected complexity. Given the sophisticated defense mechanisms against bacteria and parasites (Bartholomay *et al.*, 2004a,b; Blandin *et al.*, 2004), and the role that RNAi seems to play in RNA virus infections in mosquitoes (Keene *et al.*, 2004; Sanchez-Vargas *et al.*, 2004), it seems likely that there is a response to DNA viruses such as densoviruses. The availability of the *An. gambiae* and *Ae. aegypti* genome sequences and genomics-based approaches could provide answers to some of these questions in the near future.

Another challenge that needs to be addressed is the production of sufficient quantities of MDV for field application of the virus. This will require either large-scale mosquito rearing or large-scale cell culture. The original preparations of Viroden were produced in mosquito larvae; however, the demonstration of MDV production in C6/36 cells grown in suspension in serum-free, protein-free medium may make cell culture more feasible (Suchman and Carlson, 2004). Large-scale purification of MDV by binding to ion exchange membranes has also been investigated (Han *et al.*, 2005; Specht *et al.*, 2004). Economical production of appropriately pure preparations of MDV will be critical for effective use and acceptance in developing countries. There is much to be done to move MDV from the realm of promising prospects to that of practical tools.

Acknowledgments

We would like to thank Eugene and Victoria Azarkh for translation of several articles by Prof. Buchatsky into English. We thank Dan Konet, Joe Piper, and other members of our laboratories for sharing their work prior to publication. This work was supported by the contract NO1-AI 25489 and grant R01-AI47139 from the National Institutes of Health.

References

Afanasiev, B., Kozlov, Y., Carlson, J., and Beaty, B. (1994). Densovirus of *Aedes aegypti* as an expression vector in mosquito cells. *Exp. Parasitol.* **79:**322–339.

Afanasiev, B. N., and Carlson, J. O. (2000). Densoviruses as gene transfer vehicles. *In* "Parvoviruses: From Molecular Biology to Pathology and Therapeutic Uses" (S. Faisst and S. Rommerlaere, eds.), Vol. 4, pp. 33–58. S. Karger, Basal, Switzerland.

Afanasiev, B. N., and Carlson, J. (2001). Brevidensovirus. *In* "The Springer Index of Viruses" (J. Tidona and C. A. Darai, eds.), pp. 690–694. Springer, Berlin.

Afanasiev, B. N., Ward, T. W., Beaty, B. J., and Carlson, J. O. (1999). Transduction of *Aedes aegypti* mosquitoes with vectors derived from Aedes densovirus. *Virology* **257:**62–72.

Anderson, J. F., Andreadis, T. G., Vossbrinck, C. R., Tirrell, S., Wakem, E. M., French, R. A., Garmendia, A. E., and VanKruiningen, H. J. (1999). Isolation of West Nile virus from mosquitoes, crows and a Cooper's hawk from Connecticut. *Science* **286:**2331–2333.

Barreau, C., Jousset, F.-X., and Bergoin, M. (1996). Pathogenicity of *Aedes albopictus* parvovirus, a denso-like virus, for *Aedes aegypti* mosquitoes. *J. Invertebr. Pathol.* **68:**299–309.

Barreau, C., Jousset, F.-X., and Bergoin, M. (1997). Venereal and vertical transmission of the *Aedes albopictus* parvovirus in *Aedes aegypti* mosquitoes. *Am. J. Trop. Med. Hyg.* **57:**126–131.

Bartholomay, L. C., Cho, W. L., Rocheleau, T. A., Boyle, J. P., Beck, E. T., Fuchs, J. F., Liss, P., Rusch, M., Butler, K. M., Wu, R. C., Lin, S. P., Kuo, H. Y., *et al.* (2004a). Description of the transcriptomes of immune response-activated hemocytes from the mosquito vectors *Aedes aegypti* and *Armigeres subalbatus*. *Infect. Immun.* **72:**4114–4126.

Bartholomay, L. C., Fuchs, J. F., Cheng, L. L., Beck, E. T., Vizioli, J., Lowenberger, C., and Christensen, B. M. (2004b). Reassessing the role of defensin in the innate immune response of the mosquito, Aedes aegypti. *Insect Mol. Biol.* **13:**125–132.

Bergoin, M., and Tijssen, P. (2000). Molecular biology of Densovirinae. *In* "Parvoviruses: From Molecular Biology to Pathology and Therapeutic Uses" (P. Faisst and S. Rommerlaere, eds.), Vol. 4, pp. 12–32. S. Karger, Basal, Switzerland.

Berns, K. I. (1996). Parvoviridae: The viruses and their replication. *In* "Fundamental Virology" (K. I. Fields, B. N. Knipe, and D. M. Howley, eds.), 3rd Ed., pp. 1017–1041. Lipincott-Raven, Philadelphia.

Blandin, S., Shiao, S.-H., Moita, L. F., Janse, C. J., Waters, A. P., Kafatos, F. C., and Levashina, E. A. (2004). Complement-like protein TEP-1 is a determinant of vectorial capacity in the malaria vector *Anopheles gambiae*. *Cell* **116:**661–670.

Boublik, Y., Jousset, F.-X., and Bergoin, M. (1994). Complete nucleotide sequence and genomic organization of the *Aedes albopictus* parvovirus (AaPV) pathogenic for *Aedes aegypti* larvae. *Virology* **200:**752–763.

Buchatsky, L. P., Bogdanova, E. N., Kuznetsova, M. A., Lebedinets, N. M., Kononko, A. G., Chabanenko, A. A., and Podrezova, L. M. (1987a). Field trial of the viral preparation Viroden on the preimaginal stages of blood-sucking mosquitoes. *Meditsinskaya Parazitologiya i Parazitare. Bolezni.* **4:**69–71.

Buchatsky, L. P., Kuznetsova, M. A., Lebedinets, N. M., and Kononko, A. G. (1987b). Development and basic properties of a virus preparation: Viroden. *Vopr. Virusol.* **32:**729–733.

Buchatsky, L. P. (1989). Densonucleosis of bloodsucking mosquitoes. *Dis. Aquat. Organ.* **6:**145–150.

Buchatsky, L. P., Lebedinets, N. M., and Kononko, G. G. (1997). "Densonucleosis of Bloodsucking Mosquitoes: Student Handbook for Biology Majors." Taras Shevchenko' Kiev National University, Kiev.

Burivong, P., Pattanakitsakul, S.-N., Thongrungkiat, S., Malasit, P., and Flegel, T. W. (2004). Markedly reduced severity of Dengue virus infection in mosquito cell cultures persistently infected with *Aedes albopictus* densovirus (*Aal*DNV). *Virology* **329:**261–269.

Carlson, J., Afanasiev, B., and Suchman, E. (2000). Densoviruses as transducing vectors for insects. *In* "Insect Transgenesis: Methods and Applications" (E. Handler and A. James, eds.), pp. 139–159. CRC Press, Boca Raton.

Chen, S., Cheng, L., Zhang, Q., Lin, W., Lu, X., Brannan, J., Zhou, H., and Zhang, X. (2004). Genetic, biochemical, and structural characterization of a new densovirus isolated from a chronically infected Aedes albopictus C6/36 cell line. *Virology* **318:**123–133.

El-Far, M., Li, Y., Fediere, G., Abol-Ela, S., and Tijssen, P. (2004). Lack of infection of vertebrate cells by the densovirus from the maize worm Mythimna loreyi (MlDNV). *Virus Res.* **99:**17–24.

Fediere, G. (2000). Epidemiology and pathology of densovirinae. *In* "Parvoviruses: From Molecular Biology to Pathology and Therapeutic Uses" (G. Faisst and S. Rommerlaere, eds.), Vol. 4, pp. 1–11. S. Karger, Basal, Switzerland.

Gorziglia, M., Botero, L., Gil, F., and Esparza, J. (1980). Preliminary characterization of virus-like particles in a mosquito (*Aedes pseudoscutellaris*) cell line (Mos. 61). *Intervirology* **13:**232–240.

Gratz, N. G. (1999). Emerging and resurging vector-borne disease. *Ann. Rev. Entomol.* **44:**51–75.

Han, B., Specht, R., Wickramasinghe, S. R., and Carlson, J. O. (2005). Binding *Aedes aegypti* densonucleosis virus to ion exchange membranes. *J. Chromatogr. A* **1092:**114–124.

Jousset, F.-X., Barreau, C., Boublik, Y., and Cornet, M. (1993). A parvo-like virus persistently infecting a C6/36 clone of *Aedes albopictus* mosquito cell line and pathogenic for *Aedes aegypti* larvae. *Virus Res.* **29:**99–114.

Jousset, F.-X., Baquerizo, E., and Bergoin, M. (2000). A new densovirus isolated from the mosquito Culex pipiens (Diptera: Culicidae). *Virus Res.* **67:**11–16.

Keene, K. M., Foy, B. D., Sanchez-Vargas, I., Beaty, B. J., Blair, C. D., and Olson, K. E. (2004). RNA interference acts as a natural antiviral response to O'nyong-nyong virus (Alphavirus; Togaviridae) infection of Anopheles gambiae. *Proc. Natl. Acad. Sci. USA* **101:**17240–17245.

Kimmick, M. W., Afanasiev, B. N., Beaty, B. J., and Carlson, J. O. (1998). Gene expression and regulation from the p7 promoter of the *Aedes* densonucleosis virus. *J. Virol.* **72:**4364–4370.

Kittayapong, P., Baisley, K. J., and O'Neill, S. L. (1999). A mosquito densoviruses infecting *Aedes aegypti* and *Aedes albopictus* from Thailand. *Am. J. Trop. Med. Hyg.* **61:**612–617.

Kojima, S., Vignjevic, D., and Borisy, G. (2004). Improved silencing vector co-expressing GFP and small hairpin RNA. *Biotechniques* **36:**74–79.

Lanciotti, R. S., Roehrig, J. T., Deubel, V., Smith, J., Parker, M., Steele, K., Crise, B., Volpe, K. E., Crabtree, M. B., Scherret, J. H., Hall, R. A., MacKenzie, J. S., *et al.* (1999). Origin of the West Nile virus responsible for an outbreak of encephalitis in the Northeastern United States. *Science* **286:** 2333–2337.

Lebedeva, P. O., Zelenko, A. P., Kuznetsova, M. A., and Gudzgorban, A. P. (1972). Studies on the demonstration of a viral infection in larvae of *Aedes aegypti* mosquitoes. *Microbiol. JSU* **34:**70–73.

Ledermann, J. P., Suchman, E. L., Black, W. C., IV, and Carlson, J. O. (2004). Infection and pathogenicity of the mosquito densoviruses AeDNV, HeDNV, and APeDNV in *Aedes aegypti* mosquitoes (Diptera: Culicidae). *J. Econ. Entomol.* **97:**1827–1835.

Lewis, D. L., Hagstrom, J. E., Loomis, A. G., Wolff, J. A., and Herweijer, H. (2002). Efficient delivery of siRNA for inhibition of gene expression in postnatal mice. *Nat. Genet.* **32:**107–108.

Meynadier, G., Vago, C., Plantevin, G., and Atger, P. (1964). Virose de un type inhabituel chez le Lepedoptere *Galleria mellonella*. *Rev. Zool. Agr. Appl.* **63:**207–208.

O'Neill, S. L., Kittayapong, P., Braig, H. R., Andreadis, T. G., Gonzalez, J. P., and Tesh, R. B. (1995). Insect densoviruses may be widespread in mosquito cell lines. *J. Gen. Virol.* **76:**2067–2074.

Paterson, A., Robinson, E., Suchman, E., Afanasiev, A., and Carlson, J. (2005). Mosquito densonucleosis viruses cause dramatically different infection phenotypes in the C6/36 *Aedes albopictus* cell line. *Virology* **337:**253–261.

Rwegoshora, R. T., and Kittayapong, P. (2004). Pathogenicity and infectivity of the Thai-strain densovirus (AthDNV) in *Anopheles minimus*. *Southeast Asian J. Trop. Med. Public Health* **35:**630–634.

Rwegoshora, R. T., Baisley, K. J., and Kittayapong, P. (2000). Seasonal and spatial variation in natural densovirus infection in *Anopheles minimus* S. L. in Thailand. *Southeast Asian J. Trop. Med. Public Health.* **31:**3–9.

Sanchez-Vargas, I., Travanty, E. A., Keene, K. M., Franz, A. W. E., Beaty, B. J., Blair, C. D., and Olson, K. E. (2004). RNA interference, arthropod-borne viruses, and mosquitoes. *Virus Res.* **102:**65–74.

Specht, R., Han, B., Wickramasinghe, S. R., Carlson, J. O., Czermak, P., Wolf, A., and Reif, O.-W. (2004). Densonucleosis virus purification by ion exchange membranes. *Biotechnol. Bioeng.* **88:**465–473.

Suchman, E., and Carlson, J. (2004). Production of mosquito densonucleosis viruses by *Aedes albopictus* C6/36 cells adapted to suspension culture in serum-free protein-free media. In Vitro *Cell. Dev. Biol. Anim.* **40:**74–75.

Suchman, E. L., Kononko, A., Plake, E., Doehling, M., Kleker, B., Black, W. C., Buchatsky, L., and Carlson, J. (2006). Effects of AeDNV infection on *Aedes aegypti* (L.) lifespan and reproduction. *Biol. Control* (in press).

Vasileva, V. L., Lebedinits, N. N., Gural, A. L., Chigir, T. V., Buchatsky, L. P., and Kuznetsova, M. A. (1990). Examination of the safety of Viroden for vertebrates. *Mikrobiol. Z. (Kiev)* **52:**73–79.

Ward, T. W., Jenkins, M. S., Afanasiev, B. N., Edwards, M., Duda, B. A., Suchman, E., Jacobs-Lorena, M., Beaty, B. J., and Carlson, J. O. (2001a). *Aedes aegypti* transducing densovirus pathogenesis and expression in *Aedes aegypti* and *Anopheles gambiae* larvae. *Insect Mol. Biol.* **10:**397–406.

Ward, T. W., Kimmick, M. W., Afanasiev, B. N., and Carlson, J. O. (2001b). Characterization of the structural gene promoter of *Aedes aegypti* densovirus. *J. Virol.* **75:** 1325–1331.

ADVANCES IN VIRUS RESEARCH, VOL 68

POTENTIAL USES OF CYS-MOTIF AND OTHER POLYDNAVIRUS GENES IN BIOTECHNOLOGY

Torrence A. Gill,* Angelika Fath-Goodin,*,† Indu I. Maiti,‡ and Bruce A. Webb*

*Department of Entomology, S-225 Agricultural Science Building North
University of Kentucky, Lexington, Kentucky 40546
†ParaTechs Corp., 105c KTRDC Building, University and Cooper Drs.
Lexington, Kentucky 40546
‡Molecular Plant Virology and Plant Genetic Engineering Lab, KTRDC Building
University of Kentucky, Lexington, KY 40546

Abstract

Exploiting the ability of insect pathogens, parasites, and predators to control natural and damaging insect populations is a cornerstone of biological control. Here we focus on an unusual group of viruses, the polydnaviruses (PDV), which are obligate symbionts of some hymenopteran insect parasitoids. PDVs have a variety of important pathogenic effects on their parasitized hosts. The genes controlling some of these pathogenic effects, such as inhibition of host development, induction of precocious metamorphosis, slowed or reduced feeding, and immune

0065-3527/06 $35.00
DOI: 10.1016/S0065-3527(06)68011-1

suppression, may have use for biotechnological applications. In this chapter, we consider the physiological functions of both wasp and viral genes with emphasis on the Cys-motif gene family and their potential use for insect pest control.

I. Introduction

Exploiting the ability of insect pathogens, parasites, and predators to control damaging insect populations is a cornerstone of biological control. Agronomic manipulations that eliminate or greatly reduce these natural control agents are also widely recognized as key factors in outbreaks of secondary pests that would normally be suppressed below economic threshold levels by these biocontrol agents (Croft, 1990; Croft and Brown, 1975; Lewis *et al.*, 1997). Many strategies have been developed to augment these biocontrol agents to suppress pest populations with varying degrees of success (Lewis *et al.*, 1997). More recently, the mode of action of microbial pesticides has been investigated to identify, understand, and exploit their toxicological mechanisms. The most notable success in this area is that of the *Bacillus thuringiensis* (Bt) toxins and the proven utility of these toxins in controlling pest populations when expressed from transgenic plants (Estruch *et al.*, 1997; Kota *et al.*, 1999; Nayak *et al.*, 1997; Singsit *et al.*, 1997; Stewart *et al.*, 1996; Tian *et al.*, 1991).

Although significant efforts have been directed to production, formulation, and genetic modification of entomopathogenic viruses (Narayanan, 2004), there have been few examples of genes derived from insect viruses that have proven utility in insect control when expressed in plants (see chapter by Liu *et al.*, this volume, pp. 427–457). By contrast, genes from other pathogens, such as the coat protein genes of plant pathogenic viruses, have been successfully exploited by expression in transgenic plants for plant resistance to the viral pathogens from which the coat protein genes were derived (Abel *et al.*, 1986; Bendahmane *et al.*, 1997; Clark *et al.*, 1990; Nejidat and Beachy, 1989, 1990; Ploeg *et al.*, 1993; Reimann-Philipp and Beachy, 1993). Taken together, this brief consideration of related fields suggests that there may be opportunity for identification of genes that control insects from the entomopathogenic viruses, parasites, and predators that are used for the biological control of pest insects.

To this end, we and others have focused on hymenopteran insect parasitoids and an unusual group of viruses, the polydnaviruses (PDVs), that are obligate symbionts of some of these wasps. In this chapter we consider the physiological functions of both wasp and viral genes and their potential utility in insect control. Hence, we only describe the viral and parasitoid genes that have been sequenced, cloned, expressed, and functionally characterized. There are numerous reports of parasitoid-derived "factors" (larval and ovarian proteins, venoms, and PDVs) that have biological activities of interest, and the accumulating genomic sequence information from several systems indicates that other candidate genes are worthy of investigation (Beckage and Gelman, 2004). There is also evidence that genes from these viruses may have other biotechnological applications (see chapter by Fath-Goodin *et al.*, this volume, pp. 75–90). Here, we focus on the limited number of PDV genes for which functional studies have been reported and which have activities relevant to insect control.

II. Polydnavirus Life Cycle and Induced Pathologies

A. *Parasitoid and Polydnavirus Biology*

Endoparasitic Hymenoptera must avoid detection and subsequent attack from insect host humoral and cellular immune responses and also maintain their host insect in a physiological and developmental state that enables survival and development of the endoparasite. During parasitization, the endoparasitoid wasp may deposit one or more eggs, venoms, ovarian proteins, and viruses (Kroemer and Webb, 2004; Stoltz, 1993; Turnbull and Webb, 2002; Webb, 1998). Survival of the endoparasitoid egg requires suppression of the host's encapsulation response by these factors from the female wasp. Otherwise the immune system of the host would recognize, encapsulate, and kill the endoparasite. While many parasitoids deploy only venoms and secretory proteins from the adult and larval wasp to alter host physiology, some members of the Ichneumonidae and Braconidae rely primarily on viruses to alter host physiology. These viruses, the PDVs, exist in an obligate symbiotic mutualism with endoparasitic Hymenoptera and replicate only in the female reproductive tract. PDVs are delivered with the wasp egg to host insects during oviposition. The expression of PDV genes prevents encapsulation of the wasp egg and larva by compromising the function of hemocytes involved in this response (granulocytes and plasmatocytes). Suppression of encapsulation requires establishment of a PDV infection and expression of viral genes. PDV expression also alters other aspects of host physiology, notably development, growth, and nutritional metabolism (Bae and

Kim, 2004; D'Amico *et al.*, 2001; Gundersen-Rindal and Pedroni, 2006; Hoch *et al.*, 2002; Kaeslin *et al.*, 2005a,c; Kroemer and Webb, 2004; Malva *et al.*, 2004; Nakamatsu and Tanaka, 2004; Shelby and Webb, 1994, 1997, 1999; Shelby *et al.*, 1998).

Two genera of PDVs are recognized, the bracoviruses (BVs) and ichnoviruses (IVs). These groups are estimated to have in excess of 25,000 species (Stoltz, 1993; Turnbull and Webb, 2002). The PDV life cycle is unusual and has been described as having two separate arms in which one arm of the life cycle is responsible for virus transmission and replication (within the wasp), while the other arm is responsible for the pathogenic effects that PDVs have on the parasitized insect (usually a lepidopteran larva) (Kroemer and Webb, 2004; Stoltz, 1993; Turnbull and Webb, 2002). During the transmission and replication arm of the life cycle, in both IVs and BVs, virus replicates from proviral DNA in specialized oviduct cells known as calyx cells (Turnbull and Webb, 2002). PDV transmission is Mendelian as all wasps in which PDVs reside inherit the infection via proviral DNA which is integrated into the genome of the hymenopteran endoparasitoid. Therefore, vertical transmission occurs only through successful parasitization and emergence of an endoparasitoid larva as an adult wasp (Kroemer and Webb, 2004; Turnbull and Webb, 2002; Webb, 1998). The pathogenic arm of the PDV life cycle initiates with the replication of proviral DNA to produce the segmented, double-stranded, extrachromosomal PDV genome for packaging into virions within the calyx cells of female wasps (Gruber *et al.*, 1996; Savary *et al.*, 1997; Stoltz *et al.*, 1976; Volkoff *et al.*, 1995).

There are significant differences in replication between the two PDV groups. IVs release virus from calyx cells via budding (Volkoff *et al.*, 1995) while BVs replicate in calyx cells that lyse to free the BV virions (Stoltz *et al.*, 1976). In both cases, virions accumulate in large amounts in the lumen of the oviduct. In IVs, nucleocapsids are assembled in calyx cell nuclei and singly enveloped (Norton and Vinson, 1983; Stoltz and Vinson, 1979; Yin *et al.*, 2003) with IVs then budding through the cell membrane to acquire a second membrane (Stoltz and Vinson, 1979; Stoltz, *et al.*, 1976; Volkoff *et al.*, 1995; Wyler and Lanzrein, 2003). By contrast, BVs have a single envelope after cell lysis (Wyler and Lanzrein, 2003).

After oviposition, virus that enters the host insect with the parasitoid egg infects hemocytes, fat body cells, and other host tissues where viral expression has notable effects on host immune and development systems, but the virus itself does not replicate within the infected host insect. Thus, the physiological effects on host immune and developmental systems are caused by the infection of host cells and expression of PDV genes rather than replication of the virus. Because of the potential biotechnological applications of these genes, greater detail will be provided on the effects of PDV genes on developmental physiology. The effects of PDV on host immune responses have been reviewed and will be considered here in less detail (Kroemer and Webb, 2004; Strand and Pech, 1995a).

B. Insect Immune Responses

Insect innate immune defenses are conventionally separated into two response categories: cellular and humoral immunity. Cellular immunity entails hemocyte-mediated responses, such as encapsulation, phagocytosis, and nodulation, which involve recognition of foreign objects followed by activation of hemocyte-mediated responses that vary depending on the size of the invading organism and extent of the infection. In lepidopterans, granulocytes and plasmatocytes mediate the encapsulation response to eukaryotic organisms. Deposition of melanin usually accompanies encapsulation and nodulation responses, with the release of reactive oxygen-free radicals and cytotoxic quinones which are thought to contain and destroy the pathogen (Carton and Nappi, 1997). Hemolymph immune responses that do not directly require hemocytes comprise innate humoral immunity. The binding of lectins and other sentinel molecules, induction of antimicrobial proteins, and defensive melanization are three classes of immune responses comprising innate humoral immunity (Gillespie *et al.*, 1997). Recognition of foreign objects in the hemolymph is thought to proceed through binding of sentinel molecules to patterns that are common to various classes of pathogens. For example, some multimeric lectins bind to the surface of foreign bodies through the carbohydrate moieties conserved on bacteria, which then promotes the attachment and spreading of immunocytes (Glatz *et al.*, 2003; Tanaka *et al.*, 2003; Teramato and Tanaka, 2003).

Cecropin, attacin, and lysozyme are antimicrobial peptides that are induced in insects by microbial challenge (Lockey and Ourth, 1996a,b; Ourth *et al.*, 1994) through activation of the NF-$\kappa\beta$–mediated Toll- and IMD-signaling pathways (De Gregorio *et al.*, 2002; Kaneko and Silverman, 2005; McGettigan *et al.*, 2005). Defensive melanization responses involve deposition of eumelanin onto a pathogen and require phenoloxidase (PO) to catalyze the oxidation of tyrosine and other catecholamines to L-DOPA and dihydroxyindole, respectively. These products serve as substrates for the melanization reaction, which is important for localizing infections and producing free radicals that may be involved in killing the invading organism.

C. Effects of Parasitism on Immunity

Parasitization and notably, PDV infection, disrupts host cellular and humoral immune responses in parasitized insects. Parasitization prevents the encapsulation of parasitoid eggs, inhibits melanization of the hemolymph, and reduces the synthesis of antimicrobial peptides (Shelby and Webb, 1999).

Defensive melanization in parasitized larvae is drastically reduced due to the inhibition or reduction in monophenoloxidase activity (Carton and Nappi, 1997; Lavine and Beckage, 1995; Stoltz and Cook, 1983; Strand and Pech, 1995a,b). In the plasma of *Campoletis sonorensis* IV (CsIV)-infected larvae of *Heliothis virescens*, there is a reduction in PO activity (Shelby and Webb, 1999). Other enzymes related to the melanization process that are inhibited by CsIV infection are DOPA decarboxylase (DDC), which converts DOPA to dopamine, and dopachrome tautomerase, which converts L-DOPA to dihydroxylindole (Shelby and Webb, 1999). PO, dopachrome tautomerase, and DDC catalyze the majority of reactions in forming eumelanin from tyrosine glucoside plasma stores, and therefore suppression of activity in these enzymes can drastically affect melanization of a foreign pathogen or parasitoid (Shelby and Webb, 1999).

The antimicrobial response in lepidopteran larvae consists of the synthesis of antimicrobial peptides from larval fat body, hemocytes, and other tissues (Lockey and Ourth, 1996a,b; Ourth *et al.*, 1994). During a bacterial infection, the synthesis of cecropin, attacin, and lysozyme are induced. However, parasitized larvae of *H. virescens* infected with CsIV showed a significant reduction in cecropin and lysozyme levels (Shelby and Webb, 1999). Other antimicrobial peptides like attacins, lectins, and serine proteases may also be reduced (Shelby and Webb, 1999). In other systems, PDV-derived lectins have been shown to inactivate immunocyte spreading (Glatz *et al.*, 2004). The presence of lectin genes associated with *Cotesia* BVs (Glatz *et al.*, 2003, 2004; Tanaka *et al.*, 2003) suggests that they have a role in prevention of immunocyte recognition and cellular adhesion of hemocytes to the endoparasitoid host. CsIV infection is also known to reduce hemolymph cecropin and lysozyme activities (Shelby *et al.*, 1998).

D. Developmental Effects of Parasitism

Lepidopteran larval development is a period of rapid feeding and growth that is normally interrupted only by larval molts and which ends with the cessation of feeding in the final larval instar at the onset of pupal metamorphosis (i.e., wandering). The effects of parasitism on larval growth and development are variable. Some parasitized insects show minimal differences in larval size and timing of molts until parasite larvae complete their development (Beckage and Riddiford, 1978). Other parasitized larvae rapidly show marked changes in growth and development, which may include the onset of precocious metamorphosis (Vinson, 1990). In addition to effects on host growth and development, there are also effects on host nutritional metabolism (plasma proteins, lipids, and sugars) and endocrinology (Beckage and Gelman, 2004). Here, we consider only two systems which have been investigated in some detail at both the molecular and physiological level to illustrate the marked effect of PDVs on host developmental physiology.

In *H. virescens* parasitized by *C. sonorensis*, parasitized larvae molt one to two times depending on host size at the time of parasitization, before development is arrested and parasite larvae emerge (Shelby and Webb, 1999). Parasitized *H. virescens* larvae have reduced plasma protein concentrations which are correlated with reductions in synthesis of abundant hemolymph proteins, such as storage proteins and riboflavin-binding protein, by the fat body (Shelby and Webb, 1994, 1997). Other plasma proteins, such as ferritin, and lipophorin, remain at relatively constant levels (Shelby and Webb, 1999; Vinson, 1990). Trehalose and plasma amino acid levels are increased in a parasitized larvae (Vinson, 1990). Some proteins linked to metamorphic development are also inhibited. *H. virescens* larvae normally begin to synthesize juvenile hormone esterase (JHE) on the third day of the fifth instar, but in parasitized or CsIV-infected larvae, JHE remains undetectable (Shelby and Webb, 1999). Juvenile hormone (JH) titers remain high in parasitized larvae, which likely contributes to developmental arrest in this late larval stage (Shelby and Webb, 1997). Reductions in storage proteins (arylphorin, riboflavin-binding protein), plasma proteins, and JHE occur without significant reduction of their corresponding mRNA levels (Shelby and Webb, 1999) suggesting that the regulation of synthesis of these proteins is posttranscriptional. Reductions in storage proteins have been linked to reduced fitness, delayed larval development, and failure to attain critical weight for metamorphosis (Davidowitz *et al.*, 2003; Nijhout and Williams, 1974). The increase in plasma amino acid and trehalose levels coincident with the suppression of protein synthesis suggests that substrate limitations are not a factor in reduced protein synthesis. Rather, there seems to be a redirection of host nutrients and resources from growth and development of the host to support developing endoparasitoid larvae.

Chelonus inanitus parasitizes the eggs of *Spodoptera littoralis* with its PDV, the *C. inanitus* PDV (CiBV), which is implicated in suppressing larval growth, development, and inducing precocious onset of host metamorphosis and subsequent developmental arrest in the prepupal stage (Grossniklaus-Burgin *et al.*, 1994; Lanzrein *et al.*, 1998). *S. littoralis* eggs hatch normally and parasitized larvae are not easily distinguished from nonparasitized or X-ray parasitized larvae (parasitized larvae in which the parasite is killed by exposure to X-rays). However, the ingestion of food was reduced in penultimate and last instar-parasitized larvae (Kaeslin *et al.*, 2005a). Free sugars increased greatly at this time with parasite larvae developing rapidly (Kaeslin *et al.*, 2005a). Comparisons between nonparasitized and parasitized larvae in other systems suggest that free sugar levels are specifically regulated (Hoch *et al.*, 2002; Thompson, 1982; Thompson and Binder, 1984; Thompson *et al.*, 1990; Vinson, 1990; Vinson and Iwantsch, 1980). Lipid and glycogen levels were also increased in parasitized *S. littoralis* larvae (Kaeslin *et al.*, 2005b). These data suggest that *S. littoralis* metabolism is altered by parasitization such that feeding and development are suppressed as nutrient levels are elevated to support endoparasitoid development (Kaeslin *et al.*, 2005b).

III. Factors from Parasitoids That Alter Host Physiology

Although the focus in this chapter is on biologically active viral genes, there are other gene products derived from parasitoids that are delivered to parasitized insects and have relevant biological activities. These parasitoid-derived products are briefly considered below.

A. *Ovarian Proteins*

Ovarian proteins are synthesized by oviduct serosal cells and secreted into the lumen of the oviduct. They are introduced with parasite eggs and PDVs but have been little studied. In *C. sonorensis* ovarian proteins inhibit hemocyte spreading within 30 min of parasitization and egg encapsulation for up to 5 days (Luckhart and Webb, 1996). Ovarian proteins from CsIV alter the cytoskeleton of hemocytes and endoparasitoid eggs introduced into host larvae in the absence of ovarian proteins were rapidly encapsulated (Webb and Luckhart, 1994).

B. *Venoms*

Venoms are synthesized in a specialized organ that is connected to the reproductive tract. In wasps that lack PDVs, venoms perform functions similar to PDVs and often have similar effects on host physiology. In wasps that have both PDVs and venoms, the two factors often function synergistically. For example, *Cotesia kariyai* venoms injected during parasitization of *Pseudaletia separata* reduced circulating hemocytes (Teramoto and Tanaka, 2004). The functions provided by *C. kariyai* venoms were essential for 6-h postparasitization, and during this time hemocyte mitosis was inhibited (Teramoto and Tanaka, 2004). A specific peptide (Vn1.5) from *Cotesia rubecula* venom is essential for *C. rubecula bracovirus* (CrBV) expression and successful parasitization (Zhang *et al.*, 2004).

C. Virus-like Particles

Virus-like particles (VLPs) are present in the ichneumonid wasp *Venturia canescens*. These VLPs are very similar in structure to ichnovirus virions but lack nucleic acids. VLPs are produced in calyx cells and are essential for successful parasitization of *Ephestia kuehniella* (Lepidoptera: Phycitidae) (Edson *et al.*, 1981; Schmidt and Theopold, 1991). During oviposition of the egg, calyx fluid and VLPs are deposited. The VLPs are attached to the endoparasitoid egg surface and provide protection from the host immune response (Feddersen *et al.*, 1986; Rotheram, 1973). The VLPs have three structural peptides (VLP1, VLP2, VLP3) that are 40, 52, and 94 kDa, respectively. VLP2 has sequence similarity to the RhoGAP domain of GTPase-activating proteins (Reineke *et al.*, 2002), which activate small GTP-binding proteins (G proteins or GTPases) that regulate cellular processes (Bourne *et al.*, 1991). VLP3 has similarity to the metalloprotease neprilysin-NEP, which is an inducer of cellular adhesion and spreading of host hemocytes (Asgari *et al.*, 2002). The VLP1 peptide did not show significant sequence similarity to other peptides. Two allelic and functionally distinctive versions of this gene exist in wasp populations (Hellers *et al.*, 1996; Theopold *et al.*, 1994).

D. Endoparasitoid Larval Secretions

Endoparasitoid eggs, cells, and larvae themselves are potential sources of biologically active proteins that may manipulate host physiology. Although the egg has not been shown to secrete proteins that disrupt host physiology, both cells released at the time of egg hatch (teratocytes) and proteins synthesized by the endoparasite larvae are known to have important biological functions. The clearest example of a biologically active protein secreted by wasp larvae is that of the *C. inanitus* egg-larval endoparasitoids. During endoparasitoid larval development, the *C. inanitus* endoparasitoid secretes proteins that are involved in inducing precocious metamorphosis of the parasitized host

(Hochuli and Lanzrein, 2001; Johner *et al.*, 1999). Secreted larval proteins and their genes have been isolated from other systems, but their functions are unknown (Soldevila and Jones, 1993).

E. Proteins Secreted from Teratocytes

Teratocytes are cells derived from the serosal membrane of some endoparasitoid eggs. The serosal membrane breaks down to release its component teratocytes. These cells do not die but disperse into the hemolymph, become greatly enlarged and synthetically active, but do not divide (Jarlfors *et al.*, 1997). It is clear that these cells have important biological functions.

1. Characterization and Isolation of Biologically Active Teratocyte Proteins

Microplitis croceipes is a braconid endoparasitoid that harbors a PDV (McBV) and also produces functional teratocytes. In a series of studies, Dahlman and coworkers showed that teratocytes could be collected from eggs collected after oviposition. These teratocytes caused mortality and reduction in host growth. Teratocytes could be maintained in cell culture for some time and the proteins secreted from these cells also reduced growth of the treated insect. Injection of teratocytes into *H. virescens* larvae induced changes in hemolymph proteins such as a reduced titer of JHE, arylphorin, ecdysone, riboflavin-binding protein, and storage protein p74/76 (Jarlfors *et al.*, 1997; Zhang *et al.*, 1998). Investigations with explanted tissues showed that synthesis of these fat body proteins had reduced titers. With the exception of arylphorin, the reduction in protein synthesis appeared to occur at a posttranscriptional level as mRNA titers were unchanged (Jarlfors *et al.*, 1997). Injection of teratocyte-secreted proteins (TSPs) purified from cells maintained in culture induced a similar response (Jarlfors *et al.*, 1997). These results were then used to develop a bioassay for purification of the biologically active proteins, and the protein sequence used to isolate the gene. This gene, TSP14, was then expressed in recombinant baculoviruses and in yeast and purified recombinant protein was found to inhibit protein synthesis in the fat body cells of *Manduca sexta* and *H. virescens* larvae (Dahlman *et al.*, 2003). However, recTSP14 did not inhibit translation in all insect cells or tissues or in mammalian cell lines. In these cells, the protein did not bind to cell surfaces suggesting that the biological activity of this protein is determined by the presence or absence of specific cell-surface receptors (Dahlman *et al.*, 2003). Interestingly, the sequence of TSP14 contained a cysteine-rich motif similar to one found in an unrelated PDV, CsIV (Dahlman *et al.*, 2003; Rana *et al.*, 2002). This conservation of sequence and similar functions of CsIV and TSP14 suggests that they may have a similar physiological function in parasitized larvae.

2. *TSP14 Expressed in Plants*

The robust and dramatic effects of TSP14 on insect growth and its relatively small and highly cross-linked structure raised the possibility that it might be orally active. To test this possibility the TSP14 gene was introduced into transgenic tobacco for expression and bioassay. The expression of TSP14 in transgenic tobacco (*Nicotiana tabacum* cv Samsun NN), caused a reduction in larval survivorship, growth, development, and larval feeding relative to controls (GUS/native tobacco species). In leaf disc feeding experiments, *H. virescens* larvae were smaller and developmentally delayed when consuming TSP14 expressing lines relative to controls (Maiti *et al.*, 2003). In whole plant feeding bioassays, control lines had significantly more damage, compared to TSP14 expressing lines (Fig. 1). In addition, survivorship was only 30% among larvae feeding on TSP14 expressing lines as compared

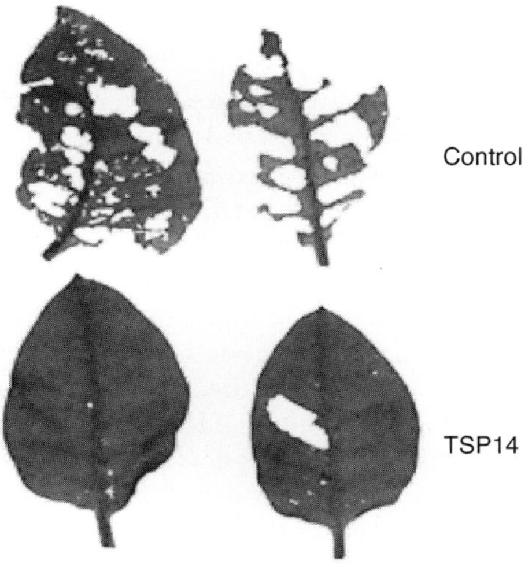

FIG 1. Reduced herbivory on TSP14 expressing transgenic tobacco leaves. Neonate *H. virescens* larvae were placed on 2-month-old tobacco plants. The larvae were allowed to feed for 17 days. At this time larvae were removed and photos of plant tissues were analyzed. The upper photos show representative leaves taken from control *Nicotiana tabacum* plants. The lower photos show representative tobacco plants expressing TSP14.

to 55% on controls. Mean weights of larvae recovered from TSP14 expressing lines (85.9 ± 16.9 mg) were less than controls (224.3 + 40.3 mg) (Maiti *et al.*, 2003). Interestingly when plants were fed to *M. sexta* larvae, which is not a permissive host to *M. croceipes*, these larvae were also sensitive to TSP14 suggesting that TSP14 may inhibit growth of other phytophagous pest insects (Maiti *et al.*, 2003).

IV. Polydnavirus Genes

Ovarian proteins, venoms, VLPs, and endoparasitoid secretions play a major role in the success of parasitization and some of these proteins could ultimately prove useful for controlling crop pest populations. However, PDVs clearly have an important biological function with a significant advantage in that these biologically active genes are encoded within the viral genome from which they can be more readily isolated for study. For the purposes of this review, the PDV genes are considered in three categories: PDV genes having at least one known function, PDV genes having a suspected function based on sequence homology to other genes, and PDV genes of unknown function. In the subsequent sections we consider these different gene groups and their biotechnological applications.

A. Viral Gene Families

The sequencing of several PDV genomes has provided new insights into both the organization of the viral genomes, and candidate genes for virus-mediated disruption of host physiology. Interestingly, a common feature of all PDV genomes is that the genome is organized into gene families and that there are relatively few predicted genes that are not present in more than one copy. Functional analyses of PDV genomes have focused on functionally important or highly abundant gene families. However, not all of these gene families are essential and none of the single-copy PDV genes have been investigated. There is every reason to expect that a number of additional genes with important biological activities will be identified. Thus, the description later of PDV genes of known, suspected, and unknown function (Table I) is a status report that is expected to change considerably as the investigation of PDV genomes matures.

1. Known Functions

Gene families that have at least one known function are described later. Most of the analyses of these genes involve investigation of gene

TABLE I
POLYDNAVIRUS OR RELATED GENES WITH BIOTECHNOLOGICAL POTENTIAL

Known function	Suspected function	Unknown function
Cys-motif	*PTP*	Early proteins (EP1, EP2, EP3)
Viral ankyrin	*Glc*	*Rep*
Vinnexin	Bracoviral C-type lectins	*Egf*
Mucin-like	*TnBV 1* and *2*	*IEP*
	TrV 1 and *4*	*N* Genes
	C. inanitus (14g1, 14g2, 12g1)	*tRNAs*
		PRRP
		M Genes *(M24, M27, M40)*

family members from a single species and often not all gene family members have been studied. However, even in this limited analyses there are some interesting and important results.

a. Cys-Motif Gene Family Genes in the Cys-motif gene family are characterized by one or more cysteine-knot structural motifs (C–C–CC–C–C) with the cysteine residues within each knot flanked by hypervariable amino acid stretches (Dib-Hajj *et al.*, 1993). CsIV and the *Campoletis chlorideae ichnovirus* (CcIV) are the only ichneumonid viral genomes known to contain Cys-motif genes. The CsIV Cys-motif gene family has 10 members (VHv1.4, VHv1.1, WHv1.0, WHv1.6, AHv1.0, AHv0.8, UHv0.8a, UHv0.8b, FHv1.4, and LHv2.8) that are named based on mRNA size and segment of origin. The genome of *Hyposoter fugitivus* IV (HfIV) contains five cysteine-rich motif genes, but names have not been assigned due to lack of functional analysis (Tanaka *et al.*, 2006, submitted for publication). All Cys-motif genes have a conserved gene structure with an intron present at the same location in every Cys-motif, all contain signal peptides, and predicted transmembrane domains along with *N*- and *O*-terminal glycosylation sites (Blissard *et al.*, 1987; Cui and Webb, 1996; Dib-Hajj *et al.*, 1993; Fath-Goodin *et al.*, 2006; Webb *et al.*, 2006) (Table II). The CcIV Cys-motif gene (CcIV 1.0) has strong amino acid sequence similarity to VHv1.4, VHv1.1, WHv1.0, and WHv1.6 of CsIV (86%, 88%, 89%, and 87%, respectively) (Zhang and Wang, 2003). The Cys-motif genes inhibit protein synthesis, host growth, and host immune responses by selective inhibition of host protein synthesis at the posttranscriptional level. We consider the pleiotropic effects of the Cys-motif gene family a reflection of their activity. The host's immune system, development

TABLE II

CURRENT PROGRESS AND GENE STRUCTURAL CHART FOR CsIV CYS-MOTIF GENES AND TSP14

Cys-motif viral genes	Cys-motifs	Signal peptide	Transmembrane domains	Nested or unique	Posttranscriptional inhibition	Slowed development through oral ingestion	Expression in transgenic tobacco	Expressed in baculovirus expression system	Introns
VHv1.4	2	Yes	0	Nested	Yes	Yes	Current	Yes	2
VHv1.1	1	Yes	0	Nested	Yes	Yes	Current	Yes	1
WHv1.6	1	Yes	0	Nested	Unknown	Yes	Current	Yes	1
WHv1.0	1	Yes	0	Nested	Unknown	Unknown	Current	Yes	1
FHv1.4	3	Yes	0	Unique	Unknown	Unknown	No	Yes	3
LHv2.8	6	Yes	0	Unique	Unknown	Unknown	No	Current	6
UHv0.8	1	Yes	1	Nested	Unknown	Unknown	No	Current	1
UHv0.8a	1	Yes	2	Nested	Unknown	Unknown	No	Current	1
AHv0.8	1	Yes	1	Unique	Unknown	Yes	No	Yes	1
AHv1.0	1	Yes	1	Unique	Unknown	Yes	No	Yes	1
TSP14	1	Yes	1	Teratocytes	Yes	Yes	Yes	Yes	1

and growth are affected because the Cys-motif genes reduce titers of host proteins that are involved in these host physiological systems. The potential of this gene family for biotechnological applications has been developed to the greatest extent as discussed below (Section V).

b. Viral Ankyrin Gene Family The viral ankyrins are the only genes present in all of the PDV genomes for which significant sequence information is available (CsIV, TrIV, HfIV, CcBV, MdBV). All of the vankyrin genes have ankyrin repeat motifs (Espagne *et al.*, 2004; Tanaka *et al.*, 2006; Webb *et al.*, 2006) that align most closely with the ankyrin repeats 3–6 of the *Drosophila melanogaster* I $\kappa\beta$ cactus (Ghosh *et al.*, 1998). The cactus protein is a *Drosophila* dorsal/NF-$\kappa\beta$-Rel protein inhibitor. I $\kappa\beta$ proteins have been linked in *Drosophila* and vertebrates to regulation of innate immune responses as well as to regulating some aspects of embryonic development (Ghosh *et al.*, 1998). CsIV, *Microplitis demolitor bracovirus* (MdBV), and CcBV vankyrin proteins lack nuclear export signals and destruction domains (regulatory elements) associated with basal degradation of I $\kappa\beta$ proteins, which suggests that these may function by irreversible binding to target proteins (Espagne *et al.*, 2004; Kroemer and Webb, 2005; Thoetkiattikul *et al.*, 2005). Such binding would prevent the activation of Nf-$\kappa\beta$ mediated transcription factors and block transcription of Nf-$\kappa\beta$ regulated immune genes (Ghosh *et al.*, 1998; Thoetkiattikul *et al.*, 2005). Vankyrin proteins H4 and N5 of MdBV, were shown to bind Rel proteins Dif and Relish and prevent Dif and Relish's binding to promoter Nf-$\kappa\beta$ sites of Cecropin A1 and drosomycin genes (Thoetkiattikul *et al.*, 2005). The CsIV vankyrin gene family members have two recognizably distinct localization patterns, namely nuclear or cytoplasmic. Interestingly, the intracellular localization of some of the vankyrin proteins changes markedly in response to immune challenge and virus infection (Kroemer and Webb, 2006, submitted for publication). These results support the hypothesis that vankyrin genes regulate host immune function during parasitization. In addition, more recent studies have shown that vankyrin protein activity enhances the expression of recombinant proteins in the baculovirus expression vector system (see chapter by Fath-Goodin *et al.*, this volume, pp. 75–90).

c. Mucin-like/Glc *Gene Family* The Glc1.8 proteins were isolated from *Pseudoplusia includens* parasitized by MdBV and localized to hemocytes where they appear to disrupt the encapsulation immune response (Trudeau *et al.*, 2000). The Glc proteins are characterized by their hydrophobic N- and C-terminal domains, which flank a heavily glycosylated central core of six tandemly arranged repeats (Trudeau *et al.*, 2000). When the *glc*1.8 gene was silenced via RNA interference (RNAi) in High Five™ cells (derived from *Trichoplusia ni* embryos: Invitrogen) infected with MdBV the cells recovered the ability to adhere and spread on foreign surfaces (Beck and Strand, 2003). Expression of *Glc*1.8 from S2 (*D. melanogaster*) and High Five™ cells inhibited phagocytosis and cell adherence. However, cells transfected with a gene encoding a Glc1.8 protein with a mutated anchor sequence retained the ability to adhere to foreign objects and phagocytose (Beck and Strand, 2005). Thus, the Glc1.8 protein is strongly implicated in the disruption of phagocytosis and the ability of immunocytes to adhere during parasitization. The *Hyposoter didymator* IV (HdIV) gene, *Hd*GorfP30 may perform a similar function in parasitized larvae. This gene is expressed 2 h postparasitization in *Spodoptera* larvae and encodes a secreted, glycosylated protein that contains mucin-like motifs similar to members of the *glc* family of MdBV (Galibert *et al.*, 2003). However, functional analysis of *Hd*GorfP30 is not complete.

d. Vinnexin Gene Family The CsIV viral innexin (vinnexin) gene family has very high homology to invertebrate gap junction proteins (Turnbull and Webb, 2002). Vinnexin genes have been identified in all sequenced ichnovirus genomes (*Tranosema rostrale* IV; TrIV, CsIV, and HfIV) (Tanaka *et al.*, 2006; Webb *et al.*, 2006). Two CsIV vinnexins that are preferentially expressed in hemocytes were shown to form functional gap junctions in *Xenopus* oocytes (Turnbull *et al.*, 2005). In invertebrate cells, innexins govern gap junction formation and regulate cell–cell communication. Therefore, vinnexin genes may disrupt cellular communication in parasitized insects possibly during the encapsulation response from a parasitized host (Kroemer and Webb, 2004). The biotechnological utility of the viral innexins is not likely to be in the area of insect control, but these novel genes that regulate gap junction formation and cellular communication may find other applications as their properties are further defined.

e. CiBV Gene Family (12g, 14g) The egg-larval parasitoid *C. inanitus* has viral genes, 12g1, 12g2, 14g1, and 14g2, that are expressed just prior to the onset of precocious metamorphosis (Bonvin *et al.*, 2004). RNAi was performed to determine if any of these late expressed genes were related to the onset of precocious metamorphosis. The activity of these genes was inhibited using RNAi transcripts derived from the 14g1 and 14g2 genes and developmental arrest was reversed (Bonvin *et al.*, 2005). 12g1 also reversed developmental arrest but the effect was less pronounced (Bonvin *et al.*, 2005). At least conceptually, these two genes might be used to induce premature metamorphosis for management of pest populations.

f. Toxoneuron nigriceps bracovirus The expression of two *Toxoneuron nigriceps bracovirus* (TnBV) transcripts during parasitization of *H. virescens* has been investigated (TnBV1 and TnBV2) (Falabella *et al.*, 2003; Lapointe *et al.*, 2005; Varricchio *et al.*, 1999). TnBV1 is a spliced gene that lacks a signal peptide, with four phosphorylation sites and a single *N*-glycosylation site. TnBV1 is expressed from prothoracic glands, and induces apoptosis in two insect cell lines (Lapointe *et al.*, 2005). TnBV2 is a spliced gene that is expressed in hemocytes and prothoracic glands. The protein lacks a signal peptide, contains several *N*-glycosylation sites, a protein kinase C phosphorylation site, and a conserved aspartyl-protease domain (Falabella *et al.*, 2003). TnBV1 may induce apoptosis of prothoracic glands, while TnBV2 is hypothesized to target cap-dependent translation or induce cleavage of cytoskeletal intermediates to disrupt translation and immune responses in *H. virescens* (Falabella *et al.*, 2003).

2. Suspected Gene Functions

The gene families considered in the previous section have experimental support for at least one biological function that appears to be relevant to the role of PDVs in parasitized insects. Many of these genes have clear potential application, while other genes have less obvious direct utility. There are a number of other gene families that have been less well studied. Most of these gene families have been identified from genome sequence data based on homology to other genes in public databases. In some cases, the identity of the homologous genes is suggestive of appropriate biological functions.

a. Protein Tyrosine Phosphatase Gene Family Protein tyrosine phosphatases (PTP) are signal transduction pathway regulators, that function by dephosphorylating tyrosine residues on regulatory proteins (Andersen *et al.*, 2001). Gene families with homology to PTPs are the most abundant viral genes in the MdBV and CcBV genomes (Espagne *et al.*, 2004; Webb *et al.*, 2006), and are also present in the genomes of TnBV, *Glyptapanteles indiensis* BV (GiBV) and *Cotesia plutellae* Bracovirus (CpBV) (Chen *et al.*, 2003; Choi *et al.*, 2005; Malva *et al.*, 2004). GiBV PTP was highly expressed at 2 h postparasitization, and decreased with time after infection (Chen *et al.*, 2003). These data suggest that PTPs may function in GiBV as inhibitors of cellular immune responses early in parasitization. Bacterial pathogens from plant and mammalian systems encode PTPs that function to inhibit innate immunity factors (Espinosa *et al.*, 2003; Sun *et al.*, 2003).

b. TrV Gene Family TrV1, TrV2, and TrV4 are members of a *T. rostrale* IV (TrIV) gene family that encode secreted peptides that are expressed postparasitization in *C. fumiferana* (Beliveau *et al.*, 2000). In this system, there are clear developmental effects associated with parasitism but effects on host hemocytes are less pronounced. TrV gene family members have homology to the CsIV Cys-motif proteins but this homology is largely limited to the signal peptide and 5′ noncoding sequence. TrV1 and TrV4 genes are expressed at high levels in premolt fifth instar larvae, which is just prior to developmental arrest (Beliveau *et al.*, 2000, 2003). Therefore, members of the TrIV gene family are thought to play a role in inducing developmental arrest in parasitized larvae.

c. Bracoviral C-Type Lectins C-type lectin-related genes have been identified and are expressed from *Chelonus* nr. *curvimaculatus* BV (CcBV), *C. kariyai* BV (CkBV), and *Cotesia ruficrus* BV (CrBV) (CrV3, Cky811, and Crf111) viral genomes during parasitization (Glatz *et al.*, 2003; Teramato and Tanaka, 2003). It is hypothesized that the C-type lectins associated with BVs are masking proteins that protect endoparasitoid larvae from host encapsulation and other cellular immune responses (Glatz *et al.*, 2003).

d. Epidermal Growth Factor Gene (EGF) Family A family of MdBV genes has been identified that has homology to vertebrate epidermal growth factors within their cysteine-rich domains. This Egf gene family has three members (egf0.4, egf1.0, and egf1.5) that are spliced (Strand *et al.*, 1997; Trudeau *et al.*, 2000). Transcription of *egf* genes coincides with physiological changes in host hemocytes between 12–24 h postparasitization of *P. includens* (Strand *et al.*, 1997). Egf proteins do not appear to prevent adhesion or phagocytosis. The Egf proteins do have sequence similarity to IEP proteins which are thought to have a role in immunoevasion or contain immunoevasive properties (Beck and Strand, 2003; Tanaka *et al.*, 2002).

3. *Unknown Gene Functions*

There are a number of PDV gene families that have no known physiological function in the parasitized host. These are now briefly considered.

a. Rep *Gene Family* The *rep* gene family is characterized by a highly conserved 540-bp repeat sequence element that in most REP proteins contains a single repeat, with a few *rep* genes contain 1–5 repeat elements in tandem arrays (Fleming and Krell, 1993; Hilgarth and Webb, 2002; Theilmann and Summers, 1987, 1988). The function of the *rep* genes has yet to be determined. Among ichnoviral genomes (HfIV, HdIV, TrIV, and CsIV) the *rep* gene family is highly conserved (Galibert *et al.*, 2003; Volkoff *et al.*, 2002) and is the most abundant gene family associated with CsIV (28 open reading frames) (Volkoff *et al.*, 2002). *Rep* gene expression profiles from *Cs*IV segments B, H, and O were expressed at highest levels by 6 h postparasitization (Theilmann and Summers, 1988). Transcripts for *rep* genes on segments H and O were detected in parasitized host and *C. sonorensis* females, however the *rep* gene on segment B (Bhv0.9), which lacked a signal peptide, was only expressed in the parasitized host (Theilmann and Summers, 1988). *Rep* genes on segment I (I0.9, I1.1, I1.2) are expressed at low levels in both parasitized host and *C. sonorensis* females (Hilgarth and Webb, 2002; Theilmann and Summers, 1988). The majority of *rep* genes in CsIV are on low-copy segments, which might suggest the targets of *rep* genes are not highly expressed.

b. M *Gene Family* Segment *M* of the HdIV genome, contains three related genes (*M24*, *M27*, and *M40*) that are expressed by 4 h postparasitization and then throughout parasitism (Volkoff *et al.*, 1999). *M* gene family members encode proteins with glycine and proline rich regions. *M* genes are theorized to exhibit host related antigenic shielding for the developing endoparasitoid larvae (Volkoff *et al.*, 1999).

c. Early Proteins Early proteins (EP1, EP2, and EP3) are expressed within 30 min of parasitization of *M. sexta* by the endoparasitoid *Cotesia congregata*. These secreted glycosylated proteins comprise 10% of total hemolymph proteins (Harwood *et al.*, 1994). Though a function has not been determined, the expression of EP proteins may affect host range (Beckage and Tan, 2002; Harwood *et al.*, 1998). A similar gene is encoded by the CkBV genome (Espagne *et al.*, 2004).

d. N *Gene Family* This family is composed of two novel genes (*N1.2* and *N1.4)*, which are located on segment *N* of CsIV (Webb *et al.*, 2006). Expression of *N* genes has been detected in parasitized *H. virescens* larvae but functional characterization has yet to be performed.

e. Proline Residue Rich Protein (PRRP) Family Five genes located on three segments of the CsIV genome are predicted to contain numerous proline residues. RT-PCR analysis has shown that all genes are expressed during parasitization, but functional analysis has yet to be performed.

f. Unassigned Open Reading Frames and Others In all of the PDV genomes there are a variety of open reading frames (ORFs) that are predicted to encode genes of unknown function. For example, within the CsIV genome 101 ORFs have been detected of which 53 reside in the *Cys-motif, vankyrin, vinnexin, rep*, and *N* gene families (Webb *et al.*, 2006). There are 61 ORFs associated with the MdBV genome with 40 residing in the *ptp, vankyrin, egf, tRNA*, and *glc* gene families (Webb *et al.*, 2006). This indicates that 48 and 21 unassigned ORFs are present in the CsIV and MdBV genomes, respectively (Webb *et al.*, 2006). The unassigned ORFs are termed unique if they do not contain homology to any sequence in the database (Webb *et al.*, 2006). Work is underway to determine whether these ORFs are expressed.

V. CsIV Cys-Motif Genes and Potential Biotechnological Applications

The Cys-motif gene family derived from CsIV is one of the most structurally and functionally characterized PDV gene families. The Cys-motif gene family contains 10 predicted members that are located on seven segments (V, W, U, F, L, A, and A2) of the CsIV genome (Bonvin *et al.*, 2005). In this section we will discuss the potential function of Cys-motif proteins as entomotoxic agents.

A. *Insecticidal Activity*

It has been known for some time that injection of CsIV into *H. virescens* larvae alters larval growth, development, and immunity but the genes responsible for these effects have not been well characterized. Recent studies by Fath-Goodin *et al.* (2006) evaluated five members of the Cys-motif protein gene family for their ability to inhibit insect growth and development. When rVHv1.1, rVHv1.4, rWHv1.6, rAHv0.8, and rAHv1.0 were expressed from recombinant baculoviruses and applied to insect diet, these proteins suppressed the growth and development of *H. virescens* larvae (Fath-Goodin *et al.*, 2006). Although these proteins are not normally delivered to the gut of their host larvae, this result was consistent with the finding that the teratocyte Cys-motif protein TSP14 was also active when administered orally (Maiti *et al.*, 2003). Of the assayed Cys-motif proteins rVHv1.1 slowed larval development the most. Therefore, this protein was the focus of a more in depth investigation (Fath-Goodin *et al.*, 2006). Oral consumption of rVHv1.1 caused a 50–70% reduction in growth of *H. virescens* larvae by day 6 relative to controls (Fath-Goodin *et al.*, 2006; Fig. 2). Furthermore, 36% of the larvae that ingested rVHv1.1 for 6 days formed nonviable pupae later during metamorphosis (Fath-Goodin *et al.*, 2006; Fig. 3), a phenotype also observed in larvae injected and orally fed with TSP14 (Fig. 3). Therefore, rVHv1.1 appeared to have effects that persisted beyond the period of direct exposure to the protein. Interestingly, when *Spodoptera exigua*, a nonpermissive host of *C. sonorensis*, consumed rVHv1.1 the reduction in weight gain was proportionally greater than that observed for *H. virescens* larvae, a permissive host (Fath-Goodin *et al.*, 2006). This result shows that the Cys-motif proteins have significant effects on insects that are not part of the normal host range of *C. sonorensis*. These feeding assays show that two species of the most economically important complexes of

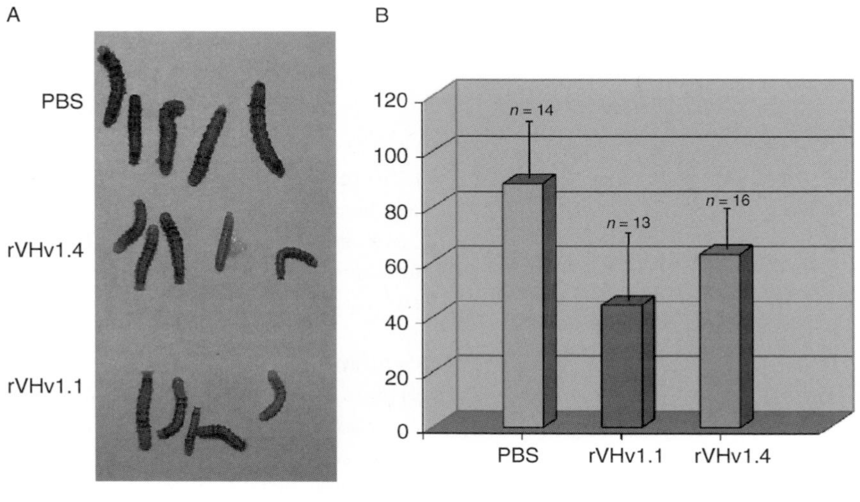

FIG 2. Effect of orally consumed rVHv1.1 and rVHv1.4 on the development of *H. virescens* larvae. (A) *H. virescens* larvae consumed diet containing rVHv1.1 and rVHv1.4 or PBS (control) for 24 h, before being placed on normal diet. The image was taken after 3 days. (B) Fresh weight of larvae fed on test and control diet 5 days after onset of the experiment. Values represent the mean ± SD. *n*, the number of larvae per treatment. (See Color Insert.)

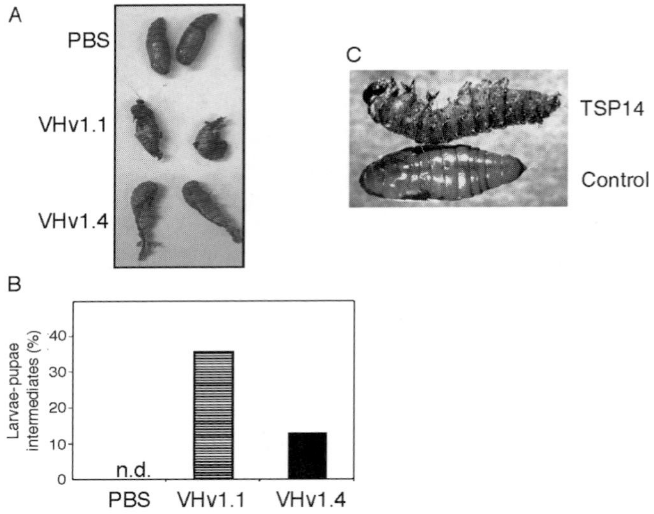

FIG 3. Cys-motif proteins induce developmental abnormalities. (A) Viable (PBS control) and nonviable (VHv1.1; VHv1.4) *H. virescens* pupae is shown. Developmental abnormalities were induced when *H. virescens* larvae consumed insect diet containing rVHv1.l or rVHv1.4. (B) The percentage of *H. virescens* larvae exhibiting developmental abnormalities after consuming rVHv1.1, rVHv1.4, or PBS with their diet is shown. (C) Similar developmental abnormalities were observed when *H. virescens* larvae were fed on TSP14.

lepidopteran pests are susceptible to the effects of the VHv1.1 protein. Furthermore, this Cys-motif protein was stable under different storage conditions and showed only moderate susceptibility to proteolytic degradation, making VHv1.1 suitable for potential use as an insecticidal agent (A. Fath-Goodin, S. Martin, and B. A. Webb, personal communication).

Most interestingly, when rVHv1.1 was injected into the *H. virescens* hemocoel, or fed at higher concentrations on insect diet, larval death was observed in many insects that was consistent with signs of baculovirus infection (Fath-Goodin *et al.*, 2006). Consumption of the VHv1.1 protein rendered these larvae more susceptible to a cryptic viral pathogen that was asymptomatic under normal rearing conditions in our colony. Since insect pathogens are important factors in regulating populations of many insects, increased susceptibility of pest insects to pathogens by Cys-motif proteins could be an important contributor to managing pest populations in agroecosystems.

To evaluate the activity of CsIV Cys-motif proteins for protection of crops from insect damage, *VHv1.1, VHv1.4, WHv1.0*, and *WHv1.6* were expressed in transgenic tobacco plants (Gill *et al.*, 2006; unpublished observation). Assessment of several independent transgenic plant lines (R1 and some R2 progeny) after exposure to *H. virescens* larvae revealed that plants expressing Cys-motif proteins show reduced feeding damage compared to the control tobacco plants (T. Gill *et al.*, unpublished observation; Fig. 4). This finding was in accordance with reduced herbivory on TSP14 expressing transgenic tobacco plants (Maiti *et al.*, 2003). These results demonstrate the potential for introduction of Cys-motif genes into plants to achieve protection against lepidopteran pests, and document that these proteins can have a major impact on insect growth and development.

Taken together, these results indicate that the Cys-motif proteins have pleiotropic effects on the physiology of insect larvae. Our results indicate that the Cys-motif proteins have similar effects with potential application as insect control agents.

FIG 4. Hemizygous Whv1.6 line suppresses insect herbivory. Ten mid-first instar *H. virescens* larvae were placed on 2-month-old transgenic tobacco plants or control plants. Larvae were allowed to consume plant material for 15 days.

B. Inhibition of Protein Synthesis

During parasitization of *H. virescens* by CsIV, cellular and humoral immunity are inhibited, and larval development is arrested. CsIV infection inhibits gene expression at the posttranscriptional level of growth-associated host proteins such as arylphorin (p74/p76), insect storage proteins, riboflavin-binding hexamer (p82), JHE, lysozyme (Shelby and Webb, 1997; Shelby *et al.*, 1998), and transcripts encoding proteins involved in the antimicrobial immune response and melanization (Shelby *et al.*, 1998). More recently, native and recombinant VHv1.1 and VHv1.4 proteins have been shown to inhibit protein synthesis in *H. virescens* testis tissue and TN 368 cells (a *Trichoplusia* cell line) (Kim, 2005). When rVHv1.1 and rVHv1.4 were tested for their ability to inhibit translation in specific tissues, these proteins reduced protein synthesis specifically in testis, hemocytes, and fat body but had little effect on other tissues (Kim, 2005). These results suggest that the ability of the Cys-motif proteins to inhibit protein synthesis may be the mechanism by which this protein family reduces the titer of hemolymph proteins involved in the host immune response and larval growth, thereby rendering these insects more susceptible to aberrant development (nonviable pupae) and disease (baculovirus infection) (Fath-Goodin *et al.*, 2006). As we continue to assess the insecticidal activity of each Cys-motif protein in CsIV and elucidate their modes of action in parasitized insects (Table II), we may find ways to improve upon the insecticidal activity of these proteins.

C. Summary

Insect resistance and herbicide tolerance are the major performance-enhancing traits in the highly successful "biotech crops," which also reduce the use of chemical insecticides and their accompanying adverse environmental impacts. These transgenic crops are credited with making the fastest-ever technological impact on agriculture, and their use continues to expand. In the insect control field, the dominance of Bt-based strategies has raised concerns about the emergence of resistant insects. Although strategies have been developed to mitigate resistance and reduce these concerns, there is an urgent need for alternative transgenic crops whose insect-resistance is based on a totally different mechanism. The Cys-motif proteins may represent an alternative technology to Bt for insect control. Application of Cys-motif proteins to diet killed or inhibited the growth of insect larvae and expression of a Cys-motif gene in transgenic tobacco protected these plants against insect feeding damage. Thus, the Cys-motif genes appear to be viable candidates for development as viable alternatives to Bt.

VI. Conclusion and Prospects for the Future Use of Polydnavirus Genes

PDVs have a variety of important pathogenic effects on the parasitized host. Some of these pathogenic effects, such as inhibition of development, induction of precocious metamorphosis, slowed or reduced feeding and immune suppression, may be applied for biotechnological purposes. Because PDVs do not replicate in the host insect, the viral genes delivered during parasitization are directly responsible for these pathogenic effects.

Only a few PDV gene families have been investigated in any detail. This chapter concentrates on the Cys-motif gene family because the encoding proteins inhibit development of tobacco budworm larvae when ingested orally. Interestingly, there is a clear similarity in structure and function between the CsIV Cys-motif protein and a protein isolated from parasitized larvae, TSP14. Our understanding of PDVs posits that many, perhaps most, of the biologically active viral genes have progenitors that are evolutionarily derived from the wasp genome (Webb and Strand, 2004). There are many examples of viruses that have acquired host genes to support and enhance virus replication and transmission and such host-derived genes are particularly common among the large DNA viruses (e.g., pox viruses). However, an important distinction is that in PDVs the acquired genes also, perhaps primarily, support the survival and development of the mutualistic wasp. Thus, PDV are considered by some investigators to be a specialized delivery system for genes that benefit the parasitic wasp eggs and larvae (Blissard *et al.*, 1989; Vinson and Iwantsch, 1980; Whitfield, 1990). In any case, it is clear that a high percentage of the genes encoded by PDVs have biological functions relevant to insect control and other biotechnological applications (see chapter by Fath-Goodin *et al.*, this volume, pp. 75–90). Given that PDV genomes have existed in their obligate mutualisms with parasitic wasps for tens of millions of years, one would expect that evolutionary processes would have selected and refined viral genes that convey benefit within the constraints of this biological system. However, as the roles of viral genes and gene families within PDV genomes become obvious it is likely that understanding the structure and organization of PDV genomes will become increasingly important to understanding and using genes derived from PDVs. It is clear that it is not only the viral genes that have undergone constrained and selective evolution within these unusual biological systems, but also the viral genome organization and mechanisms of gene delivery that have become highly modified. Thus, there may be opportunity to apply not only the genes encoded by PDV but also to apply the mechanisms of gene delivery and control used by PDVs in novel biotechnological applications.

Acknowledgments

This is publication No. 06-08-036 of the University of Kentucky Agricultural Experimental Station.

References

Abel, P. P., Nelson, R. S., De, B., Hoffmann, N., Rogers, S. G., Fraley, R. T., and Beachy, R. N. (1986). Delay of disease development in transgenic plants that express the tobacco mosaic virus coat protein gene. *Science* **232**(4751)**:**738–743.

Andersen, J. N., Mortensen, O. H., Peters, G. H., Drake, P. G., Iversen, L. F., Olsen, O. H., Jansen, P. G., Andersen, H. S., Tonks, N. K., and Moller, N. P. (2001). Structural and evolutionary relationships among protein tyrosine phosphatase domains. *Mol. Cell. Biol.* **21**(21)**:**7117–7136.

Asgari, S., Reineke, A., Beck, M., and Schmidt, O. (2002). Isolation and characterization of a neprilysin-like protein from Venturia canescens virus-like particles. *Insect Mol. Biol.* **11**(5)**:**477–485.

Bae, S., and Kim, Y. (2004). Host physiological changes due to parasitism of a braconid wasp, Cotesia plutellae, on diamondback moth, *Plutella xylostella*. *Comp. Biochem. Physiol., Part A Mol. Integr. Physiol.* **138**(1)**:**39–44.

Beck, M., and Strand, M. R. (2003). RNA interference silences Microplitis demolitor bracovirus genes and implicates glc1.8 in disruption of adhesion in infected host cells. *Virology* **314**(2)**:**521–535.

Beck, M., and Strand, M. R. (2005). Glc1.8 from Microplitis demolitor bracovirus induces a loss of adhesion and phagocytosis in insect high five and S2 cells. *J. Virol.* **79**(3)**:** 1861–1870.

Beckage, N. E., and Gelman, D. B. (2004). Wasp parasitoid disruption of host development: Implications for new biologically based strategies for insect control. *Annu. Rev. Entomol.* **49:**299–330.

Beckage, N. E., and Riddiford, L. M. (1978). Developmental interactions between the tobacco hornworm Manduca sexta and its braconid parasite Apanteles congregatus. *Entomol. Exp. Appl.* **23:**139–151.

Beckage, N. E., and Tan, F. F. (2002). Development of the braconid wasp Cotesia congregata in a semi-permissive noctuid host, *Trichoplusia ni*. *J. Invertebr. Pathol.* **81**(1)**:** 49–52.

Beliveau, C., Laforge, M., Cusson, M., and Bellemare, G. (2000). Expression of a Tranosema rostrale polydnavirus gene in the spruce budworm, *Choristoneura fumiferana*. *J. Gen. Virol.* **81**(Pt. 7)**:**1871–1880.

Beliveau, C., Levasseur, A., Stoltz, D., and Cusson, M. (2003). Three related TrIV genes: Comparative sequence analysis and expression in host larvae and Cf-124T cells. *J. Insect Physiol.* **49**(5)**:**501–511.

Bendahmane, M., Fitchen, J. H., Zhang, G., and Beachy, R. N. (1997). Studies of coat protein-mediated resistance to tobacco mosaic tobamovirus: Correlation between assembly of mutant coat proteins and resistance. *J. Virol.* **71**(10)**:**7942–7950.

Blissard, G. W., Smith, O. P., and Summers, M. D. (1987). Two related viral genes are located in a single superhelical DNA segment of the multipartite *Campoletis sonorensis* virus genome. *Virology* **160:**120–134.

Blissard, G. W., Theilmann, D. A., and Summers, M. D. (1989). Segment W of Campoletis sonorensis virus: Expression, gene products, and organization. *Virology* **169**(1)**:**78–79.

Bonvin, M., Kojic, D., Blank, F., Annaheim, M., Wehrle, I., Wyder, S., Kaeslin, M., and Lanzrein, B. (2004). Stage-dependent expression of *Chelonus inanitus* polydnavirus genes in the host and the parasitoid. *J. Insect Physiol.* **50**(11)**:**1015–1026.

Bonvin, M., Marti, D., Wyder, S., Kojic, D., Annaheim, M., and Lanzrein, B. (2005). Cloning, characterization and analysis by RNA interference of various genes of the *Chelonus inanitus* polydnavirus. *J. Gen. Virol.* **86**(Pt 4)**:**973–983.

Bourne, H. R., Sanders, D. A., and McCormick, F. (1991). The GTPase superfamily: Conserved structure and molecular mechanisms. *Nature* **349:**117–127.

Carton, Y., and Nappi, A. J. (1997). *Drosophila* cellular immunity against parasitoids. *Parasitol. Today* **13**(6)**:**218–227.

Chen, Y. P., Taylor, P. B., Shapiro, M., and Gundersen-Rindal, D. E. (2003). Quantitative expression analysis of a Glyptapanteles indiensis polydnavirus protein tyrosine phosphatase gene in its natural lepidopteran host, *Lymantria dispar. Insect Mol. Biol.* **12** (3)**:**271–280.

Choi, J. Y., Roh, J. Y., Kang, J. N., Shim, H. J., Woo, S. D., Jin, B. R., Li, M. S., and Je, Y. H. (2005). Genomic segments cloning and analysis of *Cotesia plutellae* polydnavirus using plasmid capture system. *Biochem. Biophys. Res. Commun.* **332**(2)**:**487–493.

Clark, W. G., Register, J. C., 3rd, Nejidat, A., Eichholtz, D. A., Sanders, P. R., Fraley, R. T., and Beachy, R. N. (1990). Tissue-specific expression of the TMV coat protein in transgenic tobacco plants affects the level of coat protein-mediated virus protection. *Virology* **179**(2)**:**640–647.

Croft, B. A. (1990). "Arthropod Biological Control Agents and Pesticides," p. 703. Wiley and Sons, New York.

Croft, B. A., and Brown, A. W. (1975). Responses of arthropod natural enemies to insecticides. *Annu. Rev. Entomol.* **20:**285–335.

Cui, L., and Webb, B. A. (1996). Isolation and characterization of a member of the cysteine-rich gene family from *Campoletis sonorensis* polydnavirus. *J. Gen. Virol.* **77** (Pt 4)**:**797–809.

D'Amico, L. J., Davidowitz, G., and Nijhout, H. F. (2001). The developmental and physiological basis of body size evolution in an insect. *Proc. Biol. Sci.* **268**(1476)**:**1589–1593.

Dahlman, D. L., Rana, R. L., Schepers, E. J., Schepers, T., DiLuna, F. A., and Webb, B. A. (2003). A teratocyte gene from a parasitic wasp that is associated with inhibition of insect growth and development inhibits host protein synthesis. *Insect Mol. Biol.* **12** (5)**:**527–534.

Davidowitz, G., D'Amico, L. J., and Nijhout, H. F. (2003). Critical weight in the development of insect body size. *Evol. Dev.* **5**(2)**:**188–197.

De Gregorio, E., Spellman, P. T., Tzou, P., Rubin, G. M., and Lemaitre, B. (Gregorio 2002). The Toll and Imd pathways are the major regulators of the immune response in *Drosophila. EMBO J.* **21**(11)**:**2568–2579.

Dib-Hajj, S. D., Webb, B. A., and Summers, M. D. (1993). Structure and evolutionary implications of a "cysteine-rich" *Campoletis sonorensis* polydnavirus gene family. *Proc. Natl. Acad. Sci. USA* **90**(8)**:**3765–3769.

Edson, K. M., Vinson, S. B., Stoltz, D. B., and Summers, M. D. (1981). Virus in a parasitoid wasp: Suppression of cellular immune response in the parasitoid's host. *Science* **211**(4482)**:**582–583.

Espagne, E., Dupuy, C., Huguet, E., Cattolico, L., Provost, B., Martins, N., Poirie, M., Periquet, G., and Drezen, J. M. (2004). Genome sequence of a polydnavirus: Insights into symbiotic virus evolution. *Science* **306**(5694)**:**286–289.

Espinosa, A., Guo, M., Tam, V. C., Fu, Z. Q., and Alfano, J. R. (2003). The Pseudomonas syringae type III-secreted protein HopPtoD2 possesses protein tyrosine phosphatase activity and suppresses programmed cell death in plants. *Mol. Microbiol.* **49**(2)**:** 377–387.

Estruch, J. J., Carozzi, N. B., Desai, N., Duck, N. B., Warren, G. W., and Koziel, M. G. (1997). Transgenic plants: An emerging approach to pest control. *Nat. Biotechnol.* **15** (2)**:**137–141.

Falabella, P., Varricchio, P., Gigliotti, S., Tranfaglia, A., Pennacchio, F., and Malva, C. (2003). Toxoneuron nigriceps polydnavirus encodes a putative aspartyl protease highly expressed in parasitized host larvae. *Insect Mol. Biol.* **12**(1)**:**9–17.

Fath-Goodin, A., Gill, T. A., Martin, S. B., and Webb, B. A. (2006). Effect of *Campoletis sonorensis* cys-motif proteins on *Heliothis virescens* larval development. *J. Insect Physiol.* **52**(6)**:**576–585.

Feddersen, I., Sanders, K., and Schmidt, O. (1986). Virus-like particles with host protein like antigenic determinants protect and insect parasitioid from encapsulation. *Experientia* **42:**1278–1281.

Fleming, J. G., and Krell, P. J. (1993). Polydnavirus genome organization. *Parasites and Pathogens of Insects* **1**(1)**:**189–225.

Galibert, L., Rocher, J., Ravallec, M., Duonor-Cerutti, M., Webb, B. A., and Volkoff, A. N. (2003). Two Hyposoter didmator ichnovirus genes expressed in the lepidopteran host encode secreted or membrane-associated serine and threonine rich proteins in segments that may be nested. *J. Insect Physiol.* **49**(5)**:**441–451.

Ghosh, S., May, M. J., and Kopp, E. B. (1998). NF-kappa B and Rel proteins: Evolutionarily conserved mediators of immune responses. *Annu. Rev. Immunol.* **16:**225–260.

Gill, T. A., Maiti, I. B., and Webb, B. A. (2006). The effect on herbivory of transgenic tobacco expressing Campoletis sonorensis ichnoviral(CsIV) cys-motif proteins. Unpublished data.

Gillespie, J. P., Kanost, M. R., and Trenczek, T. (1997). Biological mediators of insect immunity. *Annu. Rev. Entomol.* **42:**611–643.

Glatz, R., Schmidt, O., and Asgari, S. (2003). Characterization of a novel protein with homology to C-type lectins expressed by the *Cotesia rubecula bracovirus* in larvae of the lepidopteran host, *Pieris rapae*. *J. Biol. Chem.* **278**(22)**:**19743–19750.

Glatz, R., Roberts, H. L., Li, D., Sarjan, M., Theopold, U. H., Asgari, S., and Schmidt, O. (2004). Lectin-induced haemocyte inactivation in insects. *J. Insect Physiol.* **50**(10)**:** 955–963.

Glatz, R., Schmidt, O., and Asgari, S. (2004). Isolation and characterization of a *Cotesia rubecula bracovirus* gene expressed in the lepidopteran *Pieris rapae*. *J. Gen. Virol.* **85** (Pt 10)**:**2873–2882.

Grossniklaus-Burgin, C., Wyler, T., Pfister-wilhelm, R., and Lanzrein, B. (1994). Biology and morphology of the parasitoid *Chelonus inanitus* (Braconidae, Hymenoptera) and effects on the development of its host *Spodoptera littoralis* (Noctuidae, Lepidoptera). *Invertebrate Reprod. Develop.* **25:**143–158.

Gruber, A., Stettler, P., Heiniger, P., Schumperli, D., and Lanzrein, B. (1996). Polydnavirus DNA of the braconid wasp *Chelonus inanitus* is integrated in the wasp's genome and excised only in later pupal and adult stages of the female. *J. Gen. Virol.* **77** (Pt 11):2873–2879.

Gundersen-Rindal, D. E., and Pedroni, M. J. (2006). Characterization and transcriptional analysis of protein tyrosine phosphatase genes and an ankyrin repeat gene of the parasitoid *Glyptapanteles indiensis* polydnavirus in the parasitized host. *J. Gen. Virol.* **87**(Pt 2):311–322.

Harwood, S. H., Grosovsky, A. J., Cowles, E. A., Davis, J. W., and Beckage, N. E. (1994). An abundantly expressed hemolymph glycoprotein isolated from newly parasitized *Manduca sexta* larvae is a polydnavirus gene product. *Virology* **205**(2): 381–392.

Harwood, S. H., McElfresh, J. S., Nguyen, A., Conlan, C. A., and Beckage, N. E. (1998). Production of early expressed parasitism-specific proteins in alternate sphingid hosts of the braconid wasp *Cotesia congregata*. *J. Invertebr. Pathol.* **71**(3):271–279.

Hellers, M., Beck, M., Theopold, U., Kamei, M., and Schmidt, O. (1996). Multiple alleles encoding a virus-like particle protein in the ichneumonid endoparasitoid Venturia canescene. *Insect Mol. Biol.* **4**(5):239–249.

Hilgarth, R. S., and Webb, B. A. (2002). Characterization of Campoletis sonorensis ichnovirus segment I genes as members of the repeat element gene family. *J. Gen. Virol.* **83**(Pt 10):2393–2402.

Hoch, G., Schafellner, C., Henn, M. W., and Schopf, A. (2002). Alterations in carbohydrate and fatty acid levels of *Lymantria dispar* larvae caused by a microsporidian infection and potential adverse effects on a co-occurring endoparasitoid, *Glyptapanteles liparidis*. *Arch. Insect Biochem. Physiol.* **50**(3):109–120.

Hochuli, A., and Lanzrein, B. (2001). Characterization of a 212 kD protein, released into the host by the larva of the endoparasitoid *Chelonus inanitus* (Hymenoptera, Braconidae). *J. Insect Physiol.* **47**(11):1313–1319.

Jarlfors, U. E., Dahlman, D. L., and Zhang, D. (1997). Effects of Microplitis croceipes teratocytes on host haemolymph protein content and fat body proliferation. *J. Insect Physiol.* **43**(6):577–585.

Johner, A., Stettler, P., Gruber, A., and Lanzrein, B. (1999). Presence of polydnavirus transcripts in an egg-larval parasitoid and its lepidopterous host. *J. Gen. Virol.* **80** (Pt 7):1847–1854.

Kaeslin, M., Pfister-Wilhelm, R., and Lanzrein, B. (2005a). Influence of the parasitoid Chelonus inanitus and its polydnavirus on host nutritional physiology and implications for parasitoid development. *J. Insect Physiol.* **51**(12):1330–1339.

Kaeslin, M., Pfister-Wilhelm, R., Molina, D., and Lanzrein, B. (2005b). Changes in the haemolymph proteome of Spodoptera littoralis induced by the parasitoid Chelonus inanitus or its polydnavirus and physiological implications. *J. Insect Physiol.* **51** (9):975–988.

Kaeslin, M., Wehrle, I., Grossniklaus-Burgin, C., Wyler, T., Guggisberg, U., Schittny, J. C., and Lanzrein, B. (2005c). Stage-dependent strategies of host invasion in the egg-larval parasitoid *Chelonus inanitus*. *J. Insect Physiol.* **51**(3):287–296.

Kaneko, T., and Silverman, N. (2005). Bacterial recognition and signalling by the Drosophila IMD pathway. *Cell. Microbiol.* **7**(4):461–469.

Kim, Y. (2005). Identification of host translation inhibitory factor of *Campoletis sonorensis ichnovirus* on the tobacco budworm, *Heliothis virescens*. *Arch. Insect Biochem. Physiol.* **59**(4):230–244.

Kota, M., Daniell, H., Varma, S., Garczynski, S. F., Gould, F., and Moar, W. J. (1999). Overexpression of the *Bacillus thuringiensis (Bt) Cry2Aa2* protein in chloroplasts confers resistance to plants against susceptible and Bt-resistant insects. *Proc. Natl. Acad. Sci. USA* **96**(5):1840–1845.

Kroemer, J. A., and Webb, B. A. (2004). Polydnavirus genes and genomes: Emerging gene families and new insights into polydnavirus replication. *Annu. Rev. Entomol.* **49:** 431–456.

Kroemer, J. A., and Webb, B. A. (2005). Ikappabeta-related vankyrin genes in the *Campoletis sonorensis ichnovirus*: Temporal and tissue-specific patterns of expression in parasitized *Heliothis virescens* lepidopteran hosts. *J. Virol.* **79**(12)**:**7617–7628.

Kroemer, J. A., and Webb, B. A. (2006). Divergences in protein function and cellular localization within the Campoletis sonorensis ichnovirus vankyrin family. Submitted for publication.

Lanzrein, B., Treiblmayr, K., Meyer, V., Pfister-Wilhelm, R., and Grossniklaus-Burgin, C. (1998). Physiological and endocrine changes associated with polydnavirus/venom in the parasitoid-host system *Chelonus inanitus-Spodoptera littoralis. J. Insect Physiol.* **44**(3–4)**:**305–321.

Lapointe, R., Wilson, R., Vilaplana, L., O'Reilly, D. R., Falabella, P., Douris, V., Bernier-Cardou, M., Pennacchio, F., Iatrou, K., Malva, C., and Olszewski, J. A. (2005). Expression of a Toxoneuron nigriceps polydnavirus-encoded protein causes apoptosis-like programmed cell death in lepidopteran insect cells. *J. Gen. Virol.* **86**(Pt 4)**:**963–971.

Lavine, M. D., and Beckage, N. E. (1995). Polydnaviruses: Potent mediators of host insect immune dysfunction. *Parasitol. Today* **11**(10)**:**368–378.

Lewis, W. J., van Lenteren, J. C., Phatak, S. C., and Tumlinson, J. H. (1997). A total system approach to sustain pest management. *Proc. Natl. Acad. Sci. USA* **94:** 12243–12248.

Lockey, T. D., and Ourth, D. D. (1996a). Formation of pores in *Escherichia coli* cell membranes by a cecropin isolated from hemolymph of *Heliothis virescens* larvae. *Eur. J. Biochem.* **236**(1)**:**263–271.

Lockey, T. D., and Ourth, D. D. (1996b). Purification and characterization of lysozyme from hemolymph of *Heliothis virescens* larvae. *Biochem. Biophys. Res. Commun.* **220** (3)**:**502–508.

Luckhart, S., and Webb, B. A. (1996). Interaction of a wasp ovarian protein and polydnavirus in host immune suppression. *Dev. Comp. Immunol.* **20**(1)**:**1–21.

Maiti, I. B., Dey, N., Dahlman, D. L., and Webb, B. A. (2003). Antibiosis-type resistance in transgenic plants expressing *teratocyte secretory peptide (TSP)* gene from hymenopteran endoparasite (*Microplitis croceipes*). *Plant Biotechnol.* **1:**209–219.

Malva, C., Varricchio, P., Falabella, P., La Scaleia, R., Graziani, F., and Pennacchio, F. (2004). Physiological and molecular interaction in the host-parasitoid system *Heliothis virescens-Toxoneuron nigriceps*: Current status and future perspectives. *Insect Biochem. Mol. Biol.* **34**(2)**:**177–183.

McGettigan, J., McLennan, R. K., Broderick, K. E., Kean, L., Allan, A. K., Cabrero, P., Regulski, M. R., Pollock, V. P., Gould, G. W., Davies, S. A., and Dow, J. A. (2005). Insect renal tubules constitute a cell-autonomous immune system that protects the organism against bacterial infection. *Insect Biochem. Mol. Biol.* **35**(7)**:**741–754.

Nakamatsu, Y., and Tanaka, T. (2004). Correlation between concentration of hemolymph nutrients and amount of fat body consumed in lightly and heavily parasitized hosts (*Pseudaletia separata*). *J. Insect Physiol.* **50**(2–3)**:**135–141.

Narayanan, K. (2004). Insect defence: Its impact on microbial control of insect pests. *Curr. Sci.* **86**(6)**:**800–814.

Nayak, P., Basu, D., Das, S., Basu, A., Ghosh, D., Ramakrishnan, N. A., Ghosh, M., and Sen, S. K. (1997). Transgenic elite indica rice plants expressing CryIAc delta-endotoxin of *Bacillus thuringiensis* are resistant against yellow stem borer (*Scirpophaga incertulas*). *Proc. Natl. Acad. Sci. USA* **94**(6)**:**2111–2116.

Nejidat, A., and Beachy, R. N. (1989). Decreased levels of TMV coat protein in transgenic tobacco plants at elevated temperatures reduce resistance to TMV infection. *Virology* **173**(2)**:**531–538.

Nejidat, A., and Beachy, R. N. (1990). Transgenic tobacco plants expressing a coat protein gene of tobacco mosaic virus are resistant to some other tobamoviruses. *Mol. Plant Microbe Interact.* **3**(4)**:**247–251.

Nijhout, H. F., and Williams, C. M. (1974). Control of moulting and metamorphosis in the tobacco hornworm, *Manduca sexta* (L.): Cessation of juvenile hormone secretion as a trigger for pupation. *J. Exp. Biol.* **61**(2)**:**493–501.

Norton, W. N., and Vinson, S. B. (1983). Correlating the initiation of virus replication with a specific pupal development phase of an ichneumonid parasitoid. *Cell Tissue Res.* **231**(2)**:**387–398.

Ourth, D. D., Lockey, T. D., and Renis, H. E. (1994). Induction of cecropin-like and attacin-like antibacterial but not antiviral activity in *Heliothis virescens* larvae. *Biochem. Biophys. Res. Commun.* **200**(1)**:**35–44.

Ploeg, A. T., Mathis, A., Bol, J. F., Brown, D. J., and Robinson, D. J. (1993). Susceptibility of transgenic tobacco plants expressing tobacco rattle virus coat protein to nematode-transmitted and mechanically inoculated tobacco rattle virus. *J. Gen. Virol.* **74:** 2709–2715.

Rana, R. L., Dahlman, D. L., and Webb, B. A. (2002). Expression and characterization of a novel teratocyte protein of the braconid, *Microplitis croceipes* (cresson). *Insect Biochem. Mol. Biol.* **32**(11)**:**1507–1516.

Reimann-Philipp, U., and Beachy, R. N. (1993). Coat protein-mediated resistance in transgenic tobacco expressing the tobacco mosaic virus coat protein from tissue-specific promoters. *Mol. Plant Microbe Interact.* **6**(3)**:**323–330.

Reineke, A., Asgari, S., Ma, G., Beck, M., and Schmidt, O. (2002). Sequence analysis and expression of a virus-like particle protein, VLP2, from the parasitic wasp *Venturia canescens*. *Insect Mol. Biol.* **11**(3)**:**233–239.

Rotheram, S. M. (1973). The surface of the egg of a parasitic insect. II. The ultrastructure of the particulate coat on the egg of Nemeritie. *Proc. R. Soc. Lond. Series B* **183:** 195–204.

Savary, S., Beckage, N., Tan, F., Periquet, G., and Drezen, J. M. (1997). Excision of the polydnavirus chromosomal integrated EP1 sequence of the parasitoid wasp *Cotesia congregata* (Braconidae, Microgastinae) at potential recombinase binding sites. *J. Gen. Virol.* **78:**3125–3134.

Schmidt, O., and Theopold, U. (1991). Immune defense and suppression in insects. *Bioessays* **13:**343–346.

Soldevila, A. I., and Jones, D. (1993). Expression of a parasitism-specific protein in lepidopteran hosts of *Chelonus* sp. *Arch. Insect Biochem. Physiol.* **24**(3)**:**149–169.

Shelby, K. S., and Webb, B. A. (1994). Polydnavirus infection inhibits synthesis of an insect plasma protein, arylphorin. *J. Gen. Virol.* **75**(Pt 9)**:**2285–2292.

Shelby, K. S., and Webb, B. A. (1997). Polydnavirus infection inhibits translation of specific growth-associated host proteins. *Insect Biochem. Mol. Biol.* **27**(3)**:**263–270.

Shelby, K. S., and Webb, B. A. (1999). Polydnavirus-mediated suppression of insect immunity. *J. Insect Physiol.* **45**(5)**:**507–514.

Shelby, K. S., Cui, L., and Webb, B. A. (1998). Polydnavirus-mediated inhibition of lysozyme gene expression and the antibacterial response. *Insect Mol. Biol.* **7**(3)**:** 265–272.

Singsit, C., Adang, M. J., Lynch, R. E., Anderson, W. F., Wang, A., Cardineau, G., and Ozias-Akins, P. (1997). Expression of a *Bacillus thuringiensis cryIA(c)* gene in transgenic peanut plants and its efficacy against lesser cornstalk borer. *Transgenic Res.* **6** (2)**:**169–176.

Stewart, C. N., Jr., Adang, M. J., All, J. N., Boerma, H. R., Cardineau, G., Tucker, D., and Parrott, W. A. (1996). Genetic transformation, recovery, and characterization of fertile soybean transgenic for a synthetic *Bacillus thuringiensis cryIAc* gene. *Plant Physiol.* **112**(1)**:**121–129.

Stoltz, D. (1993). The PDV life cycle. *Parasites Pathogens Insects: Parasites* **1:**167–187.

Stoltz, D. B., and Vinson, S. B. (1979). Viruses and parasitism in insects. *Adv. Virus Res.* **24:**125–171.

Stoltz, D., and Cook, D. (1983). Inhibition of host phenoloxidase activity by parasitoid hymenoptera. *Experientia* **39:**1022–1024.

Stoltz, D. B., Vinson, S. B., and Mackinnon, E. A. (1976). *Baculovirus*-like particles in the reproductive tracts of female parasitoid wasps. *Can. J. Microbiol.* **22**(7)**:**1013–1023.

Strand, M. R., and Pech, L. L. (1995a). Immunological basis for compatibility in parasitoid-host relationships. *Annu. Rev. Entomol.* **40:**31–56.

Strand, M. R., and Pech, L. L. (1995b). Microplitis demolitor polydnavirus induces apoptosis of a specific haemocyte morphotype in *Pseudoplusia includens*. *J. Gen. Virol.* **76** (Pt 2)**:**283–291.

Strand, M. R., Witherell, R. A., and Trudeau, D. (1997). Two Microplitis demolitor polydnavirus mRNAs expressed in hemocytes of *Pseudoplusia includens* contain a common cysteine-rich domain. *J. Virol.* **71**(3)**:**2146–2156.

Sun, J. P., Wu, L., Fedorov, A. A., Almo, S. C., and Zhang, Z. Y. (2003). Crystal structure of the Yersinia protein-tyrosine phosphatase YopH complexed with a specific small molecule inhibitor. *J. Biol. Chem.* **278**(35)**:**33392–33399.

Tanaka, K., Matsumoto, H., and Hayakawa, Y. (2002). Detailed characterization of polydnavirus immunoevasive proteins in an endoparasitoid wasp. *Eur. J. Biochem.* **269**(10)**:**2557–2566.

Tanaka, K., Tsuzuki, S., Matsumoto, H., and Hayakawa, Y. (2003). Expression of *Cotesia kariyai polydnavirus* genes in lepidopteran hemocytes and Sf9 cells. *J. Insect Physiol.* **49**(5)**:**433–440.

Tanaka, K., Lapointe, R., Barney, W. E., Makkay, A. M., Stoltz, D. B., Cusson, M., and Webb, B. A. (2006). Shared and species-specific features among ichnovirus genomes. Manuscript in preparation.

Teramato, T., and Tanaka, T. (2003). Similar polydnavirus genes of two parasitoids, *Cotesia kariyai* and *Cotesia ruficrus*, of the host *Pseudaletia separata*. *J. Insect Physiol.* **49**(5)**:**463–471.

Teramoto, T., and Tanaka, T. (2004). Mechanism of reduction in the number of the circulating hemocytes in the *Pseudaletia separata* host parasitized by *Cotesia kariyai*. *J. Insect Physiol.* **50**(12)**:**1103–1111.

Theilmann, D. A., and Summers, M. D. (1987). Physical analysis of the *Campoletis sonorensis* virus multipartite genome and identification of a family of tandemly repeated elements. *J. Virol.* **61**(8)**:**2589–2598.

Theilmann, D. A., and Summers, M. D. (1988). Identification and comparison of *Campoletis sonorensis* virus transcripts expressed from four genomic segments in the insect hosts *Campoletis sonorensis* and *Heliothis virescens*. *Virology* **167**(2)**:**329–341.

Theopold, U., Krause, E., and Schmidt, O. (1994). Cloning of a VLP-protein coding gene from a parasitoid wasp *Venturia canescens*. *Arch. Insect Biochem. Physiol.* **26**(2–3)**:** 137–145.

Thoetkiattikul, H., Beck, M. H., and Strand, M. R. (2005). Inhibitor kappaB-like proteins from a polydnavirus inhibit NF-kappaB activation and suppress the insect immune response. *Proc. Natl. Acad. Sci. USA* **102**(32)**:**11426–11431.

Thompson, S. N. (1982). Effects of parasitization by the insect parasite *Hyposoter exiguae* on the growth, development, and physiology of its host *Trichoplusia ni*. *Parasitology* **84:**491–510.

Thompson, S. N., and Binder, B. F. (1984). Altered carbohydrate levels and gluconeogenic enzyme activity in *Trichoplusia ni* parasitized by the insect parasite, *Hyposoter exiguae*. *J. Parasitol.* **70:**644–651.

Thompson, S. N., Lee, R. W. K., and Beckage, N. E. (1990). Metabolism of parasitized *Manduca sexta* examined by nuclear magnetic resonance. *Arch. Insect Biochem. Physiol.* **13:**127–143.

Tian, Y. C., Qin, X. F., Xu, B. Y., Li, T. Y., Fang, R. X., Mang, K. Q., Li, W. G., Fu, W. J., Li, Y. P., and Zhang, S. F. (1991). Insect resistance of transgenic tobacco plants expressing delta-endotoxin gene of *Bacillus thuringiensis*. *Chin. J. Biotechnol.* **7** (1)**:**1–13.

Trudeau, D., Witherell, R. A., and Strand, M. R. (2000). Characterization of two novel *Microplitis demolitor* polydnavirus mRNAs expressed in *Pseudoplusia includens* haemocytes. *J. Gen. Virol.* **81**(Pt 12)**:**3049–3058.

Turnbull, M., and Webb, B. (2002). Perspectives on polydnavirus origins and evolution. *Adv. Virus Res.* **58:**203–254.

Turnbull, M. W., Volkoff, A. N., Webb, B. A., and Phelan, P. (2005). Functional gap-junction genes are encoded by insect viruses. *Curr. Biol.* **15:**291–292.

Varricchio, P., Falabella, P., Sordetti, R., Graziani, F., Malva, C., and Pennacchio, F. (1999). *Cardiochiles nigriceps* polydnavirus: Molecular characterization and gene expression in parasitized *Heliothis virescens* larvae. *Insect Biochem. Mol. Biol.* **29** (12)**:**1087–1096.

Vinson, S. B. (1990). Physiological interactions between the host genus *Heliothis* and its guild of parasitoids. *Arch. Insect Biochem. Physiol.* **13:**63–81.

Vinson, S. B., and Iwantsch, G. F. (1980). Host regulation by insect parasitoids. *Q. Rev. Biol.* **55:**143–165.

Volkoff, A. N., Ravallec, M., Bossy, J., Cerutti, P., Rocher, J., Cerutti, M., and Devauchelle, G. (1995). The replication of *Hyposoter didymator PDV*: Cytopathology of the calyx cells in the parasitoid. *Biol. Cell* **83**(1)**:**1–13.

Volkoff, A. N., Cerutti, P., Rocher, J., Ohresser, M. C., Devauchelle, G., and Duonor-Cerutti, M. (1999). Related RNAs in lepidopteran cells after *in vitro* infection with *Hyposoter didymator* virus define a new polydnavirus gene family. *Virology* **263** (2)**:**349–363.

Volkoff, A. N., Beliveau, C., Rocher, J., Hilgarth, R., Levasseur, A., Duonor-Cerutti, M., Cusson, M., and Webb, B. A. (2002). Evidence for a conserved polydnavirus gene family: Ichnovirus homologs of the CsIV repeat element genes. *Virology* **300**(2)**:**316–331.

Webb, B. A. (1998). Polydnavirus biology, genome structure, and evolution. *In* "The Insect Viruses" (L. K. Miller and L. A. Balls, eds.), pp. 105–139. Plenum Press, New York.

Webb, B. A., and Luckhart, S. (1994). Evidence for an early immunosuppressive role for related *Campoletis sonorensis* venom and ovarian proteins in *Heliothis virescens*. *Arch. Insect Biochem. Physiol.* **26**(2–3)**:**147–163.

Webb, B. A., and Strand, M. R. (2004). The biology and genomics of polydnaviruses. *In* "Comprehensive Molecular Insect Science" (L. I., Gilbert, K., Iatrou, and S. S., Gill, eds.), Vol. 6, pp. 323–360. Elsevier, San Diego, CA.

Webb, B. A., Strand, M. R., Dickey, S. E., Beck, M. H., Hilgarth, R. S., Barney, W. E., Kadash, K., Kroemer, J. A., Lindstrom, K. G., Rattanadechakul, W., Shelby, K. S., Thoetkiattikul, H., *et al.* (2006). Polydnavirus genomes reflect their dual roles as mutualists and pathogens. *Virology* **347**(1)**:**160–174.

Whitfield, J. B. (1990). Parasitoids, polydnaviruses and endosymbiosis. *Parasitol. Today* **6**(12)**:**381–384.
Wyler, T., and Lanzrein, B. (2003). Ovary development and polydnavirus morphogenesis in the parasitic wasp Chelonus inanitus. II. Ultrastructural analysis of calyx cell development, virion formation and release. *J. Gen. Virol.* **84:**1151–1163.
Yin, L., Zhang, C., Qin, J., and Wang, C. (2003). Polydnavirus of *Campoletis chlorideae*: Characterization and temporal effect on host *Helicoverpa armigera* cellular immune response. *Arch. Insect. Biochem. Physiol.* **52**(2)**:**104–113.
Zhang, C., and Wang, C. Z. (2003). cDNA cloning and molecular characterization of a cysteine-rich gene from *Campoletis chlorideae* polydnavirus. *DNA Seq* **14**(6)**:**413–419.
Zhang, D., Dahlman, D. L., and Schepers, E. J. (1998). *Microplitis croceipes teratocytes*: *In vitro* culture and biological activity of teratocyte secreted protein. *J. Insect Physiol.* **44**(9)**:**767–777.
Zhang, G., Schmidt, O., and Asgari, S. (2004). A novel venom peptide from an endoparasitoid wasp is required for expression of polydnavirus genes in host hemocytes. *J. Biol. Chem.* **279**(40)**:**41580–41585.

VIRUS-DERIVED GENES FOR INSECT-RESISTANT TRANSGENIC PLANTS

Sijun Liu, Huarong Li, S. Sivakumar, and Bryony C. Bonning

Department of Entomology, Iowa State University, Ames, Iowa 50011

Abstract

Insect viruses have evolved to counter physiological barriers to infection presented by the host insect. For the Lepidoptera (butterflies and moths), these barriers include (1) the peritrophic membrane (PM) lining the gut, which presents a physical barrier to virus infection of the midgut epithelial cells, (2) the basement membrane (BM) that overlies the gut thereby restricting secondary infection of other tissues, and (3) the immune system of the host insect. Hence, insect viruses provide a resource for genes that disrupt host physiology in a specific manner, and these genes in turn serve as a resource both for the study of physiological processes, and for disruption of these processes for pest management purposes. There are several examples of the application of genes used by an insect virus to overcome the PM barrier for production of insect-resistant transgenic plants. There are other examples of intrahemocoelic effectors, such as BM-degrading proteases that can only be used with an appropriate system for delivery of the agent from the gut into the hemocoel (body cavity) of

0065-3527/06 $35.00
DOI: 10.1016/S0065-3527(06)68012-3

the insect pest. In this chapter, we describe (1) baculovirus- and entomopoxvirus-derived genes that alter the physiology of the host insect, (2) use of these and homologous genes for production of insect-resistant transgenic plants, (3) other viral genes that have potential for use in development of insect-resistant transgenic plants, and (4) the use of plant lectins for delivery of intrahemocoelic toxins from transgenic plants. Plant expression of polydnavirus-derived genes is described by Gill *et al.* (this volume, pp. 393–426).

I. Introduction

A. *Baculoviruses and Entomopoxviruses*

There are several groups of viruses that infect insects (Miller and Ball, 1998). Among these, baculoviruses (family *Baculoviridae*) and entomopoxviruses (EPV: family *Poxviridae*, subfamily *Entomopoxvirinae*) are large double-stranded DNA viruses. EPV have been isolated from a wide range of insects including grasshoppers (Orthoptera), beetles (Coleoptera), midges (Diptera), and cockroaches (*Blattaria*) (Radek and Fabel, 2000) while baculoviruses have been isolated from wasps (Hymenoptera), butterflies and moths (Lepidoptera), and mosquitoes (Diptera).

There are two genera of baculoviruses, the *Granulovirus* (GV) and *Nucleopolyhedrovirus* (NPV). Baculoviruses have two phenotypes, the occlusion-derived virus (ODV) and budded virus (BV). ODV are occluded within a crystalline protein matrix to form the occlusion body (OB) and initiate infection in the midgut epithelium of host insects. The protein matrix of the OB for NPV and GV is composed of polyhedrin and granulin respectively, and the OBs are referred to as polyhedra (NPV) and granules (GV). BVs are produced after initiation of infection and serve to disseminate virus within the host insect. EPV virions are also assembled into OBs. The OBs of EPV are known as spheroids with the most abundant protein component being the protein spheroidin (Arif, 1995). For both baculoviruses and EPV, the OB serves to protect occluded virions from environmental stresses.

B. *Bioinformatics for Identification of Genes Involved in Virus–Host Interaction*

A bioinformatics approach has been used to identify candidate genes that are involved in insect virus–host interaction (Dall *et al.*, 2001). Comparison of insect virus genomic sequences led to identification of six groups of genes postulated to be involved in insect virus–host interaction. These genes were identified on the basis of (1) presence in viruses that infect insects and absence from viruses with other hosts, and (2) presence in at least two different taxa of insect viruses. The six groups of genes include the 11K group of genes, which harbor a core C6 motif of six cysteine residues that frequently interact with chitin. The 11K group includes the chitinases, mucins, and peritrophins. Another group is the fusolin/gp37 genes. In EPV, fusolin aggregates to form spindle bodies (SBs), which have been shown to increase infectivity of heterologous NPVs. Both of these groups of genes have potential for use in development of insect-resistant transgenic plants, as described later. Hence, this bioinformatics-based approach can provide leads for the study of specific virus genes (and homologs from other organisms), that may be useful for development of insect-resistant transgenic plants.

II. Enzymes That Target the Peritrophic Membrane

A. *Physiology of the Peritrophic Membrane*

The peritrophic membrane (PM) is an extracellular fibrous matrix that lines the midgut epithelium of some insects (Fig. 1). The PM ensures mechanical protection of epithelial cells, compartmentalizes

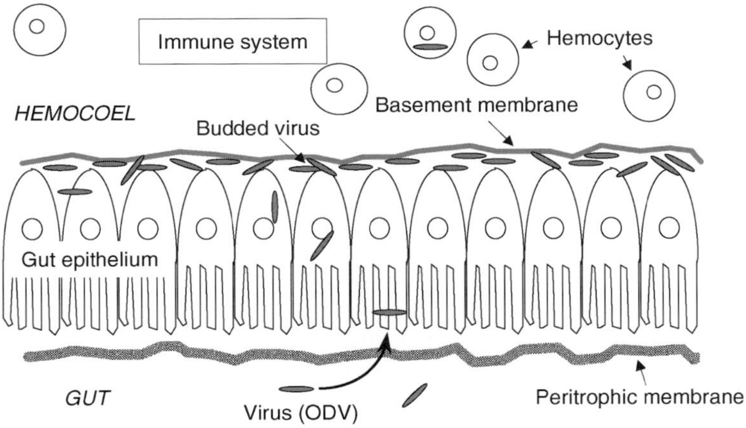

Fig 1. Host insect physiological barriers to virus infection. Barriers to virus infection include the peritrophic membrane that lines the gut, the basement membrane that overlies the gut epithelium, and the insect immune response, which may involve sloughing of infected gut cells, apoptosis of infected cells, and cellular and humoral immune factors in the hemolymph. Illustrated is infection of lepidopteran gut epithelial cells by a baculovirus, and accumulation of BV beneath the overlying basement membrane.

TABLE I
EXAMPLES OF PERITROPHIC MEMBRANE PROTEINS[a]

Protein	Insect (Order)	Reference
Intestinal insect mucin (IIM)	*Trichoplusia ni* (Lepidoptera)	Wang and Granados, 1997a,b; Wang *et al.*, 2004b
Chitin-binding proteins CBP1, CBP2		
TnPM-P42	*Trichoplusia ni* (Lepidoptera)	Guo *et al.*, 2005
Peritrophin-44	*Lucilia cuprina* (Diptera)	Elvin *et al.*, 1996
Peritrophin 50 (Ae-Aper50)	*Aedes aegypti* (Diptera)	Shao *et al.*, 2005
Ag-Aper1	*Anopheles gambiae* (Diptera)	Shen and Jacobs-Lorena, 1998
Ag-Aper14		Devenport *et al.*, 2005; Shao *et al.*, 2005

[a] See also Tellam *et al.*, 1999.

digestive enzymes in the gut, and protects the gut epithelium from microbial infection (Lehane, 1997). The PM consists of a chitinous fibrous matrix, glycosaminoglycans, and protein. The chitin content of the PM ranges from 3% to 45% of the total weight of the PM (Kramer *et al.*, 1995; Peters, 1992), while the protein content ranges from 35% to 55% of the PM by weight (Wang and Granados, 2001).

There are two types of PM. Type I PM is generated either by the whole, or by only anterior or posterior regions of the midgut epithelium, while type II PM is secreted by only a few rows of cells at the anterior region of the midgut (cardia) (Terra, 2001). Research on the molecular structure of PM has been driven by the PM as a potential target site for pest management purposes and because the PM presents a barrier to movement of parasites of medical importance. All of the PM proteins listed in Table I are chitin-binding proteins (CBP). Based on a model of the structure of the PM matrix, potential PM target sites have been proposed (Wang and Granados, 2001).

To infect insect midgut epithelial cells, bacterial and viral pathogens must negotiate the protective PM. Several insect viruses have acquired enzymes, such as enhancins and chitinases, to overcome the PM barrier.

B. Enhancin

Enhancin, which is also known as virus-enhancing factor or synergistic factor, was first isolated from GV OB. This enhancin had the

TABLE II
BACULOVIRUS ENHANCIN GENES

Virus	Number of enhancin genes	References
Granuloviruses		
Pseudaletia unipuncta GV (PuGV)	1	Tanada, 1959; Tanada *et al.*, 1973
Trichoplusia ni GV (TnGV)	1	Hashimoto *et al.*, 1991
Helicoverpa armigera GV (HaGV)	1	Roelvink *et al.*, 1995
Pseudaletia unipuncta GV-Hawaiian strain (PsunGV-H)	1	Roelvink *et al.*, 1995
Xestia c-nigrum GV (XcGV)	4	Hayakawa *et al.*, 1999
Agrotis segetum GV (AsGV)	1	GenBank accession no. AY522332
Choristoneura fumiferana GV (CfGV)	1	GenBank accession no. AAG338872
Nucleopolyhedroviruses		
Lymantria dispar MNPV (LdMNPV)	3	Bischoff, 1997; Kuzio *et al.*, 1999
Mamestra configurata NPV-A (MacoNPV-A)	1	Li *et al.*, 2003a
Mamestra configurata NPV-B (MacoNPV-B)	1	Li *et al.*, 2002
Choristoneura fumiferana MNPV (CfMNPV)	1	GenBank accession no. AF512031

ability to enhance infection by some NPV (Hukuhara *et al.*, 1987; Tanada, 1959; Tanada *et al.*, 1973). Enhancins have subsequently been identified in a number of other baculoviruses (Table II). All of the enhancins increase the susceptibility of lepidopteran larvae to heterologous baculovirus infections. A *Trichoplusia ni* GV (TnGV) enhancin enhances *Autographa californica* multiple nucleopolyhedrovirus (AcMNPV) infection by 2- to 14-fold in various insect species (Wang *et al.*, 1994). Deletions of the two *Lymantria dispar* multiple nucleopolyhedrovirus (LdMNPV) enhancins (E1 and E2) reduced the viral potency 12-fold when compared to wild-type virus (Popham *et al.*, 2001).

Enhancin-like proteins with 24–25% homology to viral enhancins have also been identified in bacteria (*Yersinia* and *Bacillus* spp.), but these proteins appear to have a different function (Galloway *et al.*, 2005).

The enhancins are metalloproteases, which contain conserved zinc-binding domains (Hashimoto *et al.*, 1991; Lepore *et al.*, 1996; Wang and Granados, 1997a). The enhancins degrade proteins within the PM, thereby disrupting the structure of the PM and increasing viral access to the midgut epithelial cells (Derksen, 1988; Lepore *et al.*, 1996;

Peng *et al.*, 1999; Slavicek and Popham, 2005). The TnGV enhancin specifically degrades intestinal insect mucin (IIM), the major PM protein of the *T. ni* midgut (Wang and Granados, 1997a). By degrading the PM, enhancin increases access of virus to the host midgut cells, thereby increasing the extent of infection of the midgut cells. It does not hasten the spread of the virus in the host insect (Li *et al.*, 2003a).

The ability of enhancin to disrupt the PM and to facilitate infection of lepidopteran larvae by baculoviruses has allowed for exploitation of this enzyme using two separate transgenic plant-based strategies. The first approach is to construct transgenic plants that express enhancin and test the effects of PM disruption on the development of insects that feed on the plant. The second approach is to use transgenic plants that express enhancin in combination with baculovirus insecticides. Disruption of the PM by enhancin would increase susceptibility of the host to baculovirus infection. Cao *et al.* (2002) expressed the enhancin genes from TnGV and *Helicoverpa armigera* GV (HaGV) in tobacco plants under the control of the constitutive *Cauliflower mosaic virus* (CaMV) 35S promoter. Expression of both of the enhancin genes in transgenic tobacco was very low; although mRNA could be detected by real time PCR (RT-PCR), enhancin could not be detected. This result suggests that either transcription or stability of the mRNA in the plants was low. Poor initial expression of *Bacillus thuringiensis* (Bt)-derived toxins in plants was attributed to cryptic polyadenylation sites, aberrant-splicing sequences, RNA instability motifs, and suboptimal codon usage (Strizhov *et al.*, 1996). Indeed, cryptic signals unfavorable to plant expression, including polyadenylation signals and splice junctions, as well as the "AUUUA" mRNA instability motif were detected in the enhancin genes (Cao *et al.*, 2002). Hence, the use of modified gene sequences could improve plant expression of enhancin. Based on the transformation efficiency, enhancin did not appear to be toxic to plants, although there may have been selection against plants in which enhancin was expressed.

However, feeding on the leaves was inhibited and the growth and development of *T. ni* larvae was significantly slowed on some of the transgenic plants carrying the enhancin gene. Control plants were completely defoliated by larvae after 8 days, while the transformed plant line E10 only suffered 30% defoliation during the same period. After 8 days, most of the larvae on the control plants were at fifth instar, with a mean head capsule size of 1.44 mm and mean body weight of 64 mg. Larvae fed on the two transformed lines E10 and E34 were second or third instars, with a mean head capsule size of 0.86–1.03 mm and mean body weight of 22–44 mg. Higher larval mortality was also observed on transgenic plants with 63% larval mortality for E10, 33% larval mortality for E34, and 25% mortality for larvae fed on control plants. Long-term feeding of lyophilized transgenic tobacco-expressing TnGV enhancin to *Pseudaletia separata* and *Spodoptera exigua* larvae also impaired insect development (Hayakawa *et al.*, 2004).

C. Chitinase

Chitin is a linear polymer of β(1,4)-linked *N*-acetylglucosamine (GlcNAc). In insects, chitin functions as a sturdy scaffold for the assembly of cuticle lining the epidermis and trachea. Chitin is also a primary component of the peritrophic membrane. Chitinases (EC 3.2.1.14) have been isolated from arthropods, bacteria, fungi, plants, nematodes, and several vertebrates (Fukamizo, 2000). Insect chitinases play an important role in chitin metabolism, which is critical for insect growth and morphogenesis (Merzendorfer and Zimoch, 2003). Chitinases belong to families 18 and 19 of the glycoside hydrolases and have been subdivided into classes on the basis of amino acid similarity (Henrissat, 1999). Family 18 chitinases are commonly found in prokaryotes, eukaryotes, and viruses, and include chitinases encoded by baculoviruses. The sequences of baculoviral chitinases are closely related to those of bacteria (Wang *et al.*, 2004a). Functionally, chitinases are classified as endo- and exochitinases. Endochitinases cleave chitin randomly at internal sites, releasing soluble oligomers of GlcNAc (Sahai and Manocha, 1993), and exochitinases catalyze the progressive digestion of GlcNAc dimers (*N,N′*-diacetylchitobioses) from the nonreducing end of the chitin microfibril (Harman *et al.*, 1993) to generate monomers of GlcNAc (Cohen-Kupiec and Chet, 1998).

Baculovirus chitinase was first identified from AcMNPV and belongs to the *chiA* class of bacterial chitinases (Ayres *et al.*, 1994; Hawtin *et al.*, 1995; Rawlings *et al.*, 1992). *Chi* genes have been found in most baculovirus genomes, with the exception of *Plutella xylostella* GV (PxGV) (Hashimoto *et al.*, 2000; Wang *et al.*, 2004a). The AcMNPV-*chiA* is expressed in the late phase of virus replication (Hawtin *et al.*, 1995) and has both endo- and exochitinase activities (Rao *et al.*, 2004). In contrast, the chitinase of *Epiphyas postvittana* NPV (EppoNPV) is an exochitinase with low-endochitinase activity (Young *et al.*, 2005). The primary function of baculoviral chitinase is in postmortem liquefaction of the infected larval host cadaver. Degradation of the insect cuticle facilitates dispersal of polyhedra (progeny viruses) into the environment. A virus-encoded cathepsin (V-cath) also contributes to the liquefaction process (Ohkawa *et al.*, 1994; Rawlings *et al.*, 1992;

Slack *et al.*, 1995). The cathepsin may degrade cuticular proteins to allow the chitinase to access the chitin matrix (Hawtin *et al.*, 1997). On the basis that cells infected with viruses without functional ChiA failed to process the V-cath precursor, it was postulated that ChiA may also serve as a molecular chaperone for proper folding of V-cath in the endoplasmic reticulum (Hom and Volkman, 2000).

Although AcMNPV-*chiA* was detected in polyhedra (OBs), it may not function in the initial infection of insect larvae: a *chiA* deletion mutant showed no difference in infectivity to second instar larvae of *T. ni* compared to the wild-type AcMNPV (Hawtin *et al.*, 1997). The lack of chitinase activity against the PM may result from the low concentration of chitinase within the polyhedra; purified, *E. coli*-expressed AcMNPV-*chiA* perforated the PM of *Bombyx mori* larvae and resulted in 100% mortality when fed at 1 μg/g of larval body weight (LW). A reduction of larval growth was also noted when larvae were fed at a sublethal dose of chitinase (0.56 μg/g LW) (Rao *et al.*, 2004). These observations demonstrate the potential for use of AcMNPV *chiA* as a transgene for production of insect-resistant transgenic plants.

In addition to harboring their own chitinase gene, baculoviruses have been used for expression of other arthropod chitinases (Han *et al.*, 2005). A nonoccluded recombinant baculovirus expressing the chitinase of the tobacco hornworm *Manduca sexta* had a decreased median time of mortality following injection into *Spodoptera frugiperda* when compared to that of wild-type virus (Gopalakrishnan *et al.*, 1995). In this case the baculovirus-expressed *M. sexta* chitinase would affect chitinous structures within the insect rather than affecting the PM (Rao *et al.*, 2004).

The potential for plant expression of chitinases for use in insect pest management has been documented. Expression of the *M. sexta* chitinase in tobacco resulted in reduced growth of larvae of the tobacco budworm, *Heliothis virescens*. When first instar larvae were reared on chitinase-positive or -negative leaves, the larvae weighed an average of 19.7 and 96.6 mg, respectively, after three weeks. The chitinase-expressing transgenic plants also resulted in increased mortality and reduced feeding damage compared to control plants. However, there were no significant differences in larval growth, survival, or foliar damage between transgenic and control plants against *M. sexta*. This result shows that susceptibility to a specific insect chitinase may vary among pest species. A synergistic effect was noted between the plant-expressed chitinase and Bt toxin against both *H. virescens* and *M. sexta* (Ding *et al.*, 1998). Given that AcMNPV ChiA was shown to increase the permeability of larval PM to methylene blue and a small neuropeptide (Rao *et al.*, 2004), it is possible that the *M. sexta* chitinase promoted the access of Bt to target receptors on the midgut epithelium.

When expressed in cotton, *M. sexta* chitinase was expressed up to 0.119 μg/mg fresh weight in the leaves and conferred strong insect resistance in the field (Wang *et al.*, 2005). The chitinase gene was also expressed *in planta* in combination with other insect resistance genes. Wang *et al.* (2005) transformed *Brassica napus*, a rapeseed crop, with the *M. sexta* chitinase and a scorpion toxin gene *BmkiT* (Bmk). Transformed plants that expressed both genes at high levels were highly resistant to larvae of the diamondback moth, *Plutella maculipenis*. Although the authors did not demonstrate a synergistic or additive effect resulting from coexpression of the chitinase and the scorpion toxin, it is possible that elimination of the PM by chitinase facilitated uptake of the neurotoxin, which would then have to be transported across the midgut epithelium to its neuronal target site receptors to exert an insecticidal effect.

There is a cautionary tale with respect to the use of transgenic plants expressing chitinase however. Transgenic potato plants expressing a chitinase from the coleopteran, *Phaedon cochleariae* showed a probiotic effect against the peach–potato aphid *Myzus persicae* and promoted population growth with a 30% reduction of the population-doubling time (Saguez *et al.*, 2005). In contrast, when pea chitinase was coexpressed with the lectin *Galanthus nivalis* agglutinin (GNA) in transgenic potatoes, a small insecticidal effect against *M. persicae* was observed (Gatehouse *et al.*, 1996). Given that GNA is highly toxic to *M. persicae* (Gatehouse *et al.*, 1996), the probiotic effect of the pea chitinase may have reduced the toxic effect of GNA to *M. persicae* on these potato plants. Aphids lack a PM and therefore may be considered nontarget insects in relation to use of the chitinase-mediated approach for insect pest management. Bacterial chitinases have been shown to be toxic to aphids however (Broadway *et al.*, 1998). Saguez *et al.* (2005) hypothesize that some arabinogalactan proteins (AGP), which are involved in plant growth, contain beta-1,4-linked GlcNAc units and hence are potential substrates for chitinase. Chitinase may therefore be generating an additional source of dietary carbohydrates within the transgenic plants for aphids.

Aphids are among the most pervasive pests of temperate agriculture, and promotion of aphid population growth by a transgene, such as chitinase, would present a considerable disadvantage for use of a transgenic plant for insect pest management. However, given that chitinases vary in their effects against different insect pests, additional work is required to determine whether all chitinases promote aphid wellbeing.

D. Chitin-Binding Proteins

EPV produce SBs in the cytoplasm of infected cells. These bodies consist of a protein called fusolin (Dall *et al.*, 1993; Hayakawa *et al.*, 1996). EPV fusolin and fusolin-like proteins also localize in the occluded virus within the spheroid (OB). EPV fusolin shares 30–84% amino acid sequence identity with baculovirus gp37 (Li *et al.*, 2000). SBs composed of gp37 have been noted for some baculoviruses (Gross *et al.*, 1993; Li *et al.*, 2000; Liu and Carstens, 1996). On the basis that EPV fusolin enhances NPV infection, it was named enhancing factor (EF) (Hayakawa *et al.*, 1996; Wijonarko and Hukuhara, 1998; Xu and Hukuhara, 1992, 1994).

The spindles of the coleopteran EPV, *Anomala cuprea* EPV (AcEPV), increased the infectivity of BmNPV to *B. mori* in excess of 10,000-fold (Furuta *et al.*, 2001; Mitsuhashi *et al.*, 1998). The spindles of AcEPV resulted in rapid damage to the PM suggesting that EPV spindles enhance NPV infectivity by providing greater access to the microvilli of the midgut (Mitsuhashi and Miyamoto, 2003). Curiously, a fusolin isolated from *P. separata* EPV (PsEPV) appeared to have an affinity for the plasma membrane of cultured cells and enhanced fusion of *Pseudoletia unipuncta* MNPV (PsunMNPV) with these cells (Hukuhara and Wijonarko, 2001; Hukuhara *et al.*, 2001).

Both gp37 and fusolin have a conserved chitin-binding domain which is also present in noncatalytic CBP. Based on homology of gp37 and fusolin to known bacterial CBP, and the chitin-binding activity of gp37 of *Spodoptera litura* MNPV (SpltMNPV), gp37 and fusolin have been characterized as CBP (Li *et al.*, 2003b). The gp37 of SpltMNPV and AcMNPV (but not OpMNPV and MbMNPV) is associated with ODV and hence, similar to fusolin, may target chitin within the PM of the host insect. Hence, for some viruses at least, gp37 and fusolin may target the chitin matrix of the peritrohpic membrane thereby enhancing access of baculoviruses and EPV to the midgut epithelium (Li *et al.*, 2003b).

There are at least two possible explanations for how gp37 and fusolin might disrupt the PM: First, fusolin or gp37 may act as chitin-binding competitors, by displacing or preventing the association of CBP with chitin and disrupting or preventing the formation of the PM as a result. This mechanism is similar to the mode of action of Calcofluor, a stilbene derivative, which facilitates baculovirus infection (Wang and Granados, 2000). Second, gp37 and fusolin may potentiate the action of viral chitinases: Studies on the CBP21 of the soil bacterium *Serratia marcescens* show that efficient degradation of chitin by the bacteria is dependent on CBP21. CBP21 binds to the insoluble crystalline substrate, causing conformational changes that facilitate hydrolysis by chitinases (Vaaje-Kolstad *et al.*, 2005). However, no chitinase homologs have thus far been identified in the two EPV sequenced to date (*Amsacta moorei* EPV and *Melanoplus sanguinipes* EPV).

Hukuhara *et al.* (1999) introduced the fusolin gene from the armyworm PsEVP into rice under the control of CaMV 35S promoter. Fusolin accounted for 0.3–0.5% of total soluble leaf proteins or 0.004–0.006% of the fresh leaf weight of the transformed rice plants. Armyworm larvae fed on transgenic rice were more susceptible to baculovirus infection. The infectious dose (ID50) to PsunMNPV was reduced 42-fold compared to that of larvae fed on nontransformed rice. The ID50 was further reduced (by 260–360-fold) when the R1 rice plants were used. These results illustrate the potential use of fusolin-expressing transgenic plants in combination with baculovirus insecticides for pest management purposes.

III. Enzymes That Target the Basement Membrane

A. *Physiology of the Basement Membrane*

BM, also referred to as basal laminae, are extracellular protein sheets surrounding all tissues of animals with the major component proteins of laminin, collagen IV, nidogen/entactin, and perlecan for vertebrates (Fig. 1) (Yurchenco and O'Rear, 1993). The major functions of the BM involve cell adhesion, cell signaling, and the structural maintenance of tissues (Page-McCaw *et al.*, 2003; Rohrbach and Timpl, 1993). There is high homology between the BMs of invertebrates and vertebrates in composition, structure, and function (Pedersen, 1991; Ryerse, 1998).

In insects, the BM must be remodeled during embryonic development, tissue and cell differentiation, and metamorphosis. This remodeling involves a number of enzymes that specifically digest components of the BM, including cathepsins, matrix metalloproteases (MMP), and ADAMTSs (*a* *d*isintegrin *a*nd *m*etalloprotease with *t*hrombospondin motifs) (Birkedal-Hansen, 1995; Cho *et al.*, 1999; Fujii-Taira *et al.*, 2000; Homma and Natori, 1996; Homma *et al.*, 1994; Maeda *et al.*, 2001; Porter *et al.*, 2005; Takahashi *et al.*, 1993; Zhao *et al.*, 1998). Expression and activation of these BM-degrading enzymes is tightly regulated to avoid uncontrolled and potentially fatal damage to other tissues. Insects may use MMP activators and inhibitors to mediate regulation of metalloprotease activity. For instance, a homolog of vertebrate tissue inhibitors of metalloproteases (TIMP) was found in *Drosophila melanogaster* that may contribute to regulation of a corresponding MMP (Vilcinskas and Wedde, 2002). The tight regulation of insect MMP homologs and other BM-degrading proteases suggest that if sufficient protease is delivered into the insect hemocoel, unregulated degradation of BM may occur. Such damage may impair insect physiological processes and kill the insect. Likewise, if sufficient protease inhibitor (TIMP) is delivered into the insect hemocoel, tissue remodeling may be disrupted. Hence, enzymes that degrade BM have potential as intrahemocoelic toxins for use in insect pest management.

The BM appears to act as a barrier to dissemination of some viruses within an infected insect. For example, the NPV of Lepidoptera are too large to diffuse through the pores in the BM that surround tissues of the host insect (Hess and Falcon, 1987; Reddy and Locke, 1990). Indeed coinjection of a baculovirus and clostridial collagenase, a protease known to degrade BM, resulted in enhanced infection of host tissues (Smith-Johannsen *et al.*, 1986). Systemic spread of virus within the host insect may occur through direct penetration of the BM into the hemocoel, possibly by an enzymatic process or where the BM is thin (Federici, 1997; Flipsen *et al.*, 1995; Granados and Lawler, 1981). An alternative hypothesis is that baculoviruses use the host tracheal system as a conduit to by-pass BM and establish systemic infection of host tissues (Engelhard *et al.*, 1994). The mechanism of penetration of the BM remains to be determined and there is debate over whether one route predominates over the other (Federici, 1997; Volkman, 1997). Interestingly, baculoviruses that infect multiple tissues within an insect (rather than being restricted to gut tissues) all harbor a fibroblast growth factor (*fgf*) gene (Detvisitsakun *et al.*, 2005). FGF may function to attract hemocytes to sites of virus infection by chemotaxis. Given that granular cells are intimately involved in the remodeling of BM and secrete a protease that digests the BM (Kurata *et al.*, 1991, 1992; Nardi *et al.*, 2001), the granular cell protease may facilitate movement of virus across the BM.

B. *Matrix Metalloproteases*

BM turnover in vertebrates is mediated by a group of zinc-binding metalloproteases known as MMPs and ADAMTS metalloproteases (Birkedal-Hansen, 1995; Maeda *et al.*, 2001; Porter *et al.*, 2005). MMPs are a family of structurally related enzymes that function in connective tissue remodeling under a variety of physiological conditions, such as embryonic growth, angiogenesis, wound healing, or reproductive processes. More than 20 different MMPs have been identified in mammalian tissues (Llano *et al.*, 2002). These MMPs have been classified into six major subfamilies, including collagenases, stromelysins, gelatinases, matrilysins, membrane-type MMPs, and other MMPs (Uria and Lopez-Otin, 2000). Most of these enzymes have several characteristic domains: a signal peptide sequence, a prodomain with a conserved Cys residue involved in maintaining the enzyme latency, a catalytic domain with a zinc-binding site, and a hemopexin-like domain that plays a role in substrate binding as well as in mediating interactions with the TIMPs. Recently, a number of MMP homologs have been identified in the genomes of insects and insect viruses, where they may function in BM turnover and remodeling in insects, or in the pathogenesis of insect viruses.

Three open reading frames (ORFs) from the genomic sequence of a grasshopper poxvirus (*Melanoplus sanguinipes* EPV) encode homologs of zinc metalloproteases (Afonso *et al.*, 1999). Two of these ORFs, MSV176 and MSV179, contain perfect copies of the His-Glu-2X-His catalytic domain consensus sequence for this group of proteases and bear a significant degree of sequence similarity to vertebrate MMPs. The ORF MSV175 has a Glu-to-Gln substitution in the active site consensus sequence. All three of these ORFs possess putative signal peptides.

Ko *et al.* (2000) identified two MMP homologs from the genomes of two baculoviruses of the genus GV. The MMP homolog from *Xestia c-nigrum* granulovirus (XcGV) has been expressed and characterized. It has the characteristics of zinc metalloproteases, shares 30% amino acid sequence identity to the catalytic domains of human stromelysin 1 and sea urchin-hatching enzyme, but it does not have a signal peptide sequence. Ko *et al.* (2000) speculated that it may be involved in tissue liquefaction and cuticle degradation of virus-killed hosts. Alternatively, this virus-encoded MMP may function in the degradation of BM in the host insect. The gene encoding the XcGV MMP was inserted into the genome of *B. mori* NPV. Infection of silkworm larvae with this engineered virus resulted in melanization, that is, blackening through deposition of melanin (Ko, R., unpublished data).

To determine the potential of MMPs as intrahemocoelic toxins, we constructed baculoviruses that express these insect virus-derived matrix metalloprotease homologs (Harrison, R. L., unpublished data). We used the baculovirus transfer vector pAcMLF9 for insertion of the protease-coding sequences into the AcMNPV genome. This vector

TABLE III

SUMMARY OF BACULOVIRUS EXPRESSION OF INSECT VIRUS-DERIVED PROTEASES

Protease	Recombinant virus	Protease activity detected?	Notes
MSV176	AcMLF9.MSV176	No[a]	Physiological pH
MSV179	AcMLF9.MSV179	No[a]	Physiological pH; budded virus unstable
MSV175	AcMLF9.MSV175	No[a]	Physiological pH
XcGV MMP	AcMLF9.spXcGVMMP	Yes[b]	pH 7.6

[a] Tested by azocoll, azocasein, and azoalbumin assay of High Five™ supernatants and cell extracts, and by gelatin zymography of supernatants. Sf21 supernatants and cell extracts were also tested by gelatin zymography and azocoll assay. It is conceivable that these assays provided inappropriate conditions for detection of these enzymes.

[b] Activity detected in 5×-concentrated High Five™ supernatants by azocoll assay.

employs the baculovirus *p6.9* promoter to drive foreign gene expression (Harrison and Bonning, 2001). Based on the lack of secretion of XcGV MMP, the bombyxin signal peptide (sp) was used for secretion of this enzyme. Expressed protease activity in the medium of High Five™ and Sf21 cells infected by the recombinant viruses was measured by gelatin or casein zymography and azocoll assay (Harrison and Bonning, 2001). Table III summarizes efforts to detect protease activity following infection of cultured insect cells.

There were indications that active proteases were produced by the recombinant baculoviruses listed in Table III, even though activity was not detected *in vitro* in all cases. Specifically, during production of virus stocks in larvae of *H. virescens*, integumental melanization (prior to death) was seen for larvae infected with viruses expressing MSV175, MSV176, and MSV179, and spXcGVMMP. In addition, the titer of the BV stock of AcMLF9.MSV179 decreased significantly over a two-week period, suggesting that this protease degrades some component of BV (Harrison, R., unpublished data).

Harrison and Bonning (2001) tested the potential of two vertebrate MMPs encoding human type IV collagenase, or collagenase A, GEL (Collier *et al.*, 1988), and a rat stromelysin-1, STR1 (Park *et al.*, 1991) as potential intrahemocoelic insect toxins. Human type IV collagenases degrade native and denatured collagens and other extracellular matrix proteins (Birkedal-Hansen, 1995). Stromelysins degrade a variety of extracellular matrix proteins, including type IV collagen and laminin. Expression of these proteases was directed from either the baculovirus *ie-1* or the *p6.9* promoter. The virus AcMLF9.STR1 caused integumental melanization of infected fifth instar *H. virescens* prior to death. This differs from wild-type baculovirus-killed lepidopteran larvae which melanize after death. Neither of the two vertebrate MMPs enhanced the insecticidal efficacy of the baculovirus. However, the level of expression of these enzymes from baculovirus-infected insect cells was relatively low (Harrison and Bonning, 2001). Posttranslational processing necessary for the secretion or activity of vertebrate MMPs may be inefficient in insect cells. Insect and insect virus homologs of MMPs may be expressed and secreted at higher levels than vertebrate proteases from baculovirus-infected insect cells, and may also digest insect basement membrane proteins more efficiently.

C. Cathepsins

Cathepsins are cysteine proteases with properties similar to papain. Papain-like cysteine proteases have three residues directly involved in catalysis: Cys^{25}, His^{169}, and Asn^{175} with Gln^{19} (papain numbering) (Barrett *et al.*, 1998; Cristofoletti *et al.*, 2005; Deraison *et al.*, 2004). Cathepsin proteases are involved in protein digestion (Terra and Ferreira, 1994), embryonic vitellin degradation (Cho *et al.*, 1999), and metamorphosis in insects (Homma *et al.*, 1994; Takahashi *et al.*, 1993).

A cathepsin L-like cysteine protease (ScathL) derived from the flesh fly, *Sarcophaga peregrina*, digests BM (Homma *et al.*, 1994). ScathL was isolated from the culture medium of NIH-Sape-4 cells, an embryonic cell line of *S. peregrina*. The proenzyme (50 kDa) undergoes autocatalytic processing to the mature enzyme (35 kDa). Cathepsin L proteases are not typically secreted, but are transported as proenzymes to lysosomes where they are activated. In NIH-Sape-4 cells, the ScathL proenzyme is both transported to lysosomes and secreted into the medium. In imaginal discs, the proenzyme normally is transported to lysosomes, but is secreted on application of 20-hydroxyecdysone to the discs (Homma *et al.*, 1994). ScathL plays an essential role in BM remodeling for the differentiation of both imaginal discs and the brains of flesh fly larvae, through the selective digestion of two BM proteins with molecular masses of 210 and 200 kDa (Fujii-Taira *et al.*, 2000; Homma and Natori, 1996).

Recombinant baculoviruses were constructed to test the insecticidal potential of ScathL (Harrison and Bonning, 2001). AcMLF9.ScathL (expressing ScathL under control of the viral *p6.9* promoter) was very effective in killing larvae of *H. virescens*. AcMLF9.ScathL killed *H. virescens* larvae approximately 30% faster than the recombinant baculoviruses AcMLF9.AaIT and AcMLF9.LqhIT2, which expressed potent insect neurotoxins, and >50% faster than the wild-type virus. Larvae infected with AcMLF9.ScathL consumed approximately 5-fold less lettuce than wild-type virus-infected larvae and 26-fold less lettuce than uninfected larvae. AcMLF9.ScathL caused integumental melanization of infected fifth instar *H. virescens* prior to death. The degree of melanization observed with AcMLF9.ScathL-infected larvae was greater than that observed with larvae infected with AcMLF9.STR1. Unlike AcMLF9.STR1, AcMLF9.ScathL also caused fragmentation and melanization of internal tissues. The presence of detached fat body lobes in the hemocoel of AcMLF9.ScathL-infected larvae suggested that baculovirus-expressed ScathL effectively degraded BM (Harrison and Bonning, 2001). These results support the idea that BM-degrading proteases can be used effectively as intrahemocoelic toxins for insect pest management.

The mechanism of insecticidal action of ScathL is under investigation. Of particular interest is the potential role of the immune system in larval mortality. Melanization is an important part of the insect immune response. Melanization is the formation and deposition of melanin, which occurs during normal postmolt hardening and darkening of insect cuticle and in response to wounding or the appearance of foreign matter in the insect (Marmaras *et al.*, 1996). Melanin is derived from the oxidation of monophenols and diphenols and the ensuing polymerization of their respective orthoquinones (Nappi and Christensen, 2005). The cellular immune response of insects consists of the phagocytosis of foreign material by hemocytes and encapsulation, in which the foreign entity is surrounded by layers of flattened hemocytes (Boucias and Pendland, 1998a,b). This cellular layer is often melanized to form an inflexible, impenetrable capsule around the foreign object. During melanogenesis, quinones and other cytotoxic reactive species are produced that may be involved in killing encapsulated parasites (Carton and Nappi, 1997; Nappi and Christensen, 2005). Quinones and the cytotoxic reactive species are also toxic to insect tissues. Therefore, melanogenesis must be localized, target-specific, and very tightly regulated in order to avoid uncontrolled and potentially fatal systemic activation in the insect's open circulatory system (Nappi and Christensen, 2005).

In insects, the integrity of the BM is important for hemocytes to respond both to cuticular wound healing and sequestration of endogenous tissues that are developing abnormally. Insects distinguish self from nonself based on the presence or absence of the correct BM structure. For example, studies on strains of *D. melanogaster* Meigen that produce melanotically encapsulated tumors revealed that the process of melanotic tumor formation started with disturbances in the BM of the fat body, followed by aggregation of hemocytes around the abnormal surface, and melanotic encapsulation of the affected area (Rizki and Rizki, 1974, 1980). Tissue grafts in *D. melanogaster* with mechanically or enzymatically generated BM damage also underwent melanotic encapsulation, but tissue grafts with an undamaged BM preparation were not encapsulated (Rizki and Rizki, 1980). Pech *et al.* (1995) tested the effects of culturing hemocytes from *Pseudoplusia includens* on an artificial BM (Matrigel). Hemocytes in mixed populations did not spread on tissue culture plates coated with Matrigel, whereas cells easily spread on uncoated plates. This result suggested that the BM can make a foreign surface appear to be self. In addition, antibodies against surface proteins of the insect pathogenic fungus *Nomuraea rileyi* cross-reacted with antigens on the surface of the fat body of the insect host, the beet armyworm, *S. exigua* (Pendland and Boucias, 1998, 2000). These results suggest that hyphae of *N. rileyi* evade the immune response of *S. exigua* by using a form of molecular mimicry in which the hyphae present a surface that resembles the BM of the host insect. These studies suggest an explanation for the melanization caused by ScathL: damage to the BM by ScathL causes the host immune defenses to recognize its own tissues as foreign. The tissues are subsequently melanized. Death of insects treated with ScathL could result from the physiological effects of basement membrane (BM) damage, or from the cytotoxic by-products of uncontrolled melanogenesis.

When fifth instar larvae of *H. virescens* were injected with recombinant ScathL at a dose of 20 μg per larva, the larvae melanized and died within 3 hours. Internal tissues and the gut were fragmented. The melanization, and fragmentation of internal tissues was similar to observations made for larvae infected with the recombinant baculovirus AcMLF9.ScathL (Harrison and Bonning, 2001). As expected from the mode of action of ScathL within the body cavity, bioassays indicate that ScathL has no insecticidal effect when ingested (unpublished data). The high-level toxicity of ScathL when delivered to the hemocoel suggests that ScathL would serve as an excellent transgene when combined with an appropriate delivery system to confer plant resistance to insect pests (Duck and Evola, 1997) (see later).

There are several examples of cysteine proteases being involved in plant defense against insects although the mechanisms of action of these proteases are unknown (Konno *et al.*, 2004; Linthorst *et al.*, 1993; Pechan *et al.*, 2000). It is conceivable that these proteases also target the peritrophic matrix or the insect BM of herbivorous insects.

IV. Delivery of Intrahemocoelic Toxins from Plants

A. Intrahemocoelic Toxins

A wide range of toxins that act within the body cavity or "hemocoel" of the insect rather than in the gut have been tested for insecticidal activity when expressed by a recombinant baculovirus (see chapter by Inceoglu *et al.*, this volume, pp. 323–360; Bonning *et al.*, 2002). The development of baculovirus insecticides that have been genetically enhanced for increased speed of kill has served to identify the most potent of a wide range of insect neurotoxins and intrahemocoelic effectors. These toxins include scorpion- and spider-derived insect selective neurotoxins. Use of the baculovirus allows delivery of the toxin from the baculovirus-infected cell into the hemocoel or to neuronal target sites. The infected caterpillar dies from the effects of the toxin delivered by the replicating virus. Until recently (with a few exceptions), it has not been possible to exploit intrahemocoelic toxins for development of insect-resistant transgenic plants because of the lack of an appropriate system to deliver the toxin through the gut epithelium and into the hemocoel of the insect. The exceptions are as follows: An intrahemocoelic effector, teratocyte secretory protein (TSP14) from the parasitic wasp *Microplitis croceipes*, was unexpectedly delivered from the gut into the hemocoel by an unknown mechanism (see chapter by Gill *et al.*, this volume, pp. 393–426). Transgenic tobacco lines stably expressing TSP14, caused retardation in growth and larval mortality when tested against first instar larvae of *H. virescens* and *M. sexta* (Maiti, 2003). There are also reports that expression of the insect-specific neurotoxin AaIT from tobacco plants (Yao *et al.*, 1996) and poplar (Wu *et al.*, 2000), and expression of spider toxins in rice (Huang *et al.*, 2001) and birch (Zhan *et al.*, 2001) confer insect resistance, although none of these studies provided direct evidence that the toxins moved from the gut into the hemocoel. Based on these exceptions to the concept that an intrahemocoelic effector will not enter the hemocoel from the gut on its own, further investigation of the mechanism of nonspecific uptake of proteins from the gut lumen is warranted.

B. Lectins

The current use of toxins for development of insect-resistant transgenic plants has been largely restricted to agents that act within the gut of the pest insect such as toxins derived from Bt, enhancin, protease

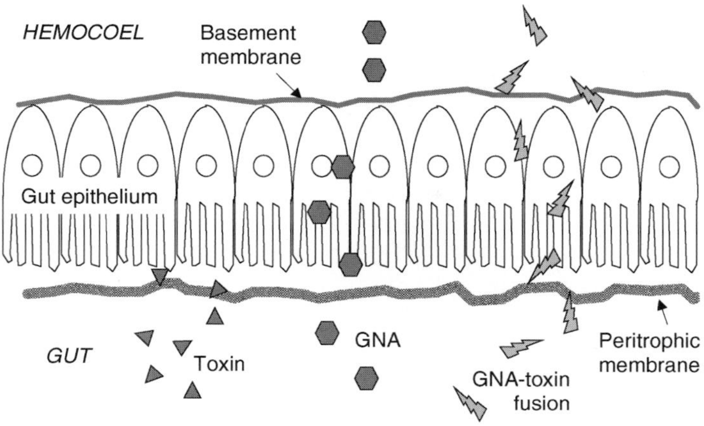

FIG 2. Lectin-mediated delivery of intrahemocoelic toxins. When ingested, most intrahemocoelic toxins have no insecticidal activity because they are unable to move across the midgut epithelium into the hemocoel to reach their site of action. Lectins, such as the snowdrop lectin *G. nivalis* agglutinin (GNA), bind to the gut epithelium and enter the hemocoel most likely by endocytosis (Fitches *et al.*, 2001b). Fusion of GNA to intrahemocoelic effectors serves to deliver the toxin into the hemocoel (Fitches *et al.*, 2002, 2004).

inhibitors, and lectins (O'Callaghan *et al.*, 2005; see Table I of associated supplemental material). It has been shown however, that some plant lectins are able to cross from the gut into the hemocoel of Lepidoptera and aphids, which comprise some of the most economically important agricultural insect pests. Drs. Elaine Fitches, John Gatehouse, and colleagues have demonstrated use of such a lectin for delivery of insect toxins into the hemocoel (Fig. 2). This research paves the way for expression and appropriate delivery of intrahemoelic toxins and enzymes, including those derived from insect viruses, from transgenic plants.

Lectins are proteins that bind carbohydrates with great specificity and sensitivity. Many plant lectins are defensive proteins that have deleterious effects on insects belonging to the orders Homoptera, Coleoptera, Lepidoptera, and Diptera (Bell *et al.*, 2001; Fitches *et al.*, 2001a; Li *et al.*, 2005; Melander *et al.*, 2003; Nagadhara *et al.*, 2004; Rao *et al.*, 1998; Setamou *et al.*, 2002; Zhou, 1998). Following ingestion by insects, the binding of insecticidal lectins to glycosylated targets on the microvilli of midgut epithelial cells causes damage to the integrity of the gut epithelium. The PM may also act as a target for lectin binding (Eisemann, 1994).

The snowdrop lectin GNA (*G. nivalis* agglutinin) is of particular interest because of its high-insect and low-mammalian toxicity (Down *et al.*, 1996; Gatehouse *et al.*, 1996; Hilder *et al.*, 1995; Rahbe *et al.*, 1995; Sauvion *et al.*, 1996; Stoger *et al.*, 1998). On ingestion by Lepidoptera and Homoptera, GNA binds to the gut epithelium and passes into the hemocoel most likely by endocytosis (Fitches *et al.*, 2001b). GNA and jackbean lectin (*Canavalia ensiformis*; Con A) were detected by immunolocalization in the gut cells, hemolymph, Malpighian tubules, and fat bodies of orally exposed insects (Fitches *et al.*, 2001b). The ability of GNA to cross the gut epithelium gives this protein the potential to act as a carrier to deliver fused peptides into the hemocoel of targeted insect pests.

Following ingestion, a recombinant fusion protein composed of GNA fused to the green fluorescent protein (GFP), was delivered to the hemolymph of larvae of the tomato moth *Lacanobia oleracea* (Raemaekers, 2000). GNA was then used successfully for delivery of a fused insect neuropeptide (*M. sexta* allatostatin, Manse-AS) (Fitches *et al.*, 2002), and a spider venom-derived neurotoxin (*Segestria florentia* toxin 1, SFI1) (Fitches *et al.*, 2004) into the hemocoel of *L. oleracea* (Fig. 2). Ingestion of the GNA-Manse-AS fusion protein resulted in inhibition of feeding and growth of the larvae. The recombinant GNA-SFI1 fusion protein killed 100% of first-instar larvae after 6 days of feeding, exhibited dose-dependent toxicity, and retarded growth when fed to late instar larvae. The GNA-SFI1 fusion protein was also highly toxic against *M. persicae* and the rice brown planthopper, *Nilaparvata lugens*. The ability of GNA to act as a carrier protein to deliver SFI1 into the hemolymph of these insects was demonstrated by immunoblotting (Down *et al.*, 2006).

These results demonstrate the potential use of GNA for delivery of virus-derived intrahemocoelic toxins such as BM-degrading proteases from transgenic plants.

V. Concluding Remarks

Insect viruses provide a useful resource for isolation and identification of potential physiological effectors that can be exploited for development of insect-resistant transgenic plants. A bioinformatics-based approach for identification of genes that are common to insect viruses can be helpful in identifying such genes (Dall *et al.*, 2001). While several such genes have been tested *in planta* for efficacy (Table IV), research is ongoing for many others. Of particular interest will be the

TABLE IV
VIRUS-DERIVED GENES USED FOR PRODUCTION OF INSECT-RESISTANT TRANSGENIC PLANTS

Gene	Origin	Target pest	Plant	References	Notes
Enhancin	TnGV, HaGV	*T. ni* *S. exigua* *P. separata*	Tobacco	Cao *et al.*, 2002	
Fusolin	PsEPV	*P. unipuncta*	Rice	Hukuhara *et al.*, 1999	Used to enhance baculovirus
TSP14	CsPDV	*H. virescens* *M. sexta*	Tobacco	Maiti, 2003	See chapter by Gill *et al.*, this volume, pp. 393–426

efficacy of baculovirus ChiA and virus-derived enzymes that target the BM as potential transgenes. As more insect virus genome sequences are determined, and our understanding of virus–host insect interaction improves, the number of virus-derived candidate genes will increase. The use of plant lectins will also allow for exploitation of virus genes that function within the hemocoel of the host insect.

ACKNOWLEDGMENTS

The authors thank Dr. Robert Harrison for critical review of this manuscript. This work was funded in part by grants from the Cooperative State Research, Education, and Extension Service, US Department of Agriculture, under Agreement Nos. 2003-35302-13558 and 2005-35607-15233, from the Consortium for Plant Biotechnology Research, as well as Hatch Act and State of Iowa funds.

REFERENCES

Afonso, C. L., Tulman, E. R., Lu, Z., Oma, E., Kutish, G. F., and Rock, D. L. (1999). The genome of *Melanoplus sanguinipes* Entomopoxvirus. *J. Virol.* **73:**533–552.

Arif, B. (1995). Recent advances in the molecular biology of entomopoxviruses. *J. Gen. Virol.* **76**(1)**:**1–13.

Ayres, M. D., Howar, S. C., Kuzio, J., Lopez-Ferber, M., and Possee, R. D. (1994). The complete DNA sequence of *Autographa californica* nuclear polyhedrosis virus. *Virology* **202:**586–605.

Barrett, A. J., Rawlings, N. D., and Wasner, J. F. (1998). "Handbook of Proteolytic Enzymes." Academic Press, London.

Bell, H. A., Fitches, E. C., Marris, G. C., Bell, J., Edwards, J. P., Gatehouse, J. A., and Gatehouse, A. M. R. (2001). Transgenic GNA expressing potato plants augment the beneficial biocontrol of *Lacanobia oleracea* (Lepidoptera; Noctuidae) by the parasitoid *Eulophus pennicornis* (Hymenoptera; Eulophidae). *Transgenic Res.* **10:**35–42.

Birkedal-Hansen, H. (1995). Proteolytic remodeling of extracellular matrix. *Curr. Opin. Cell Biol.* **7:**728–735.

Bischoff, D. S., and Slavicek, J. M. (1997). Molecular analysis of an enhancin gene in the *Lymantria dispar* nuclear polyhedrosis virus. *J. Virol.* **71:**8133–8140.

Bonning, B. C., Boughton, A. J., Jin, H., and Harrison, R. L. (2002). Genetic enhancement of baculovirus insecticides. *In* "Advances in Microbial Control of Insect Pests" (R. K. Upadhyay, ed.), pp. 109–125. Kluwar Academic/Plenum Publishers, London.

Boucias, D. G., and Pendland, J. C. (1998a). Insect immune defense system, part II: The recognition of nonself. "Principles of Insect Pathology," pp. 469–497. Kluwer Academic Publishers, Boston.

Boucias, D. G., and Pendland, J. C. (1998b). Insect immune defense system, part III: Prophenoloxidase cascade and post-attachment processes of phagocytosis. *In* "Principles of Insect Pathology," pp. 499–537. Kluwer Academic Publishers, Boston.

Broadway, R. M., Gongora, C., Kavin, W. C., Sanderson, J. P., Monroy, J. A., Bennett, K. C., Warner, J. B., and Hoffmann, M. P. (1998). Novel chitinolytic enzymes with biological activity against herbivorous insects. *J. Chem. Ecol.* **24:**985–998.

Cao, J., Ibrahim, H., Garcia, J. J., Mason, H., Granados, R. R., and Earle, E. D. (2002). Transgenic tobacco plants carrying a baculovirus enhancin gene slows the development and increase the mortality of *Trichoplusia ni* larvae. *Plant Cell Rep.* **21:**244–250.

Carton, Y., and Nappi, A. J. (1997). *Drosophila* cellular immunity against parasitoids. *Parasitol. Today* **13**(6)**:**218–227.

Cho, W. L., Tsao, S. M., Hays, A. R., Walter, R., Chen, J. S., Snigirevskaya, E. S., and Raikhel, A. S. (1999). Mosquito cathepsin B-like protease involved in embryonic degradation of vitellin in produced as a latent extraovarian precursor. *J. Biol. Chem.* **274:**13311–13321.

Cohen-Kupiec, R., and Chet, I. (1998). The molecular biology of chitin digestion. *Curr. Opin. Biotech.* **9:**270–277.

Collier, I. E., Wilhelm, S. M., Eisen, A. Z., Marmer, B. L., Grant, G. A., Seltzer, J. L., Kronberger, A., He, C., Bauer, E. A., and Goldberg, G. I. (1988). H-*ras* oncogene-transformed human bronchial epithelial cells (TBE-1) secrete a single metalloprotease capable of degrading basement membrane collagen. *J. Biol. Chem.* **263:**6579–6587.

Cristofoletti, P. T., Ribeiro, A. F., and Terra, W. R. (2005). The cathepsin L-like proteinases from the midgut of *Tenebrio molitor* larvae: Sequence, properties, immunocytochemical localization and function. *Insect Biochem. Mol. Biol.* **35:**883–901.

Dall, D., Sriskantha, A., Vera, A., Lai-Fook, J., and Symonds, T. (1993). A gene encoding a highly expressed spindle body protein of *Heliothis armigera* entomopoxvirus. *J. Gen. Virol.* **74:**1811–1818.

Dall, D., Luque, T., and O'Reilly, D. (2001). Insect-virus relationships: Sifting by informatics. *Bioessays* **23**(2)**:**184–193.

Deraison, C., Darboux, I., Duportets, L., Gorojankina, T., Rahbe, Y., and Jouanin, L. (2004). Cloning and characterization of a gut-specific cathepsin L from the aphid *Aphis gossypii*. *Insect Mol. Biol.* **13:**165–177.

Derksen, A. C. G. (1988). Alteration of a lepidopteran peritrophic membrane by baculoviruses and enhancement of viral infectivity. *Virology* **167:**242–250.

Detvisitsakun, C., Berretta, M. F., Lehiy, C., and Passarelli, A. L. (2005). Stimulation of cell motility by a viral fibroblast growth factor homolog: Proposal for a role in viral pathogenesis. *Virology* **336**(2):308–317.

Devenport, M. F. H., Donnelly-Doman, M., Shen, Z., and Jacobs-Lorena, M. (2005). Storage and secretion of Ag-Aper14, a novel peritrophic matrix protein, and Ag-Muc1 from the mosquito *Anopheles gambiae*. *Cell Tissue Res.* **320:**175–185.

Ding, X. G. B., Johnson, L. B., White, F. F., Wang, X., Morgan, T. D., Kramer, K. J., and Muthukrishnan, S. (1998). Insect resistance of transgenic tobacco expressing an insect chitinase gene. *Transgenic Res.* **7:**77–84.

Down, R. E., Gatehouse, A. M. R., Davison, G. M., Newell, C. A., Merryweather, A., Hamilton, W. D. O., Burgess, E. P. J., Gilbert, R. J. C., and Gatehouse, J. A. (1996). Snowdrop lectin inhibits development and decreases fecundity of the glasshouse potato aphid (*Aulacorthum solani*) when administered *in vitro* and via transgenic plants both in laboratory and glasshouse trials. *J. Insect Physiol.* **42:**1035–1045.

Down, R. E., Fitches, E. C., Wiles, D. P., Corti, P., Bell, H. A., Gatehouse, J. A., and Edwards, J. P. (2006). Insecticidal spider venom toxin fused to snowdrop lectin is toxic to the peach-potato aphid, *Myzus persicae* (Hemiptera: Aphididae) and the rice brown planthopper, *Nilaparvata lugens* (Hemiptera: Delphacidae). *Pest Manag. Sci.* **62:** 77–85.

Duck, N., and Evola, S. (1997). Use of transgenes to increase host plant resistance to insects: Opportunities and challenges. *In* "Advances in Insect Control: The Role of Transgenic Plants" (N. Carozzi and M. Koziel, eds.), pp. 1–20. Taylor & Francis, London.

Eisemann, C. H., Donaldson, R. A., Pearson, R. D., Cadogan, L. C., Vuocolo, T., and Tellam, R. L. (1994). Larvicidal activity of lectins on *Lucilia cuprina*: Mechanism of action. *Entomol. Exp. Appl.* **72:**1–10.

Elvin, C. M., Vuocolo, T., Pearson, R. D., East, I. J., Riding, G. A., Eisemann, C. H., and Tellam, R. L. (1996). Characterization of a major peritrophic membrane protein, peritrophin-44, from the larvae of *Lucilia cuprina*: cDNA and deduced amino acid sequences. *J. Biol. Chem.* **271:**8925–8935.

Engelhard, E. K., Kam-Morgan, L. N. W., Washburn, J. O., and Volkman, L. E. (1994). The insect tracheal system: A conduit for the systemic spread of *Autographa californica* M nuclear polyhedrosis virus. *Proc. Natl. Acad. Sci. USA* **91:**3224–3227.

Federici, B. A. (1997). Chapter 3: Baculovirus pathogenesis. *In* "The Baculoviruses" (L. K. Miller, ed.), pp. 33–60. Plenum Press, New York.

Fitches, E., Audsley, N., Gatehouse, J. A., and Edwards, J. P. (2002). Fusion proteins containing neuropeptides as novel insect control agents: Snowdrop lectin delivers fused allatostatin to insect haemolymph following oral ingestion. *Insect Biochem. Mol. Biol.* **32:**1653–1661.

Fitches, E., Ilett, C., Gatehouse, A. M., Gatehouse, L. N., Greene, R., Edwards, J. P., and Gatehouse, J. A. (2001a). The effects of *Phaseolus vulgaris* erythro- and leucoagglutinating isolectins (PHA-E and PHA-L) delivered via artificial diet and transgenic plants on the growth and development of tomato moth (Lacanobia oleracea) larvae; lectin binding to gut glycoproteins *in vitro* and *in vivo*. *J. Insect Physiol.* **47:**1389–1398.

Fitches, E., Woodhouse, S. D., Edwards, J. P., and Gatehouse, J. A. (2001b). *In vitro* and *in vivo* binding of snowdrop (*Galanthus nivalis* agglutinin; GNA) and jackbean (*Canavalia ensiformis*; Con A) lectins within tomato moth (*Lacanobia oleracea*) larvae; mechanisms of insecticidal action. *J. Insect Physiol.* **47:**777–787.

Fitches, E., Edwards, M. G., Mee, C., Grishin, E., Gatehouse, A. M. R., Edwards, J. P., and Gatehouse, J. A. (2004). Fusion proteins containing insect-specific toxins as pest control agents: Snowdrop lectin delivers fused insecticidal spider venom toxin to insect haemolymph following oral ingestion. *J. Insect Physiol.* **50:**61–71.

Flipsen, J. T. M., Martens, J. W. M., Oers, M. M. V., Vlak, J. M., and Lent, J. W. M. V. (1995). Passage of *Autographa californica* nuclear polyhedrosis virus through the midgut epithelium of *Spodoptera exigua* larvae. *Virology* **208:**328–335.

Fujii-Taira, I., Tanaka, Y., Homma, K. J., and Natori, S. (2000). Hydrolysis and synthesis of substrate proteins for cathepsin L in the brain basement membranes of *Sarcophaga* during metamorphosis. *J. Biochem. (Tokyo)* **128**(3)**:**539–542.

Fukamizo, T. (2000). Chitinolytic enzymes: Catalysis, substrate binding, and their application. *Curr. Protein Pept. Sci.* **1:**105–124.

Furuta, Y., Mitsuhashi, W., Kobayashi, J., Hayasaka, S., Imanishi, S., Chinzei, Y., and Sato, M. (2001). Peroral infectivity of non-occluded viruses of *Bombyx mori* nucleopolyhedrovirus and polyhedrin-negative recombinant baculoviruses to silkworm larvae is drastically enhanced when administered with *Anomala cuprea* entomopoxvirus spindles. *J. Gen. Virol.* **82:**307–312.

Galloway, C. S., Wang, P., Winstanley, D., and Jones, I. M. (2005). Comparison of the bacterial enhancin-like proteins from *Yersinia* and *Bacillus* spp. with a baculovirus enhancin. *J. Invertebr. Pathol.* **90:**134–137.

Gatehouse, A. M. R., Down, R. E., Powell, K. S., Sauvion, N., Rahbé, Y, Newell, C. A., Merryweather, A., Hamilton, W. D. O., and Gatehouse, J. A. (1996). Transgenic potato plants with enhanced resistance to the peach-potato aphid *Myzus persicae*. *Entomol. Exp. Appl.* **79:**295–307.

Gopalakrishnan, B., Muthukrishnan, S., and Kramer, K. J. (1995). Baculovirus-mediated expression of a *Manduca sexta* chitinase gene: Properties of the recombinant protein. *Insect Biochem. Mol. Biol.* **25**(2)**:**255–265.

Granados, R. R., and Lawler, K. A. (1981). *In vivo* pathway of *Autographa californica* baculovirus invasion and infection. *Virology* **108:**297–308.

Gross, C. H., Wolgamot, G. M., Russell, R. L., Pearson, M. N., and Rohrmann, G. F. (1993). A 37-kilodalton glycoprotein from a baculovirus of *Orgyia pseudotsugata* is localized to cytoplasmic inclusion bodies. *J. Virol.* **67**(1)**:**469–475.

Guo, W. L. G., Pang, Y, and Wang, P. (2005). A novel chitin-binding protein identified from the peritrophic membrane of the cabbage looper, *Trichoplusia ni*. *Insect Biochem. Mol. Biol.* **35:**1224–1234.

Han, J. H., Lee, K. S., Li, J., Kim, I., Je, Y. H., Kim, D. H., Sohn, H. D., and Jin, B. R. (2005). Cloning and expression of a fat body-specific chitinase cDNA from the spider, *Araneus ventricosus*. *Comp. Biochem. Physiol. B. Biochem. Mol. Biol.* **140:**427–435.

Harman, G. E., Hayes, C. K., Lorito, M., Broadway, R. M., Pietro, A. D., and Tronsmo, A. (1993). Chitinolytic enzymes of *Trichoderma harzianum* purification of chitobiosidase and endochitinase. *Phytopathology* **83:**313–318.

Harrison, R. L., and Bonning, B. C. (2001). Use of proteases to improve the insecticidal activity of baculoviruses. *Biol. Control* **20:**199–209.

Hashimoto, Y., Corsaro, B. G., and Granados, R. R. (1991). Location and nucleotide sequence of the gene encoding the viral enhancing factor of the *Trichoplusia ni* granulosis virus. *J. Gen. Virol.* **72:**2645–2651.

Hashimoto, Y., Hayakawa, T., Ueno, Y., Fujita, T., Sano, Y., and Matsumoto, T. (2000). Sequence analysis of the *Plutella xylostella* granulovirus genome. *Virology* **275:** 358–372.

Hawtin, R. E., Arnold, K., Ayres, M. D., de A. Zanotto, P. M., Howard, S. C., Gooday, G. W., Chappell, L. H., Kitts, P. A., King, L. A., and Possee, R. D. (1995). Identification and preliminary characterization of a chitinase gene in the *Autographa californica* nuclear polyhedrosis virus genome. *Virology* **212:**673–685.

Hawtin, R. E., Zarkowska, T., Arnold, K., Thomas, C. J., Gooday, G. W., King, L. A., Kuzio, J. A., and Possee, R. D. (1997). Liquefaction of *Autographa californica* nucleopolyhedrovirus-infected insects is dependent on the integrity of virus-encoded chitinase and cathepsin genes. *Virology* **238:**243–253.

Hayakawa, T., Xu, J., and Hukuhara, T. (1996). Cloning and sequencing of the gene for an enhancing factor from *Pseudoletia separata* entomopoxvirus. *Gene* **177**(1–2)**:** 269–270.

Hayakawa, T., Ko, R., Okano, K., Seong, S., Goto, C., and Maeda, S. (1999). Sequence analysis of the *Xestia c-nigrum* granulovirus genome. *Virology* **262:**277–297.

Hayakawa, T., Hashimoto, Y., Mori, M., Kaido, M., Shimojo, E., Furusawa, I., and Granados, R. R. (2004). Transgenic tobacco transformed with the *Trichoplusia ni* granulovirus *enhancin* gene affects insect development. *Biocontrol Sci. Technol.* **14**(2)**:**211–214.

Henrissat, B. (1999). Classification of chitinases modules. Birkhäuser Verlag, Basel.

Hess, R. T., and Falcon, L. A. (1987). Temporal events in the invasion of the codling moth, *Cydia pomonella*, by a granulosis virus. *J. Invertebr. Pathol.* **50:**85–105.

Hilder, V. A., Powell, K. S., Gatehouse, A. M. R., Gatehouse, J. A., Gatehouse, L. N., Shi, Y., Hamilton, W. D. O., Merryweather, A., Newell, C. A., and Timans, J. C. (1995). Expression of snowdrop lectin in transgenic tobacco plants results in added protection against aphids. *Transgenic Res.* **4:**18–25.

Hom, L. G., and Volkman, L. E. (2000). *Autographa californica* M nucleopolyhedrovirus chiA is required for processing of V-CATH. *Virology* **277:**178–183.

Homma, K., and Natori, S. (1996). Identification of substrate proteins for cathepsin L that are selectively hydrolyzed during the differentiation of imaginal discs of *Sarcophaga peregrina. Eur. J. Biochem.* **240:**443–447.

Homma, K., Jurata, S., and Natori, S. (1994). Purification, characterization, and cDNA cloning of procathepsin L from the culture medium of NIH-Sape-4, an embryonic cell line of *Sarcophaga peregrina* (flesh fly), and its involvement in the differentiation of imaginal discs. *J. Biol. Chem.* **269:**15258–15264.

Huang, J. Q., Wel, Z. M., An, H. L., and Zhu, Y. X. (2001). *Agrobacterium tumefaciens*-mediated transformation of rice with the spider insecticidal gene conferring resistance to leaffolder and striped stem borer. *Cell Res.* **11**(2)**:**149–155.

Hukuhara, T., and Wijonarko, A. (2001). Enhanced fusion of a nucleopolyhedrovirus with cultured cells by a virus enhacing factor from an entomopoxvirus. *J. Invertebr. Pathol.* **77:**62–67.

Hukuhara, T., Tamura, K., Zhu, Y., Abe, H., and Tanada, Y. (1987). Synergistic factor shows specificity in enhancing nuclear polyhedrosis virus infections. *Appl. Entomol. Zool.* **22:**235–236.

Hukuhara, T., Hayakawa, T., and Wijonarko, A. (1999). Increased baculovirus susceptibility of armyworm larvae feeding on trangenic rice plants expressing an entomopoxvirus gene. *Nat. Biotechnol.* **17:**1122–1124.

Hukuhara, T., Hayakawa, T., and Wijonarko, A. (2001). A bacterially produced virus enhancing factor from an entomopoxvirus enhances nucleopolyhedrovirus infection in armyworm larvae. *J. Invertebr. Pathol.* **78**(1)**:**25–30.

Ko, R., Okano, K., and Maeda, S. (2000). Structural and functional analysis of the *Xestia c-nigrum* granulovirus matrix metalloproteinase. *J. Virol.* **74:**11240–11246.

Konno, K., Hirayama, C., Nakamura, M., Tateishi, K., Tamura, Y., Hattori, M., and Kohno, K. (2004). Papain protects papaya trees from herbivorous insects: Role of cysteine proteases in latex. *Plant J.* **37:**370–378.

Kramer, K. J., Hopkins, T. L., and Schaefer, J. (1995). Applications of solids NMR to the analysis of insect sclerotized structures. *Insect Biochem. Mol. Biol.* **25:**1067–1080.

Kurata, S., Kobayashi, H., and Natori, S. (1991). Participation of a 200-kDa hemocyte membrane protein in the dissociation of the fat body at the metamorphosis of *Sarcophaga. Dev. Biol.* **146:**179–185.

Kurata, S., Saito, H., and Natori, S. (1992). The 29-kDa hemocyte proteinase dissociates fat body at metamorphosis of *Sarcophaga. Dev. Biol.* **153:**115–121.

Kuzio, J., Pearson, M. N., Harwood, S. H., Funk, C. J., Evans, J. T., Slavicek, J. M., and Rohrmann, G. F. (1999). Sequence and analysis of the genome of a baculovirus pathogenic for *Lymantria dispar. Virology* **253:**17–34.

Lehane, M. J. (1997). Peritrophic matrix structure and function. *Annu. Rev. Entomol.* **42:**525–550.

Lepore, L. S., Roelvink, P. R., and Granados, R. R. (1996). Enhancin, the granulosis virus protein that facilitates nucleopolyhedrovirus (NPV) infections is a metalloprotease. *J. Invert. Pathol.* **68:**131–140.

Li, G., Xu, X., Xing, H., Zhu, H., and Fan, Q. (2005). Insect resistance to *Nilaparvata lugens* and *Cnaphalocrocis medinalis* in transgenic indica rice and the inheritance of gna + sbti transgenes. *Pest Manag. Sci.* **61:**390–396.

Li, L., Donly, C., Li, Q., Willis, L. G., Keddie, B. A., Erlandson, M. A., and Theilmann, D. A. (2002). Identification and genomic analysis of a second species of nucleopolyhedrovirus isolated from *Mamestra configurata. Virology* **297:**226–244.

Li, Q., Li, L., Moore, K., Donly, C., Theilmann, D. A., and Erlandson, M. (2003a). Characterization of *Mamestra configurata* nucleopolyhedrovirus enhancin and its functional analysis via expression in an *Autographa californica* M nucleopolyhedrovirus recombinant. *J. Gen. Virol.* **84:**123–132.

Li, X., Barrett, J., Pang, A., Klose, R. J., Krell, P. J., and Arif, B. M. (2000). Characterization of an overexpressed spindle protein during a baculovirus infection. *Virology* **268:**56–67.

Li, Z., Li, C., Yang, K., Wang, L., Yin, C., Gong, Y., and Pang, Y. (2003b). Characterization of a chitin-binding protein GP37 of *Spodoptera litura* multicapsid nucleopolyhedrovirus. *Virus Res.* **96**(1–2)**:**113–122.

Linthorst, H. J. M., Oers, C. V. D., Brederode, F. T., and Bol, J. F. (1993). Circadian expression and induction by wounding of tobacco genes for cysteine proteinase. *Plant Mol. Biol.* **21:**685–694.

Liu, J. J., and Carstens, E. B. (1996). Identification, molecular cloning, and transcription analysis of the *Choristoneura fumiferana* nuclear polyhedrosis virus spindle-like protein gene. *Virology* **223**(2)**:**396.

Llano, E., Adam, G., Pendas, A. M., Quesada, V., Sanchez, L. M., Santamaria, I., Noselli, S., and Lopez-Otin, C. (2002). Structural and enzymatic characterization of *Drosophila* Dm2-MMP, a membrane-bound matrix metalloproteinase with tissue-specific expression. *J. Biol. Chem.* **277:**23321–23329.

Maeda, S., Dean, D. D., Sylvia, V. L., Boyan, B. D., and Schwartz, Z. (2001). Metalloproteinase activity in growth plate chondrocyte cultures is regulated by 1,25-(OH)2D3 and 24,25-(OH)2D3 and mediated through protein kinase C. *Matrix Biol.* **20:**87–97.

Maiti, I. B., Oey, N., Dahlman, D. L., and Webb, B. A. (2003). Antibiosis-type resistance in transgenic plants expressing a teratocyte secretory peptide (TSP) gene from a hymenopteran endoparasite (*Microplitis croceipes*). *Plant Biotech. J.* **1**:209–219.

Marmaras, V. J., Charalambidis, N. D., and Zervas, C. G. (1996). Immune responses in insects: The role of phenoloxidase in defense reactions in relation to melanization and sclerotization. *Arch. Insect Biochem. Physiol.* **31**:119–133.

Melander, M., Ahman, I., Kamnert, I., and Stromdahl, A. C. (2003). Pea lectin expressed transgenically in oilseed rape reduces growth rate of pollen beetle larvae. *Transgenic Res.* **12**:555–567.

Merzendorfer, H., and Zimoch, L. (2003). Chitin metabolism in insects: Structure, function and regulation of chitin synthases and chitinases. *J. Exp. Biol.* **206**:4393–4412.

Miller, L. K., and Ball, L. A. (1998). "The Insect Viruses." Plenum Press, New York.

Mitsuhashi, W., and Miyamoto, K. (2003). Disintegration of the peritrophic membrane of silkworm larvae due to spindles of an entomopoxvirus. *J. Invertebr. Pathol.* **82**:34–40.

Mitsuhashi, W., Furuta, Y., and Sato, M. (1998). The spindles of an entomopoxvirus of *Coleoptera* (*Anomala cuprea*) strongly enhance the infectivity of a nucleopolyhedrovirus in Lepidoptera (*Bombyx mori*). *J. Invertebr. Pathol.* **71**:186–188.

Nagadhara, D., Ramesh, S., Pasalu, I. C., Rao, Y. K., Sarma, N. P., Reddy, V. D., and Rao, K. V. (2004). Transgenic rice plants expressing the snowdrop lectin gene (gna) exhibit high-level resistance to the whitebacked planthopper (*Sogatella furcifera*). *Theor. Appl. Genet.* **109**:1399–1405.

Nappi, A. J., and Christensen, B. M. (2005). Melanogenesis and associated cytotoxic reactions: Applications to insect innate immunity. *Insect Biochem. Mol. Biol.* **35**:443–459.

Nardi, J. B., Gao, C., and Kanost, M. R. (2001). The extracellular matrix protein lacunin is expressed by a subset of hemocytes involved in basal lamina morphogenesis. *J. Insect Physiol.* **47**:997–1006.

O'Callaghan, M., Glare, T. R., Burgess, E. P. J., and Malone, L. A. (2005). Effects of plants genetically modified for insect resistance on nontarget organisms. *Annu. Rev. Entomol.* **50**:251–292.

Ohkawa, T., Majima, K., and Maeda, S. (1994). A cysteine protease encoded by the baculovirus *Bombyx mori* nuclear polyhedrosis virus. *J. Virol.* **68**:6619–6625.

Page-McCaw, A., Serano, J., Sante, J. M., and Rubin, G. M. (2003). *Drosophila* matrix metalloproteinases are required for tissue remodeling, but not embryonic development. *Dev. Cell* **4**:95–106.

Park, A. J., Matrisian, L. M., Kells, A. F., Pearson, R., Yuan, Z., and Navre, M. (1991). Mutational analysis of the transin (rat stromelysin) autoinhibitor region demonstrates a role for residues surrounding the "cysteine switch." *J. Biol. Chem.* **266**:1584–1590.

Pech, L. L., Trudeau, D., and Strand, M. R. (1995). Effects of basement membranes on the behavior of hemocytes from *Pseudoplusia includens* (Lepidoptera: Noctuidae): Development of an *in vitro* encapsulation assay. *J. Insect Physiol.* **41**(9):801.

Pechan, T., Ye, L., Chang, Y., Mitra, A., Lin, L., Davis, F. M., Williams, W. P., and Luthe, D. S. (2000). A unique 33-kD cysteine protease accumulates in response to larval feeding in maize genotypes resistant to fall armyworm and other Lepidoptera. *Plant Cell* **12**:1031–1040.

Pedersen, K. J. (1991). Structure and composition of basement membranes and other basal matrix systems in selected invertebrates. *Acta Zool.* **72**(4):181–201.

Pendland, J. C., and Boucias, D. G. (1998). Characterization of monoclonal antibodies against cell wall epitopes of the insect pathogenic fungus, *Nomuraea rileyi*: Differential binding to fungal surfaces and cross-reactivity with host hemocytes and basement membrane components. *Eur. J. Cell Biol.* **75:**118–127.

Pendland, J. C., and Boucias, D. G. (2000). Comparative analysis of the binding of antibodies prepared against the insect *Spodoptera exigua* and against the mycopathogen *Nomuraea rileyi*. *J. Inverteb. Pathol.* **75:**107–116.

Peng, J., Zhong, J., and Granados, R. R. (1999). A baculovirus enhancin alters the permeability of a mucosal midgut peritrophic matrix from lepidopteran larvae. *J. Insect Physiol.* **45:**159–166.

Peters, W. (1992). "Peritrophic Membranes." Springer, New York.

Popham, H. J. R., Bischoff, D. S., and Slavicek, J. M. (2001). Both *Lymantria dispar* nucleopolyhedrovirus *enhancin* genes contribute to viral potency. *J. Virol.* **75**(18): 8639–8648.

Porter, S., Clark, I. M., Kevorkian, L., and Edwards, D. R. (2005). The ADAMTS metalloproteinases. *Biochem. J.* **386:**15–27.

Radek, R., and Fabel, P. (2000). A new entomopoxvirus from a cockroach: Light and electron microscopy. *J. Invertebr. Pathol.* **75**(1)**:**19–27.

Raemaekers, R. J. (2000). "Expression of Functional Plant Lectins in Heterologous Systems," pp. 173–207. University of Durham, Durham, UK.

Rahbe, Y., Sauvion, N., Febvay, G., Peumans, W. J., and Gatehouse, A. M. R. (1995). Toxicity of lectins and processing of ingested in the pea aphid *Acyrthosiphon pisum*. *Entomol. Exp. Appl.* **76:**143–155.

Rao, K. V., Rathore, K. S., Hodges, T. K., Fu, X. D., Stoger, E., Sudhakar, D., Williams, S., Christou, P., Brown, D. P., Powell, K. S., Spence, J., Bharathi, M. *et al.* (1998). Expression of snowdrop lectin (GNA) in the phloem of transgenic rice plants confers resistance to rice brown planthopper. *Plant J.* **15:**469–477.

Rao, R., Fiandra, L., Giordana, B., Eguileor, M. D., Congiu, T., Burlini, N., Arciello, S., Corrado, G., and Pennacchio, F. (2004). AcMNPV ChiA protein disrupts the peritrophic membrane and alters midgut physiology of *Bombyx mori* larvae. *Insect Biochem. Mol. Biol.* **34:**1205–1213.

Rawlings, N. D., Pearl, L. H., and Buttle, J. (1992). The baculovirus *Autographa californica* nuclear polyhedrosis virus genome includes a papain-like sequence. *Biol. Chem. Hoppe Seyler* **373:**1211–1215.

Reddy, J. T., and Locke, M. (1990). The size limited penetration of gold particles through insect basal laminae. *J. Insect Physiol.* **36:**397–407.

Rizki, R. M., and Rizki, T. M. (1974). Basement membrane abnormalities in melanotic tumor formation of *Drosophila*. *Experientia* **30**(5)**:**543–546.

Rizki, R. M., and Rizki, T. M. (1980). Hemocyte responses to implanted tissues in *Drosophila melanogaster* larvae. *Roux's Arch. Dev. Biol.* **189:**207–213.

Roelvink, P. W., Corsaro, B. G., and Granados, R. R. (1995). Characterization of the *Helicoverpa armigera* and *Pseudaletia unipuncta* granulovirus enhancin genes. *J. Gen. Virol.* **76:**2693–2705.

Rohrbach, D. H. and Timpl, R. (eds.) (1993). "Molecular and Cellular Aspects of Basement Membranes." Academic Press, New York.

Ryerse, J. S. (1998). Basal laminae. *In* "Insecta" (F. W. Harrison and M. Locke, eds.), Vol. 11A, pp. 3–16. Wiley-Liss, Inc., New York.

Saguez, J., Hainez, R., Cherqui, A., Wuytswinkel, O. V., Jeanpierre, H., Lebon, G., Noiraud, N., Beaujean, A., Jouanin, L., Laberche, J. C., Vincent, C., and Giordanengo, P. (2005). Unexpected effects of chitinases on the peach-potato aphid (*Myzus persicae* Sulzer) when delivered via transgenic potato plants (*Solanum tuberosum* Linne) and *in vitro*. *Transgenic Res.* **14:**57–67.

Sahai, A. S., and Manocha, M. S. (1993). Chitinases of fungi and plants: Their involvement in morphogenesis and host parasite interaction. *FEMS Microbiol. Rev.* **11:** 317–338.

Sauvion, N., Rahbe, Y., Peumans, W. J., Damme, E. J. V., Gatehouse, J. A., and Gatehouse, A. M. R. (1996). Effects of GNA and other binding lectins on development and fecundity of the peach-potato aphid *Myzus persicae*. *Entomol. Exp. Appl.* **79:** 285–293.

Setamou, M., Bernal, J. S., Legaspi, J. C., Mirkov, T. E., and B. C. L., Jr. (2002). Evaluation of lectin-expressing transgenic sugarcane against stalkborers (Lepidoptera: Pyralidae): Effects on life history parameters. *J. Econ. Entomol.* **95:**469–477.

Shao, L., Devenport, M., Fujioka, H., Ghosh, A., and Jacobs-Lorena, M. (2005). Identification and characterization of a novel peritrophic matrix protein, Ae-Aper50, and the microvillar membrane protein, AEG12, from the mosquito, *Aedes aegypti*. *Insect Biochem. Mol. Biol.* **273:**17665–17670.

Shen, Z., and Jacobs-Lorena, M. (1998). A type I peritrophic matrix protein from the malaria vector *Anopheles gambiae* binds to chitin. Cloning, expression and characterization. *J. Biol. Chem.* **273**(28)**:**17665–17670.

Slack, J. M., Kuzio, J., and Faulkner, P. (1995). Characterization of *v-cath*, a cathepsin-L-like proteinase expressed by the baculovirus *Autographa californica* multiple nuclear polyhedrosis virus. *J. Gen. Virol.* **76:**1091–1098.

Slavicek, J. M., and Popham, H. J. (2005). The *Lymantria dispar* nucleopolyhedrovirus enhancins are components of occlusion-derived virus. *J. Virol.* **79:**10578–10588.

Smith-Johannsen, H., Witkiewicz, H., and Iatrou, K. (1986). Infection of silkmoth follicular cells with *Bombyx mori* nuclear polyhedrosis virus. *J. Inverteb. Pathol.* **48:**74–84.

Stoger, E., Williams, S., Christou, P., Down, R. E., and Gatehouse, J. A. (1998). Expression of the insecticidal lectin from snowdrop (*Galanthus nivalis* aggluitin; GNA) in transgenic wheat plants: Effects on predation by the grain aphid *Sitobion avenae*. *Mol. Breeding* **5:**65–73.

Strizhov, N., Keller, M., Mathur, J., Konca-Kalman, Z., Bosch, D., Prudovsky, E., Schell, J., Sneh, B., Koncz, C., and Zilberstein, A. (1996). A synthetic cryIC gene, encoding a *Bacillus thuringiensis* delta-endotoxin, confers *Spodoptera* resistance in alfalfa and tobacco. *Proc. Nat. Acad. Sci. USA* **93:**15012–15017.

Takahashi, N., Kurata, S., and Natori, S. (1993). Molecular cloning of cDNA for the 29 kDa proteinase participating in decomposition of the larval fat body, during metamorphosis of *Sarcophaga peregrine* (flesh fly). *FEBS. Lett.* **334:**153–157.

Tanada, Y. (1959). Synergism between two viruses of the armyworm, *Pseudaletia unipuncta* (Haworth) (Lepidoptera, Noctuidae). *J. Insect. Pathol.* **1:**215–231.

Tanada, Y., Himeno, M., and Omi, E. M. (1973). Isolation of a factor, from the capsule of a granulosis virus, synergistic for a nuclear-polyhedrosis virus of the armworm. *J. Invertebr. Pathol.* **21:**81–90.

Tellam, R. L., Wijffels, G., and Willadsen, P. (1999). Peritrophic matrix proteins. *Insect Biochem. Mol. Biol.* **29**(2)**:**87–101.

Terra, W. R. (2001). The origin and functions of the insect peritrophic membrane and peritrophic gel. *Arch. Insect Biochem. Physiol.* **47:**47–61.

Terra, W. R., and Ferreira, C. (1994). Insect digestive enzymes-properties, compartmentalization and function. *Comp. Biochem. Physiol. B.* **109:**1–62.

Uria, J. A., and Lopez-Otin, C. (2000). Matrilysin-2, a new matrix metalloproteinase expressed in human tumors and showing the minimal domain organization required for secretion, latency, and activity. *Cancer Res.* **60:**4745–4751.

Vaaje-Kolstad, G., Horn, S. J., Aalten, D. M. V., Synstad, B., and Eijsink, V. G. (2005). The non-catalytic chitin-binding protein CBP21 from *Serratia marcescens* is essential for chitin degradation. *J. Biol. Chem.* **280:**28492–28497.

Vilcinskas, A., and Wedde, M. (2002). Insect inhibitors of metalloproteinases. *IUBMB Life* **54:**339–343.

Volkman, L. E. (1997). Nucleopolyhedrovirus interactions with their insect hosts. *Adv. Virus Res.* **48:**313–348.

Wang, H., Wu, D., Deng, F., Peng, H., Chen, X., Lauzon, H., Arif, B. M., Jehle, J. A., and Hu, Z. (2004a). Characterization and phylogenetic analysis of the chitinase gene from the *Helicoverpa armigera* single nucleocapsid nucleopolyhedrovirus. *Virus Res.* **100:** 179–189.

Wang, J., Chen, Z., Du, J., Sun, Y., and Liang, A. (2005). Novel insect resistance in *Brassica napus* developed by transformation of chitinase and scorpion toxin genes. *Plant Cell Rep.* **24:**549–555.

Wang, P., and Granados, R. R. (1997a). An intestinal mucin is the target substrate for a baculovirus enhancin. *Proc. Natl. Acad. Sci. USA* **94:**6977–6982.

Wang, P., and Granados, R. R. (1997b). Molecular cloning and sequencing of a novel invertebrate intestinal mucin cDNA. *J. Biol. Chem.* **272**(26)**:**16663–16669.

Wang, P., and Granados, R. R. (2000). Calcofluor disrupts the midgut defense system of insects. *Insect Biochem. Mol. Biol.* **30:**135–143.

Wang, P., and Granados, R. R. (2001). Molecular structure of the peritrophic membrane (PM): Identification of potential PM target sites for insect control. *Arch. Insect Biochem. Physiol.* **47**(2)**:**110–118.

Wang, P., Hammer, D. A., and Granados, R. R. (1994). Interaction of *Trichoplusia ni* granulosis virus-encoded enhancin with the midgut epithelium and peritrophic membrane of four lepidopteran insects. *J. Gen. Virol.* **75:**1961–1967.

Wang, P., Li, G., and Granados, R. R. (2004b). Identification of two new peritrophic membrane proteins from larval *Trichoplusia ni*: Structural characteristics and their functions in the protease rich insect gut. *Insect Biochem. Mol. Biol.* **34**(3)**:**215–227.

Wijonarko, A., and Hukuhara, T. (1998). Detection of a virus enhancing factor in the spheroid, spindle, and virion of an entomopoxvirus. *J. Invertebr. Pathol.* **72**(1)**:**82.

Wu, N. F., Sun, Q., Yao, B., Fan, Y. L., Rao, H. Y., Huang, M. R., and Wang, M. X. (2000). Insect-resistant transgenic poplar expressing AaIT gene. Sheng Wu Gong Cheng Xue Bao **16**(2)**:**129–133.

Xu, J., and Hukuhara, T. (1992). Enhanced infection of a nuclear polyhedrosis virus in larvae of the armyworm, *Pseudoletia separata*, by a factor in the spheroids of an entomopoxvirus. *J. Invertebr. Pathol.* **60:**259–264.

Xu, J., and Hukuhara, T. (1994). Biochemical properties of an enhancing factor of an entomopoxvirus. *J. Invertebr. Pathol.* **63:**14–18.

Yao, B., Fan, Y., Zheng, Q., and Zhao, R. (1996). Insect-resistant tobacco plants expressing insect-specific neurotoxin AaIT. *Chin. J. Biotechnol.* **12**(2)**:**67–72.

Young, V. L., Simpson, R. M., and Ward, V. K. (2005). Characterization of an exochitinase from *Epiphyas postvittana* nucleopolyhedrovirus (family Baculoviridae). *J. Gen. Virol.* **86:**3253–3261.

Yurchenco, P. D., and O'Rear, J. (1993). Supramolecular organization of basement membranes. *In* "Molecular and Cellular Aspects of Basement Membranes" (D. H. Rohrbach and R. Timpl, eds.), pp. 19–47. Academic Press, New York.

Zhan, Y. G., Liu, Z. H., Wang, Y. C., Wang, Z. Y., Yang, C. P., and Liu, G. F. (2001). Transformation of insect resistant gene into birch. *J. Northeast Forestry Univ.* **29:**4–6.

Zhao, X. F., Wang, J. X., and Wang, Y. C. (1998). Purification and characterization of a cysteine proteinase from eggs of the cotton boll worm, *Helicoverpa armigera. Insect Biochem. Mol. Biol.* **28:**259–264.

Zhou, Y. T., Wu, B., and Mang, K. (1998). Inhibition effect of transgenic tobacco plants expressing snowdrop lectin on the population development of *Myzus persicae. Chin. J. Biotechnol.* **14:**9–16.

SMALL RNA VIRUSES OF INSECTS: EXPRESSION IN PLANTS AND RNA SILENCING

Karl H. J. Gordon* and Peter M. Waterhouse†

*CSIRO Entomology, Canberra, ACT 2601, Australia
†CSIRO Plant Industry, Canberra, ACT 2601, Australia

Abstract

Interest in insect small RNA viruses (SRVs) has grown slowly but steadily. A number of new viruses have been analyzed at the sequence level, adding to our knowledge of their diversity at the level of both individual virus species and families. In particular, a number of possible new virus families have emerged. This research has largely been driven by interest in their potential for pest control, as well as in their importance as the causal agents of disease in beneficial arthropods. At the same time, research into known viruses has made valuable contributions to our understanding of an emerging new field of central importance to molecular biology—the existence of RNA-based gene

0065-3527/06 $35.00
DOI: 10.1016/S0065-3527(06)68013-5

silencing, developmental control, and adaptive immune systems in eukaryotes. Subject to RNA-based adaptive immune responses in their hosts, viruses have evolved a variety of genes encoding proteins capable of suppressing the immune response. Such genes were first identified in plant viruses, but the first examples known from animal viruses were identified in insect RNA viruses. This chapter will address the diversity of insect SRVs, and attempts to harness their simplicity in the engineering of transgenic plants expressing viruses for resistance to insect pests. We also describe RNA interference and antiviral pathways identified in plants and animals, how they have led viruses to evolve genes capable of suppressing such adaptive immunity, and the problems presented by these pathways for the strategy of expressing viruses in transgenic plants. Approaches for countering these problems are also discussed.

I. Insect RNA Viruses and Their Use as Biopesticides

A. *Introduction*

Small icosahedral viruses with RNA genomes (SRVs) represent the simplest and at the same time most challenging infectious, replicating agents available for the management of pest insects. While they have minimized the number of different components (genome segments, genes, and proteins) required for self-propagation, they are challenging because a knowledge of their replication and pathology requires extensive research into the host. At the same time, host organisms have evolved a formidable array of intrinsic and adaptive defenses against such viruses. These defenses allow metazoan and plant pathways to attack viral genomes through gene control by RNA interference and silencing, and are emerging as a central issue in the use of SRVs for pest control.

Essentially, SRVs are the product of just two genes—one for a replication enzyme enabling genome amplification, the other for a structural or capsid protein enabling horizontal transmission between hosts and cells. Genomes generally comprise one or two segments of a single-stranded, message-sense RNA, of less than ~10 kb and encoding up to four (generally two to three) genes. The proteins expressed from these genes are sometimes processed further into smaller proteins with different functions. Virus particles are icosahedral capsids of less than 40 nm (±5 nm) in diameter. These particles do not have a membrane envelope and are composed of one to four different proteins derived from a single gene.

This simplicity has allowed evolution of a significant diversity of viruses in terms of host range and specificity, pathobiology, and ecology. Furthermore, with sufficient knowledge about a specific virus–host interaction, expression of viral components can be manipulated to generate attenuated viruses incapable of transmission beyond the individual target pest that acquires the control agent. All of these attributes can be related to the fact the SRVs are neither toxins nor assemblages of genes that manipulate the hosts' biology to any extent: the pathobiology that they elicit relies on the response of the host to viral infection. Design of SRV-based control agents will benefit greatly from the knowledge of host biology that is emerging from insect genome analysis and functional genomics.

The potential offered by SRVs for pest control has been little recognized, in a pest control landscape that has been dominated by baculoviruses with their obvious, sometimes spectacular pathology, and their ease of cell culture that has enabled genetic manipulation of their DNA genomes. Insect SRVs have been viewed as too difficult to work with and unlikely to make good insecticides. Few grow in cultured cells, they are hard to detect, their effects may be greatest on young larvae and therefore harder to quantify, and their genomes are considered too tiny for manipulation. The reports of devastating effects that SRVs can have upon insect populations have generally emerged from developing nations where these viruses have been employed as pest control agents (Hanzlik *et al.,* 1999). Here we discuss how the genetic simplicity of these viruses has made them attractive candidates for the basis of a powerful new paradigm for insect control. This paradigm involves production of the viruses in novel host systems —for example, the engineering of virus-producing recombinant crops protected against herbivorous or sucking arthropod pests. Among the many challenges to be overcome in realizing this new approach to pest control will be the existence in plants of the same defenses likely encountered by SRVs in insects—the RNA-silencing machinery.

B. Diversity and Phylogenetic Relationships of Insect SRVs

SRVs are widespread in nature, but have been most broadly described from plants (Fauquet *et al.,* 2005; VIIIth ICTV Report). Evidence of unsuspected diversity in aquatic environments has come from a general polymerase chain reaction (PCR)-based study of viruses concentrated by ultrafiltration from coastal marine water samples (Culley *et al.,* 2003); it is inconceivable that the traditional approach of identifying individual viruses based on a pathogenetic phenotype in a known host

TABLE I
USE AND POTENTIAL USE OF INSECT SMALL RNA VIRUSES FOR PEST CONTROL[a]

Virus family, virus	Insect host family	Incidence[b]	Crop use[c]
Tetraviruses[d]			
Darna trima virus	Lepidoptera	Field; SE Asia	Oil palm
Helicoverpa armigera stunt virus	Lepidoptera	Lab, field; Australia, USA	Crops (e.g., cotton)
Nudaurelia capensis ω *virus*	Lepidoptera	Field; South Africa	(Pine)
Parasa lepida virus	Lepidoptera	Field; SE Asia	Coconut
Thosea asigna virus	Lepidoptera	Field; SE Asia	Oil palm
Euprosterna elaeasa virus	Lepidoptera	South America	Oil palm
Dendrolimus punctatus		China	Forest
Dicistroviruses[e]			
Cricket paralysis virus	Orthoptera	Field, lab; Australasia, USA	(Pasture)
Dacus olea virus	Diptera	Lab; Greece	(Olives)
Hemitobi P virus	Hemiptera	Lab; Japan	(Rice)
Plautia stali virus	Hemiptera	Lab; Japan	(Rice)
Rhopalosiphum padi virus	Hemiptera	Lab; South Africa	(Small grain)
Aphid lethal paralysis virus	Hemiptera	Lab, field; South Africa	(Small grain)
Acute bee paralysis virus	Bees	World	Disease
Black queen cell virus	Bees	World	Disease
Iflaviruses			
Infectious flacherie virus	Lepidoptera	Lab; Japan	
DWV, VDV, SBV	Bees/mites	World	Disease
Sequenced, but unassigned small RNA viruses of insects[f]			
Acyrthosiphon pisum virus	Hemiptera	Lab; Netherlands	(Peas)
Canberra cryptic virus	Lepidoptera	Australia	Crops?
Kelp fly virus	Diptera	Australia	
Nodaviruses[g]			
Pariacoto virus	Lepidoptera	South America	
Flock house virus	Coleoptera	Field; New Zealand	(Pasture)
Other SRVs of insects[f]			
Gonometa podocarpi virus	Lepidoptera	Field; Uganda	(Pine)
Latoia viridissima virus	Lepidoptera	Field; Ivory Coast	Oil palm
Pectinophora gossypiella virus	Lepidoptera	Field; Egypt	(Cotton)

[a] Table is not a comprehensive list of insect SRVs and their uses. For more comprehensive lists see references for family groups. Table modified from Hanzlik *et al.* (1999).
[b] Where virus has been isolated.

could have revealed these viruses. In insects very fewer viruses are known, but evidence is slowly accumulating of their diversity.

With the availability of complete genomic sequences for more and more viruses, even those that are hard to isolate and work with, classification has increasingly been based on the phylogenetic/evolutionary history of each of the main genes. RNA viruses generally have been classified into three major groups or superfamilies (Goldbach, 1986; Gordon and Hanzlik, 1998; Koonin and Dolja, 1993). On the basis of shared characteristics, the insect SRVs are now grouped into four recognized families, with further lineages emerging as virologists analyze more viruses and identify new ones (Table I). Almost all of these insect virus families and lineages belong to the picorna-like and alpha-like superfamilies (see specific family reports in Fauquet *et al.*, 2005). The other major group of insect SRVs are the nodaviruses, with two genera, the alphanodaviruses that predominantly infect insects, and the betanodaviruses that predominantly infect fish (Ball and Johnson, 1998; Munday *et al.*, 2002).

C. Tetraviruses

The *Tetraviridae* is a family of icosahedral insect viruses with plus-stranded RNA genomes that includes viruses belonging to the alpha-like superfamily. They have been detected only in a single tissue of a single order of insects, the Lepidoptera (butterflies and moths) (Gordon and Hanzlik, 1998; Moore, 1991; Reinganum, 1991). Tetraviruses are horizontally transmitted via oral ingestion and are found to infect only cells lining the midgut of their caterpillar hosts. Until the discovery of the providence virus (Pringle *et al.*, 2003), none had been found to grow in cultured cells or organ explants of their hosts, even upon transfection of genomic RNA into the cells (Bawden *et al.*, 1999). The lack of cell culture systems for tetraviruses has made study of their molecular biology very difficult.

Viruses sharing the defining characteristic of the *Tetraviridae*, a $T = 4$ pseudo-symmetry particle architecture, have been classified into

[c] Parentheses indicate potential use only, either as a virus or a component of an RNA-delivery vector.

[d] Gordon and Hanzlik (1998).

[e] Christian and Scotti (1998).

[f] Not yet classified or characteristics are too poorly defined to assign to a virus family; see Christian and Scotti (1998) for viruses not described in this chapter.

[g] Ball and Johnson (1998).

two genera based on their particle morphology and genome organization (Gordon and Hanzlik, 1998). The two genera have either a mono- (*Betatetravirus*) or bipartite (*Omegatetravirus*) RNA genome. Complete genomic sequences are available for viruses in both genera. For the betatetraviruses, genomic sequences are available for the *Nudaurelia β virus* (NβV) (Gordon *et al.*, 1999), *Thosea asigna* virus (Pringle *et al.*, 1999), and *Eusprosterna elaeasa* virus (Gorbalenya *et al.*, 2002), and a partial sequence (for the capsid gene) is available for the providence virus (Pringle *et al.*, 2003). A virus obtained from the cotton bollworm, *Helicoverpa armigera*, the *Helicoverpa armigera stunt virus* (HaSV) (Gordon *et al.*, 1995; Hanzlik *et al.*, 1993, 1995), and *Dendrolimus punctatus* virus (Yi *et al.*, 2005) are the only omegatetraviruses for which complete genomic sequences are available. For the *Nudaurelia ω virus* (NωV), the genomic RNA component covering the capsid gene has been sequenced (Agrawal and Johnson, 1992); some sequence information is available for RNA1 of this virus (Agrawal and Johnson, unpublished data; Gordon, Hendry and Hanzlik, unpublished data).

Tetraviral genomic RNAs have 5′-caps and carry 3′-terminal tRNA-Val-like structures (Gordon *et al.*, 1995). Unique among the animal viruses, the tetravirus tRNA-like structures also lack the pseudo-knot found in all plant virus tRNA-like structures (Mans *et al.*, 1991). Whether mono- or bipartite, the tetraviral genomic RNAs encode a replicase and a capsid protein precursor (Fig. 1). A feature specific to the omegatetraviruses is that the smaller RNA2 carries in addition to the amino acid precursor, another overlapping open reading frame encoding a 17 K protein (P17) of unknown function (Hanzlik *et al.*, 1995), whose initiating AUG is situated in a poor context prior to the initiating AUG of the capsid gene, suggesting the latter is expressed by a leaky scanning mechanism. Although not present in the original sequence of NωV (Agrawal and Johnson, 1992), the presence of this open reading frame (ORF) was later confirmed by *in vitro* translation and resequencing (Agrawal, unpublished data).

One unusual characteristic of the viruses comprising the tetravirus family is that the conserved and monophyletic capsid architecture and protein gene phylogeny, which clearly is orthologous with the allocation of the viruses into the two genera described, contrasts with the phylogenetic diversity of replicases carried by these viral genomes. The replicases encoded by the betatetraviruses belong to three different superfamilies of RNA viruses (Gordon *et al.*, 2005). It is likely that this diversity will eventually require a redefinition of tetravirus taxonomy and also show that $T = 4$ capsid symmetry is associated with several virus families in insects. The *Omegatetravirus* virions,

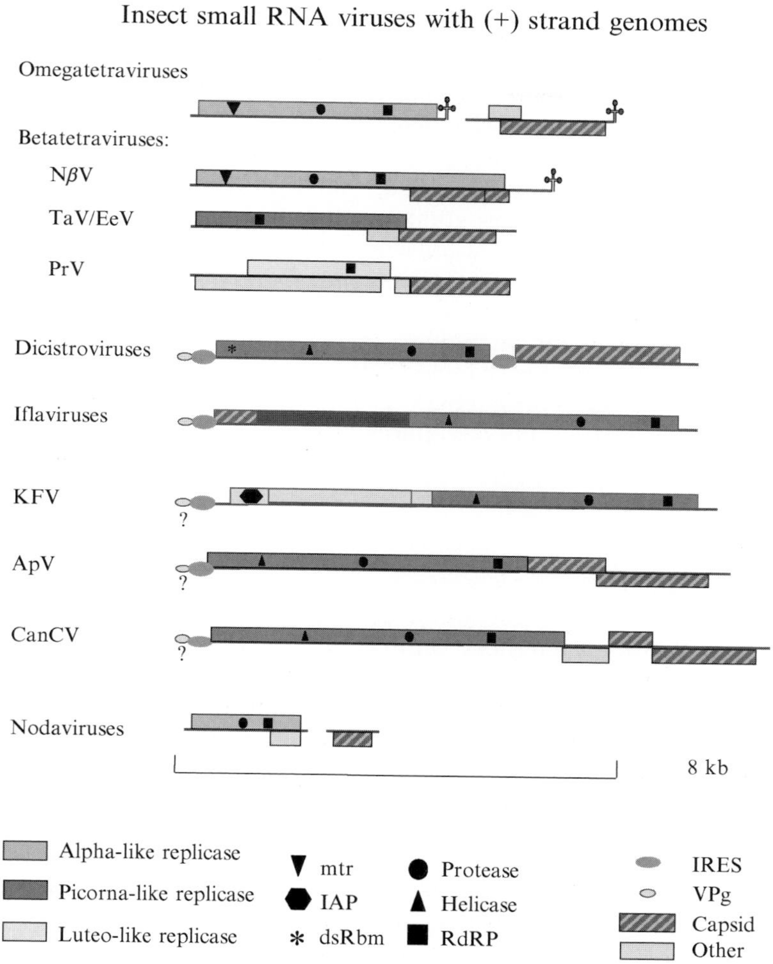

FIG 1. Genomic arrangement and encoded proteins of insect SRVs. A general schematic representation is shown for the recognized SRV families and genera. The only exception is for the betatetraviruses, where the four known members fall into three different groups as shown. Lines represent untranslated regions and boxes represent ORFs. Boxes are shaded to denote the function of the encoded protein, as indicated in the legend. The symbols in the boxes representing replication protein ORFs show locations of specific domains as indicated in the legend. mtr, methyltransferase; IAP, inhibitor of apoptosis; dsRbm, dsRNA-binding motif; RdRP, RNA-dependent RNA polymerase. Among the picorna-like viruses, VPg's and IRES function have not been confirmed in all cases and are marked with a "?" accordingly. (See Color Insert.)

that have been studied in greater detail, are 41 nm in diameter and composed of 240 copies of a 64-kDa major coat protein and a 7-kDa minor coat protein arranged in the $T = 4$ icosahedral symmetry that is characteristic of tetraviruses. The two coat proteins are derived from a 70–71 kDa precursor that is cleaved only on particle assembly to form the mature virion (Agrawal and Johnson, 1995). Cleavage of the precursor is a definitive sign of particle formation as it occurs only in provirions which are stable until a drop in pH (Canady *et al.*, 2000). Extensive structural studies have been conducted on the virions of NωV, including X-ray crystallography (Munshi *et al.*, 1996) and cryoelectron-microscopy (Canady *et al.*, 2000).

HaSV has attracted particular interest among the tetraviruses because its heliothine hosts include the most significant pests of agriculture around the world. HaSV is also the only tetravirus whose caterpillar hosts are readily available from a laboratory reared colony, and is therefore the only tetravirus upon which biological experiments can readily be conducted.

1. *Picorna-Like Viruses*

Numerous picorna-like viruses have been isolated from insect and other arthropod hosts (Christian and Scotti, 1998) and the genome sequences for many of these confirm that they are members of the picorna-like virus superfamily. Based on their genome organization, the phylogeny of their replicase protein-coding sequences, and structural data, two major taxonomic groups have emerged. The first family recognized was the *Dicistroviridae* (Mayo, 2002) with a single recognized genus, the *Cripavirus*, represented by *Cricket paralysis virus* (CrPV). Dicistroviruses have 9–10 kb genomes carrying two ORFs separated by an intragenic region (IGR), which can function as an internal ribosome entry site (IRES) to initiate translation of the downstream ORF. The IGR-IRES is more active than the 5′ IRES that initiates translation of the 5′ ORF (Wilson *et al.*, 2000). Dicistroviral capsid proteins (VP1, VP2, VP3, and possibly VP4) are encoded by the 3′ ORF, while nonstructural proteins are encoded by the 5′ ORF.

Genome sequences show that several other picorna-like viruses infecting insects resemble the canonical picornaviruses in encoding a unique large ORF with the structural proteins present in the 5′ region, and the RNA-dependent RNA polymerase (RdRP) at the 3′ end of the genome. Although this genome organization resembles that of the mammalian picornaviruses, these insect viruses are phylogenetically distinct. Insect viruses with this genome organization, and showing a clear phylogenetic relationship, include: sacbrood virus (SBV) (Ghosh *et al.*, 1999), infectious flacherie virus (IFV) (Isawa *et al.*, 1998), *Perina nuda* picorna-like virus (PnPV) (Wu *et al.*, 2002), *Ectropis obliqua* picorna-like virus (EoPV) (Wang *et al.*, 2004), deformed wing virus (DWV) (GenBank accession numbers AY292384 and AJ489744), Kakugo virus (KV) (Fujiyuki *et al.*, 2004), and *Varroa destructor virus* (VDV) (Ongus *et al.*, 2004). These viruses have been grouped into a "floating" genus (not a recognized family) termed the iflaviruses (Mayo, 2002).

2. The Unclassified Viruses—ApV, KFV, and CanCV

In addition to viruses which have been confirmed as members of the above groups on the basis of their genome sequences, there are several unclassified picorna-like viruses whose genomes show that they represent new viral groups with no close relationship to the existing families. These include *Acyrthosiphon pisum virus* (ApV) (van der Wilk *et al.,* 1997) and kelp fly virus (KFV) (Hartley *et al.,* 2005). A cryptic virus was found in a laboratory colony of *H. armigera*, maintained for production of HaSV, as an additional virus present in HaSV preparations. The isometric particles contain a single RNA strand >10 kb. However, the biological properties of the particles were less forthcoming. All attempts to elicit disease or propagate the virus in larvae by feeding or injection failed, as did attempts to propagate the virus in cultured cell lines. Sequence analysis (Hanzlik *et al.,* 2006a) of the genomic RNA showed the virus, named Canberra cryptic virus (CanCV), to have a novel genetic organization and to belong to the picorna-like supergroup of viruses. Distinctions in capsid protein characteristics, as well as unique structural properties, separate CanCV from the other picorna-like insect viruses. Like KFV, the CanCV virion also has only two major capsid proteins, but the large size of VP1 (75 kDa) distinguishes it from those of the other viruses.

The CanCV replicase contains sequence motifs shared with the RdRPs of the picorna-like superfamily. Phylogenetic analysis of the polymerase and helicase domains of CanCV (Hanzlik *et al.,* 2006a, unpublished data) and of KFV (Hartley *et al.,* 2005) yielded similar conclusions—that the three unclassified viruses ApV, KFV, and CanCV were not related to each other or members of known virus groups. These viruses were located at the ends of deeply rooted and poorly resolved branches distinct from all families in the superfamily (e.g., the dicistroviruses, iflaviruses, picornaviruses, caliciviruses, and plant viruses). Thus the distinct clades derived from insects now include: the *Dicistroviridae*, the iflaviruses, ApV, KFV, and CanCV. The high level of sequence divergence among these viruses means that the relationships between them cannot clearly be resolved.

DeMiranda and Gordon (2006) have described a number of additional members of the iflavirus genus discovered as genome sequence fragments in expressed sequence tag (EST) databases. One example was recovered from the red flour beetle (*Tribolium castaneum*), an economically important coleopteran pest that has been the subject of a genome project but for which no SRVs have previously been described. These viruses may predominantly be latent in the host, and hence asymptomatic in the apparently healthy individuals selected for cDNA library construction. This approach establishes a potentially fruitful way for analyzing virus sequence diversity in an insect pest for which a novel control agent is required. The large-scale EST sequencing that is now being applied to a greater diversity of organisms makes the task of recovering such cryptic viruses ever more feasible. However, the approach will have limitations. The sequences of tetraviruses and other viruses that lack poly(A) tails will not be obtained, and experience suggests that certain types of viruses, those related to known iflaviruses for example, are most likely to emerge. Only few new cripaviruses, and no other unclassified viruses, were identified in the Genbank dbEST.

D. SRVs as Biopesticides in the Field

As discussed by Hanzlik *et al.* (1999), SRVs are capable of causing dramatic reductions in natural insect populations and have most frequently been observed in high-density populations of lepidopteran insects in stable environments, for example, plantations of perennial crops. Most reports of such events have come from tropical regions (Table I) with the remainder from the temperate Southern Hemisphere; these reports were mainly the result of projects funded by international aid agencies. Some reports of SRV use are now accessible to the mainstream literature. To date, the best success has been with pests on oil palm and coconut plantations in Southeast Asia, West Africa, and South America (Philippe *et al.,* 1997). Here, large field applications of SRVs have been conducted typically by spraying an aqueous suspension of dispersed cadavers of SRV-killed larvae with high-volume or back-pack sprayers. Tetraviruses, and to a lesser extent picorna-like viruses, are the most frequently used in these cases (Desmier *et al.,* 1988). Viruses for pest control purposes were obtained by manual collection of cadavers after an epizootic and could be stored at room temperature for several years. Advantages of the use of SRVs for biological control as part of Integrated Pest Management (Wood, 2002) are lower costs when compared to chemical control, preservation of beneficial insects due to the specificity of the control agent, improved safety, and long-term persistence (Ginting and Desmier, 1987).

Among insect SRVs, only limited attention has been paid to nodaviruses as potential pest control agents, due largely to their host range. In New Zealand, *Flock house virus* (FHV) was found associated with field mortality of the grass grub (Bourner *et al.,* 1996). Further, *Pariacoto virus* was identified from *Spodoptera eridania*, a pest of the sweetpotato (Zeddam *et al.,* 1999). An emerging area of interest in the economic potential of nodaviruses concerns the betanodavirus genus, which contains a number of viruses that infect teleost fish. These viruses present significant problems for the health and management of fish cultures around the world (Munday *et al.,* 2002).

II. Pathology and Host Biology

A. *Lepidopteran Midgut Biology*

The larval midgut epithelium is the largest organ of the lepidopteran larva, responsible for both digestion and defense against environmental stresses, such as host plant defense chemicals or chemical pesticidal agents. It is the primary target of biological control agents such as the *Bacillus thuringiensis* toxin (Bt), as well as the primary site of infection by orally transmitted viruses. The need for the midgut to meet the often conflicting challenges of nutrient absorption and host defense entails a regenerative capacity and a major investment in immune response systems. The structure of the midgut itself is deceptively simple despite its multiple functions of digestion, ion transport, and secretion of proteins, and peritrophic membrane components. The midgut consists of a single epithelial layer of cells which is supported by a basal membrane, and surrounds the lumen where the food is digested. Midgut cells are of four main types, columnar and goblet cells that are produced by proliferation and differentiation of basal regenerative cells, and the endocrine cells; the latter two types are smaller and situated basally (Anderson and Harvey, 1966; Baldwin *et al.,* 1996; Billingsley and Lehane, 1996; Cioffi, 1979; Dow, 1986; Endo and Nishiitsutsuji-Uwo, 1981). During larval growth, the columnar and goblet cells increase significantly in both size and number. Replacement of differentiated cells that have detached from the basal membrane (Baldwin and Hakim, 1991; Engelhard *et al.,* 1991) occurs through the proliferation of basal epithelial stem cells, followed by cell differentiation (Hakim *et al.,* 2001) to yield progenitors for the differentiated cell types; there is also self-renewal of the stem cell population.

Functional genomics research in lepidopteran insects is likely to identify genes with a key role in midgut development. Of particular interest will be the study of gene expression in cultured stem cells that can be induced to proliferate and differentiate (Loeb and Hakim, 1996). An exciting development has emerged from studies of microRNAs (miRNAs). In animals, miRNAs have been implicated in the regulation of stem cell division (Hatfield *et al.*, 2005), gene expression (Houbaviy *et al.*, 2003, Suh *et al.*, 2004), and in stem cell self-renewal (Rao, 2004) and differentiation (Chen *et al.*, 2004; Kanellopoulou *et al.*, 2004). It is likely that similar observations will be made in insect guts.

B. Host Response and Pathology

Importantly for its potential use as a biological control agent, HaSV is particularly effective against young larvae (Brooks *et al.*, 2002; Christian *et al.*, 2001). Infection with virus *per os* during the first three instars of larval development is virulent and leads to rapid stunting and mortality. In contrast, no detectable symptoms occur in later larval development, signifying a high degree of developmental resistance. Neonate larvae fed very limited amounts of HaSV in a quantitative droplet-feeding assay showed rapid cessation of feeding (within 24 h), followed by death (Hanzlik and Gordon, 1998). On the other hand, presentation of the highest possible doses of HaSV ($>2 \times 10^{10}$ virus particles) to fourth or fifth instar larvae failed to elicit any detectable pathological response. The first detailed description of the interaction at the cellular level between a tetravirus (or indeed any small RNA virus) and its insect host was that between HaSV and midgut cells through all stages of larval growth by using microscopy and histopathology (Brooks *et al.*, 2002). The midgut cells of the infected larvae responded to infection with an increased rate of sloughing. In young larvae, the extent of sloughing rendered the midgut incapable of maintenance or recovery of normal function. Brooks *et al.* (2002) concluded that cell sloughing was an immune response that exists throughout larval development. While HaSV-infection resulted in the death of younger larvae, the increased resistance of older larvae was associated with a lower abundance of infection foci, that failed to expand and eventually disappeared, presumably as a result of the same process of cell sloughing. The increased rate of cell sloughing was correlated with an increase in midgut cell apoptosis. Antiapoptosis genes have not been detected by sequence analysis of tetravirus genomes. This suggests that these viruses, with small genomes, are able to rely on the speed of RNA replication to accumulate sufficient progeny before apoptosis prevents further growth in any one cell. As yet we do not know how apoptosis is linked to cell rejection or how apoptosis is induced, but it may be a factor in the disruption of the normal regulation of cell differentiation and communication during midgut regeneration.

The degree of developmental resistance toward HaSV is likely to be determined by the host midgut, given that the virus infects only this tissue. HaSV was detected only in midgut RNA of infected larvae by Northern blot (Bawden *et al.*, 1999), whether HaSV was administered orally or by hemocoelic injection of HaSV, a route that requires higher doses than *per os* to induce the same symptoms. No evidence for HaSV was seen in other tissues, even in other alimentary tissues such as the foregut and salivary gland or at advanced stages of infection. The midgut tropism of tetraviruses was originally noted by Grace and Mercer (1965) working with the Antheraea eucalypti Virus (AeV) and by a later study by Greenwood and Moore (1984) using enzyme-linked immunosorbent assays (ELISA) combined with electron microscopy on tissues from larvae infected with *Trichoplusia ni virus* (TnV). HaSV was found to infect the three most common midgut cell types, that is, the differentiated columnar and goblet cells and the regenerative basal cells, but not the fourth, more rare, endocrine cell type.

As discussed by Brooks *et al.* (2002), these observations with a tetravirus like HaSV are different to those made in the majority of studies on viral infections of lepidopterans that have focussed on cytoplasmic polyhedrosis viruses (cypoviruses; CPVs) and baculoviruses. Baculoviruses in particular are very different, most infecting their host midguts only transiently during their passage to tissues in the hemocoel. Secondary infection of tissues other than the midgut occurs within 24 h and includes the tracheal cells, fat body, hemocytes, and outer epidermis (Booth *et al.,* 1992). The midgut then appears to recover from infection, with the insects continuing to feed until near to death from virosis of their hemocoelic tissues (Engelhard and Volkman, 1995; Keddie *et al.,* 1989). The sloughing of infected midgut cells (Engelhard and Volkman, 1995; Federici, 1997; Flipsen *et al.,* 1995) evident in response to viral infection may allow for the recovery of the midgut following the initial stages of infection. Some baculoviruses, such as the Western grapeleaf skeletonizer (*Harrisina brillans*) granulovirus (Federici and Stern, 1990) are restricted to the midgut and, like HaSV, cause acute disease in the midgut with massive sloughing of infected cells prior to death from water loss and midgut failure (Federici, 1997). Cell sloughing followed by regeneration may be an effective mechanism allowing for recovery of midgut tissues from infection with other RNA viruses including CPVs and IFV (Bong and Sikorowski, 1991; Choi *et al.,* 1989; Inoue and Miyagawa, 1978; Yamaguchi 1979). Even at early larval stages, regeneration of the midgut epithelium involved healthy cells replacing the infected cells, in contrast to observations for HaSV.

In contrast to baculovirus infections, the specific and rapid spread of HaSV in the midgut is likely to account for the rapid cessation of feeding (Volkman, 1997). Such a response is precisely that required of the ideal orally transmitted viral biocontrol agent (Keddie *et al.,* 1989). This property may be caused by the ability of HaSV to elicit sloughing on a large scale in the midgut that may account for the toxicity of this virus to early instar larvae. How infection by HaSV progresses and affects the cells of the midgut likely represents a fruitful line of investigation for those interested in developing novel insect control agents. Finally, the tissue- and host-specificity of the virus (Bawden *et al.,* 1999; Christian *et al.,* 2001) are highly desirable traits for a virus with potential use for pest control (Christian *et al.*, 1993).

C. Virus–Host Interactions at the Intracellular Level

Viruses are subject to a range of cellular responses that have evolved to restrict their growth and spread. These responses include apoptosis and RNA silencing. Consequently many viruses have evolved means to overcome the host cellular response. Among these are antiapoptosis genes, first identified in baculoviruses (Clem, 2005). Antiapoptosis ("inhibitors of apoptosis," IAP) domains have been found in only two RNA viruses—KFV (Hartley *et al.,* 2005) and the arthropod-infecting dicistrovirus *Taura syndrome virus*, TSV (Mari *et al.,* 2002). It is not known whether either of these predicted IAP domains is active in inhibiting apoptosis in virus-infected cells.

RNA interference (RNAi) has emerged as a widespread process in eukaryotes. It is not only found as a defense against RNA virus replication (Ding *et al.,* 2004), but also underlies posttranscriptional gene silencing (PTGS) of genes through the degradation of mRNAs (Meins *et al.,* 2005), and the nondestructive regulation of translation by microRNAs (miRNAs) during development, for control of genes central to cell proliferation, differentiation, and death (Nakahara and Carthew, 2004). RNAi-related mechanisms are involved in the epigenetic control of gene expression and underlie genome reorganization and retrotransposon silencing (Denli and Hannon, 2003; Martienssen, 2003).

Insect genomes encode a comprehensive set of proteins required for RNAi (Meister and Tuschl, 2004). The RNAi response to double-stranded

RNA (dsRNA) and miRNAs present within cells begins with the fragmentation of such transcripts by enzymes related to RNAseIII. These enzymes fall into two classes, both with two RNAseIII catalytic domains and a dsRNA-binding domain at their C-terminus. Class III contains the Dicer (Dcr) enzymes, characterized by additional helicase and Piwi/Argonaute/Zwille (PAZ) domains. Comparative analysis of insect genomes has shown key early genes in RNAi to be variably present, with significant implications for our understanding of RNAi in these animals, as well as for the ability to harness PTGS or to apply RNAi in antiviral defense. Work in *Caenorhabditis elegans* (and later in vertebrates) has demonstrated systemic RNAi (systRNAi), in which there is widespread response within an organism to an artificially administered dsRNA trigger; such a systemic response was found to require a cellular RdRP, as well as an RNA transport channel protein, SID-1. Dipteran genomes do not encode these proteins. In contrast the bee genome encodes a single ortholog of SID-1 (Honeybee Genome Consortium, 2006); strikingly, multiple (three to date) orthologs are found in the *Bombyx mori* (Mita *et al.*, 2004; Xia *et al.*, 2004) and *T. castaneum* (Brown *et al.*, 2003) genomes, with *H. armigera* also showing at least two orthologs (Collinge and Gordon, unpublished data).

Genes for two Dcr enzymes have been found in *Drosophila melanogaster* (Meister and Tuschl, 2004), and now in the honeybee *Apis mellifera*, *Tribolium*, and *Bombyx* genomes (The Honey Bee Genome Sequencing Consortium, 2006). Nematodes and vertebrates have just one Dcr. In *Drosophila*, Dcr-1 enables miRNA generation from precursor hairpin transcripts, whereas Dcr-2 cleaves dsRNAs to yield small interfering RNAs (siRNAs) for silencing. Class II RNAseIII enzymes include Drosha, first identified in *Drosophila* and responsible for pre-miRNA processing. Drosha is also present in the honeybee, *Tribolium*, and *Bombyx* genomes. The bee genome also encodes a full set of dsRNA-binding cofactors termed R2D2, Loquacious, and Pasha that have been shown in *Drosophila* to be required for function of Dcr and Drosha. The honeybee genome also encodes the proteins identified as members of the RNA-induced silencing complex (RISC), particularly Argonaute.

D. RNAi in Insects

Gene silencing by RNAi has emerged as a research tool of great interest for the study of gene function. Analysis of insect genes encoding the components of the RNAi pathway allows for assessment of prior work on RNA silencing in insects, and of whether systRNAi can be

Li *et al.,* 2004). The FHV B2 protein binds to dsRNA and inhibits its cleavage by Dcr *in vitro* (Chao *et al.,* 2005). A cocrystal structure revealed that a B2 dimer forms a four-helix bundle that binds to one face of an A-form RNA duplex independently of sequence. These authors interpreted the results to suggest that B2 blocks both cleavage of the FHV genome by Dcr and incorporation of FHV siRNAs into the RNA-induced silencing complex.

Suppressors of RNA silencing have since been identified for a number of other animal viruses, for example, influenza and vaccinia viruses (Li *et al.*, 2004). There is intense interest in the question of how widespread such suppressors are in viral genomes (Li and Ding, 2005; Voinnet, 2005). For the picorna-like animal SRVs, a dsRNA-binding motif (PFAM motif PF00035) exists in the polymerase polyprotein of *Drosophila C virus* (DCV: residues 24–88) that may correspond to the anti-RNAi gene identified in the amino-terminal portion of the CrPV nonstructural polyprotein (Li and Ding, 2005; Wang *et al.,* 2006). No suppressors of RNAi silencing have been found in an animal virus of the alpha-like superfamily, which is consistent with the use of viruses such as Sindbis to deliver RNAi (see in an earlier section), although it may reflect the difficulty in identification of suppressors, due to their sequence diversity. The finding that insect picorna-like viruses such as CrPV and DCV encode suppressors as part of a longer polyprotein means that such a domain may yet be identified in tetraviruses as part of the replicase ORF for example, rather than as a separate RNA suppressor gene found in all members of the family. However, the strict cell tropism of tetraviruses may argue against a general ability to suppress RNA silencing. These viruses may have evolved the ability to replicate in the stem cells where they may be less vulnerable to RNAi. This situation would resemble that of some plant viruses that do not appear to encode RNAi suppressors (Roth *et al.,* 2004). Some plant viruses that are vulnerable to plant recovery via RNAi replicate in the meristem (Voinnet, 2005), and phytoreoviruses may induce tumors in order to replicate untroubled by RNAi (Waterhouse, 2006).

Latency may be another strategy by which viruses evade the host response. CanCV is an example of a virus that is so cryptic as to be apparently uninfectious. CanCV only emerges when cells are stressed, for example, through infection with another virus. Although the means by which CanCV maintains latency are unknown, it is possible that translation of virus structural proteins is dependent on cellular factors associated with stress (Hanzlik *et al.,* 2006a, unpublished). The virus genomic RNA therefore replicates but generates no infectious progeny until cell stress makes this possible.

III. Production of Insect SRVs in Nonhost Systems

A. SRVs as Biopesticides: The Production Problem

In order to produce an insect SRV outside the natural host, it is necessary to express the viral capsid proteins and biologically active genome RNAs and achieve assembly of viral particles encapsidating these RNAs. Assembly of infectious particles in nonhost systems, independently of replication, remains something of a rarity in virology, but the generation of infectious genomic RNA transcripts from full-length cDNA clones, even *in vitro*, has been possible for decades for many RNA viruses. Full-length infectious clones have been made from many plant and animal RNA viruses (Boyer and Haenni, 1994) opening up a new avenue for RNA virus genetics (Conzelmann and Meyers, 1996). The insect nodaviruses were among the first viruses for which reverse genetics by production of infectious transcripts was established (Ball and Johnson, 1999) and this work was since extended to the *Pariacoto virus* (Johnson *et al.,* 2000) and the fish beta-nodaviruses (Iwamoto *et al.,* 2001). Work on insect RNA viruses has shown that infectious transcripts can be generated from full-length reverse transcription (RT)-PCR cDNA templates for an insect picorna-like virus, the Black queen cell virus of honeybee, a member of the dicistroviruses (Benjeddou *et al.,* 2002). Infectious transcripts have also been synthesized from full-length cDNA clones of luteoviruses (Miller and Rasochova, 1997), and flaviviruses (Lai *et al.,* 1991). In order to realize the potential of HaSV as a biopesticide capable of acting rapidly and selectively on heliothine insects, it was necessary to develop an economical means of production to replace the traditional means of infecting caterpillar hosts, that is inefficient and of lower reliability.

B. Assembly of Tetraviral VLPs Using Recombinant Baculoviruses

The widely used recombinant baculovirus protein expression system has also been applied for production of viral capsid proteins allowing self-assembly of icosahedral VLPs (Kinnbauer *et al.*, 1993; Le Gall-Reculé *et al.*, 1996; Pawlita *et al.*, 1996). VLPs of RNA viruses, including the calicivirus rabbit hemorrhagic disease virus (RHDV) (Nagesha *et al.,* 1995) and poliovirus (Bräutigam *et al.,* 1993) have been produced using this system. This technology has also successfully been applied to tetraviruses, including NωV (Agrawal and Johnson, 1995; Canady *et al.,* 2000) and to HaSV (Gordon and Hanzlik, 1998; Hanzlik *et al.,* 1999, 2006b, unpublished). While baculovirus expression of viral capsid proteins has allowed production of tetravirus VLPs that resemble virions, and allowed detailed structural studies, and analysis of the mechanism of particle formation and capsid protomer processing (Bothner *et al.*, 2005; Helgstrand *et al.*, 2004), it has not been possible to generate infectious particles by this route (Gordon *et al.*, unpublished). This result is because of the difficulty of generating sufficient transcripts for RNA2 and RNA1 that have identical termini to those found on native viral RNAs.

The encapsidation of HaSV or NωV-related sequences was found to be highly specific under the conditions of production, that is, where overall levels of capsid protein are moderate to low (Agrawal and Johnson, 1995; Hanzlik and Gordon, unpublished). This makes tetraviruses such as HaSV ideal test candidates to address whether infectious virus particles can be assembled in a suitable nonhost system, as described in a later section.

C. Assembly of Infectious HaSV in Plant Cells

The first step toward asking whether infectious HaSV could be produced in plants was accomplished by transfection of protoplasts of *Nicotiana plumbaginifolia* with three plasmids designed to express the three main components, RNA1, RNA2, and P71 of the HaSV virion (Gordon *et al.*, 2001). Each cDNA was placed under the control of the *Cauliflower mosaic virus* 35S promoter and followed by a *cis*-acting ribozyme so that the resultant transcripts corresponded precisely to the two genomic RNAs (Fig. 2). No replication of HaSV in protoplasts was detected in pulse-labeling and blotting experiments. Assembly of infectious particles in the protoplasts could only be detected by electron microscopy and bioassay of host insect larvae, which became diseased and produced virus particles confirmed as HaSV. As observed for many other RNA viruses (Boyer and Haenni, 1994), transcripts carrying nonviral sequences at either or both termini of the RNAs yielded no infectious particles. For infectious particles to be assembled, expression of the capsid protein was required from a separate plasmid (Fig. 2). The requirement for separate expression of the capsid protein was consistent with the lack of evidence for viral replication, and resulted from the low levels of capsid protein translated from genomic RNA2 in plant cells. This novel plasmid-based system confirmed that full-length clones of HaSV represented infective genomes and established a procedure for the reverse genetics of a tetravirus.

An interesting aspect of this study was demonstration that insects could be infected by particles assembled in protoplasts transfected with only the RNA1 and coat protein plasmids. The symptomatic larvae contained only RNA1 and failed to yield infectious progeny virus, suggesting that RNA1 is capable of self-replication. Omission of the viral genome component encoding P71 (RNA2) resulted in assembly of "one-way" virions that delivered only RNA1 to cells where RNA1 self-replicated to produce a lesser pathology in larvae, characterized by a lower level of stunting. Histochemical analysis of midgut sections from the diseased larvae confirmed a pathology characteristic of HaSV infection, but without production of progeny virus (E. M. Brooks, T. Hanzlik, and K. Gordon, unpublished data). Self-replication of the genomic RNA strand encoding the replicase in the absence of other genome components has been observed for a number of positive-sense RNA viruses that have multipartite genomes including the nodaviruses (Hendry, 1991). What appears to be unusual is that the self-replication of HaSV RNA1 leads to the observed disease symptoms, possibly through interference with proper cell function in neonates and young larvae. That RNA self-replication can persist in larval gut cells was shown by continuing reporter gene activity in gut cells over 10 days after ingestion of protoplasts transfected with pR1, pCAP, and a gene for RNA2 modified to express reporter genes in place of the coat protein (see in a later section).

D. SRV Capsids as RNA Vectors for Foreign Genes

The demonstration that particles of a tetravirus could be assembled in a nonhost system, in the absence of replication, and deliver a biologically effective RNA into the midgut cells of the target insect led us to ask whether the same particles could deliver an RNA encoding a nonviral gene into the same target cells. If this were possible, it would represent a novel approach to the problem of precise, efficient transfection of specific RNAs, for example, for RNAi, into specific cells of animals, even allowing the harsh environment of the lepidopteran gut lumen to be traversed. The first question asked (Hanzlik *et al.,* 2006b, unpublished) was whether an exogenous gene could replace the P71 ORF on RNA2, for amplification by the RNA1-encoded replicase. Fusion of the GFP-coding sequence to the start of the P71 ORF, followed by cotransfection with the plasmids for RNA1 and the capsid protein into protoplasts, yielded fluorescent particles in larvae fed the protoplasts as neonates (Fig. 2C).

It was then asked whether gene expression could be achieved in the absence of viral replication in the target host through omission of RNA1, which encodes the viral replicase. The resulting need to produce

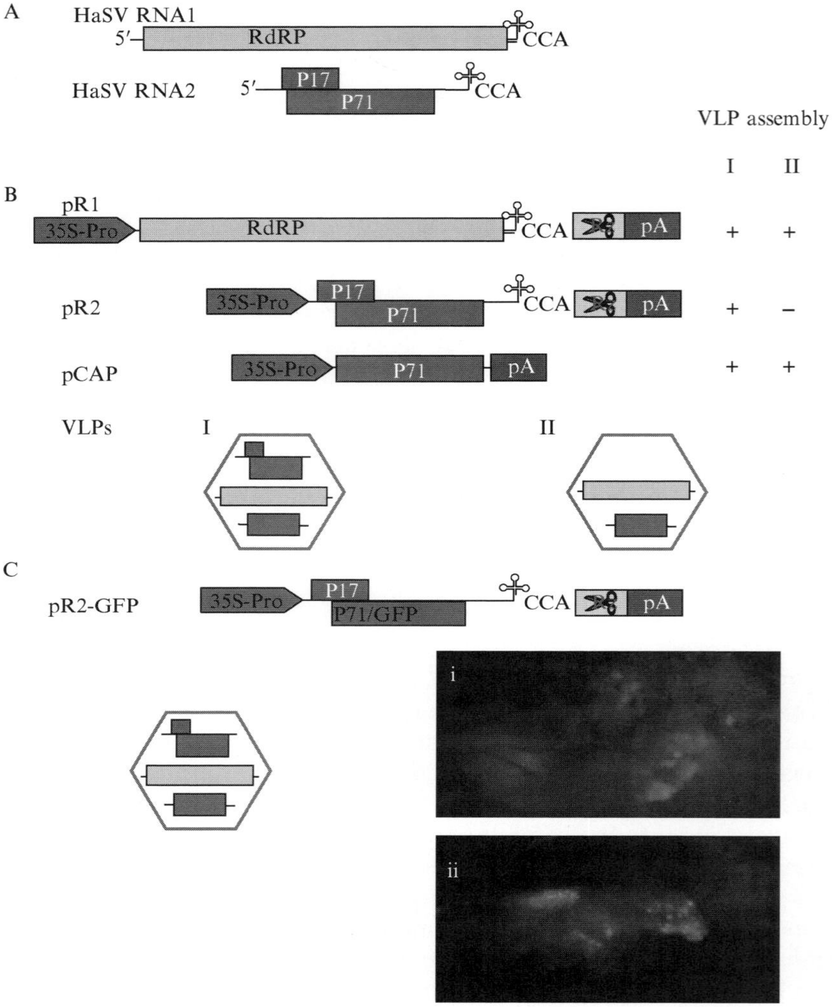

FIG 2. HaSV RNAs and their use for transgene expression. (A) Genome organization of HaSV RNAs 1 and 2, showing ORFs and 3′-terminal tRNA-like structures. (B) Schematic diagrams showing plasmids for expression of HaSV RNAs in plant protoplasts, and the resulting virus-like particles (VLPs). The plasmids were as described in Gordon *et al.* (2001). Genomic RNA1 is expressed from pR1, and pR2 which produces genomic RNA2, was used for two derivative plasmids. Plasmid pCAP carries only the HaSV P71 ORF. The CaMV 35S promoter (35S-Pro), *cis*-acting ribozyme (Rz), and the CaMV 35S polyadenylation sequence (pA) are indicated for each construct. Boxes represent ORFs for

much larger numbers of particles than was possible in protoplasts led to use of the baculovirus expression system, based on generating recombinant *Autographa californica* multiple nucleopolyhedrovirus (recNPV). Several approaches were used (Hanzlik *et al.,* 2006b, unpublished). Assembly of VLPs produced using a mixture of recNPVs containing either the capsid gene alone or in combination with an upstream reporter gene on the same RNA was investigated. This approach yielded particles able to deliver the chimeric RNA expressing the upstream reporter gene. (The capsid gene on this chimeric RNA was silent and served only to ensure encapsidation of the RNA containing the reporter gene.) One disadvantage of this two-recNPV approach is that the resultant VLPs contained large amounts of the capsid protein mRNA, resulting in reduced ability to vector the RNA carrying the reporter gene. To simplify this approach and assemble both the reporter gene and the capsid gene on the same construct for expression in a single recNPV, a new dicistronic RNA was designed. This RNA used an IRES, a segment of RNA that enables ribosomes to translate downstream ORFs (Masoumi *et al.,* 2003). IRESs of insect origin usually function only in specific cell types (Masoumi *et al.,* 2003); the IRES used in this study was obtained from the 5′ region of the novel picorna-like virus isolated from *H. armigera*, the CanCV (Hanzlik *et al.,* 2006b, unpublished).

Assembly of VLPs containing a single RNA species carrying both the capsid and reporter genes proved to be an effective delivery strategy. While an IRES was used for translation of a downstream reporter gene on a dicistronic RNA, this IRES would not be necessary for other effector RNA sequences, for example, to promote gene silencing by RNA interference of insect genes. Alternatively, P71 could be produced

gene products as noted within them. Combinations of plasmids transfected into protoplasts for feeding to larvae are shown to the right, and schematic diagrams of the resulting VLPs (combinations I and II) assembled in protoplasts, below. Combination I corresponds to assembly of VLPs containing the complete HaSV genome, and combination II to VLPs of the subvirus, lacking genomic RNA2. (C) Expression of green fluorescent protein (GFP) in larvae. The capsid gene on plasmid pR2 was replaced by a P71/GFP fusion as described by Hanzlik *et al.* (2006b, unpublished), allowing assembly in protoplasts of VLPs containing RNA1, the RNA2-GFP fusion, and the P71 mRNA as shown in the schematic. Following transfection of the plasmids shown, protoplasts were incubated for 3 days before aliquots were fed to larvae; larval midguts were examined for fluorescence after 11 days. Panels (i) and (ii) show epifluorescence images of anterior midgut regions from GFP-expressing larvae, with the anterior of each image oriented to the left. Control larva, fed protoplasts transfected with only pR1 and pCAP, showed no GFP fluorescence. (See Color Insert.)

from a separate, nonencapsidated codon-modified transcript, allowing the effector sequences to precede nontranslated P71-derived sequences whose only function is to enable encapsidation. The successful production of HaSV particles in plant cells is suggestive of the use of this system as a novel control approach for protection of crops against heliothine caterpillars. The technology could have wider application through manipulation of the tropism exhibited by the tetraviral capsid.

E. Transgenic Plants Expressing Virus Genomes and Capsid Genes

Plants have been successfully transformed with capsid genes from nonplant viruses to address whether these capsid proteins are able to self-assemble into VLPs. Of particular interest for generating plants expressing insect picorna-like viruses is that the Norwalk virus capsid protein (NVCP) assembles VLPs in transgenic tomato plants (Huang *et al.*, 2005). This virus is related to the abundant picorna-like viruses of insects. These workers asked whether VLPs of this calicivirus could be assembled, mimic the form of authentic virions and display neutralizing antibody epitopes. NVCP expressed in plants assembled 38-nm virion-size icosahedral ($T = 3$) VLPs, similar to those produced in insect cells. The VLPs stimulated serum immunoglobulin G (IgG) and IgA responses in mice and humans when they were delivered by ingestion of fresh potato tuber. Although the predominant VLP form in tomato fruit was a small 23-nm particle also observed in insect cell-derived NVCP, transgenic NVCP tomato fruit yielded stable VLP-antigen preparations that stimulated excellent IgG and IgA responses against NVCP when fed to mice. Interestingly, 23-nm particles predominated in the transgenic tomato, at the expense of full-size 38-nm particles. Both particles are present in insect cell-derived VLPs, representing $T = 1$ (23 nm) and $T = 3$ (38 nm) icosahedral symmetry (White *et al.*, 1997). The 38-nm VLPs had previously been observed in tobacco leaf and potato tuber (Mason *et al.*, 1996) but the 23-nm particles were not evaluated in that study; the later study provided evidence that these 23-nm VLPs were highly immunogenic (Huang *et al.*, 2005).

F. Attempts to Produce HaSV in Transgenic Tobacco

The successful assembly of the three components of HaSV (the capsid protein and the two RNA molecules of the virus genome) into infectious HaSV particles in protoplasts led to the ambitious attempt to generate transgenic plants capable of assembling this insect SRV for protection against chewing pests (Hanzlik and Gordon, 1998;

Larkin *et al.*, 1996). For this, the expression cassettes comprising DNA copies of the viral RNAs driven by the CaMV 35S promoter were assembled into a binary vector, pHaSV1, to produce transgenic plants carrying the complete HaSV genome as for the protoplast work. The genes were placed under the control of a plant-specific constitutive promoter derived from *Cauliflower mosaic virus*. In later work, promoters derived from another plant virus, subclover stunt virus (SCSV) which show high levels of expression in various dicotyledonous crop plants including cotton (Schunmann *et al.*, 2003), were used, instead of the CaMV 35S promoter. Generation of transgenic plants using *Agrobacterium*-mediated transfer of foreign DNA to the plant genome was followed by molecular studies on the presence and integrity of transgenes and expression of RNA and protein components of the virus. Plants transformed with combinations of genes theoretically capable of yielding infectious virus were screened by bioassay of neonate larvae on leaf material analogous to studies on plants expressing Bt toxins. Initially, tobacco cultivar Wisconsin 38 (W38) was used. Problems with bioassaying control material from this cultivar led to the later use of cultivar Samsun.

A plant binary vector carrying all three genes to transform the model plant tobacco generated several primary (*T*o) transformants that showed good to very good stunting of *Heliothis virescens* larvae fed on the early leaves. The larvae fed on these plants contained large quantities of HaSV and appeared to have been infected from eating the transgenic plants; this gave promising protection against feeding damage. Extraction of RNA from larvae and Northern blotting confirmed HaSV infection as the cause of stunting observed in the larvae. Further analysis of the progeny of these transformants showed that expression of the viral transgenes was not stable, resulting in loss of the ability to infect larvae with HaSV, and that the inserted genes were unstable, and contained deletions. Examination of further tobacco transformants showed that although some lines contained a complete set of HaSV genes, no plant was found in which all the genes remained active. These results led to the eventual conclusion that virus production was highly unstable and could not be confirmed. Had the HaSV-transgenes also have been modified or inactive in the original transformants analyzed, then the source of infection of the larvae may have been HaSV present at low levels in the insect colony from which these larvae were derived. Bioassays were, however, carefully designed to assess the levels of virus in uninfected control larvae (Gordon *et al.*, 2001).

An alternative approach was to engineer homozygous plants expressing individual HaSV components that could be crossed to yield plants expressing all three genes, and therefore capable of making complete stunt virus. Tobacco plants carrying only the gene for the capsid protein gave expression of this protein (Fig. 3). The maximum expression level for protein detected was at low levels—ca. 0.02% of total soluble plant protein, with most of this being the 64 kDa processed form of the capsid protein. A faint band at p71 was detected, showing the extent of processing to be $\gg$ 95%. This result indicated that all the capsid protein detected was assembled into VLPs, and was supported by demonstration that the protein was resistant to protease digestion (Fig. 3B). Electron microscopic analysis of extracts from these plants bound to the anti-HaSV monoclonal antibody on grids showed the presence of characteristic HaSV particles. Further analysis of capsid protein expression in these plants showed that the gene was not stable in the longer term, possibly due to silencing.

This work showed that transgenic tobacco plants carrying just the gene for the capsid protein could express the protein at levels that although low, still allowed for assembly into virus particles. Moreover, the low levels observed initially were still greater than the levels detected in protoplasts where production of infectious virus was verified by bioassay. Engineering of transgenic plants expressing the two virus genomic RNA strands proved to be more problematic. It was not possible to generate large numbers of plants containing RNA2 alone; this result appeared to be due to interference of the p17 gene with transformation efficiency. Expression of RNA2 was demonstrated indirectly and transiently by Western blotting detection of very low amounts of capsid protein translated from this RNA. No evidence for expression of RNA1 transcripts from transgenes was obtained, suggesting that this gene is particularly prone to silencing in plants. This problem occurred although the RNA1 expression cassette was not included in a tandem or inverted repeat unit and was present only at very low copy numbers in the genome. Silencing likely explains why attempts to express the combination of RNA1 and the capsid gene corresponding to the subvirus in transgenic plants were unsuccessful.

Although the approach has generated interest since first proposed (Hilder, 2003; Ranjekar *et al.*, 2003; Sayyed and Wright, 2002; Service, 1996), these findings indicate that further work is required to achieve stable expression of the HaSV components in transgenic plants. Questions requiring investigation to achieve this aim center on whether HaSV gene expression is subject to silencing by cosuppression. Beyond this, it is likely that the efficiency of the production system would be improved through use of a synthetic capsid gene with codon usage modified to resemble that in plants (R. Drake and

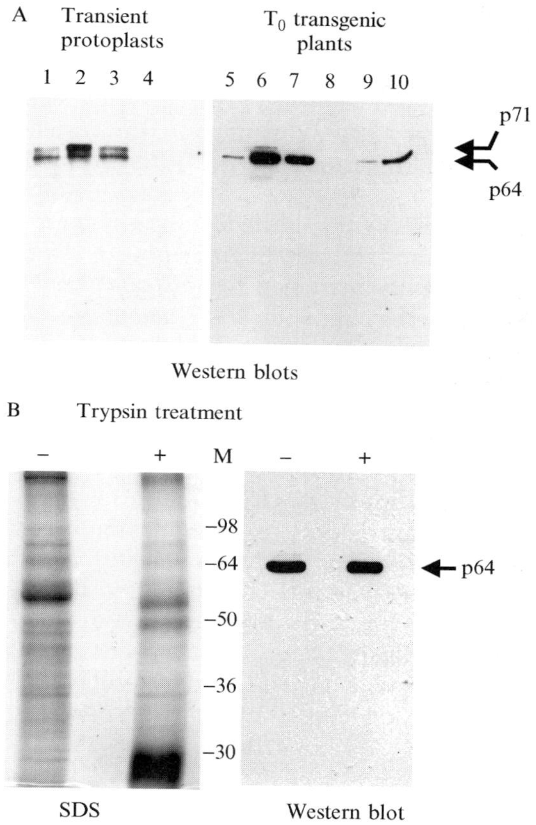

FIG 3. Expression and assembly of HaSV capsid protein in transgenic plants. (A) Western blots showing HaSV capsid protein expressed in transgenic plants and assembly of VLPs. Proteins were separated on a 12% SDS gel, blotted and probed with antiserum against purified HaSV (see Hanzlik *et al.,* 1995 for details). Lanes 1–3 show extracts from protoplasts transfected with plasmids expressing the HaSV capsid protein (see Gordon *et al.,* 2001 for details). Lane 4 is a control protoplast extract. Lanes 5–7 are from three different transgenic plant lines expressing the HaSV capsid protein. Lane 8 is a nontransgenic control, and lanes 9 and 10 contain 1 & 10 μg, respectively of purified HaSV. The positions of the p71 precursor and p64 major capsid protein following assembly-dependent self-cleavage are shown. (B) Coat protein in transgenic plants assembles VLPs that are protease resistant (Gordon, Hanzlik, and Larkin, unpublished). The left two lanes show 20 μg of total soluble protein from transgenic plants expressing the HaSV capsid protein, before (−) and after (+) treatment with protease. The extracts were run on 12% SDS gels and stained with Coomassie blue. The right two lanes show Western blotting of extracts treated in the same way and blotted with the anti-HaSV antiserum. Markers are denoted by M. The position of the p64 main capsid protein following assembly-dependent self-cleavage is shown.

K. Gordon, unpublished data); such a gene would offer two theoretical advantages. Sequence changes that led to altered base composition and reduced secondary structure could lead to less rapid silencing of the gene in plants. Moreover, without nucleotide sequences with extensive homology to the native capsid gene, the transcript should also not be encapsidated, and therefore not interfere with production of infectious virus.

Future work on engineering plants to express insect SRVs is likely to benefit enormously from the use of *Arabidopsis thaliana* as a model system. This plant has been shown to be a good host for a number of major pests, including the lepidopterans *Spodoptera exigua, Spodoptera littoralis, Pseudoplusia includens, Trichoplusia ni, Pieris rapae, Plutella xylostella, Helicoverpa zea, Heliothis virescens*; the coleopterans *Phyllotreta zimmermani* and *Psylliodes convexior*; the homopterans *Brevicoryne brassicae* and *Myzus persicae*; the dipteran *Bradysia impatiens*; and the thysanopteran *Frankliniella occidentalis* (Grant-Peterson, 1993; Grant-Petersson and Renwick, 1996; Mauricio, 1998; McConn *et al.*, 1997; Rashotte and Feldmann, 1996; Reymond *et al.*, 2000; Santos *et al.*, 1997; Singh *et al.*, 1994; Stotz *et al.*, 2000). *Arabidopsis* is also a host for *H. armigera* (East, Larkin, Gordon, and Hanzlik, unpublished data). *Arabidopsis* is not only readily transformed, but benefits from the availability of a complete and well-annotated genome. Furthermore, it is becoming the model system for the study of silencing mechanisms in dicotyledonous plants, and many mutations in silencing genes are available (see in the following section). It will therefore be possible to select the best plants in which to test insect virus assembly and to explore options for overcoming inactivation of such transgenes. Complicating selection of which *Arabidopsis* strain to use for such work is the observation of natural resistance to some insects conferred by the TASTY locus in the Columbia strain (Jander *et al.*, 2001).

G. *Other Approaches for Analysis of SRV Assembly in Plants*

1. *Replication of Nodavirus in Plants*

The first indication that an insect nodavirus could replicate in cells other than insect cells was the finding that FHV could replicate in barley protoplasts and in inoculated leaves of several plant species (Selling *et al.*, 1990). On inoculation with FHV RNA, newly synthesized FHV particles were detected in whole plants of barley, cowpea, chenopodium, tobacco, and *Nicotiana benthamiana*, as well as in protoplasts derived from barley leaves. The virions produced in plants contained newly synthesized RNA as well as capsid protein. These results showed that the intracellular environment in these plants allowed translation of viral RNA, RNA replication, and virion assembly for this insect virus. This study showed that in *N. benthamiana*, virions resulting from inoculation with RNA were detected not only in inoculated leaves but also in other leaves of inoculated plants, suggesting that virions could move in this plant species. Such movement probably occurred by a passive transport through the vascular system rather than by active systemic transport involving mechanisms that have evolved for plant viruses.

The absence of systemic movement of FHV in plants led others to test the ability of movement proteins of plant viruses to provide movement functions for systemic spread of FHV in plants (Dasgupta *et al.,* 2001). The movement proteins (MPs) of *Tobacco mosaic virus* or *Red clover necrotic mosaic virus* (RCNMV) mobilized cell-to-cell and systemic movement of FHV in transgenic *N. benthamiana* plants. Comparison of FHV produced in leaves of nontransgenic and MP-transgenic plants showed that the amount of FHV was more than 100-fold higher in the inoculated leaves of transgenic plants than in the inoculated leaves of nontransgenic plants. FHV also accumulated in the noninoculated upper leaves of both MP-transgenic plants, with the RCNMV MP proving more efficient in mobilizing FHV to noninoculated leaves. These results demonstrated that plant viral MPs could enable cell-to-cell and long-distance movement of an animal virus in plants, suggesting approaches for the development of novel, RNA-virus-derived vectors for transient expression of foreign genes in plants.

2. *BMV Expression Using Agroinfiltration*

To investigate packing of Brome mosaic virus (BMV) RNAs by the capsid protein in the absence of replication, Annamalai and Rao (2005) used a T-DNA-based *Agrobacterium*-mediated transient expression system (termed agroinfiltration, Schob *et al.,* 1997) in *N. benthamiana* leaves to express either individual or desired pairs of the three genomic RNAs. Either individual or desired pairs of the three genomic RNAs were expressed and the packaging of these RNAs into virions by the transiently expressed coat protein (CP) analyzed. In the absence of a functional replicase, assembled virions contained nonreplicating viral RNAs (RNA1 or RNA2 or RNA3 or RNA1 + RNA3, or RNA2 + RNA3) as well as cellular RNAs. By contrast, virions assembled in the presence of a functional replicase contained only viral RNAs. These workers concluded that packaging of BMV genomic RNAs is not replication-dependent, whereas expression of a functional viral replicase plays an active role in increasing the specificity of RNA packaging.

IV. RNA Silencing in Plants and the Effect on Expression of Transgenes

Replication of an RNA virus during the infection of its plant host induces RNA silencing against the viral genome, resulting in the accumulation of viral siRNAs. This siRNA-generating pathway (Fig. 4) is probably a natural antiviral defense mechanism in plants (Waterhouse *et al.*, 2001), although some of the details of the biogenesis of the viral siRNAs are yet to be fully resolved. A key enzyme in the pathway is an RNaseIII-like enzyme, called Dicer (DCR) that processes dsRNA into siRNAs. The dsRNA replication intermediates of a virus are probably the substrates for this siRNA production, although direct processing of duplex structures formed within single-stranded viral RNAs could also contribute to the siRNA pool. There is good evidence that a plant-encoded RdRP is also involved in antiviral defense (Mourrain *et al.*, 2000; Muangsan *et al.*, 2004) suggesting that RdRP-mediated synthesis of secondary viral dsRNA also plays a role in viral siRNA accumulation. Mammals seem to have only one Dcr gene, and most insects and fungi possess two. However, plants have a basic set of four dicer-like genes (Dcl1 to Dcl4) with some species, such as rice, having as many as six genes (Margis *et al.*, 2006; Watson *et al.*, 2005). The DCL proteins are probably associated with a set of five cofactor dsRNA-binding proteins (DRB1 to DRB5). DCL1 produces miRNAs, which regulate plant development, from hairpin-like precursor RNAs. DCL3 produces ~24 nt siRNAs that suppress the activation of transposable elements in the plant genome by directing chromatin structure modifications (Xie *et al.*, 2004). DCL2 is reported to be involved in the production of siRNAs from a replicating virus (Xie *et al.*, 2004), although DCL4 appears to be the major player in this process (Fusaro *et al.*, 2006). The mi- and siRNAs produced by DCL1, -2, and, -4 are used to guide protein complexes (RISC), containing the key endonuclease, Argonaute, to cognate single-stranded RNAs. Once the target-RNAs are cleaved by the RISC they are rapidly degraded by nonspecific nucleases.

The presence of the DCL2/DCL4-mediated viral defense pathway in plants is possibly a major reason why attempts to express insect SRVs in plants have been largely unsuccessful. These attempts have a number of parallels with one of the early descriptions of unintended gene

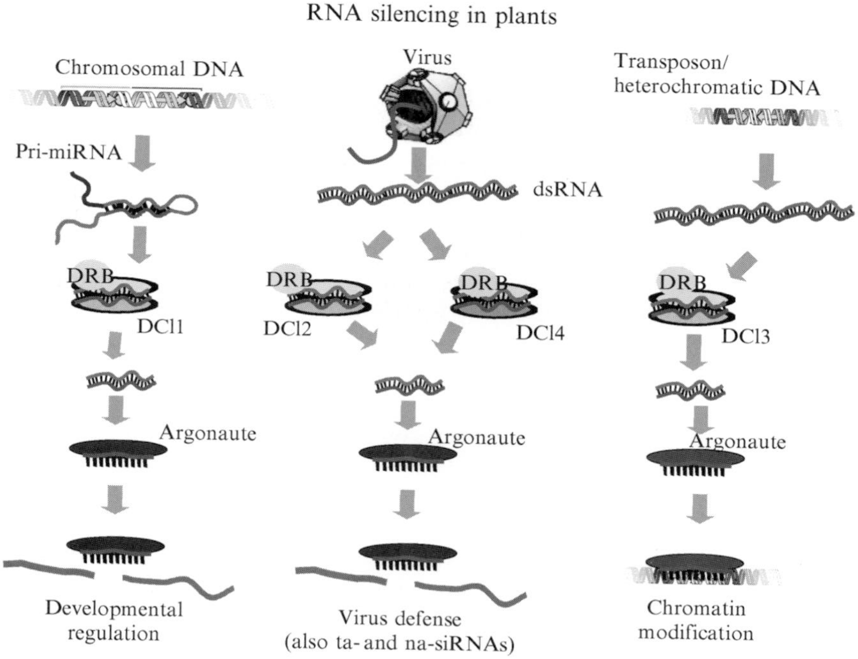

FIG 4. Pathways leading to RNA silencing in plants. The miRNA pathway is shown on the left, the antiviral RNA-silencing pathway in the center, and the RNAi pathway leading to chromatin modification on the right. The miRNA pathway processes transcripts from genes yielding Pri-miRNA precursors; these are processed by DCL1 and the resulting 21–24 nt miRNAs guide the RNA-induced silencing complex (RISC) to the appropriate mRNA for miRNA-directed cleavage and degradation. RISC contains Argonaute, the central endoribonuclease involved in RNA silencing. For silencing of viral sequences, dsRNAs derived from the virus are cleaved by DCL2 or DCL4 (see text) and yield 21 nt siRNAs that guide the RISC to target viral sequences. *Trans*-acting (ta-) and na-siRNAs (see text) operate by similar mechanisms. Transcripts from highly repetitive heterochromatin (e.g., retrotransposons) are processed by DCL3 to yield 24–26 nt siRNAs that then direct intranuclear complexes containing a different member of the Argonaute family (AGO-4) to the chromatin sequence itself for modification (e.g., methylation). (See Color Insert.)

silencing in plants (Angell and Baulcombe, 1997). In this study, the researchers initially set out to overexpress transgenes in plants using a DNA construct encoding a plant virus, *Potato virus X* (PVX). The construct was composed of the strong 35S promoter directing transcription of the entire PVX genome, into which had been inserted the coding region of the β-glucuronidase (GUS) reporter gene. The DNA construct was transformed into tobacco plants with the expectation that the RNA transcribed from it would generate a replicating PVX virus that would, in turn, produce high levels of GUS expression. What happened was almost the complete opposite. The viral transgene was transcribed into RNA, which initially replicated and expressed GUS, but the plants showed no symptoms of virus infection, the GUS expression and PVX transcript levels reduced, and the plants were resistant to super-infection by PVX. It now seems clear that the replicating transgene-encoded virus triggered the DCL2/4 pathway which produced siRNAs against the virus and its embedded GUS sequences. These, in turn, guided the degradation of the virus and GUS RNA, and the genomic RNA of any super-inoculated PVX.

From the PVX work, it seems likely that insect SRV constructs transformed into plants will suffer the same fate. If the SRV replicates it will induce the DCL2/4 pathway and bring about its own destruction. Even for a virus such as HaSV that does not appear to replicate in any cultured cells, a single round of dsRNA production may be sufficient to trigger silencing. One possible way to overcome the induction of this pathway would be to ensure that the transgene-encoded SRV is unable to replicate in the plant to generate dsRNA. This may be difficult as the SRV genome, which has to be functional in the insect, may always have at least a limited degree of replication in plants. Even if it is possible to alter the SRV genome sequence so that it is functional in an insect but not in a plant, it will also be important to select for plants in which the DNA construct has integrated as a single copy into a single locus as there appears to be a direct correlation between induction of RNA silencing and integration of transgenes that have occurred as inverted repeats (Muskens *et al.*, 2000; Wang and Waterhouse, 2000). Presumably this is because read-through transcription produces a hairpin RNA that initiates the defense pathway. However, there are also examples where the pathway is induced by a single transgene insertion (Lechtenberg *et al.*, 2003); in these cases it is thought that the transcript is somehow triggering a host-encoded RdRP to generate a complementary RNA strand. Such a plant response is likely to have led to the transcriptional silencing of HaSV sequences that were not in themselves capable of replication, for example, those encoding the capsid gene only.

There are at least two further strategies that may allow the expression and assembly of SRVs in plants. Both strategies accept that the integration of the SRV constructs will produce dsRNA that can activate the DCL2/4 pathway but take actions to prevent the pathway from destroying the SRV transcripts. One strategy employs the expression of proteins to suppress this pathway, the other uses plants with mutations in the genes encoding the pathway.

Plant viruses infecting their hosts activate the RNA-silencing pathway, but almost all of them express some sort of suppressor protein that negates this response (Chapman *et al.*, 2004). Potyviruses encode a protein, HC-Pro that interacts with RISC to prevent it from cleaving the target RNA. The p21, p19, and 2B proteins from *Beet yellows virus, Tomato bushy stunt virus*, and *Cucumber mosaic virus* (CMV), respectively, also do not prevent the production of the siRNAs but rather interfere with their action. The coat protein of *Turnip crinkle virus* appears to prevent the action of DCL2 in generating siRNAs, and the ORF0 protein of poleroviruses seems to act as an F-Box protein in targeting the ubiquitin-mediated protein degradation pathway against one or more proteins involved in the RNA-silencing pathway (Pazhouhandeh *et al.*, 2006). One of these viral suppressor proteins could be produced in plants from a transgene alongside the transgenes encoding the SRV genomic RNA and coat protein. Expression of the suppressor protein may prevent siRNAs from being produced against the SRV RNA or impair their effectiveness. However, there is a major drawback associated with this approach. Transgenic expression of all of these suppressor proteins, except the CMV 2B protein, cause severe developmental defects in the plant because most, if not all of them, also interfere with the action of miRNAs. Transgenic expression of CMV 2B protein might be a possible solution, except that this is the weakest suppressor of RNA silencing and may not provide sufficient protection for the SRV RNA.

An alternative strategy is to put SRV encoding transgenes into plants in which one or more of the genes involved in the RNA silencing pathways have been mutated. This approach is especially attractive as the virus defense pathway can be nullified without affecting the main DCL1-mediated miRNA pathway and hence has less impact on the development of the plant. Under glasshouse conditions, *Arabidopsis* plants mutant for Dcl2 have little or no obviously deleterious phenotype. *Arabidopsis* plants mutant for Dcl4 have an accelerated vegetative phase change (Xie *et al.*, 2005), which is caused because DCL4 also processes an endogenous dsRNA into *trans*-acting (ta)-siRNAs that regulate this phase change, but this is a relatively mild phenotype. *Arabidopsis* plants triple mutant for dcl2, dcl3, and dcl4 are fertile plants that have only the dcl4 mutant phenotype (Fig. 5). One drawback of this approach is that DCL2 is involved in production of a newly discovered class of siRNAs, called na-siRNAs that enable the plant to respond to environmental stresses such as high salt conditions

FIG 5. Dicer-deficient *Arabidopsis* plants. Phenotypes of *A. thaliana* ecotype Columbia, carrying knockout mutations in the dicer genes. Top row (from left): wildtype; single dicer-like 2 (dcl2) mutant; single dcl4 mutant. Bottom row (from left): double dcl2 and -4 mutant; triple dcl2, -3, and -4 mutant.

(Borsani *et al.*, 2005). The plants would also be expected to be more susceptible to viral infection. Therefore, SRV-producing plants with a dcl2/dcl4 mutant background may not perform so well in field conditions. However, the availability of these RNA-silencing mutant *Arabidopsis* lines makes the practicality of expressing functional SRV in these different backgrounds easily testable.

V. Conclusions

There is an irony about the fate of the first attempt to harness an insect SRV for pest control based on the engineering of transgenic plants. The same host adaptive immune responses encountered by the virus in its natural host are also to be met in plants. While the tetravirus HaSV has evolved the ability to deal with RNA silencing in its natural host, its biological characteristics have not armed it for survival in plants, even if the latter only requires nonreplicative virus production from transgenes. It may be possible to apply the accumulating knowledge about plant RNA-based defense systems (whose existence was scarcely suspected when the first experiments with HaSV were undertaken), to allow for production of HaSV in DCR mutants of *Arabidopsis* for example. However, it is moot that this will actually achieve a practical approach for pest control because plants able to express the complete virus may well be rendered vulnerable to other pathogens that threaten crops in the field.

As a further irony, the existence of the RNA-based immune systems may provide an opportunity to use SRVs as agents to trigger another immune response—not against themselves, but against genes central to the insect's survival. Demonstration that an insect SRV assembled in a plant cell can deliver a nonviral, biologically active RNA opens the possibility that such VLPs could deliver dsRNAs designed to silence genes in the target insect pest. This approach is likely to lead to renewed interest in SRV VLPs for plant protection. If feasible, this strategy would avoid the regulatory concerns and RNA-silencing problems that are currently associated with expression of viral genomes in transgenic plants.

Acknowledgments

We thank our colleagues, especially Terry Hanzlik, Phil Larkin, and Jean-Louis Zeddam, with whom we have worked on insect SRVs and RNA silencing, for access to unpublished material and valuable discussions. The work on HaSV described in this chapter was supported by Syngenta, the Cotton Research and Development Corporation, and the Grains Research and Development Corporation.

References

Adelman, Z. N., Blair, C. D., Carlson, J. O., Beaty, B. J., and Olson, K. E. (2001). Sindbis virus-induced silencing of dengue viruses in mosquitoes. *Insect Mol. Biol.* **10:**265–273.

Agrawal, D. K., and Johnson, J. E. (1992). Sequence and analysis of the capsid protein of Nudaurelia capensis omega virus, an insect virus with $T = 4$ icosahedral symmetry. *Virology* **190:**806–814.

Agrawal, D. K., and Johnson, J. E. (1995). Assembly of the $T = 4$ *Nudaurelia capensis* ω virus capsid protein, post-translational cleavage, and specific encapsidation of its mRNA in a baculovirus expression system. *Virology* **207:**89–97.

Amdam, G. V., Simoes, Z. L., Guidugli, K. R., Norberg, K., and Omholt, S. W. (2003). Disruption of vitellogenin gene function in adult honeybees by intra-abdominal injection of double-stranded RNA. *BMC Biotechnol.* **3:**1.

Anderson, E., and Harvey, W. R. (1966). Active transport by the Cecropia midgut. II Fine structure of the midgut epithelium. *J. Cell Biol.* **31:**107–134.

Angell, S. M., and Baulcombe, D. C. (1997). Consistent gene silencing in transgenic plants expressing a replicating potato virus X RNA. *EMBO J.* **16:**3675–3684.

Annamalai, P., and Rao, A. L. (2005). Replication-independent expression of genome components and capsid protein of brome mosaic virus in planta: A functional role for viral replicase in RNA packaging. *Virology* **338:**96–111.

Baldwin, K. M., and Hakim, R. S. (1991). Growth and differentiation of the larval midgut epithelium during moulting in the moth, *Manduca sexta*. *Tissue & Cell* **23:**411–422.

Baldwin, K. M., Hakim, R. S., Loeb, M. J., and Sadrud-Din, S. Y. (1996). Midgut development. *In* "Biology of the Insect Midgut" (S. Y. Lehane and M. J. Billingsley, eds.), pp. 31–54. Chapman & Hall, London.

Ball, L. A., and Johnson, K. L. (1998). Nodaviruses. *In* "The Insect Viruses" (K. L. Miller and L. K. Ball, eds.), pp. 225–267. Plenum Press, London.

Ball, L. A., and Johnson, K. L. (1999). Reverse genetics of nodaviruses. *Adv. Virus Res.* **53:**229–244.

Bawden, A. L., Gordon, K. H. J., and Hanzlik, T. N. (1999). The specificity of the Helicoverpa armigera stunt virus infectivity. *J. Invertebr. Pathol.* **74:**156–163.

Benjeddou, M., Leat, N., Allsopp, M., and Davison, S. (2002). Development of infectious transcripts and genome manipulation of Black queen-cell virus of honey bees. *J. Gen. Virol.* **83:**3139–3146.

Beye, M., Hartel, S., Hagen, A., Hasselmann, M., and Omholt, S. W. (2002). Specific developmental gene silencing in the honeybee using a homeobox motif. *Insect Mol. Biol.* **11:**527–532.

Beye, M., Hasselmann, M., Fondrk, M. K., Page, R. E., and Omholt, S. W. (2003). The gene csd is the primary signal for sexual development in the honeybee and encodes an SR-type protein. *Cell* **114:**419–429.

Billingsley, P. F., and Lehane, M. J. (1996). Structure and ultrastructure of the insect midgut. *In* "Biology of the Insect Midgut" (M. J. Lehane and M. J. Billingsley, eds.), pp. 3–29. Chapman & Hall, London.

Bong, C. F. J., and Sikorowski, P. P. (1991). Histopathology of cytoplasmic polyhedrosis virus (Reoviridae) infection in corn earworm, *Helicoverpa zea* (Boddie), larvae (Insecta: Lepidoptera: Noctuidae). *Can. J. Zool.* **69:**2127–2130.

Booth, T. F., Bonning, B. C., and Hammock, B. D. (1992). Localization of juvenile hormone esterase during development in normal and recombinanat baculovirus infected larvae of the moth *Trichoplusiani*. *Tissue and Cell* **24:**267–282.

Borsani, O., Zhu, J., Verslues, P. E., Sunkar, R., and Zhu, J. K. (2005). Endogenous siRNAs derived from a pair of natural *cis*-antisense transcripts regulate salt tolerance in *Arabidopsis*. *Cell* **123:**1279–1291.

Bothner, B., Taylor, D., Jun, B., Lee, K. K., Siuzdak, G., Schultz, C. P., and Johnson, J. E. (2005). Maturation of a tetravirus capsid alters the dynamic properties and creates a metastable complex. *Virology* **334:**17–27.

Bourner, T. C., Glare, T. R., O'Callaghan, M., and Jackson, T. A. (1996). Towards greener pastures—pathogens and pasture pests. *NZJ Ecol.* **20:**101–107.

Boyer, J.-C., and Haenni, A.-L. (1994). Infectious transcripts and cDNA clones of RNA viruses. *Virology* **198:**415–426.

Bräutigam, S., Snezhkov, E., and Bishop, D. H. (1993). Formation of poliovirus-like particles by recombinant baculoviruses expressing the individual VP0, VP3, and VP1 proteins by comparison to particles derived from the expressed poliovirus polyprotein. *Virology* **192:**512–524.

Brooks, E. M., Gordon, K. H. J., Dorrian, S. J., Hines, E. R., and Hanzlik, T. N. (2002). Infection of its lepidopteran host by the *Helicoverpa armigera stunt virus* (*Tetraviridae*). *J. Invertebr. Pathol.* **80:**97–111.

Brown, S. J., Denell, R. E., and Beeman, R. W. (2003). Beetling around the genome. *Genet. Res. Camb.* **82:**155–161.

Bucher, G., Scholten, J., and Klingler, M. (2002). Parental RNAi in *Tribolium* (Coleoptera). *Curr. Biol.* **12:**R85–R86.

Canady, M. A., Tihova, M., Hanzlik, T. N., Johnson, J. E., and Yeager, M. (2000). Large conformational changes in the maturation of a simple RNA virus, Nudaurelia capensis omega virus. *J. Mol. Biol.* **299:**573–584.

Chao, J. A., Lee, J. H., Chapados, B. R., Debler, E. W., Schneemann, A., and Williamson, J. R. (2005). Dual modes of RNA-silencing suppression by Flock house virus protein B2. *Nat. Struct. Mol. Biol.* **12:**952–957.

Chapman, E. J., Prokhnevsky, A. I., Gopinath, K., Dolja, V. V., and Carrington, J. C. (2004). Viral RNA silencing suppressors inhibit the microRNA pathway at an intermediate step. *Genes Dev.* **18:**1179–1186.

Chen, C.-Z., Lodish, H. F., and Bartel, D. P. (2004). MicroRNAs modulate hematopoietic lineage differentiation. *Science* **303:**83–86.

Choi, H. K., Kobayashi, M., and Kawase, S. (1989). Changes in infectious flacherie virus specific polypeptides and translateable m-RNA in the midgut of the silkworm, *Bombyx mori*, during larval molt. *J. Invertebr. Pathol.* **53:**128–131.

Christian, P. D., and Scotti, P. D. (1998). Picornalike viruses of insects. *In* "The Insect Viruses" (P. D. Miller and L. K. Ball, eds.), pp. 301–336. Plenum Press, London.

Christian, P. D., Hanzlik, T. N., Dall, D. J., and Gordon, K. H. (1993). Insect viruses: New strategies for pest control. *In* "Molecular Approaches to Fundamental and Applied Entomology" (K. H. Oakeshott and J. Whitten, eds.), pp. 128–163. Springer-Verlag, New York.

Christian, P. D., Dorrian, S. J., Gordon, K. H. J., and Hanzlik, T. N. (2001). Pathogenicity and characteristics of the *Helicoverpa armigera* stunt virus. *J. Biol. Control* **20:**65–75.

Cioffi, M. (1979). The morphology and fine structure of the larval midgut of a moth (Manduca sexta) in relation to active ion transport. *Tissue and Cell* **11:**467–479.

Clem, R. J. (2005). The role of apoptosis in defense against baculovirus infection in insects. *Curr. Top. Microbiol. Immunol.* **289:**113–129.

Conzelmann, K.-K., and Meyers, G. (1996). Genetic engineering of animal RNA viruses. *Trends Microbiol.* **4:**386–393.

Culley, A. I., Lang, A. S., and Suttle, C. A. (2003). High diversity of unknown picorna-like viruses in the sea. *Nature* **424:**1054–1057.

Dasgupta, R., Garcia, B. H., and Goodman, R. M. (2001). Systemic spread of an RNA insect virus in plants expressing plant viral movement protein genes. *Proc. Natl. Acad. Sci. USA* **98:**4910–4915.

DeMiranda, J. R., and Gordon, K. H. J. (2006). Novel viruses as a by-product from EST databases. Manuscript in preparation.

Denli, A. M., and Hannon, G. J. (2003). RNAi: An ever-growing puzzle. *Trends Biochem. Sci.* **28:**196–201.

Desmier de Chenon, R., Mariau, D., Monsarrat, P., Fédière, G., and Sipayung, A. (de Chenon 1988). Research into entomopathogenic agents of viral origin in leaf-eating Lepidoptera of the oil palm and coconut. *Oléagineux* **43:**107–117.

Ding, S. W., Li, H., Lu, R., Li., F., and Li, W. X. (2004). RNA silencing: A conserved antiviral immunity of plants and animals. *Virus Res.* **102:**109–115.

Dong, Y., and Friedrich, M. (2005). Nymphal RNAi: Systemic RNAi mediated gene knockdown in juvenile grasshopper. *BMC Biotechnol.* **5:**25.

Dow, J. A. T. (1986). Insect midgut function. *Adv. Insect. Physiol.* **19:**188–303.

Endo, Y., and Nishiitsutsuji-Uwo, J. (1981). Gut endocrine cells in insects: The ultrastructure of the gut endocrine cells of the lepidopterous species. *Biomed. Res.* **2:**270–280.

Engelhard, E. K., and Volkman, L. E. (1995). Developmental resistance in fourth instar *Trichoplusia ni* orally inoculated with *Autographica californica* M nuclear polyhedrosis virus. *Virology* **209:**384–389.

Engelhard, E. K., Keddie, B. A., and Volkman, L. E. (1991). Isolation of third, fourth and fifth instar larval midgut epithelia of the moth *Trichoplusia ni*. *Tissue and Cell* **23:**917–928.

Farooqui, T., Vaessin, H., and Smith, B. H. (2004). Octopamine receptors in the honeybee (Apis mellifera) brain and their disruption by RNA-mediated interference. *J. Insect Physiol.* **50:**701–713.

Fauquet, C. M., Mayo, M. A., Maniloff, J., Desselberger, U., and Ball, L. A. (eds.) (2005). "Virus Taxonomy, VIIIth Report of the ICTV." Elsevier/Academic Press, London.

Federici, B. A. (1997). Baculovirus pathogenesis. *In* "The Baculoviruses" (B. A. Miller, ed.), pp. 33–59. Plenum Publ. Corp., New York.

Federici, B. A., and Stern, V. M. (1990). Replication and occulusion of a granulosis virus in larval and adult midgut epithelium of the Western Grapeleaf Skeleltonizer, *Harrisina brillians*. *J. Invertebr. Pathol.* **56:**401–414.

Flipsen, J. T. M., Marten, J. W. M., van Oers, M. M., Vlak, J. M., and van Lent, J. W. M. (1995). Passage of the *Autographica californica* nuclear polyhedrosis virus through the midgut epithelium of *Spodoptera exigua* larvae. *Virology* **208:**328–335.

Fujiyuki, T., Takeuchi, H., Ono, M., Ohka, S., Sasaki, T., Nomoto, A., and Kubo, T. (2004). Novel insect picorna-like virus identified in the brains of aggressive worker honeybees. *J. Virol.* **78:**1093–1100.

Fusaro, A. F., Matthew, L., Smith, N. A., Curtin, S. J., Dedic-Hagan, J., Ellacott, G. A., Watson, J. M., Wang, M.-B., Brosnan, C., Carroll, B. J., and Waterhouse, P. M. (2006). RNAi-inducing hairpin RNAs in plants act through the viral defence pathway. *EMBO Reports* (in press).

Ghosh, R. C., Ball, B. V., Willcocks, M. M., and Carter, M. J. (1999). The nucleotide sequence of sacbrood virus of the honeybee: An insect picorna-like virus. *J. Gen. Virol.* **80:**1541–1549.

Ginting, C. U., and Desmier de Chenon, R. (1987). Nouvelles perspectives biologiques pour le contrôle d'un ravageur très important du cocotier en Indonésie: *Parasa lepida* Cramer, Limacodidae par l'utilisation de virus. *Oléagineux* **42**(3)**:**107–115.

Goldbach, R. W. (1986). Molecular evolution of plant RNA viruses. *Ann. Rev. Phytopathol.* **24:**289–310.

Gorbalenya, A. E., Pringle, F. M., Zeddam, J.-L., Luke, B. T., Cameron, C. E., Kalmakoff, J., Hanzlik, T. N., Gordon, K. H. J., and Ward, V. K. (2002). The palm subdomain-based active site is internally permuted in viral RNA-dependent RNA polymerases of an ancient lineage. *J. Mol. Biol.* **324:**47–62.

Gordon, K. H. J., and Hanzlik, T. N. (1998). Tetraviruses. *In* "The Insect Viruses" (L. K. Miller and L. K. Ball, eds.), pp. 269–299. Plenum Press, London.

Gordon, K. H. J., Johnson, K. N., and Hanzlik, T. N. (1995). The larger genomic RNA of Helicoverpa armigera stunt tetravirus encodes the viral RNA polymerase and has a novel 3′-terminal tRNA-like structure. *Virology* **208:**84–98.

Gordon, K. H. J., Williams, M. R., Hendry, D. A., and Hanzlik, T. N. (1999). Sequence of the genomic RNA of Nudaurelia beta virus (*Tetraviridae*) defines a novel virus genome organization. *Virology* **258:**42–53.

Gordon, K. H. J., Williams, M. R., Baker, J. S., Gibson, J., Bawden, A. L., Millgate, A., Larkin, P. J., and Hanzlik, T. N. (2001). Replication-independent assembly of an insect virus (*Tetraviridae*) in plant cells. *Virology* **288:**36–50.

Gordon, K. H. J., Gorbalenya, A. E., Hanzlik, T. N., Hendry, D. A., Pringle, F. M., Ward, V. K., and Zeddam, J.-L. (2005). *Tetraviridae*: Taxonomic structure of the family. *In* "Virus Taxonomy, VIIIth Report of the International Committee on Taxonomy of Viruses" (J.-L. Fauquet, C. M. Mayo, M. A. Maniloff, J. Desselberger, and U. Ball, eds.), Elsevier/Academic Press, London.

Grace, T. D. C., and Mercer, E. H. (1965). A new virus of the Saturniid *Antheraea eucalypti* Scott. *J. Invertebr. Pathol.* **7:**241–244.

Grant-Peterson, J. (1993). The effect of allelochemical differences in *Arabidopsis thaliana* on the responses of the herbivores *Trichoplusia ni* and *Pieris rapae*. PhD thesis. Department of Entomology, Cornell University, Ithaca, NY.

Grant-Petersson, J., and Renwick, J. A. A. (1996). Effects of ultraviolet B exposure of *Arabidopsis thaliana* on herbivory by two crucifer-feeding insects. *Environ. Entomol.* **25:**135–142.

Greenwood, L. K., and Moore, N. F. (1984). Determination of the location of an infection in *Trichoplusia ni* larvae by a small RNA-containing virus using enzyme-linked immunosorbent assay and electron microscopy. *Microbiologica* **7:**97–102.

Guidugli, K. R., Nascimento, A. M., Amdam, G. V., Barchuk, A. R., Omholt, S., Simoes, Z. L., and Hartfelder, K. (2005). Vitellogenin regulates hormonal dynamics in the worker caste of a eusocial insect. *FEBS Lett.* **579:**4961–4965.

Hakim, R. S., Baldwin, K. M., and Loeb, M. J. (2001). The role of stem cells in midgut growth and differentiation. *In vitro Cell. Dev. Biol.—Animal* **37:**338–342.

Hanzlik, T. N., and Gordon, K. H. J. (1998). Beyond Bt: Plants with stunt virus genes. *In* "Pest Management-Future Challenges: Proceedings of Sixth Australasian Applied Entomological Research Conference" (K. H. J. Zalucki, M. Drew, and R. White, eds.), pp. 206–214. University of Queensland, Brisbane.

Hanzlik, T. N., Dorrian, S. J., Gordon, K. H. J., and Christian, P. D. (1993). A novel small RNA virus isolated from the cotton bollworm, *Helicoverpa armigera*. *J. Gen. Virol.* **74:**1105–1110.

Hanzlik, T. N., Dorrian, S. J., Johnson, K. N., Brooks, E. M., and Gordon, K. H. J. (1995). Sequence of RNA2 of the *Helicoverpa armigera stunt virus* (*Tetraviridae*) and bacterial expression of its genes. *J. Gen. Virol.* **76:**799–811.

Hanzlik, T. N., Zeddam, J.-L., Gordon, K. H. J., and Christian, P. D. (1999). A new view of small RNA viruses of insects. *Pesticide Outlook* **10:**22–26.

Hanzlik, T. N., Campbell, P. M., Masoumi, A., Ewins, J., and Gordon, K. H. J. (2006a). Structural and genomic organization of the *Canberra Cryptic Virus,* a novel insect RNA virus belonging to the picorna-like virus superfamily. Submitted to *Virology.*

Hanzlik, T. N., Dorrian, S. J., Williams, M. R., Brooks, E. M., Hines, E. J., Masoumi, A., Larkin, P. J., and Gordon, K. H. J. (2006b). A novel gene, derived from a virus, for transfecting and expressing RNA in gut cells of an insect. Submitted to *Nature Biotech.*

Hartley, C. R., Greenwood, D. R., Gilbert, R. J. C., Masoumi, A., Gordon, K. H. J., Hanzlik, T. N., Fry, E. E., Stuart, D. I., and Scotti, P. D. (2005). Kelp fly virus: An insect small RNA virus with a unique genome, novel particle structure and unusual functional capabilities. *J. Virol.* **79:**13385–13398.

Hatfield, S. D., Shcherbata, H. R., Fischer, K. A., Nakahara, K., Carthew, R. W., and Ruohola-Baker, H. (2005). Stem cell division is regulated by the microRNA pathway. *Nature* **435:**974–978.

Helgstrand, C., Munshi, S., Johnson, J. E., and Liljas, L. (2004). The refined structure of Nudaurelia capensis omega virus reveals control elements for a T = 4 capsid maturation. *Virology* **318:**192–203.

Hendry, D. A. (1991). Nodaviridae of invertebrates. *In* "Atlas of Invertebrate Viruses" (J. R. Adams and J. R. Bonami, eds.), pp. 227–276. CRC Press, Boca Raton.

Hilder, V. (2003). GM plants and protection against insects: Alternative strategies based on gene technology. *Acta Agric. Scand., Sect. B Soil Plant Sci.* **53**(Suppl. 1)**:**34–40.

Houbaviy, H. B., Murray, M. F., and Sharp, P. A. (2003). Embryonic stem cell-specific microRNAs. *Developmental Cell* **5:**351–358.

Huang, Z., Elkin, G., Maloney, B. J., Beuhner, N., Arntzen, C. J., Thanavala, Y., and Mason., H. S. (2005). Virus-like particle expression and assembly in plants: Hepatitis B and Norwalk viruses. *Vaccine* **23:**1851–1858.

Inoue, H., and Miyagawa, M. (1978). Regeneration of midgut epithelial cells in the silkworm, *Bombyx mori*, infected with viruses. *J. Invertebr. Pathol.* **32:**373–380.

Isawa, H., Asano, S., Sahara, K., Lizuka, T., and Bando, H. (1998). Analysis of genetic information of an insect picorna-like virus, infectious flacherie virus of silkworm: Evidence for evolutionary relationships among insect, mammalian and plant picorna (-like) viruses. *Arch. Virol.* **143:**127–143.

Iwamoto, T., Mise, K., Mori, K.-I., Arimoto, M., Nakai, T., and Okuno, T. (2001). Establishment of an infectious RNA transcription system for Striped jack nervous necrosis virus, the type species of the betanodaviruses. *J. Gen. Virol.* **82:**2653–2662.

Jander, G., Cui, J., Nhan, B., Pierce, N. E., and Ausubel, F. M. (2001). The TASTY locus on chromosome 1 of *Arabidopsis* affects feeding of the insect herbivore *Trichoplusia ni*. *Plant Physiol.* **126:**890–898.

Johnson, K. L., Price, B. D., Eckerle, L. D., and Ball, L. A. (2004). Nodamura virus nonstructural protein B2 can enhance viral RNA accumulation in both mammalian and insect cells. *J. Virol.* **78:**6698–6704.

Johnson, K. N., Zeddam, J.-L., and Ball, L. A. (2000). Characterization and construction of functional cDNA clones of Pariacoto virus, the first *Alphanodavirus* isolated outside Australasia. *J. Virol.* **74:**5123–5132.

Kanellopoulou, C., Muljo, S. A., Kung, A. L., Ganesan, S., Drapkin, R., Jenuwein, T., Livingston, D. M., and Rajewsky, K. (2004). Dicer-deficient mouse embryonic stem cells are defective in differentiation and centromeric silencing. *Genes Devel.* **19:**489–501.

Keddie, B. A., Aponte, G. W., and Volkman, L. E. (1989). The pathway of infection of *Autographica californica nuclear polyhedrosis virus* in an insect host. *Science* **243:**1728–1730.

Kennerdell, J. R., and Carthew, R. W. (2000). Heritable gene silencing in *Drosophila* using double-stranded RNA. *Nat. Biotechnol.* **18:**896–898.

Kinnbauer, R., Taub, J., Greenstone, H., Roden, R., Dürst, M., Gissmann, L., Lowry, D. R., and Schiller, J. T. (1993). Efficient self-assembly of human papillomavirus type 16 L1 and L1-L2 into virus-like particles. *J. Virol.* **67:**6929–6936.

Koonin, E. V., and Dolja, V. V. (1993). Evolution and taxonomy of positive-strand RNA viruses: Implications of comparative analysis of amino acid sequences. *Crit. Rev. Biochem. Mol. Biol.* **28:**375–430.

Lai, C. J., Zhao, B., Hori, H., and Bray, M. (1991). Infectious RNA transcribed from stably cloned full-length cDNA of dengue type 4 virus. *Proc. Natl. Acad. Sci. USA* **88:** 5139–5143.

Larkin, P. J., Gordon, K. H. J., and Hanzlik, T. N. (1996). Virally-armed plants: New genetic defence against insects. *In* "Proceedings of the Third Asia-Pacific Conference on Agricultural Biotechnology: Issues and choices," pp. 29–32. Hua Hin Prachuap-khirikhan, Thailand: National Center for Genetic Engineering and Biotechnology. ISBN 9747-7577-52-6.

Lechtenberg, B., Schubert, D., Forsbach, A., Gils, M., and Schmidt., R. (2003). Neither inverted repeat T-DNA configurations nor arrangements of tandemly repeated transgenes are sufficient to trigger transgene silencing. *Plant J.* **34:**507–517.

Le Gall-Reculé, G., Jestin, V., Chagnaud, P., Blanchard, P., and Jestin, A. (1996). Expression of muscovy duck parvovirus capsid proteins (VP2 and VP3) in a baculovirus expression system and demonstration of immunity induced by the recombinant proteins. *J. Gen. Virol.* **77:**2159–2163.

Li, H. W., and Ding, S. W. (2005). Antiviral silencing in animals. *FEBS Lett.* **579:** 5965–5973.

Li, H. W., Li, W. X., and Ding, S. W. (2002). Induction and suppression of RNA silencing by an animal virus. *Science* **296:**1319–1321.

Li, W. X., and Ding, S. W. (2001). Viral suppressors of RNA silencing. *Curr. Opin. Biotech.* **12:**150–154.

Li, W. X., Li, H., Lu, R., Li, F., Dus, M., Atkinson, P., Brydon, E. W., Johnson, K. L., Garcia-Sastre, A., Ball, L. A., Palese, P., and Ding, S. W. (2004). Interferon antagonist proteins of influenza and vaccinia viruses are suppressors of RNA silencing. *Proc. Natl. Acad. Sci. USA* **101:**1350–1355.

Liu, C., Shuai, X. R., Cheng, T. C., Xu, H. F., Li, C. F., Dai, F. Y., Xia, Q. Y., and Xiang, Z. H. (2004). Study on the RNA interference of Bombyx mori embryo. *Prog. Biochem. Biophys.* **31:**322–327.

Loeb, M. J., and Hakim, R. S. (1996). Insect midgut epithelium *in vitro*: An insect stem cell system. *J. Insect Physiol.* **42:**1103–1111.

Mans, R. M. W., Pleij, C. W. A., and Bosch, L. (1991). tRNA-like structures: Structure, function and evolutionary significance. *Eur. J. Biochem.* **201:**303–324.

Margis, R., Fusaro, A. F., Smith, N. A., Finnegan, E. J., and Waterhouse, P. M. (2006). The evolution and diversification of Dicers in plants. *FEBS Lett.* **580:**2442–2450.

Mari, J., Poulos, B. T., Lightner, D. V., and Bonami, J. R. (2002). Shrimp Taura syndrome virus: Genomic characterization and similarity with members of the genus Cricket paralysis-like viruses. *J. Gen. Virol.* **83:**915–926.

Martienssen, R. A. (2003). Maintenance of heterochromatin by RNA interference of tandem repeats. *Nat. Genet.* **35:**213–214.

Mason, H., Ball, J. M., Shi, J-J., Jiang, X., Estes, M. K., and Arntzen, C. J. (1996). Expression of Norwalk virus capsid protein in transgenic tobacco and potato and its oral immunogenicity in mice. *Proc. Natl. Acad. Sci. USA* **93:**5335–5340.

Masoumi, A., Hanzlik, T. N., and Christian, P. D. (2003). Functionality of the 5′- and intergenic IRES elements of cricket paralysis virus in a range of insect cell lines, and its relationship with viral activities. *Virus Res.* **94:**113–120.

Mauricio, R. (1998). Costs of resistance to natural enemies in field populations of the annual plant *Arabidopsis thaliana*. *Am. Nat.* **151:**20–27.

Mayo, A. M. (2002). Virus Taxonomy: Houston 2002. *Arch. Virol.* **147:**1071–1076.

McConn, M., Creelman, R. A., Bell, E., Mullet, J. E., and Browse, J. (1997). Jasmonate is essential for insect defense in *Arabidopsis*. *Proc. Natl. Acad. Sci. USA* **94:** 5473–5477.

Meins, F., Si-Ammour, A., and Blevins, T. (2005). RNA silencing systems and their relevance to plant development. *Annu. Rev. Cell. Dev. Biol.* **21:**297–318.

Meister, G., and Tuschl, T. (2004). Mechanisms of gene silencing by double-stranded RNA. *Nature* **431:**343–349.

Miller, W. A., and Rasochova, L. (1997). *Barley yellow dwarf viruses. Annu. Rev. Phytopathol.* **35:**167–190.

Mita, K., Kasahara, M., Sasaki, S., Nagayasu, Y., Yamada, T., Kanamori, H., Namiki, N., Kitagawa, M., Yamashita, H., Yasukochi, Y., Kadono-Okuda, K., Yamamoto, K. *et al.* (2004). The genome sequence of silkworm, *Bombyx mori*. *DNA Res.* **11:**27–35.

Moore, N. F. (1991). The *Nudaurelia* β family of insect viruses. *In* "Viruses of Invertebrates" (N. F. Kurstak, ed.), pp. 277–285. Marcel Dekker, New York.

Mourrain, P., Beclin, C., Elmayan, T., Feuerbach, F., Godon, C., Morel, J. B., Jouette, D., Lacombe, A. M., Nikic, S., Picault, N., Remoue, K., Sanial, M. *et al.* (2000). *Arabidopsis SGS2* and *SGS3* genes are required for posttranscriptional gene silencing and natural virus resistance. *Cell* **101:**533–542.

Muangsan, N., Beclin, C., Vaucheret, H., and Robertson, D. (2004). Geminivirus VIGS of endogenous genes requires SGS2/SDE1 and SGS3 and defines a new branch in the genetic pathway for silencing in plants. *Plant J.* **38:**1004–1014.

Munday, B. L., Kwang, J., and Moody, N. (2002). Betanodavirus infections of teleost fish: A review. *J. Fish Dis.* **25:**127–142.

Munshi, S., Liljas, L., Cavarelli, J., Bomu, W., McKinney, B., Reddy, V., and Johnson, J. E. (1996). The 2. 8 A structure of a T = 4 animal virus and its implications for membrane translocation of RNA. *J. Mol. Biol.* **261:**1–10.

Muskens, M. W. M., Vissers, A. P. A., Mol, J. N. M., and Kooter, J. M. (2000). Role of inverted DNA repeats in transcriptional and post-transcriptional gene silencing. *Plant Mol. Biol.* **43:**243–260.

Nagesha, H. S., Wang, L. F., Hyatt, A. D., Morrissy, C. J., Lenghaus, C., and Westbury, H. A. (1995). Self-assembly, antigenicity, and immunogenicity of the rabbit haemorrhagic disease virus (Czechoslovakian strain V-351) capsid protein expressed in baculovirus. *Arch. Virol.* **140:**1095–1108.

Nakahara, K., and Carthew, R. W. (2004). Expanding roles for miRNAs and siRNAs in cell regulation. *Curr. Opin. Cell Biol.* **16:**127–133.

Ongus, J. R., Peters, D., Bonmatin, J.-M., Bengsch, E., Vlak, J. M., and van Oers, M. M. (2004). Complete sequence of a picorna-like virus of the genus *Iflavirus* replicating in the mite *Varroa destructor*. *J. Gen. Virol.* **85:**3747–3755.

Pawlita, M., Müller, M., Oppenländer, M., Zentgraf, H., and Herrmann, M. (1996). DNA encapsidation by viruslike particles in insect cells from the major capsid protein VP1 of b-lymphotropic papovavirus. *J. Virol.* **70:**7517–7526.

Pazhouhandeh, M., Dieterle, M., Marrocco, K., Lechner, E., Berry, B., Brault, V., Hemmer, O., Kretsch, T., Richards, K. E., Genschik, P., and Ziegler-Graff, V. (2006). F-box-like domain in the polerovirus protein P0 is required for silencing suppressor function. *Proc. Natl. Acad. Sci. USA* **103:**1994–1999.

Philippe, R., Veyrunes, J.-C., Mariau, D., and Bergoin, M. (1997). Biological control using entomopathogenic viruses. Application to oil palm and coconut pests. *Plantations, Recherche, Développement* **1:**39–45.

Pringle, F. M., Gordon, K. H., Hanzlik, T. N., Kalmakoff, J., Scotti, P. D., and Ward, V. K. (1999). A novel capsid expression strategy for *Thosea asigna virus* (*Tetraviridae*). *J. Gen. Virol.* **80:**1855–1863.

Pringle, F. M., Johnson, K. N., Goodman, C. L., McIntosh, A. H., and Ball, L. A. (2003). Providence virus: A new member of the *Tetraviridae* that infects cultured insect cells. *Virology* **306:**359–370.

Quan, G. X., Kanda, T., and Tamura, T. (2002). Induction of the white egg 3 mutant phenotype by injection of the double-stranded RNA of the silkworm white gene. *Insect Mol. Biol.* **11:**217–222.

Ranjekar, P. K., Patankar, A., Gupta, V., Bhatnagar, R., Bentur, J., and Kumar, P. A. (2003). Genetic engineering of crop plants for insect resistance. *Curr. Sci.* **84:**321–329.

Rao, M. (2004). Conserved and divergent paths that regulate self-renewal in mouse and embryonic stem cells. *Dev. Biol.* **275:**269–286.

Rashotte, A., and Feldmann, K. (1996). Epicuticular waxes and aphid resistance in *Arabidopsis cer* mutants and ecotypes. *Plant Physiol. Suppl.* **111:**87.

Reinganum, C. (1991). *Tetraviridae. In* "Atlas of Invertebrate Viruses" (C. Adams and J. R. Bonami, eds.), pp. 553–592. CRC Press, Boca Raton.

Reymond, P., Weber, H., Damond, M., and Farmer, E. E. (2000). Differential gene expression in response to mechanical wounding and insect feeding in *Arabidopsis*. *Plant Cell* **12:**707–719.

Roignant, J. Y., Carre, C., Mugat, B., Szymczak, D., Lepesant, J. A., and Antoniewski, C. (2003). Absence of transitive and systemic pathways allows cell-specific and isoform-specific RNAi in *Drosophila*. *RNA* **9:**299–308.

Roth, B. M., Pruss, G. J., and Vance, V. B. (2004). Plant viral suppressors of RNA silencing. *Virus Res.* **102:**97–108.

Santos, M. O., Adang, M. J., All, J. N., Boerma, H. R., and Parrott, W. A. (1997). Testing transgenes for insect resistance using *Arabidopsis*. *Mol. Breeding* **3:**183–194.

Sayyed, A. H., and Wright, D. J. (2002). Genetic diversity of Bt resistance: Implications for resistance management. *Pakistan J. Biol. Sci.* **5:**1330–1344.

Schob, H., Kunz, C., and Meins, F. (1997). Silencing of transgenes introduced into leaves by agroinfiltration: A simple, rapid method for investigating sequence requirements for gene silencing. *Mol. Gen. Genet.* **256:**581–585.

Schroeder, R. (2003). The genes orthodenticle and hunchback substitute for bicoid in the beetle Tribolium. *Nature* **422:**621–625.

Schunmann, P. H. D., Llewellyn, D. J., Surin, B., Boevink, P., De Feyter, R. C., and Waterhouse, P. M. (2003). A suite of novel promoters and terminators for plant biotechnology. *Funct. Plant Biol.* **30:**443–452.

Selling, B. H., Allison, R. F., and Kaesberg, P. (1990). Genomic RNA of an insect virus directs synthesis of infectious virions in plants. *Proc. Natl. Acad. Sci. USA* **87:**434–438.

Service, R. (1996). Arming plants with a virus. *Science* **271:**145.

Singh, R., Ellis, P. R., Pink, D. A. C., and Phelps, K. (1994). An investigation of the resistance to cabbage aphid in Brassica species. *Ann. Appl. Biol.* **125:**457–465.

Stotz, H. U., Pittendrigh, B. R., Kroyman, J., Weniger, K., Fritsche, J., Bauke, A., and Mitchell-Olds, T. (2000). Induced plant defense responses against chewing insects: Ethylene signaling reduces resistance of *Arabidopsis* against Egyptian cotton worm, but not diamondback moth. *Plant Physiol.* **124:**1007–1017.

Suh, M. R., Lee, Y., Kim, J. Y., Kim, S. K., Moon, S. H., Lee, J. Y., Cha, K. Y., Chung, H. M., Yoon, H. S., Moon, S. Y., Kim, V. N., and Kim, K. S. (2004). Human embryonic stem cells express a unique set of microRNAs. *Dev. Biol.* **270:**488–498.

The Honey Bee Genome Sequencing Consortium (2006). The genome of a highly social species, the honey bee *Apis mellifera*. *Nature* (in press).

Uhlirova, M., Foy, B. D., Beaty, B. J., Olson, K. E., Riddiford, L. M., and Jindra, M. (2003). Use of Sindbis virus-mediated RNA interference to demonstrate a conserved role of Broad-Complex in insect metamorphosis. *Proc. Natl. Acad. Sci. USA* **100:** 15607–15612.

van der Wilk, F., Dullemans, A. M., Verbeek, M., and Van den Heuvel, J. F. J. M. (1997). Nucleotide sequence and genomic organization of *Acyrthosiphon pisum virus*. *Virology* **238:**353–362.

Voinnet, O. (2005). Induction and suppression of RNA silencing: Insights from viral infections. *Nat. Rev. Genet* **6:**206–220.

Volkman, L. E. (1997). Nucleopolyhedrovirus interactions with their insect hosts. *Adv. Virus Res.* **48:**313–348.

Wang, M. B., and Waterhouse, P. M. (2000). High-efficiency silencing of a β-glucuronidase gene in rice is correlated with repetitive transgene structure but is independent of DNA methylation. *Plant Mol. Biol.* **43:**67–82.

Wang, X. C., Wu, C. Y., and Lo, C. F. (2004). Sequence analysis and genomic organization of a new insect picorna-like virus, *Ectropis obliqua* picorna-like virus, isolated from *Ectropis obliqua*. *J. Gen. Virol.* **85:**1145–1151.

Wang, X. H., Aliyari, R., Li, W. X., Li, H. W., Kim, K., Carthew, R., Atkinson, P., and Ding, S. W. (2006). RNA interference directs innate immunity against viruses in adult *Drosophila*. *Sciencexpress* 23.03.2006.

Waterhouse, P. M. (2006). Defense and counterdefense in the plant world. *Nat. Genet.* **38:**138–139.

Waterhouse, P. M., Wang, M. B., and Lough, T. (2001). Gene silencing as an adaptive defence against viruses. *Nature* **411:**834–842.

Watson, J. M., Fusaro, A. F., Wang, M. B., and Waterhouse, P. M. (2005). RNA silencing platforms in plants. *FEBS Lett.* **579:**5982–5987.

White, L. J., Hardy, M. E., and Estes, M. K. (1997). Biochemical characterization of a smaller form of recombinant Norwalk virus capsids assembled in insect cells. *J. Virol.* **71:**8066–8072.

Wilson, J. E., Powell, M. J., Hoover, S. E., and Sarnow, P. (2000). Naturally occurring dicistronic cricket paralysis virus RNA is regulated by two internal ribosome entry sites. *Mol. Cell. Biol.* **20:**4990–4999.

Wimmer, E. A. (2003). Applications of insect transgenesis. *Nat. Rev. Genet.* **4:**225–232.

Wood, B. J. (2002). Pest control in Malaysia's perennial crops: A half century perspective tracking the pathway to integrated pest management. *Integr. Pest Manage. Rev.* **7:**173–190.

Wu, C. Y., Lo, C. F., Huang, C. J., Yu, H. T., and Wang, C. H. (2002). The complete genome sequence of *Perina nuda* picorna-like virus, an insect-infecting virus with a genome organization similar to that of the mammalian picornaviruses. *Virology* **294:**312–323.

Xia, Q., Zhou, Z., Lu, C., Cheng, D., Dai, F., Li, B., Zhao, P., Zha, X., Cheng, T., Chai, C., Pan, G., Xu, J. *et al.* (2004). A draft sequence for the genome of the domesticated silkworm (*Bombyx mori*). *Science* **306:**1937–1940.

Xie, Z., Johansen, L. K., Gustafson, A. M., Kasschau, K. D., Lellis, A. D., Zilberman, D., Jacobsen, S. E., and Carrington, J. C. (2004). Genetic and functional diversification of small RNA pathways in plants. *PLoS Biol.* **2:**E104.

Xie, Z., Allen, E., Wilken, A., and Carrington, J. C. (2005). DICER-LIKE 4 functions in trans-acting small interfering RNA biogenesis and vegetative phase change in *Arabidopsis thaliana*. *Proc. Natl. Acad. Sci. USA* **102:**12984–12989.

Yamaguchi, K. (1979). Natural recovery of the fall web worm, *Hyphantia cunea*, to infection by a cytoplasmic-polyhedrosis virus of the silkworm *Bombyx mori*. *J. Invertebr. Pathol.* **33:**126–128.

Yi, F., Zhang, J., Yu, H., Liu, C., Wang, J., and Hu, Y. (2005). Isolation and identification of a new tetravirus from Dendrolimus punctatus larvae collected from Yunnan Province, China. *J. Gen. Virol.* **86:**789–796.

Zeddam, J.-L., Rodriguez, J. L., Ravallec, M., and Lagnaoui, A. (1999). A Noda-like virus isolated from the sweetpotato pest *Spodoptera eridania* (*Cramer*) (Lep. Noctuidae). *J. Invertebr. Pathol.* **74:**267–274.

INDEX

C

D

E

F

G

J

K

L

M

N

O

P

Q

R

S

T

U

V

W

X

Y

Z

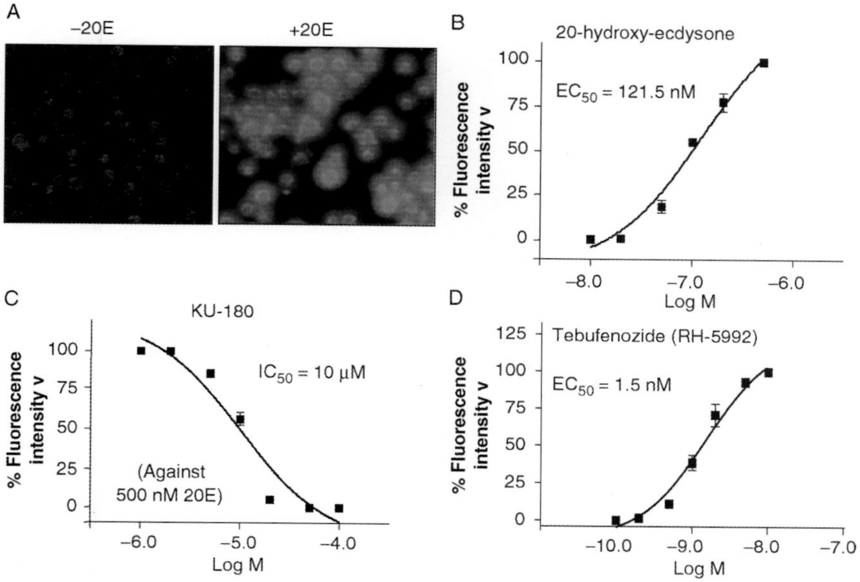

DOURIS *ET AL.*, FIG 6. HTS system for ecdysone mimetics based on transformed silkmoth-derived Bm5 cells. (A) Fluorescence photographs of transformed cells before (left) and after challenge with 1 μM 20E (right). (B) and (D) Dose–response curves of the natural insect molting hormone 20E and the synthetic ecdysone agonist tebufenozide (RH-5992) as determined by measurements using a fluorescence microplate reader. (C) Identification of an ecdysone antagonist (KU-180) after screening of a library of dibenzoyl hydrazine compounds (provided by Dr. Y. Nakagawa, University of Kyoto, Japan). Shown is the inhibition of the response by 500 nM 20E using different concentrations of KU-180. The median effective concentration (EC_{50}) of the agonist compounds (Panels B and D) and the median inhibitory concentration (IC_{50}) of the antagonist compound (Panel B) are indicated.

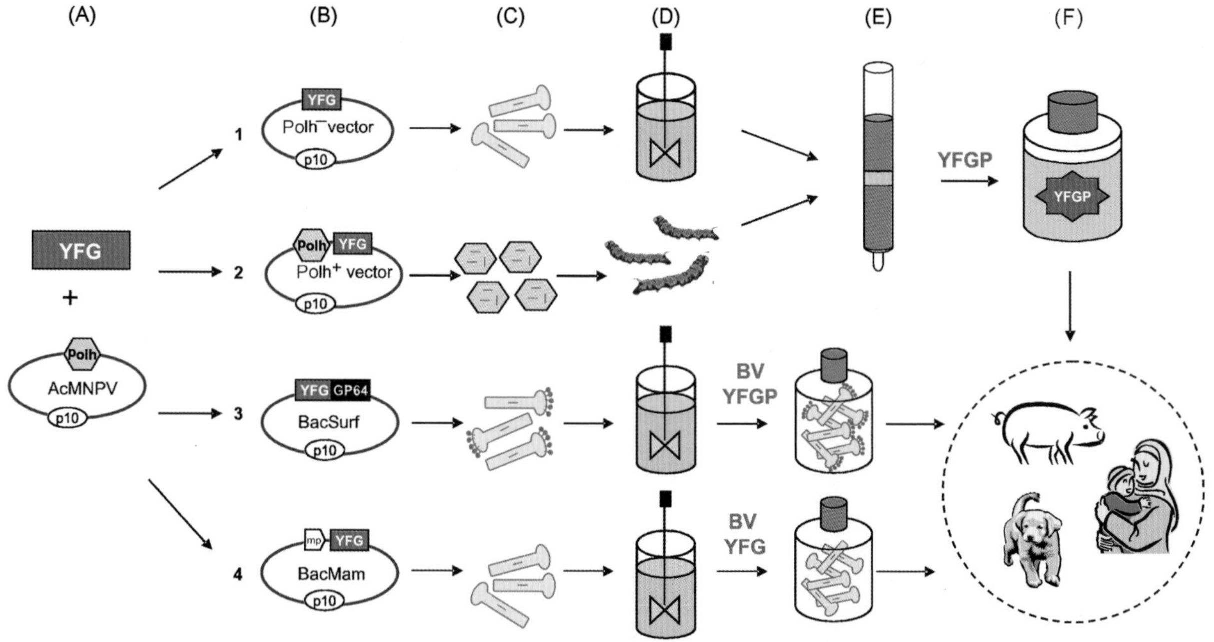

VAN OERS, FIG 1. Flow chart showing four different methods to make a vaccine based on your favorite gene (*YFG*) in the baculovirus expression system: (1) protein expression in insect cell bioreactors using the *polyhedrin* locus for expression, (2) protein expression in insect larvae leaving the *polyhedrin* gene intact; expression is driven either by a duplicated *p10* promoter or by the original *p10* promoter, (3) baculovirus surface display methods where YFG is fused to GP64, and (4) DNA vectors with a mammalian promoter (*mp*) for synthesis of your favorite gene product (YFGP) in the target species. Subsequent steps in the process are: (A) selection and PCR amplification of *YFG*, (B) cloning into the appropriate baculovirus vector, (C) generation of recombinant BV particles or occlusion bodies, (D) production in insect cells in bioreactors or larvae, (E) purification of recombinant YFGP or collection of BVs loaded with either YFGP or YFG, and (F) delivery of prophylactic or therapeutic vaccines.

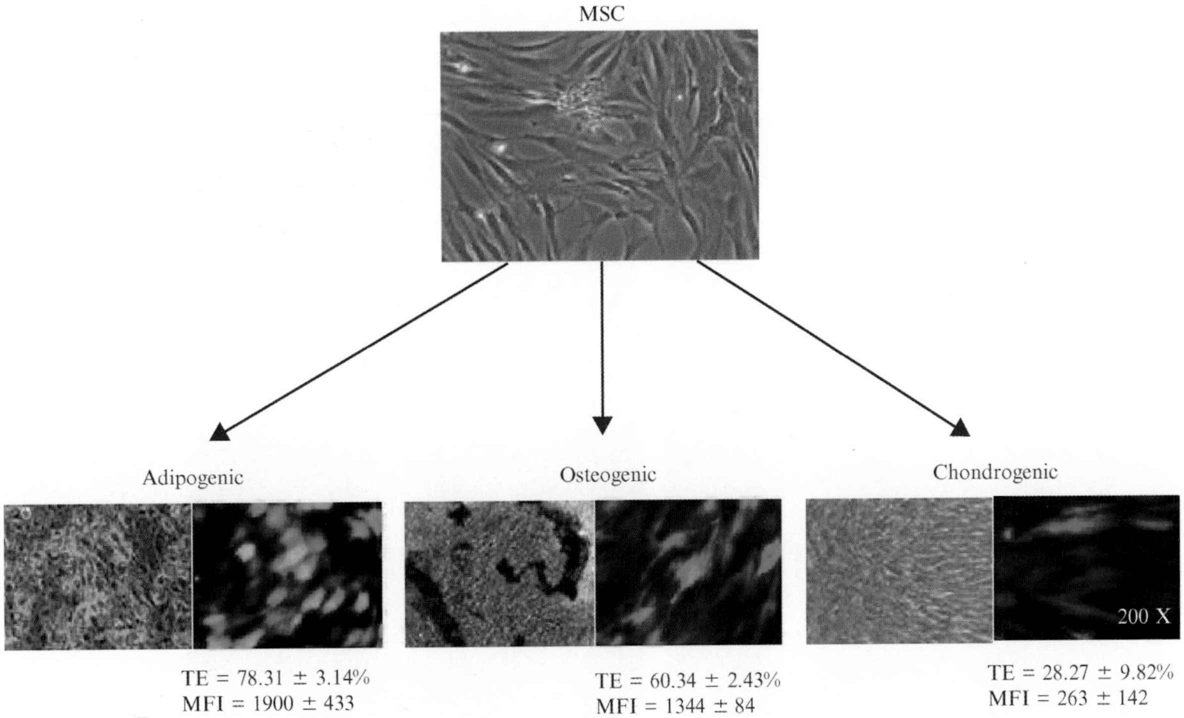

Hu, Fig 2. Baculovirus transduction of adipogenic, osteogenic, and chondrogenic progenitors. Human mesenchymal stem cells (MSC) were induced to differentiate into adipogenic, osteogenic, and chondrogenic pathways. The adipogenic, osteogenic, and chondrogenic progenitors were revealed by staining with oil red-O, von Kossa, and safranin-O staining, respectively, two weeks postinduction (left panels of each lineage pathway). The progenitors were transduced with baculoviruses expressing enhanced green fluorescent protein (EGFP) 2 weeks postinduction and exhibited different degrees of EGFP expression (right panel of each lineage pathway). The transduction efficiency (TE) and mean fluorescence intensity (MFI) were measured by flow cytometry.

A

B

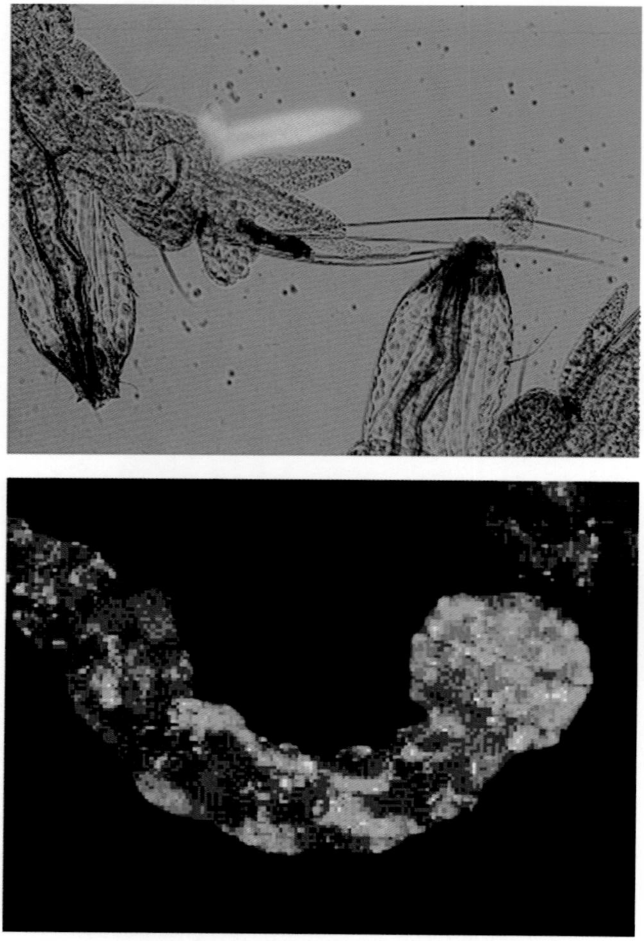

CARLSON *ET AL.*, FIG 6. *Ae. aegypti* larvae infected with GFP transducing virus. (A) Infected anal papilla of an *Ae. aegypti* larva infected by the pANS1-GFP–transducing virus. (B) Larva infected with the pASN1-GFP–transducing virus showing dissemination of virus.

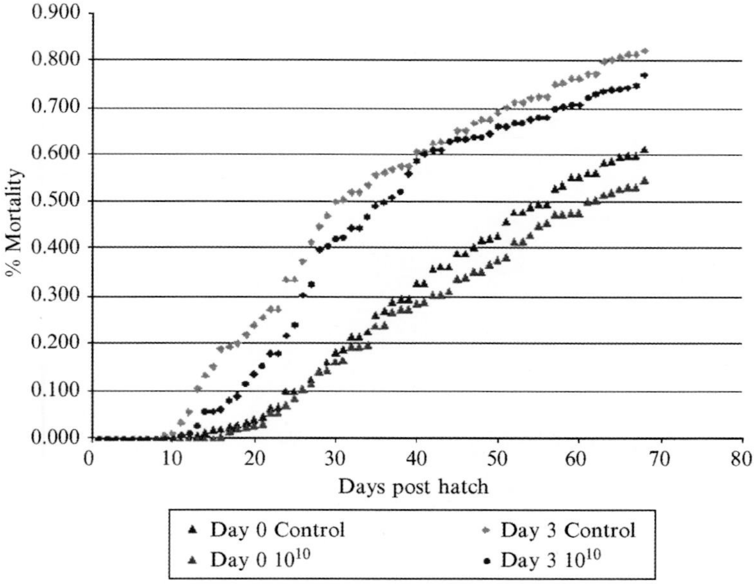

Carlson *et al.*, Fig 7. Effect of AeDNV infection on survival of adult *Ae. aegypti*. Two hundred larvae were infected with 10^{10} geq/ml of AeDNV either immediately after hatching or 3 days after hatching. The mortality percentage is shown for virus and control treatments.

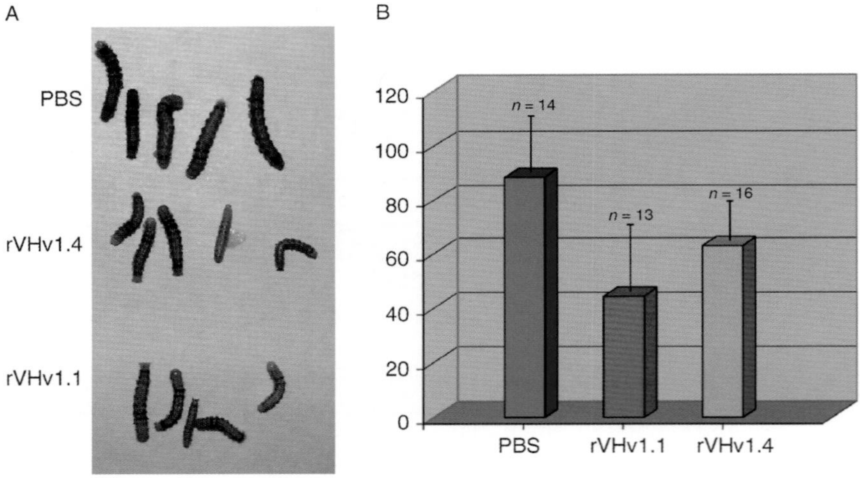

GILL *ET AL.*, FIG 2. Effect of orally consumed rVHv1.1 and rVHv1.4 on the development of *H. virescens* larvae. (A) *H. virescens* larvae consumed diet containing rVHv1.1 and rVHv1.4 or PBS (control) for 24 h, before being placed on normal diet. The image was taken after 3 days. (B) Fresh weight of larvae fed on test and control diet 5 days after onset of the experiment. Values represent the mean ± SD. *n*, the number of larvae per treatment.

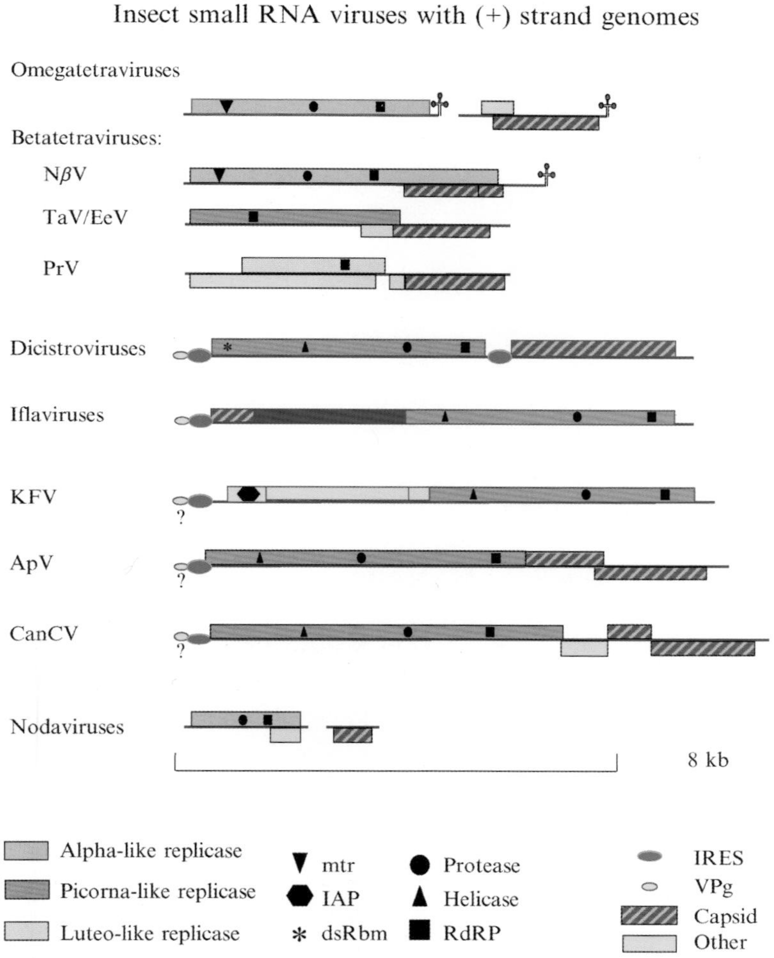

Gordon and Waterhouse, Fig 1. Genomic arrangement and encoded proteins of insect SRVs. A general schematic representation is shown for the recognized SRV families and genera. The only exception is for the betatetraviruses, where the four known members fall into three different groups as shown. Lines represent untranslated regions and boxes represent ORFs. Boxes are shaded to denote the function of the encoded protein, as indicated in the legend. The symbols in the boxes representing replication protein ORFs show locations of specific domains as indicated in the legend. mtr, methyltransferase; IAP, inhibitor of apoptosis; dsRbm, dsRNA-binding motif; RdRP, RNA-dependent RNA polymerase. Among the picorna-like viruses, VPg's and IRES function have not been confirmed in all cases and are marked with a "?" accordingly.

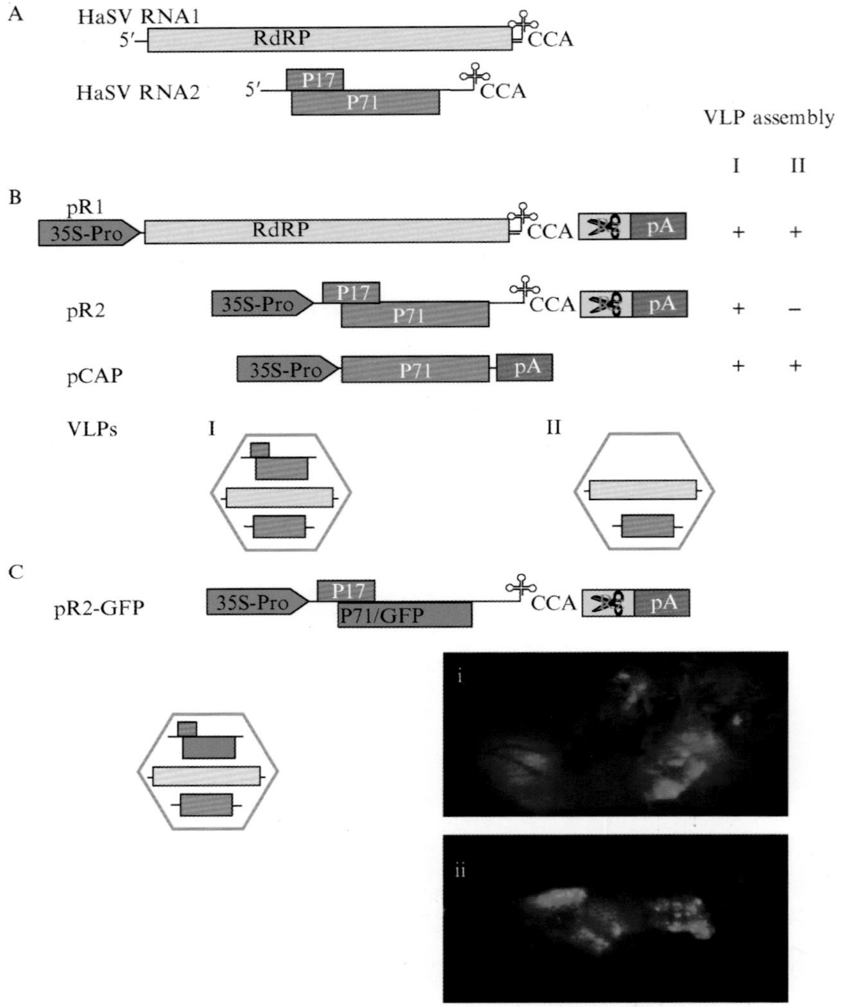

Gordon and Waterhouse, Fig 2. HaSV RNAs and their use for transgene expression. (A) Genome organization of HaSV RNAs 1 and 2, showing ORFs and 3′-terminal tRNA-like structures. (B) Schematic diagrams showing plasmids for expression of HaSV RNAs in plant protoplasts, and the resulting virus-like particles (VLPs). The plasmids were as described in Gordon *et al.* (2001). Genomic RNA1 is expressed from pR1, and pR2 which produces genomic RNA2, was used for two derivative plasmids. Plasmid pCAP carries only the HaSV P71 ORF. The CaMV 35S promoter (35S-Pro), *cis*-acting ribozyme (Rz), and the CaMV 35S polyadenylation sequence (pA) are indicated for each construct. Boxes represent ORFs for gene products as noted within them. Combinations of plasmids transfected into protoplasts for feeding to larvae are shown to the right, and schematic diagrams of the resulting VLPs (combinations I and II) assembled in protoplasts, below. Combination I corresponds to assembly of VLPs containing the complete HaSV genome, and combination II to VLPs of the subvirus, lacking genomic RNA2. (C) Expression of green fluorescent protein (GFP) in larvae. The capsid gene on plasmid pR2 was replaced by a P71/GFP fusion as described by Hanzlik *et al.* (2006b, unpublished), allowing assembly in protoplasts of VLPs containing RNA1, the RNA2-GFP fusion, and the P71 mRNA as shown in the schematic. Following transfection of the plasmids shown, protoplasts were incubated for 3 days before aliquots were fed to larvae; larval midguts were examined for fluorescence after 11 days. Panels (i) and (ii) show epifluorescence images of anterior midgut regions from GFP-expressing larvae, with the anterior of each image oriented to the left. Control larva, fed protoplasts transfected with only pR1 and pCAP, showed no GFP fluorescence.

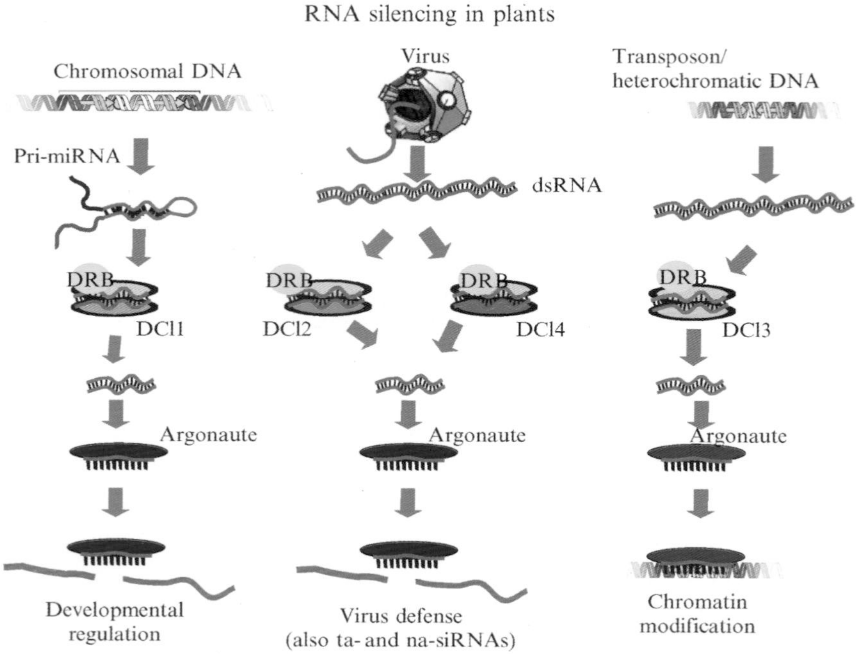

GORDON AND WATERHOUSE, FIG 4. Pathways leading to RNA silencing in plants. The miRNA pathway is shown on the left, the antiviral RNA-silencing pathway in the center, and the RNAi pathway leading to chromatin modification on the right. The miRNA pathway processes transcripts from genes yielding Pri-miRNA precursors; these are processed by DCL1 and the resulting 21–24 nt miRNAs guide the RNA-induced silencing complex (RISC) to the appropriate mRNA for miRNA-directed cleavage and degradation. RISC contains Argonaute, the central endoribonuclease involved in RNA silencing. For silencing of viral sequences, dsRNAs derived from the virus are cleaved by DCL2 or DCL4 (see text) and yield 21 nt siRNAs that guide the RISC to target viral sequences. *Trans*-acting (ta-) and na-siRNAs (see text) operate by similar mechanisms. Transcripts from highly repetitive heterochromatin (e.g., retrotransposons) are processed by DCL3 to yield 24–26 nt siRNAs that then direct intranuclear complexes containing a different member of the Argonaute family (AGO-4) to the chromatin sequence itself for modification (e.g., methylation).